Progress in Material Handling and Logistics

Editors: John A. White, Ira W. Pence

Material Handling Research Center
Georgia Institute of Technology

Progress in Material Handling and Logistics

Volume 2

Material Handling '90

Springer-Verlag Berlin Heidelberg NewYork
London Paris Tokyo Hong Kong Barcelona

Dr. John A. White

Regents' Professor and Eugene C.
Gwaltney Professor
School of Industrial
and Systems Engineering
Georgia Institute of Technology
Atlanta, GA 30332-0206
USA

Dr. Ira W. Pence, Jr.

Director
Material Handling Research Center
Georgia Institute of Technology
Atlanta, GA 30332-0206
USA

Robert J. Graves, Ph. D.

Professor
Industrial Engineering
University of Massachusetts
Amherst, MA 01003
USA

Leon F. McGinnis, Ph. D.

Professor
School of Industrial
and Systems Engineering
Georgia Institute of Technology
Atlanta, GA 30332-0206
USA

Mickey R. Wilhelm, Ph. D.

Professor and Chairman
Industrial Engineering
University of Louisville
Louisville, KY 40292
USA

Richard E. Ward, Ph. D.

Staff Vice President-Education
The Material Handling Institute, Inc.
The Material Handling Industry
of America, Inc.
Charlotte, NC 28217
USA

ISBN-13: 978-3-642-84358-7 e-ISBN-13: 978-3-642-84356-3
DOI: 10.1007/978-3-642-84356-3

Library of Congress Cataloging-in-Publication Data
Progress in material handling and logistics.
Includes bibliographical references.
1. Materials handling--Automation. 2. Flexible manufacturing systems.
I. White, John A. II. Pence, Ira W. III. Georgia Institute of Technology.
Material Handling Research Center.
TS180.P74 1989 658.7'81 89-26085

Softcover reprint of the hardcover 1st edition 1991

2161/3020-543210 – Printed on acid-free paper

ACKNOWLEDGEMENTS

A large number of organizations contributed to the success of the 1990 Material Handling Research Colloquium. CIC-MHE is grateful for the encouragement and support that it received and is proud to acknowledge all of the Colloquium's co-sponsors.

Institutional Co-Sponsors:

- *The Material Handling Industry of America (MHIA), a subsidiary of The Material Handling Institute (MHI), which sponsors and supports all of CIC-MHE's activities. Were it not for MHIA's backing the Colloquium would never have gotten off the ground.*
- *The Material Handling Research Center (MHRC), the only research center in the United States devoted entirely to the science of material movement, storage, and control. MHRC is headquartered at the Georgia Institute of Technology, but conducts research on several campuses throughout North America.*
- *The National Science Foundation (NSF), an independent federal agency created by the National Science Foundation Act of 1950 (P.L. 81-507), sponsors and supports advances in all areas of science and engineering. NSF's grant in support of the Colloquium helped defray travel costs of all academic participants in the Colloquium.*

Corporate Co-Sponsors and Hosts:
(all of whom are member companies of MHIA)

Accu-Sort Systems
Allen Bradley Company
Blue Giant Equipment Corporation
Littion IAS
Mannesmann Demag Corporation
Penco Products
Rapistan Corporation
Republic Storage Systems
Stanley-Vidmar, Inc.
Unarco Material Handling

PREFACE

The contents of this book are based on invited papers submitted for presentation and discussion at the 1990 Material Handling Research Colloquium held in Hebron, Kentucky, June 19-21, 1990. The Colloquium was sponsored and organized by the College Industry Council for Material Handling Education (CIC-MHE) with additional co-sponsorship and funding provided by numerous organizations (see acknowledgements).

The purpose of the Colloquium was to foster open discussion about the current state of material handling research at universities from across the United States and Canada. It was an opportunity to share specific research directions and accomplishments. But more importantly, it was an opportunity to discuss the implications of the basic constraints to solving industry relevant problems in the field of material handling and closely related activities; the efficacy of the approaches being taken at the present time; and the directions believed to be of most value to the industry and to advancing the knowledge and science base of the material handling engineering discipline.

The sponsoring organization, the College Industry Council for Material Handling Education was founded in 1952. The council is composed of college and university educators, material handling equipment manufacturers, distributors, users and consultants, representatives of the business press plus professional staff and members of other organizations concerned with material handling education.

CIC-MHE receives its financial support from The Material Handling Industry of America (MHIA) whose parent corporation is The Material Handling Institute (MHI), a national trade association incorporated in 1945. All three organizations have their offices at 8720 Red Oak Blvd., Suite 201, Charlotte, North Carolina 18217.

All CIC-MHE activities are educationally oriented and designed to promote a better understanding of material handling. Through the work of its various committees, CIC-MHE provides quality material and teaching aids for use in materia handling education. The Council has long encouraged material handling researc and is well known in the United States for its sponsorship of educational seminars an training courses for the improvement of material handling teaching at the college an university level. Through its various activities and diverse representations c membership, CIC-MHE facilitates the interchange of information between industr academia and the user community. The 1990 Colloquium held in Hebron, Kentuck was a first for CIC-MHE, but because of its success, it undoubtedly will be the first many such gatherings.

All of the above companies hosted some portion of the Colloquium. In addition, Litton IAS provided its corporate training facility for the first two days of the Colloquium.

Since so many individuals influenced the Colloquium in so many ways, one is always at risk when trying to name a few. However, in this case I would be remiss by not thanking my co-editors, who also served with me as an organizing committee; Dr. Ira W. Pence, Jr., Director of the MHRC, for the production of the proceedings; Dr. Louis Martin-Vega of NSF, both for his part in sponsoring the Colloquium, and also for taking the time to participate over all three days of the Colloquium; Dennis Caster and Connie Gayle of Litton IAS, who together coordinated our use of Litton's training center; and lastly Bobbie Curtis of MHIA, who was invaluable in the coordination of all the logistics associated with the Colloquium.

These companies, these orgaizations, and these individuals all working together made it possible.

Robert J. Graves, Ph.D.
1990-91 President
College Industry Council for Material Handling Education

The editors would like to thank the staff of the Material Handling Research Center for their support. Special thanks to Ms. Joene Owen who organized the material, translated the various formats and styles submitted into a consistent presentation, and conferred with the authors on omissions, figures, references, etc. We are also indebted to our students for their assistance with some of the less well known word processing packages.

We wish to acknowledge Albrecth von Hagen and Erdmuthe Raufelder of Springer-Verlag who provided help and guidance in the preparation of this work.

Dr. John A. White　　　　Dr. Ira W. Pence., Jr.

CONTENTS

SECTION THREE: STORAGE SYSTEMS AND LOGISTIS

SECTION FOUR: ECONOMICS AND AUTOMATION

SECTION ONE
PERSPECTIVE ON 1990s

The environment for material handling in the 1990s will undergo some radical changes. The swing toward more flexibility and faster responsiveness in our manufacturing and service industries and the continued emphasis on individuality within our society will increase the breadth of product offering, reduce lot sizes, and place a greater emphasis on tracking and control.

A new definition of manufacturing as the process of converting resources into economic satisfaction is emerging, one in which the strategic importance of material handling is more widely recognized. Many of the old boundaries between functions will fall, and the elements of the material handling system will be integrated and treated as a single network.

New cost accounting methods are being developed to reflect time absorption as opposed to direct labor hours as the fundamental economic measure, and these will impact the justification of systems that integrate labor, hardware, and software for planning, operation, and maintenance.

Management structures will emphasize lean flat organizations with cross-trained employees and greater responsibility for those actually doing the work. Education, both to broaden understanding and to provide specific skills, will be a key to success.

Although the environment may alter their roles, government, industry, consultants, and research will remain the four essential sectors that will shape material handling in the 1990s. In this section, a leader from each of the first three sectors presents his perspective of the current state and what the future holds. John A. White addresses the role of government in fostering the development of material handling in an environment in which an increasing number of issues compete for federal tax dollars, tax dollars that citizens are reluctant to increase.

As a consultant deeply involved with the problems which users encounter in designing and installing material handling systems, James A. Tompkins discusses the need for new answers and alternatives versus the current emphasis of research and suggests changes needed for the 1990s.

Industry's needs and interests, both as user and supplier, are often not well aligned with those of the research organizations. This often leads to unfulfilled expectations by both. Nowhere is this more true than in the transfer of technology. Stephen L. Parsley reviews industry's perception of the existing research and development process, highlights weaknesses that need to be corrected, and suggests improvements.

The researchers are represented through the comments made by nearly forty thought leaders from the major research universities in the U.S. and Canada, as summarized by the editors. The lack of a formal categorization of material handling problems and the lack of identification of Material Handling as a distinct discipline inhibit the structured development of analysis techniques. Great progress has been made in some areas, but others lag. The combined thoughts provide insight into the major research areas for the 1990s.

Ira W. Pence, Jr.

MATERIAL HANDLING RESEARCH: NEEDS AND OPPORTUNITIES

John A. White
National Science Foundation

ABSTRACT

Despite the impact of material handling on industrial competitiveness, it has not attracted the attention of researchers to the extent it should. In this chapter, a brief retrospective assessment of material handling is provided, followed by a consideration of research needs and opportunities. The connections between material handling education and research are explored and the traditional sources of support are examined. Throughout the chapter, both education and research needs are addressed. The chapter does not attempt to provide a comprehensive review of material handling research. Its focus is on future needs and opportunities, not past successes and failures. Finally, it focuses on the prospects for future support of material handling research.

1 INTRODUCTION

Material handling is not a research intensive field. In fact, material handling has attracted very little research attention, especially when one considers its role in industrial competitiveness. As a function and as a field of technology, it has not attracted the attention of either academic or industrial researchers. As a function, material handling has not attracted attention equivalent to, say, inventory control, production planning and scheduling, or quality control. As a technology, manufacturers of material handling equipment do not devote as much attention to product research and development as, say, manufacturers of processing equipment. Even though relatively more R&D attention is given to material handling in Europe and Japan than in the U.S., it still does not attract the degree of attention as, say, machine tool technology.

Despite the absence of a culture with a long tradition of research, material handling has benefited from a tradition of industrially supported education. The College Industry Council on Material Handling Education, the Material Handling Education Foundation, the Material Handling

Institute, and the Material Handling Industry of America have provided continual support of material handling education for more than 25 years. The support has included: fellowships for graduate students; travel support for faculty to attend material handling conferences and trade shows; summer institutes for material handling teachers; slides, films, videotapes, case studies, and other teaching materials for use in material handling courses; and the highest ranking material handling education award -- the Reed-Apple Award.

The foundry industry and the machine tool industry have a history of providing support for undergraduate and graduate education in their subjects. Hence, material handling is not unique in its support from industry. However, few industrial sectors have provided support for collegiate education comparable to that provided by the material handling industry.

Despite the support that has been given to strengthening material handling education, the subject is still relatively invisible among academic researchers. In fact, it might be the case that the relatively strong support provided by industry for material handling education, coupled with the lack of a research focus by the industry, contributed to material handling being a neglected area of research.

For most of its existence, material handling suppliers tended to be small, closely held corporations. In fact, in the 1950s, 60s, and 70s very few of the conveyor and rack suppliers employed more than 250 people. Many of the manufacturers served local and regional markets. Truly, many material handling suppliers were garage operations. Relatively little capital investment was needed to become a supplier of pallets, racks, and conveyors. Through the 1950s, the most sophisticated material handling equipment were industrial truck, cranes and hoists, automatic palletizers, and tow line conveyors.

The introduction of electronic controls and the subsequent introduction of automation to the material handling industry introduced a degree of complexity to the industry that few were prepared to accommodate. The emergence of automated guided vehicles, automatic identification, and automated storage and retrieval led to a flurry of entries, with a subsequent "shake down" to a few survivors. Even so, the survivors have not tended to make significant investments in research and development.

2 THE STATE OF THE FIELD

While material handling has attracted relatively little attention from industrial researchers, it is difficult to understand why it has not attracted more attention from academic researchers. By any measure, one would expect it to be an attractive field for academic research. Why? For those interested in modeling complex, highly integrated systems, material handling is a logical candidate. For those interested in electronic and mechanical control issues, material handling is a natural choice. For those interested in focusing on broad systems issues, as well as those interested in studying narrow specialized technological issues, material handling offers considerable challenges. From software to hardware complexity, from human-machine combinations to fully automated environments, from design to operating conditions material handling offers a plethora of research opportunities.

In addition to the richness of research issues, one would expect material handling to receive more attention from academic researchers since the establishment of a bonafide research laboratory is generally within the capabilities of most research universities. Through a combination of industrial contributions and the acquisition of various components of a material handling system, it is possible to assimilate enough material handling technology to equip a laboratory that is adequate for both instructional and research purposes. In the absence of full-scale equipment, it is possible to develop reasonable representations of material handling equipment through the use of analog and iconic models; scale model equipment, for example, has been used by several universities to demonstrate automated storage and retrieval operations, as well as conveyor operations.

Yet another aspect of material handling that should attract the attention of academic researchers is its interface with other technologies, e.g. machine tools, robots, sensors, and controls. Further, the cross-disciplinary research opportunities are abundant; among the disciplines whose expertise is especially suited to material handling research are agricultural engineering, chemical engineering, civil engineering, computer science, electrical engineering, industrial engineering, management science, manufacturing engineering, mechanical engineering, mining engineering, operations research, petroleum engineering, and transportation engineering. Among the professional and technical societies which focus on some aspect of material handling (and/or logistics) are the American Society of Mechanical Engineers, the American Institute of Mining Engineers, the Council for Logistics Management, the Institute of Industrial Engineers, the International Materials Management Society, the Operations Research Society of America, the Society of Manufacturing Engineers, the Society of Logistics Engineers, the Society of Packaging and Handling Engineers, the Institute of Management Sciences, and the Warehouse Education and Research Council; undoubtedly, there are many more, but this listing establishes the claim of a breadth of interest and a diffusion of focus, which diminishes its visibility as a bonafide research field.

The relative importance of material handling as a research field is not much different in Europe and Japan than in the U.S.; however, recent investments by the European Community in material handling related research has been noteworthy. The PROMETHEUS and the ESPRIT programs, in particular, have focused resources and attention on logistics, transportation, and material handling research problems. The PROMETHEUS program addresses the subject of an intelligent highway system, including pathway and vehicle design and control. The ESPRIT program focuses on artificial intelligence, automation, robotics, guided vehicles, and computer science issues. The level of industry and government funding in these two programs is far greater than that devoted to these subjects in the U.S.

An assessment of material handling research suggests the following shortcomings:

- a single discipline, industrial engineering, has tended to dominate the research agenda;
- inadequate attention has been given to joint material handling/manufacturing research problems;
- insufficient focus has been given to bulk handling, and essentially no attention has been given to the joint bulk/discrete handling problem, e.g. when should bulk materials become unitized in the flow process?
- the research has been dominated by abstract analysis, relatively little attention has been given to synthesis;
- very little "hands on" focus occurs, much of the research focuses on an abstraction of the real problem and little attention is given to validating the abstraction, performing experimental research, or addressing hardware research issues;
- for a field that is intimately related to industrial competitiveness, relatively little industry interaction is apparent in the focus, conduct, or validation of the research; and
- the research is dominated by incremental thinking, with considerable focus on analyzing single technologies or making minor changes in assumptions for queueing network representations of flow processes.

3 MATERIAL HANDLING NEEDS

Among the major needs facing the material handling research community are the following: education, human resources, and research. In particular, a broader exposure of material handling design considerations is needed across many disciplines at both the undergraduate and graduate levels; material handling education needs to include both bulk and unit handling, hardware and software focus, design and operating issues, analysis and synthesis, abstract and experimental studies, human-machine and fully automated environments, and productivity/quality/economic issues. In terms of human resources, greater efforts are needed in broadening the participation of women, minorities, and the disabled in material handling education and research. Increased recognition must be given to the challenges facing industry of the coming shortfall in an educated work force, including equipment operators and technicians, managers, analysts, and engineers.

From a review of the chapters contained in this volume, as well as those included in [2] and a review of the research literature, it is possible to identify a number of material handling research needs. Rather than duplicate an earlier listing [1], the following builds on that listing.

- Attention must be given to what the French term ultimate technologies, i.e., technology pushed to its perceived performance limits. In the case of material handling, research is needed at the limits of speed, material/load size, weight, accuracy, control, stability, and precision. Research is needed at the nanoscale and at the megascale extremes. Moving, storing, and controlling material in harsh and/or non-traditional environments would be included in the ultimate technology category.
- The nuclear engineering and fluid dynamics communities have been performing research on multi-phase processing, in which combinations of gases, solids, and liquids are co-mingled in the flow stream. Similarly, material handling research is needed to understand the actual flow dynamics of material moving in/on different flow media, e.g., air film handling, gravity chutes and hoppers, pneumatic tubes, slurries, screw conveyors, and belt/roller conveyors.
- A unified theory of material handling, based on scientific principles of handling, should be attempted. The so-called Principles of Material Handling are empirically based or rely on anecdotal evidence, rather than scientifically based; they are of minimal utility in terms of their scientific rigor.
- Research is needed across all material handling generations (manual, mechanized, automated, integrated, and intelligent), especially manual, integrated, and intelligent handling.
- User data bases are needed in order to perform research on the development of industry standards on containers, pallets, load sizes, conveyor widths and heights, etc.
- Little is known about the true impact of material handling on productivity, quality, safety, and overall profitability of the firm; additional evidence of such impacts could prove quite useful in terms of establishing the value of well-designed material handling systems.
- High fidelity case studies would prove useful for both education and research.
- Models of interconnected sets of technologies are needed; thus far, most of the attention has been given to modeling individual technologies.
- Automated palletizing of mixed loads continues to be a research challenge worthy of attention from both industry and academia.
- Automatic identification of material at the nanoscale, as well as direct identification of material through the use of new kinds of sensors, merit the attention of the research community.
- Automated storage/retrieval of individual items continues to be a research challenge; the available technologies are finding limited application due to cost/performance considerations.

- For both manufacturing and distribution applications, the virtual machine concept has not been approached in terms of developing an integrated systems design tool; before such a tool can be developed, considerable work is needed in comprehending scheduling and operating rules and in modeling the various hardware, software, and human components and interfaces.

In a sense, a material handling simulator is needed. Consider a flight simulator, which combines hardware and software with human operators/managers. Now, consider providing the equivalent of a flight simulator for a warehouse or a manufacturing plant as a teaching machine for undergraduate students! What are the research impediments to a full realization of such a teaching machine? It must be high fidelity and incorporate software representations of state-of-the-art manufacturing and warehousing technology. It must be highly flexible, incorporate modular technology components, provide for human-machine interfaces, incorporate learning, and be of value to both system designers and operating managers. Finally, it must be affordable to academic institutions.

4 MATERIAL HANDLING RESEARCH SUPPORT

Currently, material handling and logistics research receive approximately $5 million of Federal support. Most of this support comes from the Department of Defense, with logistics support being a significant component. Support from the National Science Foundation for material handling and logistics research totals approximately $1 million. NSF's support includes that which occurs within the Engineering Research Centers at Carnegie Mellon University, the University of Maryland, and Purdue University, as well as the Industry/University Cooperative Research Centers at Georgia Tech, Oklahoma State University, and a combination of California universities (UCLA, Southern California, and others). Also, NSF supports a number of Presidential Young Investigators and Research Initiation Awards where the research is focused on material handling and/or logistics. Finally, a few individual investigator awards address material handling and/or logistics research problems.

Is the research support from NSF adequate or commensurate with the significance of the research impact? No! But, the same can be said of practically every area supported by NSF's Engineering Directorate. The total budget for the Engineering Directorate is approximately $200 million. To put into perspective the amount, recognize that it would be "rounded off" by the Joint Chiefs of Staff. Similarly, the Engineering Research Expenditures at MIT are greater than $250 million.

A reasonable 5-year goal for NSF support of material handling and/or logistics research is $3 million. However, to achieve that goal, it will be essential for the education/research community to approach the matter in a coordinated manner. Piecemeal and incremental approaches will not yield significant increases in support. The community must become more unified. Further, it must develop a broader constituency, including professional societies, the National Research Council, industrial researchers and leaders, and the academic deans. Material handling must be more than a subset of industrial engineering; multiple disciplines must be involved. Further, handling of material in multiple media must be included, e.g. solid, liquid, and gas. Bulk handling, as well as unit handling, should be included within the research scope. Greater attention must be given to technological issues. Finally, both the educational and the research needs must be addressed within the context of human resources development.

5 CONCLUSION

One of the best predictors of Federal research support for an engineering discipline is the number of doctoral degrees produced. While the linkage between doctoral student support and degree production is such that increasing the latter without the former presents a formidable challenge, until material handling can establish itself more fully as a research field, it will continue to find it difficult to obtain Federal support. Not for this reason alone, it is especially important for material handling researchers to develop strong linkages to industry. Quite likely, industry will prove to be the major supporter of material handling research in the near future.

In addition to increasing the production of doctoral degrees (and consequent doctoral research on material handling problems), it is essential that a research journal be developed that focuses specifically on material handling and logistics. The disappearance of the *Material Flow* journal provides chilling evidence of the difficulty material handling faces in establishing itself as a bonafide research field.

6 REFERENCES

1. White, John A., "Material Handling in Integrated Manufacturing Systems," Design and Analysis of Integrated Manufacturing Systems, W. Dale Compton (editor), National Academy Press, Washington, D.C., pp. 46-59 (1988).
2. White, John A. and Ira W. Pence (editors), *Progress in Materials Handling and Logistics*, Vol. 1, Springer-Verlag, New York, NY (1989).

THE FOCUS OF MATERIAL HANDLING RESEARCH IN THE UNITED STATES

James A. Tompkins
Tompkins Associates, Inc.

ABSTRACT

Material handling research can be categorized as:

Category I: Material handling equipment design.
Category II: Operational optimization of material handling equipment.
Category III: Design of material handling systems.
Category IV: Design of manufacturing and distribution systems which consider material handling a critical component.

The emphasis of material handling research has been on Categories II and III. To increase material handling research credibility and relevance, more emphasis is required on Categories I and IV.

1 INTRODUCTION

Material handling research may be categorized as follows:

Category I: *The design of material handling equipment:* This research is undertaken primarily by manufacturers of material handling equipment and is pursued primarily by mechanical and electrical engineers. The thrust is to improve equipment operating characteristics and is very applications oriented.

Category II: *The operational optimization of material handling equipment:* This research is undertaken primarily by industrial engineering professors and is focused on optimization of the utilization of material handling equipment. The majority of the work done here is on AS/RS and AGVS. The primary deliverable from this work is research proposals and publications.

Category III: *The design of material handling systems:* This research is also primarily undertaken by industrial engineering professors. The focus has been on facilities planning and on the application of industrial engineering tools to material handling (i.e. simulation, waiting-line analysis, economic justification, etc.). The facilities planning research tends to view the design of facilities from a material handling perspective and has had some acceptance by industry. The application of industrial engineering tools has varied from case study applications to highly theoretical work.

Category IV: *The design of manufacturing and distribution systems while realizing material handling is a critical component*: The thrust of this work is not the design of the material handling system, but rather the design of the manufacturing and/or distribution system while material handling plays a critical systems support role. Often, material handling research has forgotten that material handling is not an end unto itself. The objective should not be to optimize the material handling system. The objective is not to reduce the costs of material handling. The objective is to design material handling systems that result in the minimal total systems cost of manufacturing and/or distribution.

Of the 29 chapters published in the this book, this author categorizes:

7% as falling in the design of material handling equipment category,

38% being in the material handling equipment operational optimization category,

34% being in the material handling systems design category, and

21% falling in the material handling is a component of manufacturing and distribution systems category.

This mix is better than what one encounters in the material handling research literature. An estimate of what typically appears is:

Virtually no published work on the design of material handling equipment,

40% in the material handling equipment operational optimization category,

50% in the material handling systems design category, and

10% in the material handling is a component of manufacturing and distribution systems category.

This mix is a problem. Industry has the greatest need for research in the first and last categories and the material handling research community is doing most of its work in the second and third categories. There is a tremendous need to get mechanical and electrical engineers more involved with the design of material handling equipment (Category I) and to get industrial engineers more involved with the research focusing on material handling being a component of manufacturing and distribution systems (Category IV). Since the author of this chapter is an industrial engineer, the remainder of this chapter shall address the needs for Category IV research. Nevertheless, the need for the Category I research is very important and work should be done to identify and pursue the most relevant, industry-wide material handling equipment design research needs.

2 MATERIAL HANDLING AS A COMPONENT OF MANUFACTURING AND DISTRIBUTION SYSTEMS DESIGN

As the culture of manufacturing and distribution systems change, so too must the material handling research that is done to support the design of manufacturing and distribution systems. In fact, material handling plays a key role in achieving the manufacturing and distribution cultural changes. The cultural changes that must be addressed when undertaking material handling research are:

1. **Customer Orientation**: Successful operations must have a high regard for the customer, must know the customer's requirements and must consistently meet these requirements. There must be an awareness of the real customer. Many organizations have gotten confused on the topic of customer orientation. They start defining "internal customers" and attempt to meet these internal needs. This is a problem! Of course, it is important to work with people within your organization. But, the internal team should work in unison to meet the "real customers" needs. The customer must be viewed as a friend, as an ally to achieve success. The golden rule should be followed in establishing supplier-customer relationships, i.e. treat your vendors like you want to be treated by your customers. To establish the proper relationships, fewer, more long-term relationships must be established. These relationships must be nurtured and allowed to grow over time. Truly, all beyond competitiveness relationships shall be based upon material handling systems that are based upon meeting and exceeding customer requirements.

2. **Time Compression**: The reduction of lead times, the reduction of lot sizes, the reduction of setup times, the shortening of product lives, the shortening of product development cycles and the increased turnover of inventory will all result in a compression of time. Due to this time compression, there will be an increased need for responsiveness to the quickened pace. Future shock and the resulting cultural changes will

significantly alter the material flow requirements, and thus material handling systems must be designed to meet these responsiveness and pace requirements.

3. **Systems Planning**: Managers in the past have been considered successful when they have managed uncertainty, when they have successfully fought fires. In today's demanding environment, the fire fighter is obsolete. A manager today will not be successful by managing uncertainty, but rather by managing certainty. Systems and operations must be put into effect to that will allow management to proactively plan and control the operations. Strategic plans will be established and implemented. Strategic plans will be established for material handling systems and these plans, once implemented, will result in operations that perform according to plan. These material handling strategic master plans will be integrated with business strategic master plans to result in an overall plan for successful operations.

4. **Adaptability**: Material handling systems must be designed to handle the variety and volume requirements that must be addressed to meet customer requirements and time compression. Systems must be planned to handle the flexibility and modularity requirements that will occur due to the day-to-day material flow requirement variations. The cultural adjustment in material handling systems that will result from more adaptable material flow requirements will be very significant.

5. **Cost Competitiveness**: To make the proper impact, a vision of long-term cost competitiveness must be established. This vision will not be based upon an annual cost reduction of 3-5%. The culture of cost competitiveness will demand annual cost reductions in the 10-15% category. These cost reductions will differ from the traditional 3-5% cost reductions in that they will not simply focus on improving operations, but rather, will focus on the total redesign of operations. The objective will not be to refine the material handling systems, but to revolutionize the material handling systems. This broader expectation will result in the need for a material handling system innovativeness not typically seen in the past.

To perform relevant material handling research we must redirect our energies. We must encourage material handling equipment design research (Category I) and material handling as a component of manufacturing and distribution systems research (Category IV). It is only by emphasizing these two areas that the benefits of material handling research will be realized.

3 ACKNOWLEDGEMENT

This chapter is based in part upon the new book, *Winning Manufacturing: The How-To Book of Successful Manufacturing,* published by the Institute of Industrial Engineers and available from Tompkins Associates, Inc.

THE STATE OF MATERIAL HANDLING RESEARCH

Stephen L. Parsley
Litton Industrial Automation

1 INTRODUCTION

Litton Industrial Automation Systems has been a strong supporter of material handling research. New product innovations have been the backbone of Litton's ability to retain market leadership. In addition to totally new ideas, Litton is constantly looking for improvements of existing products and new ways of applying existing products.

Our interest extends beyond new product innovation into the area of evolving standards for performance and analysis of material handling systems. While most standards that are proposed by various organizations are accepted if they are good, bad standards serve to confuse the market and cause our customers to make bad decisions. For this reason, we stay constantly in touch with those organizations that are doing research in system analysis. (We also stay in touch with those organizations that make "rule of thumb" type of application recommendations.)

We stay in touch with the research institutions for other reasons as well. For example, associations with academic institutions gives us a familiarity with a pool of people who could be available for special research. Additionally, this set of contacts allows us to more thoroughly review the qualifications and interest of perspective employees that we may hire out of the university. Finally, we find that by allowing our engineers and scientists to maintain a regular contact with academia, a positive effect on creativity is experienced.

2 RESEARCH VS. DEVELOPMENT

From the introduction, it should be apparent that Litton has some very specific ideas about research and development. Further, to Litton, these are two distinctly different issues. Research is the process of making new theories or novel ideas into good theories and ideas.

At this point, it may be appropriate to define a "good" theory. A theory is considered "good" if it satisfies two requirements:

1. It must accurately describe a large class of observations on the basis of a model that contains only a few arbitrary elements.
2. It must make definite predictions about the results of future observations.

Example: Aristotle said all matter was made up of four elements; earth, wind, fire, and water. The problem with his theory is that there was no criteria established with which to test its validity.

Newton, on the other hand, proposed the theory that two separate bodies are attracted to each other by a force that is proportionate to their masses, and inversely proportionate to the distance between them.

The latter theory is even simpler than the first in that it contains fewer arbitrary elements, and it is considered a good theory in that it provides a means of validation. Still it is subject to invalidation and thus modification, as was shown by Einstein with the theory of relativity.

A good theory is always provisional. After all, a good theory is only a hypothesis. You can never prove it, but you may disprove it. No matter how many times you repeat the experiment, you can never be sure that the next time the results won't contradict the theory.

With each test, however, the theory either survives or it fails. With survival, we gain confidence in the theory. With failure, we are faced with the choice of abandoning the theory or modifying it based on what we have learned. (Another approach, is to question the competence of the person who carried out the failing experiment.)

Development is the process of converting "good" theories into practice. Unlike research whose goal is knowledge and the development of confidence in a hypothesis, the goal of development is to produce a product or service based on that hypothesis. While a research program may also be driven by budget and schedule considerations, it is far more common to see the developer focused on finance and timing issues.

If the developer doesn't foresee a return from the effort that will justify the cost of development, the work is rarely pursued. Failure at this point can be caused by many factors, not the least of which is the possibility that the idea cannot be converted into something useful. On the other hand, many research projects never leave the lab because the developer was unable to assimilate all the knowledge from the researcher, enabling the developer to appraise the true potential of the idea or concept. When this occurs, we blame it on inadequate technology transfer.

3 TECHNOLOGY TRANSFER

Technology transfer is the process of communicating a good theory to a developer in such a way that the developer accepts the validity of the theory and is able to put the theory to practical use. In some cases, the process of technology transfer may involve documentation of the theory, documentation of the tests and the findings, or creation of a prototype. In all cases, however, technology transfer is a process of communication; the transfer of information from the researcher to the developer. The techniques for improving communications are entirely appropriate for evaluating and improving the technology transfer process. When applying these techniques, however, some unique barriers to communication appear that are not under the direct control of the researcher or the developer.

3.1 Proprietary Agreement

Within most research centers in this country, research is funded by a combination of sources. With less funding coming from the government, research institutes are turning more to private industry for their support (*see also pages 6 and 8, ed.*). With this reliance on industry, however, proprietary agreements are becoming common place. With such agreements in place, research institutions are unable to exchange data with other institutions who may be funded by the former's competitors.

3.2 Publish or Perish

Another problem often found in academic institutions is the publish or parish syndrome. Sharing of data may be a problem, but sharing of publication acknowledgements is even more of a problem. In academia, publications affect one's compensation and promotion. Publication volume plays an important role in the reward system for anyone on a tenure track. Sharing publications, particularly across academic boundaries, dilutes the value of the reward.

Adding to the problem is the need for those publications that contribute to a researcher's career path to conform to a specific technical content and structure standard. They are extremely time consuming to produce, and if routinely takes months for the referee process to be completed. The dwell between start and end often deletes the momentum of even the most exciting research programs.

Furthermore, the finished product, while perhaps an effective tool for communication with other researchers, may not effectively communicate information to a developer. It has been said that every question in a technical report will cut its audience in half. This may be an optimistic estimate. The simple fact is that too little time is available for publication of technical articles, but even less is available for the publication of articles for the layman.

3.3 Not Invented Here

Most material handling research is performed within the engineering community. As a result, the researchers are well acquainted with the term, "Not invented here".

This becomes a significant barrier to communication partly because the inability to communicate enthusiasm often eliminates the ability to communicate information, even at the researcher level.

At this point, it may be appropriate to point out some of the buzz phrases that contribute to the "Not invented here" syndrome.

1. Simplicity breeds suspicion.
2. The principle of design inertia: any change looks terrible at first.
3. No design is bad enough that it can't get worse.
4. The key to flexibility is indecision.

In order for technology transfer to be effective, there needs to be early involvement between the researcher and the developer. As already mentioned, the inability to communicate enthusiasm is often caused by the researcher trying to pass an idea or theory in an unsolicited fashion to a developer. If the developer does not have the same enthusiasm as the researcher, the developer will not take full advantage of the idea's potential. Furthermore, his disinterest may have a deleterious effect on the researcher's enthusiasm and further efforts.

We often joke that the best way to get an idea across is to get the user to think it was his idea. This is exactly what must happen in order to avoid the "not invented here" suspicion that accompanies transfers to developers who may not fully understand why you want them to use a new theory.

3.4 Not Wanted Here

A related problem is one that we classify as the "not needed here" problem. In reviewing material handling research, it is common to find research that is duplicating what has already been done at other institutions or has reached the commercial development arena. To blame this on the researcher is unfair. Information on what is in development may not be flowing from the developer back to the research institution, misleading him as to what research is needed. One reason for this, is that some of the things that are currently being done in industry probably are proprietary in nature, and industry does not wish to share this information with a researcher where it might enter the public domain.

Another part of the "not needed here" problem is that some researchers are pursuing problems that are already obsolete. The lack of a national research clearing house has resulted in significant redundancy. The research community is full of examples where parallel research has been conducted, only to have one of the research institutions move rapidly ahead of the other through discovery or other factors. With proprietary agreements inhibiting cross boundary communications, it is also well documented that some research becomes obsolete due to work done elsewhere.

A clearing house would facilitate the elimination of some of these problems. I think that with sufficient involvement on the part of industry, a clearing house could act to preserve research resources and help them to avoid falling into an unproductive trap.

3.5 Profit Motive

Another barrier to effective technology transfer is the profit motive. With reductions in state and federal funding forcing research institutions to look to private industry for support, compromises have been necessary. In the past, publicly funded research could be done in the public domain, openly sharing data with other researchers or agencies. To attract industry sponsorship, however, proprietary agreements were introduced to given the sponsoring agency a competitive advantage through an "early look" at the results.

In the material handling industry, a good idea is worth six months. In other words, a supplier who gets a good idea from a research institution will have six relatively competition free months in which to exploit the idea and recoup his development expense. After that time, a good idea will be cloned and possibly improved upon by competitors within the industry.

Few organizations will invest in the development of something that has little or no proprietary market advantage. We often see this with research programs that are not picked up by the sponsoring companies. When this occurs, we sometimes rationalize that the technology transfer process failed. More often, we accuse the disinterested party of not being smart enough to see the product's potential.

The fact is, too often, development occurs at the research institution, and the developed work becomes public domain before technology transfer has had a chance to work. When this occurs, it reduces the potential value of the idea to that of who can produce the potential products this idea may spawn at the lowest cost. In many cases, the organizations that have the foresight to sponsor research and support research organizations are large, high quality suppliers who do not typically have the lowest manufacturing costs. The proprietary agreement gives these members of the development community a chance to be competitive and to recoup some of their investment.

3.6 Poor Marketing

Finally, a barrier to technology transfer is ineffective marketing or delivery. As mentioned before, the loss of momentum between research and development often comes from the researcher assuming that the developer needs or wants the theory. It can also come from having developed a theory that is not broad enough to be of practical use. For example, many cycle-time programs have been written to study specific cases of ASRS performance, with the program itself being held up is a potential standard for analysis. Unfortunately, because of the diversity with which ASRS equipment is applied, no one program has been adopted, even by the research community, as a standard. Still, hundreds of hours of researcher time is absorbed each year analyzing specific cases.

4 PRODUCTIVE RESEARCH

Litton is interested in productive research. Productive research is work that will seed our own development of products and services which are aimed at helping us compete in current or future markets. We are interested in receiving the "why" projects from academia. Given the results of this type of research, we can then focus on developing a value added product. Thus, we prefer research projects that focus on processes versus products.

While we realize prototyping is an effective way of supplementing research, putting the "how" or the development burden on the researcher is usually inefficient. We find that the somewhat idealistic environment of the research lab often causes the researcher to misconstrue or negate the goal. For this reason, we believe that the researcher is better utilized developing and testing new theories.

Furthermore, we are interested in pursuing and funding research that addresses dominant problems. We are not interested in research that is going to get an additional 2 or 3 percent of performance out of an already mature technology. We are interested in research that may enable us to apply an existing or mature family of technology in a previously unserved market.

Finally, we are interested in research that is conducted with technology transfer as a goal. In this sort of a project, the research proposal will include a framework for technology transfer even before the research begins. This framework will not only include a listing of the deliverables such as documentation, prototypes, etc., that will be used to communicate theory to the developer, but it will also identify the developers and a schedule for their involvement so that research momentum can be passed on.

5 ISSUES FOR THE FUTURE

Material handling research issues for the future dominantly revolve around issues for improving the effectiveness of technology transfer. In our opinion, research is solely limited by this. For this reason, we will be supporting the encouragement of two types of research reports. First, the academic documentation that is needed to pass information between researchers must be continued. While this type of report has long been critiqued as too complex for the layman, this level of information is needed so that research continuity can be preserved at the researcher level. With it, however, we need to see more time spent on developing practitioner/developer documentation.

Another challenge is the need for a national material handling and manufacturing research clearing house. A current project to affiliate several major research universities into a multi university consortium for material handling research is a start toward this goal. In a world economy where competitiveness is at stake it is essential that research be free to operate unencumbered by traditional restrictions. While respecting the concept of academic rewards through the publish or perish system, a national clearing house for research needs to be structured so as to provide rewards to researchers to replace those rewards that are lost through a cooperative sharing of ideas and a loss of exclusive research idea ownership.

This research clearing house must also find a method of combining resources to supplement those projects that would otherwise be delayed because of a lack of effective researchers. Furthermore, this clearing house could serve to distribute the work of rapidly proof testing theories to produce good theories prior to moving them into the technology transfer process.

Finally, a national research clearing house should be charged with the responsibility of tracking the progress of research, serving as a barometer for encouraging new research efforts, and facilitating the expeditious turnaround of technical publications.

6 CONCLUSIONS

Overall, research is important to companies like Litton, who depend on innovation and creativity to retain market leadership positions. Today, our emphasis has shifted to improving the effectiveness of the technology transfer process. We are interested in acquiring research at the transfer stage, performing the development ourselves to add proprietary value to the end product.

Barriers to communication have been around since the beginning of time, but only now are we beginning to realize how these same barriers are impeding the technology transfer process. These barriers can be reduced through early involvement between the researcher and the developer. Other steps need to be taken as well, however, and a National Clearing House for the classification, monitoring, exchange and documentation of Material Handling and Manufacturing research may be a partial answer.

Finally, industry is interested in research on dominant problems. The research program designed to extract another 2-3% of performance out of a mature technology is rarely worth investigation, while work to evaluate the potential of using this same mature technology in an entirely new and previously unthought of way is of significant value.

As our world changes, the successfulness, both of research and of the technology transfer process, is becoming increasingly vital to sustaining U.S. competitiveness in world markets. In the 1990s, industry must play a stronger role in encouraging and guiding research and take the initiative in establishing effective technology transfer.

RESEARCH IN THE 1990s

The Editors

1 INTRODUCTION

The Material Handling Research Colloquium convened June 19-21 in Hebron, Kentucky and brought together over 40 researchers from nearly 25 institutions. The objectives of these two and one-half days of intensive meetings were the exchange of viewpoints, the assessment of the state of material handling research, and the identification of key issues confronting the material handling research community. The goal was to anticipate the directions of material handling research in the 1990s through an examination of current activity and the identification of key barriers. The following sections of this chapter provide some insight into the kinds of problems being addressed, as well as the methodologies and approaches being used by researchers.

The agenda for the Colloquium did not consist of a typical presentation by each researcher. Rather, a subset of the papers was selected for presentation, with lengthy discussion periods interspersed. This format proved quite successful, and was interesting in that the ensuing discussions addressed issues and concerns that were only peripherally related to the technical aspects of the papers presented! An industry participant in the Colloquium observed that the discussions revealed much more about the concerns of the researchers than did their papers.

It would not be possible to give a complete account of all the lively discussions, which often carried over into breaks, meals, and evenings. There were, however, several recurring "global" themes. Note that only the nature of the issues that were raised has been recorded and not suggested resolutions, for no consensus was reached in that regard.

2 TECHNOLOGY TRANSFER

Many of the discussions related to the general issue of technology transfer, and in particular to the impact that research results have on the practice of material handling. By and large, the material handling research community is concerned that its efforts not only address relevant problems, but also make significant contributions to solving them. A number of factors were identified as potentially detrimental to accomplishing these goals, including: the distortion induced by the academic recognition and reward structure, which emphasizes primarily refereed publications; the disparity in the goals and time frames for accomplishing those goals between universities and industry; and the difficulties many academicians face in establishing a fruitful collaborative relationship with industry. The point was made that the research community may focus on the communication within the research community to the detriment of communication of results to the potential users outside the research community.

Another salient point made during the discussions was that the researchers themselves could benefit from technology transfer from other, closely related disciplines. The researchers in logistics and transportation (movement between factories) have been working on a number of problems which have direct analogies to the problems considered by those researchers who usually concern themselves with movement within the factory or warehouse.

Related to the technology transfer issue was the observation that material handling research almost always addresses problems observed in existing systems, or in existing technologies. These problems already have solutions, so the most that research can offer is an incremental improvement. Research is likely to have a greater impact if it addresses new technologies, not already in use, or addresses new applications. Therein lies a great challenge for the research community.

The discussions also revealed that technology transfer in material handling research may take place without a specific identifiable research "product," such as a piece of software or a new device. For example, concepts formalized in a research context may be presented to material handling system developers or users, resulting in a change in design approach or management strategy. While the impact of the research is indirect, it is, nonetheless, real. The problem for the research community is to recognize when the opportunity for such technology transfer exists.

3 SCIENCE BASE

Another topic of many discussions was the science base, or the lack of one, for material handling systems. Evidence that the science base is meager and poorly organized includes: the absence of any standard terminology for material handling; the paucity of rich descriptions of material handling problems or systems; the absence of any unifying system for classifying material handling problems or systems; and the near total absence of any real empirical research in material handling. A compelling comparison was drawn between queuing theory, for which a very robust classification system exists, and storage systems, which appear to exhibit similar degrees of structure, yet for which no such classification system exists.

In all fairness, it should be recognized that the broader arena of manufacturing systems also suffers from this malaise, i.e., a very weak and poorly organized science base. Perhaps an important contribution of the Colloquium has been to bring this issue to the fore.

One difficulty that may be unique to material handling systems research is the availability of a suitable "laboratory" in which to perform empirical research. Given space and budget constraints in the university setting, it is practically impossible to create a suitably realistic or sufficiently complex material handling system using full scale equipment. While scale-model system replicas have some appeal, they do not appear to adequately capture the behavior of full scale systems. Operating systems in such small scale environments rarely, if ever, can be used for future experimentation. The development of a solid science base of material handling systems awaits a creative resolution of this difficulty.

4 IDENTITY

A third theme that arose with some frequency can best be summarized as community spirit. The Colloquium attendees recognized, and the other chapters in this book attest to, the narrowness of the individual perspective. For example, most of the chapters address operational and systems related issues, rather than, say, devices, information, communication, container design, etc. Also, the chapters all address very traditional material handling applications rather than, e.g., moving people or luggage in an airport, medicine in a hospital, mail, hazardous materials, bulk materials, or materials in space. The material handling research community needs to be expanded to recognize, identify, and assimilate research on these broader issues.

One could, in fact, question whether a true material handling research "community" exists. There is no focused, general attendance forum for material handling research. In fact, one can find material handling research being presented at almost every engineering discipline's academic conference, as well as under names like operations research, management science, warehousing, and logistics. A similar situation holds for the publication of material handling research; such research is scattered across literally dozens of journals. A related problem seems to be the inordinately long publication review cycle, reflecting the difficulty that journal editors may have finding qualified and willing referees.

5 FLEXIBLE SYSTEMS

Another common discussion point throughout the Colloquium was focused on the term "flexibility" and its relationship to many of the areas of concern to the participants. While detailed definitions of flexibility were not offered, it was noted that flexibility in material handling systems design and flexibility in material handling systems operation seem to be commonly recognized as desirable. While there has been some recent work on "flexibility metrics," which are designed to measure the flexibility of a system in terms of the number of alternatives available at any point in design or in operation, there appears to be much left to be addressed regarding this concern.

Colloquium attendees clearly came prepared to discuss "big" issues in material handling research. One should not infer, however, that questions specific to the research presentations were not also raised. Presentation of specific models or methods of analysis generated questions about the process within which they might be used. Presentations of design methods spawned discussions around the issue of implementation, e.g., should academic researchers be building prototype software tools? This discussion also relates to testing environments in laboratory settings as mentioned previously.

6 FOCUS GROUPS

On the last day of the Colloquium, the participants were divided into three focus groups:

1) Facility design/Unit loads/Flow control
2) Storage systems/Order picking
3) Economics/Automation

The charge to each focus group was to address the following questions and to bring forth specific recommendations:

1) If material handling research is to advance, what are the principle issues that need to be addressed?
2) What are the important questions in material handling research?

Summaries from the focus group discussions follow.

6.1 Facility Design/Unit Load/Flow Control

6.1.1 Facility Design

The facilities design researchers focused their comments on the trend toward increasing complexity. The facilities design process was viewed as much more than just layout; the modern requirement is to actively integrate with other design processes, specifically those of manufacturing unit process design, assembly process design, material handling system design, and shop floor control system design. The facilities design activity, therefore, was viewed as one increasingly required to address design issues heretofore considered as "givens" in the more traditional facilities design layout context.

The changing nature of facilities design has had an impact upon research directions. The development of design aids for facilities design must push toward incorporating consideration of these additional design issues. In this sense, integrated design systems are needed to support this new, broadly-based concept of the facilities design process. The problems are both methodological and computational.

Concurrent with the broadening of the facilities design process and its consequent demand upon the research community is the need for more commonality in the representation of design decisions. A common framework for design representation would aid the research community in organizing both the existing knowledge and the advances to new knowledge as the complexity of design increases.

The demand for broader design system coverage and its impact on research activity is not the only dimension of increased complexity for the facilities design process. The traditional quantitative layout design criterion typically is focused on flow-distance as a measure of recurring material handling costs. Traditional layout design algorithms have attempted to minimize this criterion. Increasingly, a criterion of flexibility of layout design is being applied and this can significantly impact the direction of layout design research. The specification of units of measure for the flexibility criterion and the development of layout design algorithms based upon this criterion are therefore also part of the changing nature of the facilities design research area.

6.1.2 Unit Load

A fundamental question raised relative to unit loads concerned the definition and character of a unit load. This question can arise at a variety of levels of unit load definition, e.g., the definition of kits vs. component totes for assembly and the definition of pallet loads vs. bins for transport. Increased understanding of this problem would appear to be based upon better understanding of the issues of integration of unit load concepts with the needs and capabilities of the production system, the material handling system, and the distribution system. Furthermore, understanding this problem requires understanding the criteria to be used and the factors that constrain the formation of unit loads.

Further research is needed to improve our understanding of the role of automation in the unitization process. A significant question was noted during the discussion; namely, what should guide the design decision to specify manual, robotic, or automated unitizing approaches?

When conditions are favorable for unitizing loads with a robot or other automated device, then further research work is needed in algorithms for the unitizing of loads. Better understanding of conditions of imperfect geometric fit (the inexact matching of load geometry to the carrier geometry) may enable the development of algorithms with improved performance characteristics. Better design tools for three dimensional pattern development may also allow for the implementation of improved algorithms.

In addition to the need for improved understanding of the role of automation is the need to quantify the effects of container standardization on design decisions. Benefits of container standardization would appear to accrue to manual as well as to robotic and to automated unitizing approaches, but to differing degrees. If container standardization is to occur, then the definition of standard container sizes is a research area which should have a substantial impact upon the unit load research community.

6.1.3 Flow Control

The continuum of approaches to material handling can be characterized as follows:

Manual..>Mechanical..>Automated..>Integrated..>Intelligent

Issues of flow control are certainly important at every step of this continuum, but recent research has concentrated primarily on the automated and integrated segments of this continuum.

A significant research issue in flow control is that of interfacing and integrating between distinct material handling systems or between a material handling system and other systems. The issue of interfacing renders research work on integrated control systems very difficult and impedes progress on the testing and implementation of conceptual models for flow control systems.

Within the domain of conceptual control models, there are a number of important issues. One important issue is that of conceptual model validation. If the conceptual model and a simulation are both intended to represent a hypothesized system and both fail to capture key attributes of the physical system to be controlled, then apparently valid, but truly erroneous, results may be reported. Another issue for flow control researchers was that of identifying appropriate objectives and constraining elements for control system design. If hierarchical architectures are to be used, what is the appropriate assignment of responsibilities to levels in the system and how does this design decision interface with the control hardware capability at each level? What role does a criterion of flexibility play in this design/model decision and how should it be measured? And, is a purely hierarchical paradigm appropriate in systems that incorporate material movement?

Advancing along the continuum from control system design for automated and integrated systems to control system design for intelligent systems presents a formidable challenge to researchers. What might "smart" material handling devices be designed to do and how might an intelligent control system support this role? The design issues and the testing/validation issues for such control systems represent significant new research frontiers for the 1990s.

6.2 Storage Systems/Order Picking

The interface between material flow and information flow must not be forgotten when designing storage systems or management information systems, or for that matter, any material handling system. For example, in designing automated storage and retrieval systems, or in designing alternative methods for orderpicking, etc., the specific problem will define the information system required to support it. In turn, the information system design may have an impact on the design decisions regarding the batching of orders, the equipment selection, etc.

Therefore, the flow of information is an important consideration for material handling system model development. It was noted that commercial computer-aided software engineering (CASE) products may be of assistance to material handling systems modelers, at least with respect to information; however, the CASE systems usually only consider information flow while ignoring material flow. It was thought that some standardization of terms, approach, problems, etc., is needed for the sake of communication of research activities. Toward this point, a number of paradigms were offered.

One paradigm argues for context specificity when building models of material handling systems in research. But, unfortunately, there is not a standard language for succinctly describing the context of material handling problems. There is an enormous number of research papers dealing with automatic storage and retrieval systems that have been published in the refereed literature. Yet, there is no "Dewey decimal system" for classifying AS/RS papers as there is for classifying queuing models. A classification system for material handling models is needed to support the context specific research paradigm.

Another paradigm was proposed in the context of material handling system design. In this paradigm, a metamodel (a model of models) framework would be created to support design modeling and design integration. The metamodel would ensure consistency between, say, layout models, which are essentially geometric, and control models, which are logical and algorithmic. The metamodel paradigm also addresses the lack of attention paid to modeling the design process itself. Most modeling-oriented research focuses on operational or control aspects of the system being designed, and ignores the design process.

The discussion of the two paradigms revealed that they really define opposite ends of a spectrum. The question arose as to who will interpret, or translate, the metamodels into context specific applications to solve real material handling design or operational problems. One answer is that the human, as an element of the system, will do so. As a consequence, more research into the development of design tools, like the engineering design workstation approach, is needed to bridge the gap. Such tools assist with the definition and visualization of the problem and employ graphics technologies. Other answers appear to be possible and await further research.

No matter which modeling paradigm, or combination of paradigms, is used, great care must be taken by researchers that the objectives of material handling systems models, e.g., picker utilization, minimization of the product of flow volume times distance, etc., align with the objectives of material handling system applications in industry. A misalignment of model objectives from those of the application environment is a major reason that research results are not put into practice.

The conclusion of this aspect of the discussion was that the dynamic nature and high state of flux of many industries today demands rapid development of answers to material handling system design problems. At the practical level, optimum solutions from models may not be as desirable as feasible solutions obtained in a timely manner. Therefore the material handling industry and the research community need to work together more closely in order to more quickly develop solutions to the right problems.

6.3 Economics/Automation

Of all the topics presented and discussed at the Colloquium, this was the most diverse and stood the greatest risk of fostering scattered comment and opinions.

This turned out not to be the case, either during the open discussion periods or during the small group discussions on the final day of the Colloquium.

During the open discussion, attention immediately focused on the general absence of research related to the electromechanical aspects of material handling equipment. In terms of attendance at the Colloquium, it was acknowledged that those in attendance were persons doing research on subjects related to planning, systems design, and the logic of control systems. But, two other points were also made. The first being that material handling tends to be perceived as a "people" problem and an operations problem. Therefore, it tends to attract industrial engineers more readily than engineers from other disciplines. The second point was that there is indeed research and development being done in areas which directly impact material handling but which doesn't get labeled as material handling research. Seismic loading of (storage) rack structures is one example, work on mechanical power transmission and servo mechanisms is another, and the whole area of data communications as it relates to remote control is yet a third. This discussion then led to another regarding what is included in the field of material handling and what is not. For example, when is a production planning and control problem also a material handling problem?

There appears to be very little concrete research and development work in economic justification taking place at this time, despite the lament from all corners that existing cost accounting practices and justification measures are inadequate. There was some debate over the issue of how best to deal with uncertainty when evaluating large, complex systems. The conclusion of these discussions was that if indeed economic justification is an important question, it is also a fertile and open area for further research work.

As the question of economic justification was debated, one very interesting point was made. It was suggested that persons concerned about justification problems might be better advised to think of the problem as being as much a confidence problem as a truly economic problem. It is a confidence problem in the sense that word spreads quickly when a new automation project encounters difficulties. The challenge for researchers is to develop more reliable methods for predicting the economic impact of material handling automation.

A strong recommendation was made by this discussion group for a sharing of empirically validated simulators among the research community. The argument even went so far as to suggest a standardized test bed for classes of modeling problems. This point quickly led to a number of related issues. The first being the need for more empirical research and validation with reference to both simulation models and analytical models. A second, related point was the need to examine potential benefits that could be derived from high-fidelity case studies; these case studies are called high fidelity because of the inclusion of extensive system performance data.

On the subject of information, it was agreed that it is almost taken for granted that information flow is an inseparable part of the material flow problem. However, despite this relationship and its importance, there appears to be little or no effort to establish standard methodologies or procedures for defining the information requirements of a material flow problem. Communication among researchers and material handling practitioners could be significantly enhanced if such standard methodologies were to be widely adopted within the field of material handling system planning and design during the 1990s.

7 CONCLUSION

The Material Handling Research Colloquium appears to have succeeded in achieving its stated goals. The participants presented and discussed their viewpoints on a wide range of subjects. Although there were differences of opinion, there was a remarkable coincidence of views. The participants offered their assessments of the state of material handling research and these general assessments have been presented in the previous paragraphs. Finally, a number of key issues confronting the material handling research community were identified and described. In addition to the formal communication and documentation goals, it appears that the Colloquium strengthened the sense of community of those who are active in material handling research. The accomplishment of these formal and informal results for the first Material Handling Research Colloquium would therefore argue strongly for future colloquia to foster further progress in the field of material handling research.

SECTION TWO
FACILITIES DESIGN AND UNIT LOADS

Facilities layout research over the last twenty years has produced a number of advances, both in theory and in computational procedures. However, these advances have not, in general, found their way into common use. Sometimes this is because the information never becomes known beyond the narrow academic discipline where it originated, sometimes it is because the theory or technique requires software that is not commercially available. Whatever the reasons, the software tools commercially available to aid in facilities layout are technologically "obsolete", yet are widely used because nothing better is available.

Elements of facility layout, like the design of an AGV network, can be performed using sophisticated computer-aided tools, like the engineering workstation described by Leon F. McGinnis. New theories, like that for directly producing a layout from the material flow requirements proposed by Benoit Montreuil, and new heuristics, like that Wilbert E. Wilhelm suggests for locating machines along a production line, hold promise for additional tools for the 1990s.

The Unit Load principle, one of the cornerstones of material handling system design, has come under increasing attack from the proponents of the just-in-time (JIT) philosophy. Decreasing set-up costs, increasing product variety, and fewer handling operations in a JIT environment, are some of the factors that reduce the benefits of handling products in unit loads.

Three chapters address the issue of Unit Load size. Pius J. Egbelu considers unit load size in a batch production environment where loads are assigned to different machines. At a more detailed level, Jessica O. Matson considers single and multiple carton sizes for shipping products. Jaime Trevino presents two solution approaches to a nonlinear integer programming formulation that minimizes the sum of storage, transport, handling, and pallet costs.

In the last chapter of this section, Russel D. Tsai presents a dynamic-programming-based heuristic for solving the three-dimensional pallet configuration problem in real time.

Gunter P. Sharp

BATCH PRODUCTION WITH UNIT LOAD DESIGN AND SCHEDULING CONSIDERATION

Pius J. Egbelu
The Pennsylvania State University

ABSTRACT

Despite the fact that most batch jobs are moved through shops in unit loads, very limited work has been reported on how to select the best unit load size and the routing of these unit loads through the machines to minimize the batch completion time. In this chapter, the problem of determining the best unit load size for a job and the routing of these unit loads through the machines in a multistage system with material handling consideration is addressed. The design objective is the minimization of the batch completion time. Both stationary and mobile based material handling systems are considered. Results indicate that the unit load and routing requirements for mobile and stationary systems are different.

1 INTRODUCTION

Statistics indicate that about 78% of all manufacturing activities in the U.S.A. fall under the classification of a batch production system [13]. In batch production, parts are manufactured in lots that may range from a few parts, say ten, to as many as thousands. With improved manufacturing technology through advanced machine design and just-in-time application coupled with the need to lower inventory cost, the trend is toward smaller lot sizes. One of the problems that must be addressed to ensure efficient manufacture for a batch job is that of material movement between the machines on the job route. Decision must be made whether the entire batch is to be moved as a unit between machines or broken down into sublots or unit loads. If the decision to move the batch in sublots is accepted, then the second stage decision of determining the optimal number of parts to form a unit load must be made.

As is common in most manufacturing systems, a production stage visited by the batch may consist of multiple, heterogeneous machines that differ in performance due to differences in age, make and model. Machines in a particular stage may occupy a particular section of the shop floor or the stage location may simply be logical. Although machines in a logical stage are functionally similar, they do not necessarily occupy a unique section of the shop floor.

Machines within a logical stage may be located at different sections of a shop or different buildings in a manufacturing complex. Therefore unit load movement from one stage to another must decide to which machine in the next production stage the unit load has to be sent if production efficiency is to be maintained. Such decision must not only consider the production efficiency of the machines at the succeeding stage but also the distances between the source machines and the potential succeeding machines.

In this chapter, the problem of manufacturing a batch job consisting of multiple identical parts through a multistage production system with unit load design and material handling consideration is addressed. The machines in a given stage are assumed to be functionally similar but may differ in performance due to age, make, and model. The objective is to minimize the batch manufacturing time through the optimal selection of the unit load size and the assignment of the unit loads to machines at each stage. The routing of the unit loads necessary to achieve the minimum manufacturing time is also to be determined.

2 RESEARCH JUSTIFICATION

It is a well known result in scheduling that job splitting can reduce the shop time of a job [2,5,10]. This is because when job splitting is allowed, the same job can witness processing at multiple stages and machines simultaneously and thus accelerate the completion time of the job. Since, in most scheduling models the effect of material handling on production time is generally ignored [2,5,6,10,12], it follows from a scheduling viewpoint that the optimal policy is to move a unit part to the next operation as soon as processing is completed on the part in the preceding stage. Furthermore, it is not common in scheduling models to consider a production stage as consisting of two or more machines with nonuniform performance [2,5,6,10,12]. In these respects, the problem presented in this chapter differs significantly from the traditional scheduling models. Because small size unit loads may overburden the material handling system to the extent of becoming the bottleneck resource, in the problem considered here, it is difficult to speak with any authority what the best unit load size would be a priori. The increased material handling time imposed on the system due to small size unit load specification may be sufficiently large to neutralize, or worse still, surpass any gains made from reduced operation time due to job splitting. Another source of distinction between the traditional job scheduling problem and the problem addressed here is the number of jobs that are involved. Whereas traditional scheduling problems are concerned with multiple distinct jobs assumed to exist in lot sizes of one [2,5,6,10,12], the problem considered in this chapter is concerned with a single job assumed to exist in multiple units. Whereas the traditional scheduling problem represents higher level planning, the single job planning problem is more detailed and represents a lower level planning than is called for at the multiple job scheduling level. Detailed and lower level planning are an essential component of all production systems, although they may be done informally in most manufacturing organizations. Although detailed planning for a job is as critical in affecting manufacturing efficiency as the aggregate scheduling of multiple jobs, it has not received the necessary research attention it deserves. The work reported in this chapter is aimed partially at stimulating more research interest in the area of lower level planning and to call attention to the importance of lower level planning in ensuring the realization of the objectives set out at the higher level planning stage.

3 SURVEY OF LITERATURE

Ever since the concept of unit load was popularized by Apple [1] in his book on material handling, several research works that deal with the design of unit loads to improve manufacturing and material flow efficiency have appeared. Tanchoco, Wysk, Davis, and Agee [20] in 1979 discussed the economic effect of unit loads and their relationship to manufacturing efficiency. This was followed by another paper in 1981 by Tanchoco and Agee [18] on the interaction between unit loads and other elements of warehousing systems. For example, a paper by Tanchoco et al. [19] addressed the problem of designing a unit load conveyor. In that paper, the unit load size is one of the decision variables in the conveyor selection model. The paper pursued the objective of minimizing the conveyor system cost through the optimal selection of unit load size. The work by Tanchoco, Davis, Egbelu, and Wysk [21] dealt with the problem of unit load size specification in a lot sizing decision environment. The model they developed related unit load decision to production lot size decision.

The problem of determining the best unit load size, in order to minimize manufacturing cost in a batch production system with automated guided vehicle based handling systems, has been addressed by Egbelu [8,9]. Steudel and Moon [17] also addressed a similar problem as Egbelu [8,9] except that their focus was on minimization of mean flow time of jobs. The solution approach used by Egbelu and Steudel and Moon was a combination of optimization and simulation techniques.

The problem of timing and launching of unit loads into a synchronous automated guided vehicle based batch flow line was addressed by Egbelu and Roy [7]. Their work was aimed at minimizing the production time of a batch while minimizing the queue holding time of unit loads. Truscott [22], in 1985, addressed the problem of setup and operation scheduling in a multi-stage batch manufacturing system with the objective of minimizing the number of machine setups and maximize facility utilization. He also discussed the problem of unit load size selection. Unfortunately, no algorithm was provided on how to determine the best unit load size. Both the work by Egbelu and Roy [7] and Truscott only considered production systems that contain one machine at each production stage. In this study, multiple machines per stage is considered and the machines need not be identical. Furthermore, this chapter also differs from the Egbelu and Roy and Truscott in the sense that both mobile material handling equipment (e.g. vehicles) with round trip consideration and fixed material handling equipment (e.g. conveyor) are modeled. An algorithm for selecting the best unit load size in order to minimize batch production time is developed as part of the chapter. The unit load selection problem was only identified as a necessary factor in the earlier papers but no algorithms were actually developed.

Other works that have pursued the determination of the best unit load size in a production scheduling context are those of Santos and Magazine [16] and Baker [3]. Santos and Magazine addressed the single facility scheduling problem with multiple jobs. The objective of their paper was to determine the optimal number of sublots to split a given job in order to optimize on some measure of performance. Thus, one of the decision variables in their model is the number of sublots for each job. Several single facility models under various conditions were proposed for what they hoped would stimulate a new direction in scheduling research. Baker's work on the other hand dealt with a two stage and three stage flow shop with no material handling consideration. Mathematical models for determining the optimal number of unit loads and the size of each unit load were developed. Like the work by Egbelu and Roy and Truscott, only one facility per production stage is considered in Baker's paper. A unique feature of Baker's work is that the sublots are not constrained to be of equal size. Its drawback is that it is applicable only to small size problems and no material handling effect is considered. The work reported in this chapter does consider material handling and multiple machines per stage. Furthermore, it is capable of dealing with a large problem because of the optimal seeking heuristic algorithm employed.

4 MODEL DEVELOPMENT

The following assumptions are made in the development of the models presented in this section.

i) Parts in a unit load are identical.
ii) Parts in a unit load constitute both the handling and processing unit or entity.
iii) Once unitized, a unit load cannot be broken down into smaller unit loads later in the process or combined with other unit loads to form a larger unit load.
iv) Unit loads are transported between machines one at a time.
v) All unit loads are of equal size unless it is impossible to form a full load with the last unit load due to the batch size.
vi) Once completed at a machine in a preceding stage, the unit load immediately proceeds to the succeeding production stage as soon as the material handling system becomes available.
vii) Material transfer between every machine pair in adjacent production stages is possible subject to obeying the routing requirements of the job. If material transfer between any pair of machines is prohibited, the associated transfer time between them is set to infinity.
viii) The operation sequence required by the batch is known.
ix) Each machine has a unit of material handling device that serves it to move completed unit loads to their next destination.

Assumption (ix) implies that each machine has a dedicated transportation device assigned to serve its needs in the case of a mobile system. For the case of a stationary system, assumption (ix) implies machines in adjacent stages are all connected by the transportation system. The following variables are used at various sections of this chapter. Additional local variables are defined in different sections of the chapter as the need arises.

Q = number of parts contained in the batch (i.e., the batch size).
q = number of parts in a full unit load or sublot. This parameter is also referred to as the unit load size.
N_q = number of unit loads generated from a batch of size Q if a full unit load contains q parts.

$\overline{N}_q$ = the size of any partial unit load present within the set of the N_q unit loads generated from the batch size Q, given that a full unit load contains q parts.

P_{ij} = processing time per part on the jth machine of stage i.

Z_{ij} = setup time for machine j in stage i.

F_{ij} = preparatory time per unit load on machine j of stage i. This time is independent of unit load size.

$t_{ij(i+1)k}$ = travel time from machine j in stage i to machine k in stage (i+1).

M = number of machine stages.

$\overline{M}(i)$ = set of machines in stage i.

m_i = $|\overline{M}(i)|$, the number of machines in the ith stage.

$A^n_{ij(i+1)k(q)}$ = the arrival time of the nth unit load at machine k in stage (i+1), given that the unit load was sent to machine $k \varepsilon \overline{M}(i+1)$ from machine $j \varepsilon \overline{M}(i)$ and the size of the unit load is q.

$S^n_{ij(i+1)k(q)}$ = operation start time on the nth unit load on machine $k \varepsilon \overline{M}(i+1)$, given that unit load n was sent to machine $k \varepsilon \overline{M}(i+1)$ from machine $j \varepsilon \overline{M}(i)$ and the size of the unit load is q.

$C^n_{ij(i+1)k(q)}$ = the completion time of the nth unit load on machine $k \varepsilon \overline{M}(i+1)$, given that unit load n was sent to machine $k \varepsilon \overline{M}(i+1)$ from machine $j \varepsilon \overline{M}(i)$ and the size of the unit is q.

$\overline{T}^n_{(i+1)k}$ = the earliest ready time of machine $k \varepsilon \overline{M}(i+1)$ after the arrival of unit load n at machine station $k \varepsilon \overline{M}(i+1)$. Note that $\overline{T}^n_{(i+1)k} = A^n_{ij(i+1)k(q)}$ if the machine is idle when unit load n arrives at station $k \varepsilon \overline{M}(i+1)$.

$R^n_{ij(i+1)(\lambda,k)}$ = the earliest time the material handling device can pick up unit load n from machine $j \varepsilon \overline{M}(i)$ for delivery to machine $k \varepsilon \overline{M}(i+1)$, given that the unit load that preceded n on machine $i \varepsilon \overline{M}(i)$ was sent to machine $\lambda \varepsilon \overline{M}(i+1)$.

4.1 The Unit Load and Processing Time Submodel

Given that the batch size is Q and a full unit load size is $q \leq Q$, the number of unit loads, N_q, obtainable from Q is given by Equation (1).

$$N_q = [\,Q/q\,]^+, \quad 1 \leq q \leq Q \tag{1}$$

where $[a/b]^+$ refers to the smallest integer greater than or equal to the fraction (a/b). If Q is not a perfect multiple of q, the last unit load will only be partially filled. The size, $\overline{N}_q < q$ of any partial unit load is as expressed in Equation (2).

$$\overline{N}_q = Q - (N_q - 1)\,q \tag{2}$$

The total processing time, $\overline{P}_{(i+1)k(q)}$ of a unit load of size q on machine $k \; \varepsilon \; \overline{M}(i+1)$ is defined according to Equation (3).

$$\overline{P}_{(i+1)k(q)} = F_{(i+1)k} + qP_{(i+1)k} \tag{3}$$

For a partial unit load, the corresponding processing time is given by Equation (4).

$$\overline{P}_{(i+1)k(\overline{N}_q)} = F_{(i+1)k} + \overline{N}_q P_{(i+1)k} \tag{4}$$

Given that the nth unit load is processed on machine $k \; \varepsilon \; \overline{M}\,(i+1)$, the earliest operation start time $S^n_{ij(i+1)k(q)}$ on unit load n on machine $k \; \varepsilon \; \overline{M}\,(i+1)$ is defined by Equation (5).

$$S^n_{ij(i+1)k(q)} = \max\left\{A^n_{ij(i+1)k(q)},\ \overline{T}^n_{(i+1)k}\right\} \tag{5}$$

where machine $j \; \varepsilon \; \overline{M}\,(i)$ is the source of unit load n from the preceding stage and $\overline{T}^n_{(i+1)k}$ is the completion time of the unit load that immediately preceded unit load n on machine $k \; \varepsilon \; \overline{M}\,(i+1)$. For i=0, $A^n_{ij(i+1)k(q)} = 0$, for all $k \; \varepsilon \; \overline{M}(i+1)$. Also if unit load n is the first to be processed on machine $k \; \varepsilon \; \overline{M}\,(i+1)$, then $\overline{T}^n_{(i+1)k} = Z_{(i+1)k}$. The completion time of unit load n on machine $k \; \varepsilon \; \overline{M}\,(i+1)$ is given by Equation (6).

$$C^n_{ij(i+1)k(q)} = \begin{cases} S^n_{ij(i+1)k(q)} + \overline{P}_{(i+1)k(q)}; & \text{for a full load} \\ S^n_{ij(i+1)k\overline{N}_q} + \overline{P}_{(i+1)k(\overline{N}q)}; & \text{for a partial load} \end{cases} \tag{6}$$

The completion time $C^n_{ij(i+1)k(q)}$ will occasionally be donated simply as

$C^{n}_{..(i+1)k(q)}$ when reference to the source machine $j \varepsilon \overline{M}(i)$ is unimportant.

4.2 The Material Handling Submodel

Common material handling devices used for unit load transfer between machines in adjacent production stages can be classified broadly into two main categories, namely,

i) stationary or fixed position equipment (e.g., conveyors)
ii) mobile equipment (e.g., trucks, AGVs, etc.)

Under normal operating conditions, stationary equipment serving a machine can be assumed to be continuously available to receive a unit load for transportation to its next destination as soon as processing is completed on that unit load. This is not true with respect to a mobile equipment. For a mobile material handling system, the equipment must make a complete round trip between the source station and the destination station before the equipment becomes available again for another load pickup at the source station. Although the material flow is unidirectional for the problem considered in this chapter, the movement of the material handling system itself is unidirectional only for a stationary equipment and bidirectional for a mobile equipment. For bidirectional systems, the distance between pairs of machines in adjacent stages can be symmetric or nonsymmetric.

Given the operating characteristics of material handling systems as described above, the arrival times of unit loads at a station and the ready times of the material handling system can be expressed algebraically. Given that unit load n was completed at time $C^{n}_{..ij}(q)$ on machine $j \varepsilon \overline{M}(i)$, the arrival time of the load at station $k \varepsilon \overline{M}(i+1)$ is given by Equation (7).

$$A^{n}_{ij(i+1)k(q)} = \begin{cases} C^{n}_{..ij(q)} + t_{ij(i+1)k}\ ; \text{ stationary equipment} \\ \max\left\{C^{n}_{..ij(q)}, R^{n}_{ij(i+1)(\lambda,k)}\right\} + t_{ij(i+1)k};\ \text{mobile equip.} \end{cases} \tag{7}$$

If unit load n was the first to be processed on machine $j \varepsilon \overline{M}(i)$, then $R^{n}_{ij(i+1)(\lambda,k)} = 0$ since $\lambda=0$. From (7), it is obvious that the departure time of unit load n from station $j \varepsilon \overline{M}(i)$ is equal to $D^{n}_{ij(i+1)k(q)}$, where

$$D^{n}_{ij(i+1)k(q)} = \begin{cases} C^{n}_{..ij(q)};\ \text{for stationary equipment} \\ \max\left\{C^{n}_{..ij(q)}, R^{n}_{ij(i+1)(\lambda,k)}\right\};\ \text{for mobile equipment} \end{cases} \tag{8}$$

The ready time of the material handling system at station $j \varepsilon \overline{M}$ (i) after the pickup of unit load n from station $j \varepsilon \overline{M}$ (i) is given by Equation (9).

$$R^{m}_{ij(i+1)(k,\lambda)} = \begin{cases} D^{n}_{ij(i+1)k(q)}; \text{ for stationary equipment} \\ D^{n}_{ij(i+1)k(q)} + 2t_{ij(i+1)k}; \text{ for mobile equipment and symmetric distances} \\ D^{n}_{ij(i+1)k(q)} + t_{ij(i+1)k} + t_{(i+1)kij} \text{ for mobile equipment and nonsymmetric distances} \end{cases} \tag{9}$$

where m refers to the unit load that will follow n on machine $j \varepsilon \overline{M}$ (i) and λ refers to any machine in stage (i+1) to which load m may be routed.

5 SOLUTION APPROACH

From the relationships expressing the arrival time, operation start time, and the completion time of a given unit load at a given machine station and stage, it is obvious that these parameters depend also on the unit load departure time from the source machine at the preceding stage. The departure time itself is a function of the completion time and the material handling system ready time at the preceding machine. Thus, given a set of machines visited by a unit load, the completion time at each machine visited can be computed recursively using Equations (5), (6), (7) and (8). The presence of the recursive relationships makes it very attractive for the use of dynamic programming based heuristics as a solution approach to the problem. The application of dynamic programming based algorithm is also suggestive by the fact that the problem can be decomposed into stages and states. Each production stage corresponds to a dynamic programming stage while the machines in each stage serve as states. Given that the number of sublots or unit loads to be processed is known, the problem of finding the shortest shop time for the batch is equivalent to that of finding the shortest completion time path for each unit load and picking the longest of these paths as the shop time.

The shortest path for each unit load is found using an algorithm similar to that commonly employed for project planning under CPM (critical path method) or PERT (project evaluation and review technique) [11,15].

The best unit load size can be found by searching over the feasible range. For each value of unit load size generated by the search routine, the number of unit loads from the batch size of Q is computed. Given the number of unit loads and the size of each load, the dynamic programming algorithm is invoked to return the minimum manufacturing time corresponding to that decision. Thereafter, a new search value for the unit load size is generated to invoke the dynamic programming algorithm again. For each search value of unit load size generated, the corresponding minimum manufacturing time for the entire batch is determined. The search value with the minimum overall batch manufacturing time becomes the adopted and the best unit load size decision.

Because the search on the unit load size is unidimensional and the batch completion time function is unimodal on unit load size, the one dimensional search algorithm known as the Golden Section [4,14] is used in this chapter. The dynamic programming based heuristic algorithm is embedded within the unidimensional search algorithm.

5.1 The Unidimensional Search Algorithm

The following additional variables are used in the algorithms.

$\bar{a}$ = size of the smallest unit load that is permitted in the shop.
$\bar{b}$ = size of the largest unit load that is permitted in the shop.
$S_{(q)}$ = the best schedule determined when a unit load of size q is used. The best schedule is defined as the schedule that yields the smallest determined completion time with a unit load size of q.
$T_{(q)}$ = the completion time associated with schedule $S_{(q)}$.

The detailed unidimensional search algorithm is as described below.

1. If $|\overline{M}(M)| > 1$, create a dummy last stage M=M+1 and set the following parameters associated with the dummy stage as follows: $|\overline{M}(M)| = 1$; $P_{M1} = 0$; $Z_{M1} = 0$; $F_{M1} = 0$; and $t_{(M-1)jM1} = t_{M1\,(M-1)j} = 0$ for all $j \varepsilon \overline{M}$ (M-1).
2. Set $L = \bar{a}$ if $\bar{a}$ is given. Otherwise, set L = 1.
3. Set $U = \bar{b}$ if $\bar{b}$ is given. Otherwise, set U = Q.
4. Set $q_1 = [\,L + 0.38\,(U\text{-}L)\,]^+$.
5. With $q = q_1$, calculate N_{q1} and $\overline{N}_{q1}$ according to Equations (1) and (2).
6. Using the parameters q_1, N_{q1}, and $\overline{N}_{q1}$ calculated in steps (4) and (5), solve for the minimum completion time schedule using the dynamic programming based algorithm (described in the next section). Let $S_{(q1)}$ and $T_{(q1)}$ be the output of the dynamic programming algorithm associated with the input parameters q_1, N_{q1}, and $\overline{N}_{q1}$.
7. Set $q_2 = [L + 0.62\,(U\text{-}L)]^-$, where $[X]^-$ is the highest integer less than or equal to X. If $q_2 = q_1$, go to 12.
8. With $q = q_2$, calculate N_{q2} and $\overline{N}_{q2}$ according to Equations (1) and (2).
9. Using the parameters q_2, N_{q2}, $\overline{N}_{q2}$, solve for the minimum completion time schedule as described in step (6) above. Again, let $S_{(q2)}$ and $T_{(q2)}$ be the output of the dynamic programming algorithm.
10. If $T_{(q1)} \leq T_{(q2)}$, set $U = q_2$, $q^* = q_1$, $T_{(q^*)} = T_{(q1)}$, $S_{(q^*)} = S_{(q1)}$, $N_{q^*} = N_{q1}$ and $\overline{N}_{q^*} = \overline{N}_{q1}$. Otherwise, set $L = q_1$, $q^* = q_2$, $T_{(q^*)} = T_{(q2)}$, $S_{(q^*)} = S_{(q2)}$, $N_{q^*} = N_{q2}$, and $\overline{N}_{q^*} = \overline{N}_{q2}$.
11. Go to 4.
12. With the interval of uncertainty narrowed, evaluate all unit load sizes,

$q = [L, U]$ and set $q* = \bar{q}$, where $T(q*) = T(\bar{q}) \leq T(q)$ for all $q = [L, U]$. The best solution corresponds to $q*$, N_q*, $\bar{N}_q*$, $S(q*)$, and $T(q*)$.

5.2 The Dynamic Programming Algorithm

The input into the dynamic programming based heuristic algorithm are the parameters q_1, N_{q1}, and $\bar{N}_{q1}$, or $q_2, N_{q2}, \bar{N}_{q2}$ defined in steps 4, 5, 7, and 8 of the unidimensional search algorithm. These parameters will simply be identified as q, N_q, $\bar{N}_q$ in this algorithm. The algorithm proceeds as follows:

1. Set $T_{max} = 0$ and the unit load number $n = 1$.
2. a) Set the stage number $i = 1$.
   b) Set $\bar{q}$ = the size of unit load n.
3. If $i \geq 2$, go to 4. Otherwise, for each machine $j \in \bar{M}(i)$, compute:
   a) $A^n_{00ij(\bar{q})} = 0.$
   b) $S^n_{00ij(\bar{q})} = \bar{T}^n_{ij}$, where $\bar{T}^n_{ij} = Z_{ij}$ if machine $j \in \bar{M}(i)$ has not previously been assigned a unit load.
   c) $P_{ij(\bar{q})} = F_{ij} + \bar{q}P_{ij}$.
   $C^n_{00ij(\bar{q})} = S^n_{00ij(\bar{q})} + P_{ij(\bar{q})}$
   d) Record the results computed in (a) through (c) and store in a working array against machine $j \in \bar{M}(i)$.
   e) Go to 6.
4. For each machine $k \in \bar{M}(i)$ and over all preceding machines $j \in \bar{M}(i-1)$, compute:

   a)
$$D^n_{(i-1)jik(\bar{q})} = \begin{cases} C^n_{\cdot\cdot(i-1)j(\bar{q})} & \text{if stationary handling system used} \\ \max\left\{C^n_{\cdot\cdot(i-1)j(\bar{q})} ; R^n_{(i-1)ji(\lambda,k)}\right\} & \text{otherwise} \end{cases}$$

   Note that $R^n_{(i-1)ji(\lambda,k)} = 0$ if no unit load had been previously assigned to machine $j \in \bar{M}(i-1)$. All machines start initially with no assignment. Machine λ is any machine in $M(i)$, including the zero or dummy machine.

   b) $A^n_{(i-1)jik(\bar{q})} = D^n_{(i-1)jik(\bar{q})} + t_{(i-1)jik}$

c) $$R^{m}_{(i-1)ji(k,\lambda)} = \begin{cases} C^{n}{}_{..(i-1)j(\bar{q})} & \text{if stationary handling system used} \\ A^{n}{}_{(i-1)jik(\bar{q})} + t_{ik(i-1)j} & \text{otherwise} \end{cases}$$

where $R^{m}{}_{(i-1)ji(k,\lambda)}$ is the ready time of the material handling equipment at machine j ε $\overline{M}$(i-1) and m is the unit load that will follow n immediately on machine j ε $\overline{M}$(i-1). The index λ refers to any machine in $\overline{M}$(i) that load m may be sent to.

d) $$S^{n}_{(i-1)jik(\bar{q})} = \max\left\{A^{n}_{(i-1)jik(\bar{q})} \; ; \; \overline{T}^{n}_{ik}\right\}$$

where $\overline{T}^{n}_{ik} = Z_{ik}$ if no unit load had been previously assigned to machine k ε $\overline{M}$(i).

e) $$\overline{P}_{ik(\bar{q})} = F_{ik} + \bar{q}P_{ik}$$

f) $$C^{n}_{(i-1)jik(\bar{q})} = S^{n}_{(i-1)jik(\bar{q})} + \overline{P}_{ik(\bar{q})}.$$

5. a) Determine the source machine h ε $\overline{M}$(i-1) that yields the earliest completion time of unit load n on machine k ε $\overline{M}$(i). Algebraically, this is equivalent to identifying the machine h ε $\overline{M}$(i-1) such that

$$C^{n}_{(i-1)hik(\bar{q})} = \min_{\forall j \,\varepsilon\, \overline{M}(i-1)}\left\{C^{n}_{(i-1)jik(\bar{q})}\right\}.$$ Break ties arbitrarily.

b) Record against machine k ε $\overline{M}$(i) in a working array, the best source machine h ε $\overline{M}$(i-1) and the associated parameters

$$R^{m}_{(i-1)hi(k,\lambda)},\; A^{n}_{(i-1)hik(\bar{q})},\; S^{n}_{(i-1)hik(\bar{q})} \text{ and } C^{n}_{(i-1)hik(\bar{q})}.$$

6. Set i = i+1. If i ≤ M, go to 3.

7. Set $T_{max} = C^{n}_{(M-1)hM1}$ since there is only one machine in the last stage.

8. Trace back the earliest completion time path for unit load n. For each machine k ε $\overline{M}$(i), i=1,2,...,M in the path, record in the schedule file

$S_{(\bar{q})}$ the results $A^n_{(i-1)hik(\bar{q})}$, $S^n_{(i-1)hik(\bar{q})}$, $C^n_{(i-1)jik(\bar{q})}$, h, and the material handling system ready time $R^m_{ik(i+1)(h',\lambda)}$ computed at machine h' $\varepsilon\ \overline{M}(i+1)$ where h' $\varepsilon\ \overline{M}(i+1)$ follows machine k $\varepsilon\ \overline{M}(i)$ immediately on the path.

9. Set n = n+1. If $n \leq N_q$, go to 2.

10. A complete schedule $S_{(\bar{q})}$ for all unit loads has been constructed. Set $T_{(\bar{q})} = T_{max}$ and return to the unidimensional search algorithm.

6 ILLUSTRATIVE PROBLEM

The above algorithm has been applied in the planning of a job containing 100 pieces of identical parts. Five operations are to be performed on each part, with each operation requiring a separate production stage. The stages are numbered according to the operations. There are three machines each in stages 1, 4, and 5. Stages 2 and 3 each has four machines. Processing time data per part on the machines are given in Table 1. There is a preparatory time on each load at each machine station. These times are given in Table 2. Setup time requirements for the machines are summarized in Table 3. The data of Tables 1, 2, and 3 are drawn from the uniform distribution with parameters of (a,b) = (0.5, 3.0), (a,b) = (1.0, 3.5), and (a,b) = (20., 50.) respectively. Data on material handling times between pairs of machines in adjacent production stages are given in Table 4. For the case of a mobile material handling system, it is assumed that the travel times are symmetric (i.e., $t_{ij(i+1)k} = t_{(i+1)kij}$). Given the problem scenario described above, what is the best unit load size and unit load routing policies that will minimize the production makespan for the job.

The problem was solved under the two cases of material handling systems, namely, stationary and mobile systems as discussed earlier in the chapter. Firstly, the results of the case of a mobile system is presented. This is followed by the case of a stationary system. Finally, the results of the two cases are compared and inferences are drawn.

Using the algorithm presented in the chapter, for the case of mobile material handling system, the best makespan was found at a unit load size of six parts. The corresponding makespan is 218.44. With a batch size of 100 parts and a full unit load size of 6 parts, a total of 17 unit loads are generated. The seventeenth unit load is only partially filled and contains four parts.

Table 5 is a summary of the distribution of the 17 unit loads on the machines at each stage. For example, of the 17 unit loads processed through stage 1, 5 were processed by machine 1, while machines 2 and 3 each processed six units. The assignments in the other stages can be similarly read.

Table 1 Part processing times

		Stage No.				
		i = 1	i = 2	i = 3	i = 4	i = 5
Machine	m_{i1}	1.98	2.64	2.12	1.98	2.35
	m_{i2}	1.49	1.22	1.97	2.38	2.93
	m_{i3}	2.42	0.77	2.27	1.53	2.00
	m_{i4}	-	1.18	0.64	-	-

Legend: m_{ij} = j^{th} machine in stage i.

Table 2 Unit load preparatory times at machine stations

		Stage No.				
		i = 1	i = 2	i = 3	i = 4	i = 5
Machine	m_{i1}	1.80	2.47	2.61	2.35	2.54
	m_{i2}	1.92	3.30	3.07	2.56	2.68
	m_{i3}	2.26	1.97	2.52	1.66	2.72
	m_{i4}	-	1.02	2.05	-	-

Table 3 Setup times at machine stations

		Stage No.				
		i = 1	i = 2	i = 3	i = 4	i = 5
Machine	m_{i1}	39.01	48.65	29.46	41.15	21.06
	m_{i2}	32.21	43.36	31.20	24.88	28.00
	m_{i3}	20.90	46.35	26.42	41.09	49.00
	m_{i4}	-	21.45	37.15	-	-

Table 4 Material handling times between machines in adjacent stages*

To Machine

From Machine	i = 2				i = 3				i = 4			i = 5		
	m_{21}	m_{22}	m_{23}	m_{24}	m_{31}	m_{32}	m_{33}	m_{34}	m_{41}	m_{42}	m_{43}	m_{51}	m_{52}	m_{53}
$m_{(i-1)1}$	6.43	9.14	10.67	6.26	9.57	11.64	5.18	6.27	10.52	6.44	11.80	10.96	7.49	5.78
$m_{(i-1)2}$	11.10	7.73	5.93	9.49	5.54	11.49	9.50	11.08	10.06	8.03	9.97	10.98	11.97	9.06
$m_{(i-1)3}$	5.22	6.57	11.55	9.21	7.94	10.85	8.39	9.34	6.69	5.88	7.68	11.99	10.34	11.19
$m_{(i-1)4}$	-	-	-	-	8.81	11.70	8.90	6.11	9.47	9.29	8.71	-	-	-

* Symmetric times for mobile handling systems.

Table 5 Unit load - machine assignment (mobile system)

Stage No.

		i = 1	i = 2	i = 3	i = 4	i = 5
Machine	m_{i1}	5	2	5	8	5
	m_{i2}	6	4	2	4	4
	m_{i3}	6	5	4	5	8
	m_{i4}	-	6	6	-	-
	Total	17	17	17	17	17

Table 6 Unit load flow volume between machines (mobile system)

To Stage

		i=2				i=3				i=4			i=5		
	Machine	m_{21}	m_{22}	m_{23}	m_{24}	m_{31}	m_{32}	m_{33}	m_{34}	m_{41}	m_{42}	m_{43}	m_{51}	m_{52}	m_{53}
From	$m_{(i-1)1}$				5		1	1		4	1				8
	$m_{(i-1)2}$		2	4		3			1	2			4		
	$m_{(i-1)3}$	2	2	1	1	2		3		1		3	1	4	
	$m_{(i-1)4}$						1		5	1	3	2			
Stage	Total	17				17				17			17		

The material flow volume (in unit loads) between machines in the schedule that yielded the makespan of 218.44 is summarized in Table 6. For example, of the six unit loads processed by machine 3 of stage 1, machines 1 and 2 in stage 2 each received two unit loads while machines 3 and 4 in stage 2 each received 1 unit load. Again the remainder of the table can be similarly read. Since the timing of the unit load flow is critical in achieving the resulting makespan, a knowledge of the transition pattern followed by each unit load is necessary. The routing for each of the 17 unit loads is summarized in Fig. 1. By tracing the machines visited by each unit load through all the stages, the routing for each unit load is uniquely identified. Fig. 1 also gives the unit load sequencing on each machine. For example, the unit load sequencing on machine 4 in stage 2 is: 1, 3, 6, 9, 12, 16 and the routing for unit load 9 is m_{11}, m_{24}, m_{34}, m_{42}, m_{51}.

Fig. 2 is the corresponding Gantt Chart for Fig. 1. When used jointly, Figs. 1 and 2 give a complete production plan for the batch of 100 parts when a mobile material handling system is used.

Whereas a unit load size of six parts is the best decision when a mobile material handling system is used, the best unit load size to use under a stationary

Machine								
m_{11}	3	6	9	12	16			
m_{12}	2	4	7	10	11	14		
m_{13}	1	5	8	13	15	17		
m_{21}	13	17						
m_{22}	5	7	11	15				
m_{23}	2	4	8	10	14			
m_{24}	1	3	6	9	12	16		
m_{31}	2	5	8	11	15			
m_{32}	6	13						
m_{33}	4	10	14	17				
m_{34}	1	3	7	9	12	16		
m_{41}	1	2	4	6	8	11	13	15
m_{42}	5	9	12	16				
m_{43}	3	7	10	14	17			
m_{51}	3	5	9	12	16			
m_{52}	7	10	14	17				
m_{53}	1	2	4	6	8	11	13	15

Fig. 1 Unit load routing and machine sequencing (mobile system)

material handling system is three parts. This represents a 50% reduction in unit load size compared to the mobile system. For a batch of 100 parts, this represents a total of 34 unit loads. The last unit load is a single part load. Results on unit load assignment to machines are given in Table 7. Table 8 contains the results on material flow (in unit loads) frequency between machines in consecutive stages. Similar to Fig. 1 in the case of a mobile material handling system, Fig. 3 contains the complete production plan for the 34 unit loads generated from the batch. Fig. 3 provides the detailed information on the routing of the unit loads and the unit load sequence on the machines. The Gantt Chart to the plan is not shown here for the sake of brevity. The associated makespan for the plan of Fig. 3 is 201.96 time units. As one would expect, this is lower than the makespan of 218.44 obtained for the case of a mobile material handling systems.

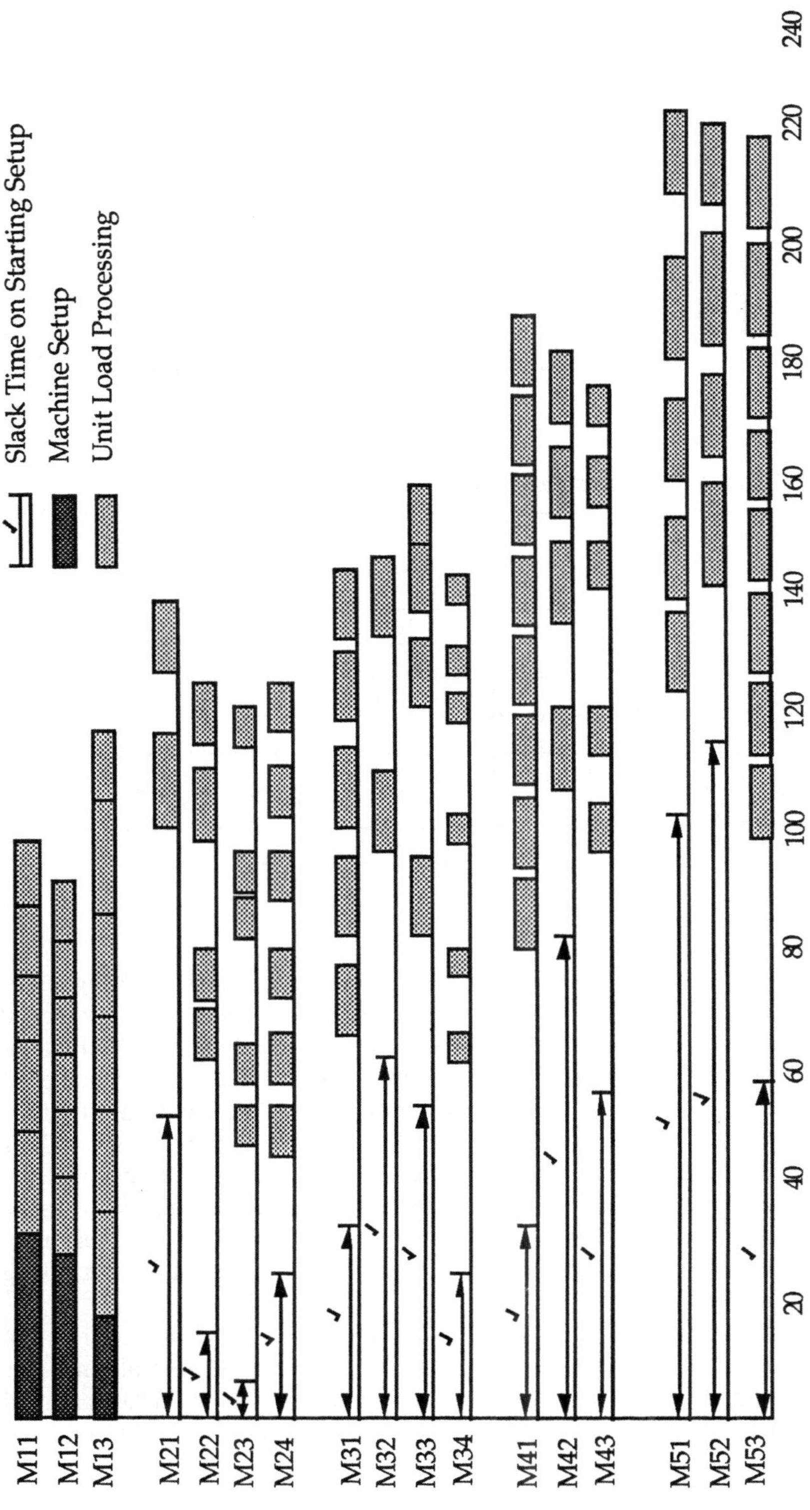

Fig. 2 *Gantt chart based on mobile handling system*

Table 7 Unit load - machine assignment (stationary system)

Stage No.

		i = 1	i = 2	i = 3	i = 4	i = 5
Machine	m_{i1}	10	5	10	14	11
	m_{i2}	14	8	5	9	9
	m_{i3}	10	10	6	11	14
	m_{i4}	-	11	13	-	-
	Total	34	34	34	34	34

Table 8 Unit load flow volume between machines (stationary system)

To Stage

		i=2				i=3				i=4			i=5		
	Machine	m_{21}	m_{22}	m_{23}	m_{24}	m_{31}	m_{32}	m_{33}	m_{34}	m_{41}	m_{42}	m_{43}	m_{51}	m_{52}	m_{53}
From	$m_{(i-1)1}$		2		8	2		2	1	10					14
	$m_{(i-1)2}$		3	10	1	7	1			1	4		7	2	
	$m_{(i-1)3}$	5	3		2	1	3	4	2	1	4	1	4	7	
	$m_{(i-1)4}$						1		10	2	1	10			
Stage	Total	34				34				34			34		

In analyzing the differences between the two material handling cases considered, based on the results reported here as well as other experimental results not shown in this chapter, certain interesting preliminary observations were made. If the round trip times of a mobile material handling system are not dominated by the unit processing times on machines, the best unit load size for a stationary handling system cannot exceed the best unit load size found under a mobile handling system. Furthermore, the best makespan found under a stationary handling system is as good or better than that obtained under a mobile handling system. As Figs. 2 and 3 indicate, the pattern of unit load flow between machines is not affected significantly by the type of material handling systems used.

Whereas, transporting parts in large unit loads may be attractive from the point of view of minimizing material handling cost, this is generally not true when the objective is to minimize the batch completion time. Whereas small unit loads accelerate the batch through the shop, large unit loads tend to tie up the batch at each workstation. As the unit load preparatory times tend to zero,

Machine														
m_{11}	5	8	11	13	16	19	23	26	30	31				
m_{12}	2	4	7	9	12	15	17	20	22	24	28	29	33	34
m_{13}	1	3	6	10	14	18	21	25	27	32				
m_{21}	10	18	21	27	32									
m_{22}	6	8	14	20	22	25	29	30						
m_{23}	4	7	9	12	15	17	24	28	33	34				
m_{24}	1	2	3	5	11	13	16	19	23	26	31			
m_{31}	6	8	14	17	20	22	25	27	30	32				
m_{32}	11	24	28	29	33									
m_{33}	4	12	15	18	21	34								
m_{34}	1	2	3	5	7	9	10	13	16	19	23	26	31	
m_{41}	1	2	4	6	8	11	14	17	20	22	25	27	30	32
m_{42}	7	12	15	18	21	24	28	29	33					
m_{43}	3	5	9	10	13	16	19	23	26	31	34			
m_{51}	3	7	10	13	15	18	21	24	28	31	33			
m_{52}	5	9	12	16	19	23	26	29	34					
m_{53}	1	2	4	6	8	11	14	17	20	22	25	27	30	32

Fig. 3 Unit load routing machine sequencing (stationary system)

the unit load size tends to unity when a stationary handling system is used. For the case of a mobile handling system, the unit load size depends on both the unit load preparatory time and the magnitude of unit processing times relative to the round trip handling time. If the unit processing times dominate the round trip handling times, the same unit load size and production plan will be optimal for both the stationary and mobile material handling systems.

7 CONCLUSIONS

In this chapter, the problem of determining the minimum production time plan for a batch job consisting of multiple parts is addressed. This chapter is developed around the notion that batch splitting is a common industrial practice whose objective is to facilitate batch movement and early completion time. When split, several subbatches, known as unit loads, are generated from a batch. These unit loads constitute the moving entities between the machines visited by the batch. Models describing the flow of unit loads through the production stages have been developed. Given that the number of unit loads is known as well as their sizes, the models provide a basis for recursively determining the best machine assignment to each unit load at each machining stage. The assignment is accomplished using a dynamic programming based heuristic algorithm.

The dynamic programming algorithm is embedded in a unidimensional search algorithm that is based on the Golden Section search. The search algorithm generates potential unit load sizes with which the dynamic programming algorithm is evaluated. An example problem illustrates the application of the models in planning a typical batch production instance.

Although the chapter assumes that all unit loads generated from a batch are of equal size, it is believed that the procedure presented can be applied directly to several batch production systems since uniform unit load sizes seem to be the norm rather than the exception. The current model does, however, allow for the last unit load to be only partially full if the batch size is not perfectly divisible by the size of a full unit load.

The model accommodates both stationary and mobile material handling systems. Round trip travel times are explicitly recognized for the case of mobile system. The model is easy to use and when coded into a computer program, can run very large problem sizes in negligible time.

8 REFERENCES

1. Apple, J.M., *Material Handling Systems Design*, John Wiley and Sons, New York (1972).
2. Baker, K.R., *Introduction to Sequencing and Scheduling*, John Wiley and Sons, New York (1974).
3. Baker, K.R., "Lot Streaming to Reduce Cycle Time in a Flow Shop," Working Paper No. 203, The Amos Tuck School of Business Administration, Dartmouth College, Hanover, NH (1988).
4. Beightler, C.S., D.T. Phillips, and D.J. Wilde, *Foundation of Optimization*, Prentice-Hall (1979).
5. Conway, R.W., W.L. Maxwell, and L.W. Miller, *Theory of Scheduling*, Addison-Wesley Publishing Co. (1967).
6. Day, J.E. and M.P. Hottenstein, "Review of Sequencing Research," *Naval Research Logistics Quarterly*, Vol. 17 (1970).
7. Egbelu, P.J. and N. Roy, "Material Flow Control in AGV/Unit Load Based Production Lines," *International Journal of Production Research*, Vol. 26, No. 1, pp. 81-94 (1988).
8. Egbelu, P.J., "Unit Load Design for Minimum Cost Manufacture in an Automated Guided Vehicle-Based Transport System," *Proceedings, SME ULTRATECH Conference*, Vol. 1, Long Beach, CA, pp. 3.19-3.37 (1986).
9. Egbelu, P.J., "Integration of Unit Load Selection in the Design of an AGV-Based Automated Manufacturing Shop," *Recent Developments in Production Research*, (Ed. Anil Mital), Elsevier Publishers (1987).
10. French, S., *Sequencing and Scheduling: An Introduction to the Mathematics of the Job Shop*, John Wiley & Sons, NY (1982).

11. Gido, J., *An Introduction to Project Planning*, 2nd Ed., Industrial Press, Inc. (1985).
12. Graham, R.L., R.L. Lawler, J.K. Lenstra, and A.H.G. Rinooy Kan, "Optimization and Approximation in Deterministic Sequencing and Scheduling: A Survey," *Annals of Discrete Mathematics*, Vol. 5, pp. 287-326 (1979).
13. Groover, M.P., *Automation, Production Systems, and Computer-Integrated Manufacturing*, Prentice-Hall, Inc. (1987).
14. Himmelblau, D.M., *Applied Nonlinear Programming*, McGraw-Hill Book Co. (1972).
15. Phillips, D.T., A. Ravindran, and J.J. Solberg, *Operations Research: Principles and Practice*, John Wiley and Sons (1976).
16. Santos, C. and M. Magazine, "Batching in Single Operation Manufacturing Systems," *Operations Research Letters*, Vol. 4, No. 3, pp. 99-103 (1985).
17. Steudel, H.J. and H.K. Moon, "Selection of Unit Load Size for Automated Guided Vehicles (AGVs) in Large Cellular Manufacturing Environments," *Proceedings 9th International Conference on Production Research*, Cincinnati, OH, pp. 2198-2209 (1987).
18. Tanchoco, J.M.A. and M.H. Agee, "Plan Unit Loads to Interact with All Components of Warehouse Systems," *Industrial Engineering*, Vol. 13, No. 6, pp. 36-48 (1981).
19. Tanchoco, J.M.A., R.P. Davis, R.A. Wysk, and M.H. Agee, "Component Selection and Operating Specifications for Belt Conveyor Systems,"*Proceedings 1979 Fall Industrial Engineering Conference* (1979).
20. Tanchoco, J.M.A., R.A. Wysk, R.P. Davis, and M.H. Agee, "Economic Unit Load and Its Impact on Material Flow Systems Planning," *Proceedings 1979 Fall Industrial Engineering Conference* (1979).
21. Tanchoco, J.M.A., R.P. Davis, P.J. Egbelu, and R.A. Wysk, "Economic Unit Loads (EUL) for Multi-product Inventory Systems with Limited Storage Space," *Material Flow*, Vol. 1, pp. 141-148 (1983).
22. Truscott, W.G., "Scheduling Production Activities in Multi-stage Batch Manufacturing Systems," *International Journal of Production Research*, Vol. 23, No. 2, pp. 315-328 (1985).

A GROUP TECHNOLOGY APPROACH FOR CONTAINER SIZE SELECTION

Jessica O. Matson
Girish N. Naik
The University of Alabama

ABSTRACT

The problem considered in this chapter is the selection of carton sizes for shipping items produced in a small batch manufacturing environment. Rather than selecting an optimum container design for each item, the problem is that of selecting a single container or a limited number of containers for the variety of items and range of order sizes that are characteristic of the particular small batch manufacturer. The focus is on two issues: cost considerations and the carton size selection method. The relevant cost factors are identified and used as the criteria for evaluating the selected carton sizes. A grouping technique is used to assist in the determination of carton sizes, and the results of the application of the methodology for a set of historical data are evaluated.

1 INTRODUCTION

The selection of the containers to be used in a production system can have a major impact on the cost of handling. The container selection problem involves tradeoffs among material costs, labor costs, transportation costs, storage costs, and control costs. The determination of the optimal container design(s) depends upon various factors such as item shapes, item weights, and production quantities. In addition, there are numerous interactions with the facility and material handling system which must be considered, including the method of shipping, the handling equipment, the storage method, and the facility configuration, among others [10].

The problem considered in this chapter is the selection of shipping containers (i.e., cartons or boxes) for a small batch manufacturing environment. Since it is commonly estimated that as much as 75% of all discrete parts manufacturing is in lot sizes of 50 pieces or less, the container size selection problem is a significant one. In a mass production environment with large stable production quantities of a limited number of items, it may be possible to consider the relevant factors and evaluate the cost tradeoffs so as to select an optimal container for each item produced by the firm. However, such an approach is not realistic for small batch manufacturers who typically produce a large variety of items in small or medium volumes. The problem becomes that of selecting a single container or a limited number of containers for the variety of items and range of order sizes that are characteristic of the particular small batch manufacturer.

The selection of shipping container(s) for such an environment is a unit load design problem. The main design issue is container size. Tanchoco and Agee [10] note that "the specification of carton size is perhaps the most critical element in the design of a 'unit load' system." Tompkins and White [11] also state that, if the unit load design cannot be determined simultaneously with the handling system design, then the "size of the tote box or the pallet should be among the first determinations made, rather than among the last."

Although important, issues such as stacking height and common carrier dimensions are not considered. The production of a variety of items in small quantities often means that containers will not be stacked in traditional pallet loads. In addition, the shipping method will not typically be full trailer or rail car shipments.

2 RELATED LITERATURE

The container selection problem has been examined previously by a number of researchers. Some of the earliest reported efforts coincided with the development of the digital computer. In 1958, Carrabino [1] used simulation to determine the optimum single size of cargo containers for a specific sea shipping route and compared the costs of containerization versus palletization. King and Goode [6] reported in 1960 on the application of operations research techniques in the development of cost models for determining the optimum sizes of containers. They considered material and labor costs and focused on specifying the optimum number of packages per carton and the optimum number of cartons per case. In 1967, Normandin [7] used an enumerative procedure with increments of one-half inch in length and width and one-quarter inch in depth for finding the best container size for a mix of baked goods. The objective was to minimize the volume occupied by the containers.

Other work related to the container size selection problem is in the category of packing and loading problems. For a given master container size, Peleg and Peleg [8] used an enumerative approach to evaluate all feasible containers (cartons) for optimum space utilization of the given master container. Udy et al. [12] described the use of a graphics package to model interference, container, and packing problems. The focus was on how to orient the set of convex objects within the container to maximize space utilization and to avoid interference. Container size was not a decision variable.

In work on the pallet loading problem, Steudel [9], Hodgson [4], and Hodgson, Hughes, and Martin-Vega [5] developed procedures for placing rectangular-shaped cartons on a rectangular pallet or in a rectangular container. The objective was to select the pallet or container size which would maximize the utilization of the pallet perimeter [9] or maximize the area covered by the cartons [4,5].

Friedman and Kipnees [2] provide a number of rules of thumb for container size selection. They state that "some packaging authorities contend that the best shaped box for handling and storage is one in which length is approximately one and one-half times the width, and depth is a little less than width." The half-cube, where length is equal to depth and is twice the width, is the most economical shape in relation to cubic capacity. The perfect cube, while not suitable for interlocked stacking, sustains the least damage in handling.

3 PROBLEM APPROACH

This chapter investigates a different approach for determining the container size(s) to be used for shipping. The benefits of group technology techniques are well known for a number of problems related to small batch manufacturing, including machine selection and facility layout. It would seem worthwhile to investigate group technology methods for selecting the container(s) to be used for shipping.

However, before group technology methods can be applied, the criteria for evaluating the results must be determined. Many of the previous studies focused on space utilization as the single criterion to be evaluated. While space utilization is important, it is not the only factor which should be considered. This chapter proposes the development of a cost model which incorporates all relevant cost factors.

The focus of this chapter is thus two-fold. The first step is the determination of all relevant cost factors for inclusion in a total cost model. The second step is the demonstration of the application of group technology methods, using a set of actual data, for selecting container sizes.

4 CONTAINER COST FACTORS

The design of the container has a direct or indirect effect on material cost, shipping costs, in-plant handling costs, storage costs, and control costs. Two decisions which impact these costs are the size(s) of the containers to be used and the number of different container sizes to be permitted. If only a single container size is to be permitted, then that container must be large enough to hold the largest item to be shipped. As the number of different containers permitted increases, smaller container sizes can be added. It can be shown that, by adding a second, smaller container size, the total number of containers required will be greater than or equal to the number of containers when restricted to a single (large) container size. Thus, as the number of different (smaller) sizes increases, the total number of containers may increase, resulting in increases in carton costs, handling costs and control costs. However, as the number of different sizes increases, one might also expect increases in carton utilization, resulting in a reduction in shipping and storage costs. The size of the cartons influences the type of handling equipment and storage method to be used, two additional factors which affect costs. The major cost categories are described below, and a total cost model is proposed.

4.1 Material Costs

Material costs include the container itself and any internal packing material. Container costs are based on the amount and type of material used to produce the container. Corrugated cartons are a commonly used container for shipping, and consideration in this chapter is restricted to corrugated cartons. The carton dimensions, carton style, and wall construction are three of the main factors which influence corrugated carton costs. The larger the carton dimensions, the greater the carton volume is, and the greater the carton cost. Carton style is determined by the type of flap or closure desired. Various styles require varying amounts of material. If a single carton design is to be used, the wall construction must provide adequate strength and protection for the heaviest items. If multiple carton designs are permitted, varying grades (strengths) of wall construction can be used. Quantity discounts may also be an important factor in carton costs. For example, the carton order quantity may be sufficient to qualify for a quantity discount if only one container design is to be permitted in the system; if several different container sizes are permitted, the smaller quantities of each size may not qualify for a discount. The unit cost of a particular style of carton is thus a function of wall construction (WC), carton length (L), carton width (W), carton height (H), and quantity ordered (Q). For example, a regression model was developed for estimating the unit cost of one style of cartons, based on prices obtained from several corrugated carton distributors. Including the quantity discount, the model is given by

$$CC = \begin{cases} WC\,(a_0 + a_1\,LWH) & \text{for } Q < \text{discount quantity} \\ WC\,(a_0 + a_1\,LWH) - d & \text{for } Q \geq \text{discount quantity} \end{cases}$$

where CC is unit carton cost; a_0 and a_1 are constants based on carton dimensions and costs; WC is a constant whose value depends on single wall or double wall construction; and d is the unit discount.

Internal packing may be required to ensure the safety of the items in the container or to prevent the container from crushing when not full packed. The amount of internal packing required is proportional to the unutilized space in the carton. It might be expected that the average utilization of all cartons would be increased by allowing more than one container size in the system. (An increase usually will occur, but not always.) Using styrofoam packing material as an example, the internal packing costs can be estimated by

$$IP = k_p(1-U)LWH$$

where

IP = cost of internal packing for a given carton
k_p = cost per cubic foot for internal packing material
U = carton utilization $(0<U<1)$

Combining the carton and internal packing costs, a general expression for material costs (MC) is given by MC = CC + IP.

4.2 In-plant Handling Costs

In-plant handling includes carton preparation, carton loading, and in-plant transportation. Carton preparation costs reflect the effort involved in taking a carton from the unmade (flat) position to ready-for-loading status. The process can be manual or part of an automated packing line. Both labor and equipment cost estimates may be needed, depending upon the system. An example cost model for manual carton preparation, assuming that the time required is dependent upon the size of the carton, is given by

$$PC = b_0 + b_1 LWH$$

where PC = preparation cost per carton and b_0 and b_1 are constants based upon the time required and the applicable wage rate.

Carton loading costs reflect the time required to fill the carton with items to be shipped. Again, the loading can be accomplished either manually or with an automated system. For a given item, the time per piece required for filling the carton may be influenced by the carton size and by the item's nesting characteristics. For example, consider an item with an order quantity of 100 pieces. If a small carton (capacity = 100 pieces) were to be used, considerable effort might be expended in placing the pieces in the carton to ensure their fit. If a larger carton (capacity = 150 pieces) were to be used, the time per piece required for placement would not be as great. Similarly, if parts are to be nested, the effort required for placement to achieve nesting would be much greater than the effort that would be required if the pieces could simply be tossed into the carton. Assuming that pieces are to be loaded manually, tightly packed and nested if possible, and assuming that placement ease increases with increasing carton volume, an example of a cost model for carton loading is given by

$$LC = N(c_0 - c_1 LWH)$$

where LC = loading cost per carton of a given item, N = the number of pieces of the item to be loaded into the carton, and c_0 and c_1 are constants based upon the time required and the applicable wage rate.

In-plant transportation costs are dependent upon the number of moves, the labor and equipment requirements, and other factors such as distance to be moved. The in-plant transportation costs may also be affected by the size of the container(s) selected. If a single size carton is permitted, larger cartons will result in a smaller number of cartons to be moved/handled both empty and packed. If multiple carton sizes are permitted, the sizes may be selected to achieve higher average carton utilization so that more goods (and less packing material) are being moved in a trip. Another consideration is that large containers may prohibit manual handling; mechanical or automated systems may be required. If TR represents the in-plant transportation costs, the total in-plant handling costs (HC) are given by HC = PC + LC + TR.

4.3 Storage Costs

Two categories exist for storage costs: costs for unmade cartons and costs for packed cartons. When multiple carton sizes are permitted, the storage space requirements for unmade cartons will likely be greater than if a single size were to be used because the number of different items to be stored is greater. The storage area required for a flat carton is approximated by (L+W)(2L+H). Storage costs for packed containers will be influenced by the utilization of the cartons. If multiple carton sizes are permitted, average carton utilization should be greater and space requirements may be reduced. A cost model can be developed based on the cost per square foot or cost per cubic foot of space occupied and the time period that the carton is stored. Total storage costs can be designated by ST.

4.4 Shipping Costs

Depending upon the shipping method, the cost may be a function of packed carton weight, volume of space occupied, and/or shipping distance. Again, allowing multiple carton sizes may result in greater carton utilization and less volume occupied. There may be restrictions on carton dimensions and carton weights for particular shipping methods. A cost model incorporating the appropriate costs can be developed, and costs for shipping can be denoted by SH.

4.5 Control Costs

These costs include inventory control and damage costs. The unit carton inventory control costs, including costs for purchasing and record keeping, will be greater if multiple sizes of cartons are permitted. Depending upon the inventory system, record keeping and control may be more difficult as the number of container sizes increases. Multiple sizes of cartons increase the need for a more sophisticated identification and control system.

Damage costs are also included in the category of control and include the costs of scrap and rework for damaged parts. When multiple container sizes are allowed, damage costs per carton might be expected to decrease. By using the best container for a given item, the cartons will have better weight distribution and will be less likely to receive damage in an accident. Also, the use of smaller cartons means that fewer pieces may be damaged if the carton is dropped. Damage costs are influenced by the characteristics of the item, i.e., whether it is fragile or sturdy, its value, its weight, and the quantity. Combining inventory control costs and damage costs, total control costs per container are denoted by RC.

4.6 Total Costs

Summing the identified costs, a total cost (TC) model is given by

$$TC = MC + HC + ST + SH + RC$$

This model is used as the basis for evaluating different container size possibilities.

5 APPLICATION OF GROUP TECHNOLOGY

The basic philosophy of group technology (GT) is to identify and bring together similar items to take advantage of their similarities. The typical application of GT is in the formation of part families and machine cells. It seems reasonable to investigate whether or not a GT approach would be helpful in container size selection.

There are two possible ways of using the results of a grouping technique. First, each group of items can be studied to help select the appropriate container size for that group. Secondly, the technique can also be used to assign products to groups. Both uses were investigated for a set of 92 items for which a year's production and demand history was known. Other characteristics of the items, such as weight, shape, and value, were also available. Cost models were determined or estimated for each of the cost categories described above and were used as the basis for the evaluation. The following section describes the methodology for an application of GT to container size selection for the set of 92 items.

Before applying the grouping technique, a number of decisions were required. For example,

- Will a given container be limited to a single item type or will it be permitted to hold multiple item types? For this problem, a given container was limited to a single item type.
- What weight limitations exist for a full container? There were no weight restrictions assumed for the problem investigated.
- Will items be tightly packed and nested if possible, or can they be tossed into the carton (random placement)? For this problem, the items were required to be tightly packed and nested if possible.
- How many different container designs (sizes) will be permitted? For purposes of this chapter, at most two container sizes were permitted. However, the methodology presented here can be easily extended to three or more container sizes.

5.1 Methodology

The following steps describe the procedure to be followed.

1. Determine the factors to consider in the grouping method. Many factors are available for consideration. The factors investigated in this study included item weight, item shape, item dimensions, nesting capability, fragility, average order size, item value, and number of orders, among others. Questions related to the selection of factors are: How many factors should be considered? Which factors are most important?
2. Apply ABC analysis to determine the most popular items. In the application considered, ABC analysis led to the following classifications, based on the previous year's demand.
   - Category A: 18 items 70% volume
   - Category B: 26 items 23% volume
   - Category C: 48 items 7% volume
3. Apply the grouping technique; determine the number of groups. A basic question is: Which grouping method should be used? The technique selected in this application was the Similarity Coefficients Method (SCM) [3], and the number of groups was two.
4. For each group identified in step 3, determine the first, second, and third critical dimensions using the following procedure. For group j, order item dimensions such that $L \geq W \geq H$ for each item in group j. Then,
   - critical dimension 1 (CD1) is the largest item length value,
   - critical dimension 2 (CD2) is the largest item width value, and
   - critical dimension 3 (CD3) is the largest item height value.

   Repeat the procedure for each group.

5. Using CD1, CD2, and CD3 for a given group, determine the carton size. Consideration can be limited to standard cartons, or a custom size can be specified. The carton size (LxWxH) for a group is the smallest carton such that

$$L > CD1, W > CD2, \text{ and } H > CD3.$$

6. If carton sizes for two groups are very similar, consider limiting consideration to the largest carton size. Otherwise, assign category C items to one of the two groups based on the item dimensions. That is, a C item should be placed in the smaller container if it will fit.
7. Evaluate the costs and utilization. Consider sensitivity analysis on carton sizes.

5.2 Results

The procedure was applied to the group of 92 items using the previous year's demand history. Results of the procedure were mixed. As discussed below, some of the results were expected; others seemed counter-intuitive.

1. Factors. A number of different factors were considered. Among those considered, the combinations of four factors which yielded the best results for both container size selection and item assignment were item weight, item shape, item volume (LWH), and average order quantity. These factors resulted in the specification of a container size which had lower total costs than the other combinations of factors considered.

2. Grouping Technique. The Similarity Coefficients Method was the primary method investigated. Several other techniques were considered, but the ease in applying SCM and the quality of the results made SCM the preferred method. SCM was applied both with and without ABC analysis. The results obtained when incorporating ABC analysis (and restricting consideration to A and B items for the determination of container sizes) were superior to the results obtained when all items were considered in determining container sizes. In addition, the analysis was more manageable when ABC analysis was used.

3. Application of Results. There are two alternative approaches for applying the results.
   - Alternative 1: Use the grouping method only to determine container sizes. Then assign individual items to a group on the basis of minimum costs.
   - Alternative 2: Use the grouping method to determine container sizes and assignment to groups.

   The grouping procedure was found to be an effective way of selecting "good" container sizes for the problem investigated. For Alternative 1, the container sizes selected by the grouping procedure (24"x24"x12" and 20"x20"x19") yielded costs only 4.1% higher than the costs for the optimum container sizes (24"x24"x12" and 14"x14"x14"). (The optimum sizes were determined by enumeration of container combinations, evaluation of costs for each item for each container size, and assignment of each item

Table 1 Cost and container size comparison

SIZES		COSTS	
Container 1	Container 2	Alternative 1	Alternative 2
24"x24"x12"	20"x20"x19"	$59,422	$ 63,137
24"x24"x12"	16"x16"x16"	$57,129	$ 73,457
24"x24"x12"	14"x14"x14"	$57,108	$ 79,105
24"x24"x12"	12"x12"x12"	$61,244	$105,083
24"x24"x12"	12"x12"x 6"	$62,004	$242,044

to the container size which resulted in minimum total cost for that item). For Alternative 2, the costs were 10.6% greater than the costs for optimum size and assignment. An analysis of the sensitivity of total costs to container size indicated that total costs are not very sensitive to changes in container size when optimum assignments are made (Alternative 1). However, for assignments based on the grouping procedure (Alternative 2), costs are highly sensitive to container size. Table 1 illustrates the sensitivity of costs to container size.

4. Number of container sizes. For the items considered, costs are decreased from $69,706 to $59,422 by adding a second container size for shipping, when optimum assignments are made (Alternative 1). Extending the analysis to three groups and selecting a third container size resulted in only a slight decrease in costs for Alternative 1. Under Alternative 2, however, costs increased significantly. Thus, the grouping procedure used does not appear to be very useful for assigning items to groups when several container sizes are to be permitted.

6 EXTENSIONS AND CONCLUSIONS

This chapter has demonstrated the development of cost models and the application of group technology methods for container size selection. Further analysis is needed with different data sets and considering different factors, grouping methods, and cost models before any conclusions can be made concerning the general applicability of a group technology approach to container size selection. If such an approach can be shown to be useful, a wide range of opportunities exist for its application.

7 REFERENCES

1. Carrabino, J.D., "Determination of Optimum Sizes and Economic Feasibility of Shipping Containers Using Operations Analysis Techniques," *Journal of Industrial Engineering*, pp. 513-520 (November 1958).
2. Friedman, W.F. and J.J. Kipnees, *Distribution Packaging*, Robert E. Krieger Publishing Company: New York, (1977).
3. Gallagher, C.C. and W.A. Knight, *Group Technology Production Methods and Manufacture*, Ellis Harwood Series, Holsted Press, (1986).

4. Hodgson, T.J., "A Combined Approach to the Pallet Loading Problem," IIE Transactions, pp. 175-182 (September 1982).
5. Hodgson, T.J., D.S. Hughes and L.A. Martin-Vega, "A Note on a Combined Approach to the Pallet Loading Problem," *IIE Transactions*, pp. 268-271 (September 1983).
6. King, E.P. and R.J. Goode, "Application of Operations Research Techniques to a Packaging Problem," *Proceedings of the 14th Midwest Quality Control Conference*, pp. 177-182 (September 1959).
7. Normandin, G., "Computer Picks the One 'Best' Container Size," *Modern Materials Handling*, pp. 60-62 (May 1967).
8. Peleg, K. and E. Peleg, "Container Dimensions for Optimal Utilization of Storage and Transportation Space," *Computer Aided Design*, pp. 175-181 (July 1976).
9. Steudel, H.J., "Generating Pallet Loading Patterns: A Special Case of the Two-Dimensional Cutting Stock Problem," *Management Science*, pp. 997-1004 (October 1979).
10. Tanchoco, J.M.A. and M.H. Agee, , "Unit Load Interfaces," *Proceedings of the 1981 National Material Handling Forum*, (April 1981).
11. Tompkins, J.A. and J.A. White, *Facilities Planning*, John Wiley and Sons: New York, (1984).
12. Udy, J.L., R.J. Balling, S.E. Benzley and M.D. Landon, "Computation of Interferences Between Three Dimensional Objects and the Optimal Packing Problem," *Advances in Engineering Software,* Vol.10, No.1, pp. 8-14 (1988).

COMPUTER-AIDED FACILITY DESIGN REVISITED: A PROTOTYPE DESIGN WORKSTATION FOR AGV SYSTEMS

Leon F. McGinnis
Georgia Institute of Technology

ABSTRACT

In the mid-sixties, there was great hope that computers would revolutionize facility design. Research focused on facility layout and resulted in a number of computer programs for the block layout problem. Sadly, in the intervening twenty years, while computers have become prominent in facility design, facility design research has failed to produce a significant impact on the practice of facility design. Perhaps the reason is that the research has focused on the wrong issues. This chapter describes a *strategy* for computer-aided facility design, and describes the current state of a prototype design workstation developed from that strategy.

1 INTRODUCTION

The *facility design problem* is to specify the design of a facility. Behind that simple definition hides a nightmare of complexity. This enormously complex problem is at the heart of both traditional and modern industrial engineering, yet facility design remains essentially a craft, depending on individual expertise and ad hoc analysis.

The thesis of this chapter is that *facility design research has failed to impact the practice of facility design because it has focused on models of particular problems in facility design, largely ignoring the facility design process*. The models themselves, by and large, are not used, thus the research has little impact on practitioners.

Research on models is, and will always be, important for advancing our understanding. It is not sufficient, however, simply to develop new models. As researchers, we must also address the process in which the models are intended to be used. In this chapter, it is argued that consideration of the facility design process can lead to challenging and fruitful research.

To set the scene, we first discuss the facility design problem and the facility design process, focusing on the difficulty of facility design and the role and use of models. Then a *computational model of facility design* is proposed. The computational model provides a baseline for subsequent discussions of appropriate requirements for computer aiding of facility design.

To demonstrate that the concepts advanced in this chapter can be made operational, we present a case study. An integrated design workstation is described, which supports the design of unit load automated guided vehicle systems in a manufacturing facility. The workstation represents a first step toward a fully integrated facility design environment, and has been in active use for over two years. During that period, over a dozen industry case studies have shown the benefit of integrating the design models with the design process.

2 THE FACILITY DESIGN PROBLEM

Before we can address the facility design problem, we must define it. As Francis and White [16] point out, it is very difficult to draw the boundary around a facility design problem. For example, what is **not** the facility? Can we treat the product as distinct from the facility? Can we treat the planning and control system as distinct from the facility? Can we consider the facility independently of the people who work in it?

The obvious answer is "No," because *the facility designer cannot ignore the interaction of these considerations*. Fig. 1 illustrates one view of the facility design problem. Each box in Fig. 1 represents a set of design decisions. The dashed line encloses those design decisions that directly determine the physical configuration of the facility, i.e., the facility design. The facility design problem, as usually defined, involves specifying those decisions enclosed by the dashed line.

As Fig. 1 illustrates, the facility design consists of five sets of design decisions. *Machine grouping* decisions specify what pieces of process equipment are required, and the relationships between them. For example, all the milling machines may be grouped in a milling department, or they may be distributed among manufacturing cells. The *stores design* describes the sizes and types of centralized stores, which may be for materials, work-in-process, or finished goods, and the relationships between stores and processes. *Buffer design* specifies the work-in-process staging capability for each processing operation. The *material flow system design* specifies the technology for material handling, and the layout of the material handling systems. Finally, the *layout* depicts the physical configuration of the facility, including the processing equipment, stores, buffers, and material handling. (Although it often is not treated in detail either in IE research or courses, the layout also must address the extremely critical issue of *utilities*.) Layout decisions include the allocation of space, e.g., for aisles and bathrooms, breakrooms, etc., as well as the arrangement of space.

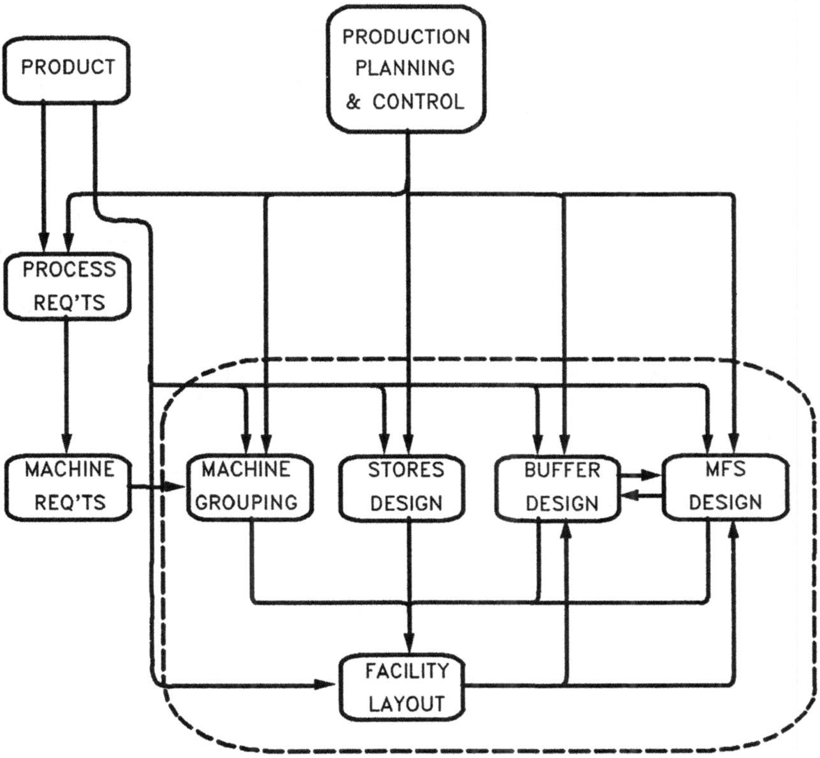

Fig. 1 Facility design decisions

The facility design problem is quite challenging, for several reasons. First, it is unlikely that a single individual will make all the design decisions. Since one set of decisions often has a major impact on other sets of decisions, having different decision makers introduces the potential for communication failures. For example, changes to the product design, such as redesigning for manufacturability, often should be accompanied by changes to the processing requirements, the machine groupings, and the material flow system. Unless there is excellent communication and coordination, these changes may not be made.

Second, there is a great deal of uncertainty in most facility design problems regarding the product. Often, when the facility is being designed, there is only a general idea about the product(s) it will produce. This problem is growing more important as product life cycles become shorter.

A corresponding problem is uncertainty in the production requirements, i.e., the amounts to be produced. Again, this is especially difficult when there are many products in production, each at a different point it its life cycle.

A third type of problem arises because of the fuzzy boundary between design and implementation. It is not unusual for the "final" facility design to omit many important details, which are subsequently determined during the implementation. For example, the details of the material flow system may not be specified until after the facility is erected, and much of the process equipment is already in place. As another example, the details of the control systems are almost never worked out in the facility design.

Facility design research has addressed all five sets of decisions, but almost always as independent problems. That is, research addressing one set of decisions treats all the other decisions as known and given, employs some aggregate or approximate representation of the other design elements, or ignores them altogether. For example, material flow system design research generally treats the layout of the processing equipment as a given, block layout methods generally consider material handling only through the proxy of material flow distance, and machine grouping research generally ignores both the material flow system design and the layout.

3 THE FACILITY DESIGN PROCESS

There are a number of well-known prescriptions for doing facility design. One of the best known is the Systematic Layout Planning method of Muther [32], which identifies 10 specific steps in three distinct phases. Apple [1] lists twenty steps, and Reed [34] lists ten. While the specifications of the various authors vary somewhat, they also share common elements.

The essential elements of the facility design process are summarized in Fig. 2. The solid box represents the facility design itself, i.e., the set of documents or databases in which the facility design is recorded. The facility design, and corresponding set of documents and databases, will change as various aspects of the design are elaborated or modified throughout the design process. At some point, the design is considered to be complete, is *frozen*, and becomes the reference data for implementation, i.e., construction of the facility.

The design is created and modified by the designer(s), represented by the broken line box. The designer must be able to comprehend the current state-of-the-design in order to make decisions to add to or modify the design. In making these *design decisions*, the designer integrates the current state-of-the-design as he/she understands it, the requirements and constraints, as he/she understands them, domain specific knowledge, and the results of supporting models and analyses. Note that there may be several designers working either simultaneously or sequentially on different aspects of the facility design. Comprehending the current state-of-the-design often is difficult because of the *lack of integration* of the design decisions.

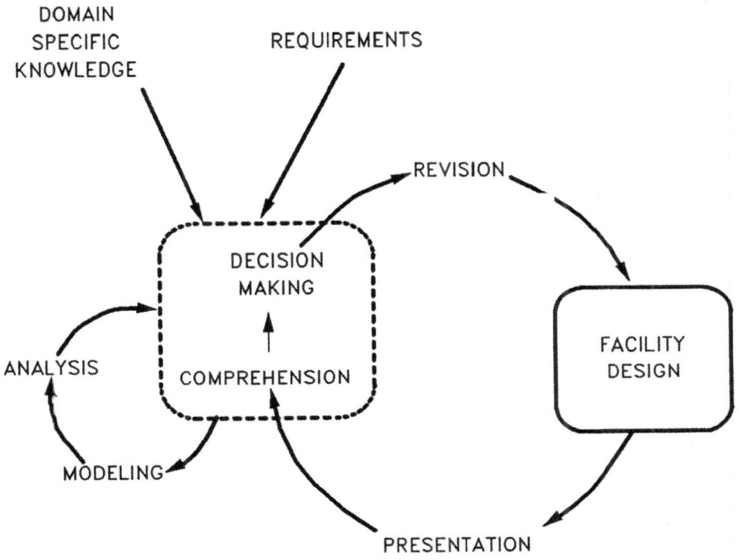

Fig. 2 Facility design process

It would not be unusual for each set of design decisions identified in Fig. 1 to reside in an independent database. Thus, for example, in developing a block layout, the designer might have to repeatedly consult four different databases in order to assess the requirements for the layout and also to estimate the potential impact of a specific layout alternative. Where Fig. 2 shows a single box for the facility design, in reality the designer might be working with as many as five relatively independent databases.

Domain specific knowledge may come from experience with prior designs, reference data such as standards and codes, prior education, research, or some other source. For example, the designer may use knowledge about the company's typical practices with regard to palletizing, along with experience in lift truck applications to determine the necessary floor space to allow for block stacking the finished product. Domain specific knowledge may be documented, and thus available to a number of designers. On the other hand, a considerable amount of very useful domain specific knowledge is not recorded or documented.

A variety of analytic models have been developed which might be used to aid the designer in decision making. For example, in designing buffers, the designer might use queueing network analysis (see, e.g., [36], [37], [38], and [41]). Analytic models may be used for sizing stores (see, e.g., [9], [10], and [14]). Block layout heuristics may be used in developing the facility layout (see, e.g., [2], [23], and [29]). In every case, however, the designer will be forced to create a separate database for each model, and must maintain the model database in parallel to the facility design.

The presentation of the current design to the designer takes two forms. *Geometric data* about the relative locations of equipment, aisles, walls, columns, etc., is presented in *line drawings*, or sometimes in *physical models*. All other aspects of the design are presented as *alphanumeric data*, in the form of prose, data tables, equipment lists, etc. In most cases, however, the designer will be required to make any necessary references between data types. For example, a designer examining a layout drawing, and wishing to know what parts will move between two processing operations, will be forced to search through a separate database to find those parts corresponding to the two operations.

As Hales [19, 20] points out, the complexity of modern facility design dictates the use of computer tools to manage the design database. CAD systems are commonly used for creating and maintaining the line drawings. Database systems may be used for managing equipment lists. For example, a relational database might be used to permit equipment lists to be viewed by department, by manufacturer, by type, or other classification scheme. **In the practice of facility design, by far the most frequent use of computers is for managing the facility design information.**

Facility design research, on the other hand, has focused almost exclusively on modeling. More specifically, research has tended to focus on block layout models of the problem. From the early work resulting in block layout programs, such as CRAFT [2] and CORELAP [23], to more recent work on analytic models (see, e.g., [15], [28], [29], and [30]), the focus has been on the shapes and relative locations of areas whose sizes and relationships are assumed to be known.

Despite the substantial body of research on layout analysis, computer aided facility design, in practice, amounts to the use of templates for speeding up the data entry to CAD systems. In rare cases, the CAD system may have the capability to generate a summary report for space allocation, equipment lists, or some similar type of information. However, **there are no commercially available CAD packages that provide integrated modeling.** The most reasonable explanation for the lack of modeling integration is that the model developers have never addressed the data problems associated with modeling.

4 A COMPUTATIONAL MODEL OF FACILITY DESIGN

Fig. 3 summarizes the current state of the computational aspect of the facility design process. As in Fig. 2, the broken line box represents the designer, who must interact with multiple design databases, and with supporting models. Each of the design databases may require a different mode of interaction. For example, requirements and specifications may require the use of a word processor (broadly defined, a typewriter or pencil could be the realization of

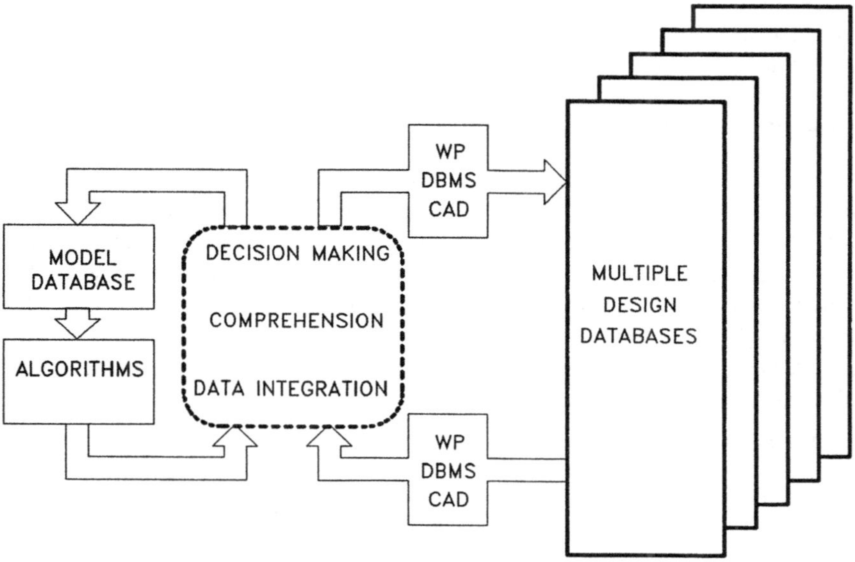

Fig. 3 As-Is computational model of facility design

the word processor!). Equipment and material lists may require the use of a database management system (DBMS), and a computer-aided drafting (CAD) package may be required to access facility layouts.

In this computational model, data integration is performed by the designer, both for the design data coming from different design databases, and also for the results obtained from supporting models. The designer must use a number of different interfaces, each requiring a significant degree of expertise for proficiency. In using supporting models, the designer must create and maintain the model database. The end result is that **the designer expends a great deal of time and effort simply getting, moving, translating, and integrating data.**

If the foregoing is an accurate assessment of current practice in facility design, then it may provide some insight into the lack of use of models in facility design. Models are simply tools to aid the designer. Any tool must provide an advantage, or it is not useful. If the amount of effort required to use the tool is not justified by an improved result (either in time, energy, or quality), then the tool will not be used. If the job can be done without using the tool (even if the result is not ideal), and especially if the effort to use the tool appears significant, then there is a low probability that the tool will be used.

Researchers who develop models for facility design problems are essentially "sellers" of tools. If they are to be successful, they must not only develop the methodology necessary for the tools, but also the methodology that insures the tools will be accepted and used. The latter is significantly absent in most facility design research.

5 DESIGN AIDING STRATEGIES

There are two clear strategies for research aimed at improving the practice of facility design: facility design data integration; and facility design modeling integration. Data integration methodologies can reduce the information processing burden on the facility designer, freeing time and energy for the creative aspects of facility design. Data integration also is a prerequisite for widespread use of facility design models. Modeling integration will reduce the overhead associated with models, making them easier to use, and thereby eliminating one of the obstacles to their use.

5.1 Design Data Integration

What does data integration mean in the context of facility design? It might mean simply the capability to simultaneously view data in two or more distinct databases. Such a capability can be provided in a computing environment that supports windowing, e.g., on a Unix workstation or a DOS microcomputer system running MS-Windows™. While this kind of *programming solution* would be a giant step forward for most facility designers, it falls far short of true data integration.

The ideal of design data integration would be an *intelligent design database*. In the intelligent design database, a change to one design decision would automatically flag all the affected design elements, thereby notifying the appropriate designer of the need to re-evaluate, and perhaps to change, previous design decisions. For example, consider a decision to change the specification of a machining center to give it greater capability. Greater capability at a machining center could impact part routes, which impacts material flow requirements. In turn, this might change the buffer requirements, and even the facility layout.

An intelligent design database might be thought of as a *design compiler*. The output of the design compiler is either a set of messages identifying inconsistencies in the current state-of-the-design or an error-free design, meaning that there are no unresolved design change references. In other words, the potential impact of each design decision has been evaluated and resolved by the appropriate designer.

The programming solution can be achieved without research. The design compiler, on the other hand, clearly requires research. For example, what are "important" relationships between design decisions? Is there one design database, or are there a number of distinct databases, with some sort of communication between them? How will the change/notify/evaluate/modify process be managed? Will the process converge naturally, or must convergence be forced through specific rules or disciplines? Understanding these kinds of questions, much less answering them, will require much experimentation, and the formulation of appropriate theories.

5.2 Design Modeling Integration

It is not enough to have facility design models; the models must be sufficiently convenient to use, so that the effort to use them is not perceived by the designer to be excessive. Thus, facility design models must be integrated with the design database. There are at least three important elements of this integration.

One of the great difficulties with current modeling practice is that the designer must create the model database, independently of the design database. One requirement for design model integration is *automatic model generation*, in other words, providing the necessary interface software so that the appropriate design data may be extracted and translated to the required model format. The role of the designer then becomes one of managing and controlling model generation, rather than hand coding the model database.

A second difficulty with many current facility design models is that the results of the model are not necessarily in a readily understood format. A second requirement for design modeling integration is the model must provide its *results in context*. For example, the results of a queueing network analysis of machine utilization must be provided with appropriate names so that the designer need not reference another database to understand the results. A collateral requirement is that models addressing the same issue should present their results in comparable formats. For example, a queueing network analysis of machine utilizations and a simulation analysis of the same problem should generate results that are directly comparable.

A third requirement for design modeling integration is the development of *structured model libraries*. Ideally, the computing environment will present the designer a selection of models appropriate to the design data currently being viewed. For example, if the designer is viewing a block layout, the appropriate models would be block layout models, and not process planning models.

Some progress toward design modeling integration could be made without design data integration. However, this probably is not a wise strategy. Consider, for example, the issue of automatic model generation. Without design data integration, automatic model generation would be almost completely *ad hoc*, requiring a new interface between every database/model pair.

5.3 Ideal Facility Design Computing Environment

Ideally, the facility designer would work through a single, consistent user interface. All the relevant design data could be accessed and viewed at any time, and control could be switched easily and seamlessly between any of the applications software.

A conceptual framework for an ideal design environment is illustrated in Fig. 4. The user need not be concerned with the specific format in which the data are stored; the interface system provides seamless access to any of the data. The user can simultaneously view the layout and product information, and simultaneously modify both. At any time, the user can invoke a model to evaluate a design or to suggest design alternatives. The data integration system, which is transparent to the user, enforces consistency, and manages all necessary extraction or translation of data formats.

At a minimum, the ideal environment illustrated in Fig. 4 must permit the facility designer to continue to do all the things that can be done today, without such an integrated environment. If no additional design functionality is created, the design environment would still be a great step forward, because it would eliminate much of the noncreative effort associated with accessing, translating, and manipulating data.

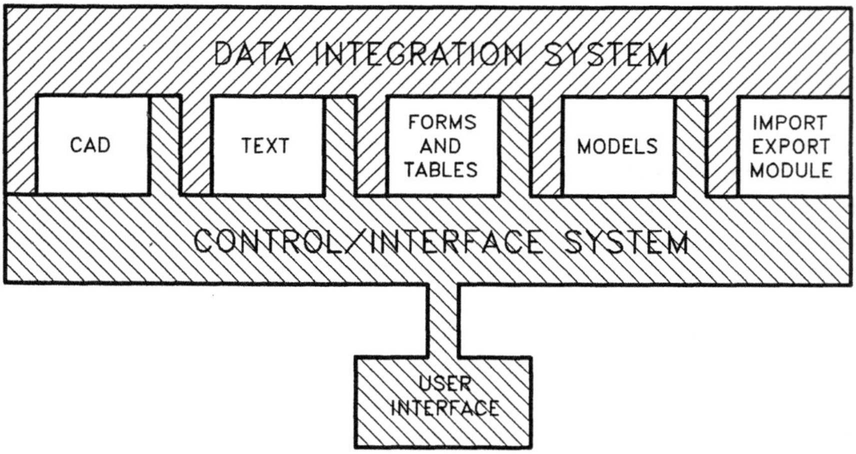

Fig. 4 Ideal facility design environment

6 CASE STUDY: AN AGVS WORKSTATION

The concepts of data integration and model integration described above have been implemented in a prototype computing environment for designing unit load automated guided vehicle systems (AGVS) in a manufacturing facility. This prototype, referred to as the AGVS engineering workstation, or EWS, is not claimed to completely integrate either data or models. However, as we shall describe below, it does incorporate significant data and model integration in the context of AGVS design.

AGVS applications were selected as the testbed for an integrated design environment for several reasons. At the time the project started, AGVs were a "hot" technology, and there was little experience in designing applications, at least in the U.S. Thus, it was thought that there would be considerable interest from practitioners, and corresponding opportunities for field testing. Second, as material handling systems, AGVs are relatively complex: they interface with the manufacturing processes at many dispersed locations; they must respond to service requests that appear at random; the vehicles operate asynchronously and individually; there is potential for randomly occurring interference between the vehicles; and the problems of managing and controlling an AGV system are not well-solved. It seemed clear that the implementation process would generate many interesting research questions. Finally, the complexity of AGVS design derives from the dependencies between the AGV system and all the other elements of the facility design, as depicted in Fig. 1.

The AGVS EWS project has been conducted within the Material Handling Research Center at Georgia Tech, under the joint sponsorship of the Member Companies of the Center and the Member Companies of the Computer Integrated Manufacturing Systems program. As a result, from the beginning, there has been a strong emphasis on technology transfer, and a corresponding need to understand and support the design process.

The objective of the EWS is to provide an integrated computing environment in which a knowledgeable designer can quickly develop a system design for an AGVS application. The level of detail in the design includes the placement and orientation of the guidepath segments, the placement of the load/unload stations, the number of vehicles, and some aspects of the vehicle dispatching and routing. The EWS functionality includes creating and manipulating a design database, accessing a range of supporting models, specifying and modifying the AGVS design, and generating appropriate reports.

In the remainder of this section, we will briefly discuss the data requirements for an AGVS application design, the principles that have guided our development of the EWS, the architecture that has emerged from the research project, and the current status of the implementation. The section closes with a brief discussion of additional research in the AGVS EWS arena.

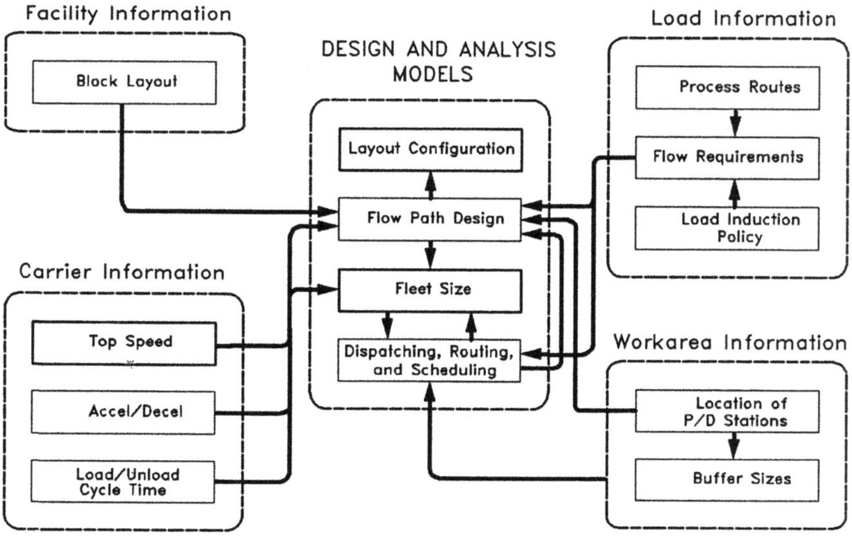

Fig. 5 AGVS design decision interdependencies (from [6])

6.1 AGVS Application Data Requirements

An AGV system design is heavily dependent on other facility design decisions, and, in turn, may have a significant impact on other facility design decisions. Some of these interdependencies are illustrated in Fig. 5.

Block layout clearly affects the AGV system design, if for no other reason than because it specifies the aisle structure. For example, if the aisles are not sufficiently wide for two vehicles to pass one another, then either the AGV system will have only unidirectional travel in the aisle, or it will require bidirectional vehicle travel in the aisles. While bidirectional vehicles are common, the bidirectional capability is used primarily off the aisle for load/unload operations. Controlling bidirectional travel in the aisle is not practical for large systems with many vehicles.

Machine grouping decisions, together with the facility layout, will create constraints on the placement of pickup/deposit (P/D) stations. In addition, the buffer sizing decisions may have a significant effect on the AGV system performance requirements. Generally, as the size of the buffers in a work area increases, there is less pressure on the AGV system for responsiveness.

The load information follows from the process requirements, the production volumes, and the unit load size. For example, in a batch manufacturing environment, the size of the container to be moved by the vehicles clearly has an impact on the number of trips required. Likewise, the production schedule can impact the AGV system, by the way it distributes move requirements over time and space.

The AGV system design must specify the attributes of the vehicles. Critical factors for the system design include the speed, acceleration, and load/unload time. Obviously, the capacity of the vehicle, in terms of size and weight, also is important. Other characteristics that need to be considered at some point include factors such as the curve turning envelop, i.e., the space swept out by the vehicle when it turns a corner. We consider that to be relevant to *detailed system design* and do not consider it in the EWS at this time.

Fig. 5 identifies four key AGV system design elements. *Flow path design* specifies the location and orientation of *path segments*; in effect, the flow path design dictates where vehicles may travel. For bidirectional path segments, the orientation is not relevant. An integral part of flow path design is the number and placement of P/D stations, and the identity of the workareas that they serve.

Fleet sizing is the second major AGV system design decision, and specifies the number of vehicles (of each type, if more than one type is used) required to meet the flow requirements. Clearly, the fleet size is affected by the flow path design, and also by the system control algorithms. In particular, the nature of the vehicle dispatching can have a major impact, especially when buffer sizes become a major consideration.

Dispatching, routing and scheduling represent the major system management and control functions. The algorithms specified for these functions will determine which vehicle is assigned to service each transport request, how a vehicle shall be routed for both unloaded and loaded travel, and such factors as scheduling battery charging and preventive maintenance and the management of congestion.

In Fig. 5, *layout configuration* refers to the details of the guidewire arrangement for systems that use an active rf signal and frequency-shifting for vehicle guidance and control (see, e.g., [8] and [21]). For these systems, decisions are required for the number of signal sources, their frequencies, and the arrangement of their wire loops.

As Fig. 5 clearly illustrates, AGV system design requires integrating many of the key elements of facility design. Our approach has focused on the decisions required to specify the AGV system, i.e., flow path design, fleet sizing, dispatching, and guidewire layout, and for those decisions, we have developed integrated modeling capabilities. For the decisions related to load information, block layout, and workarea information, we have not developed integrated models at this time, although we have implemented the necessary data creation and manipulation capabilities.

6.2 Implementation Principles

It was recognized from the beginning that the AGVS EWS project would be a large multiyear effort. Thus, while data and modeling integration would be the beacons that would guide research in the project, it was important to set out some other principles that would guide the associated implementation.

Our concern for technology transfer lead to the first principle, which has proven to be an important source of research issues: *the implementation must be user focused, with frequent "live" testing.* Live testing sometimes can be done in the research laboratory, by bringing in designers with real-world design problems. The limitation is that in such an environment, the researchers are close at hand to "coach" the designers on the software, and to provide "work-arounds" when software idiosyncrasies appear. Thus, it also is important to put test versions of the computing environment in the field.

The desire to test the implementation in the field lead to a very important initial implementation decision: *the computing platform must be one which is likely to be available to a large number of facilities designers.* Thus, the decision was made to target the implementation on microcomputer systems based on the Intel 80x86 family of microprocessors running under DOS, as these platforms are widely available today, and the price-performance curves for them are a favorable indication of continued widespread availability over the next five to ten years. As we shall see, this implementation decision has not proven to be overly limiting.

A second result of the concern for technology transfer was a recognition that *there must not be a long learning curve for the knowledgeable designer.* In other words, someone who is familiar with manufacturing and AGV systems should be able to become relatively proficient with the EWS in a reasonable amount of time, say two to four hours. This means that the interface should require a minimum of arcane computer knowledge, and that the user should not have to deal with flaky software.

Our recognition that the EWS would likely become a very large software system lead us to the second principle: *the implementation must make maximum possible use of commercial and public domain software.* In a research environment, we simply could not afford to reinvent the wheel. Clearly, this might be a poor decision if the objective were to develop a commercial product, since it might make the cost prohibitive. However, since we are most interested in demonstrating concepts, it is an acceptable tradeoff. In other words, commercialization is not a key concern.

Our intention to prove the concepts of data and modeling integration lead to the third principle: *the AGVS EWS must remain a research platform.* Since data integration, model development, and model integration are research topics, the EWS must continue to support research efforts. Thus, we often may be forced to sacrifice computing efficiency for implementation efficiency and flexibility.

6.3 EWS Architecture and Implementation

The architecture of the AGVS EWS is illustrated in Fig. 6. The designer works with a single *common user interface*, through which all data manipulation and modelling operations are performed. There are two subsystems through which data may be viewed or manipulated: a *CAD system* and an *editor*. All data, whether requirements, design, or model related, reside in a *neutral database*. The neutral database contains both *persistent data* and *transient data*. Persistent data are carried over from one design session to another, but transient data are destroyed at the end of a session. Models are characterized as *fast analysis, optimization, and simulation*. To be considered a fast analysis, a result must be returned within a matter of seconds. Optimization routines typically take longer, and the nature of the returned result is a suggested alternative for some design decision. Simulation analyses also typically take much longer than a few seconds, and the nature of the returned result is an evaluation of a design alternative.

At the core of the EWS, and critical to its success to date, is the concept of the neutral database. The neutral database, in its current realization, is simply a collection of flat ASCII files, plus the corresponding CAD drawing files. In essence, each distinct type of information about the AGVS design problem has an associated ASCII file, with a specific file extension. The one exception is the CAD drawing files, which are in a proprietary format.

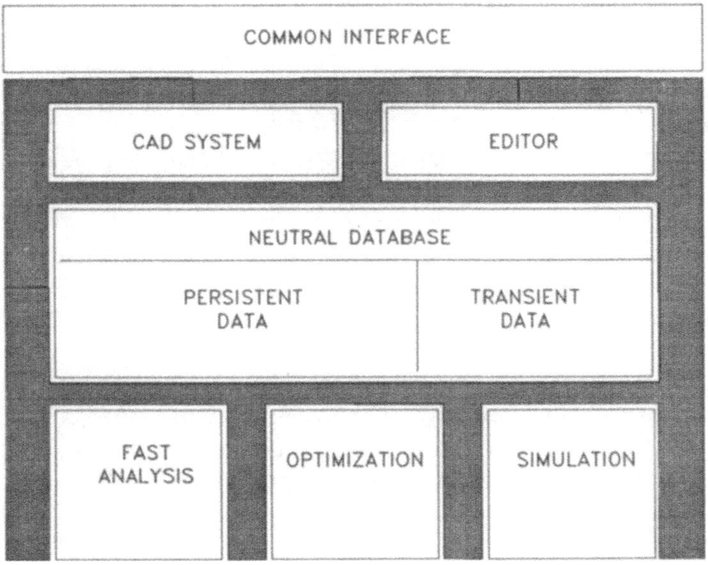

Fig. 6 AGVS EWS architecture

This is not necessarily a computationally efficient nor functionally powerful way to manage a design database. Yet it has one crucial redeeming value: since it has a minimum of structure, it is very easy to change the content and format of the data. Likewise, very simple methods, with very low computational overhead, suffice to control access, updating, and coordination between the separate files. These attributes are critical in the research environment, where the ultimate requirements are not really known.

The terminology, neutral database, reflects the fact that the core database in the EWS may be in a format that is not directly useful for **any** application software. Thus, for every application program which accesses the neutral database, either to view or to modify data, some interface logic may be required. Because the neutral database has such a simple structure, the necessary interface logic also is very simple, and therefore relatively easy to implement. There is a very modest software overhead in reading and writing to the neutral database.

Since the neutral database takes the form of a set of files in a DOS environment, some mechanism is needed to insure that the specific files being read or written to in a session constitute a legitimate database. The necessary control is implemented through a special file called the *project profile*. A project profile contains some alphanumeric information about the project, such as the basic units of measure and the designer's identification, along with a list of the specific files to be used during the design session. Table 1 lists the contents of the project profile. The files listed in the project profile, along with the project profile itself represent persistent data. In addition, during a session a number of other temporary files are created.

A session may be started with a "dummy" project profile, allowing the designer to create a new design database; in this database, each of the individual files listed in Table 1 will have the same name as the project profile, but with the appropriate file extensions. The project profile might remain "transparent" to the designer. On the other hand, the profile gives the designer considerable flexibility to create variations of a design. For example, several alternative guidepaths might be created, with corresponding .cps, .seg, .wsl, .wsi, .pth, and .dwg files. All of the guidepaths could be used with the original .vht, .vhl, .wsf, .pti, and .ptr files, provided the designer has maintained consistency between the P/D station labels in the CAD drawings and those in the workstation flow data.

The project profile represents a tradeoff. While it is relatively simple, and provides a great deal of flexibility, it also requires some intelligence on the part of the designer. It is possible for a designer to make spectacular mistakes by incorrectly combining files in the project profile. When there are gross inconsistencies, such as inconsistent naming of workareas, or of P/D stations, the interface logic will detect and flag them. However, if the same names have been used, but have different meanings, it is possible that no

Table 1 Content of project profile

NAME	CONTENT	COMMENTS
LENGTH UNIT	[length unit]	IN - inches, FT - feet, M - meters
TIME UNIT	[time unit]	SEC - seconds, MIN - minutes, HRS - hours
PLANNING PERIOD	[period length]	Period duration in TIME UNITS
MODELER'S NAME	[name]	Identifier for user
PROJECT	[project id]	Identifier for project
PROJECT DATE	[date]	In format: MM-DD-YYYY
VEHICLE TYPES	[filename].VHT	created in DIALOG submenu
VEHICLE LABELS	[filename].VHL	created in DIALOG submenu
CONTROL POINT LIST	[filename].CPS	created by CAD program functions
SEGMENT LIST	[filename].SEG	created by CAD program functions
P/D STATION LIST	[filename].WSL	created by CAD program functions
WORKSTATION INFO	[filename].WSI	created by CAD program functions
WORKSTATION FLOWS	[filename].WSF	created in DIALOG submenu
PART TYPE INFO	[filename].PTI	created in DIALOG submenu
PART TYPE ROUTES	[filename].PTR	created in DIALOG submenu
SHORTEST PATHS	[filename].PTH	created by CAD program functions
GUIDEPATH LAYOUT	[filename].DWG	AutoCad format, non-neutral

detectable error will occur, even though the combined database is meaningless.

The CAD functions in the EWS are provided by the proprietary software package AutoCad™, from Autodesk, Inc. [3]. AutoCad™ was chosen for two reasons: it had, at the time, the largest installed base of any CAD package for microcomputers; and it has an extremely flexible and customizable menu structure. The latter has been quite important in the implementation of interface logic.

The editor functions in the EWS are provided, not by a single editor, but by a number of editor modules coordinated by the interface logic. Some of the editor modules were developed by the research team, and some are adaptations or applications of modules from commercial sources. For example, for editing from/to flow data, we have made good use of spreadsheet routines from the QuickBasic™ library provided by Microsoft [27].

Fast analysis functions have been implemented in a variety of programming languages, including Fortran, Pascal, and Basic. The architecture of the EWS facilitates the development of analytic routines using whatever programming language is most appropriate. We have found that the most appropriate language is almost always the one with which the researcher has the greatest facility! Since the EWS is a team-oriented research project, a variety of languages have been used.

At the present, the optimization functions in the EWS are very similar to the fast analysis functions with respect to implementation, i.e., they have been implemented by the research team using a variety of programming languages. In addition, however, the EWS architecture makes it relatively easy to incorporate standard optimization software. For example, Liu [24] developed a mixed integer programming model for optimizing the final design of the guidepath network, and solved it using the Lindo™ package [35]. The architecture of the EWS permits the incorporation of such a model by making the modification to the neutral database necessary to represent the decision alternatives and model solution, and modifying the interface logic to generate the input files to Lindo™, and to view the model solutions.

Integrating fast analysis and optimization models in the EWS framework has been relatively straightforward, because analytic models have a mathematical statement. Thus, the translation from the design database to the model database usually is a simple transformation of the data into the structure required by the model. Simulation models of manufacturing applications, on the other hand, have no obvious mathematical statement. As a consequence, simulation model integration is a much more difficult problem than mathematical model integration.

Our approach to simulation model integration has been based on the concept of a *simulation code generator*, or SCG (see, e.g., [13] for discussions of simulation generators). At the present, our target simulation language is SIMAN™, selected primarily because, at the time the project started, it was the only widely-available simulation language for microcomputers with an animation capability. Today, a variety of commercial packages could be used.

The initial implementation of the SCG was intended only as a proof-of-concept demonstrations, and was based on an AGVS simulation model originally developed by Davis [12]. The SCG, which is coded in Basic, reads the appropriate neutral database files, and creates model and experimental frame files with the same structure as the model in [12]. For example, the method used to represent vehicle routes is the same, although the specific realization depends upon the problem data. This model structure is quite limited in terms of the material flow requirements it can support, and in terms of the system control methodology represented. A substantial research effort has been devoted to the development of a more generic and robust model structure, which is still in its initial implementation phase.

The common interface "seen" by the designer is simply an AutoCad™ menu. We have customized the menu, so that most of the functionality of the EWS is seen either as a main menu item, or as part of some submenu. Obviously, a great deal of the functionality must reside outside of AutoCad™, even though the AutoLisp™ language does provide a significant ability for creating new functions within AutoCad™. Access to externally implemented functions is accomplished by using the "shell" command in AutoCad™ to exit AutoCad™, execute the external function, and then return to AutoCad™. The shell command essentially hides all of the program changes from the user.

When one of the AutoCad™ menu items is selected and starts an external process, such as an editor, the external process may have its own interface. In other words, the designer is no longer working with the AutoCad™ screen format or menu structure. However, when the designer quits the external function, control is returned to the AutoCad™ screen. While this may be inconsistent with the emerging practice in "graphical user interfaces," it is a reasonable trade-off between implementation elegance and implementation efficiency, especially for a research platform.

The interface logic is the "glue" which integrates the neutral database with all the editing and analysis functions. In an ideal system, the interface logic would be realized in distinct software modules, so that it could be documented and maintained. In the real world, unfortunately, we have yet to discover such a clean solution.

In terms of reading and writing the neutral database files, each stand-alone software module contains its own code to implement the interface. Because the EWS architecture is so simple, this means that *any neutral database file can be "used" by at most one module at a time*. Thus, for example, we would find it impossible to generalize the EWS to a networked version without a major restructuring of the interface logic.

In terms of coordinating the various functions, the interface logic is distributed. When an "external" function is selected in the AutoCad™ menu, the interface logic is a combination of AutoLisp™ macros that execute a shell command, and, perhaps, a DOS batch file that executes a sequence of programs and then exits the shell. There are AutoLisp™ macros that automatically convert certain aspects of the drawing file into a format that is useful for analysis, then check the data for consistency. For example, the guidepath network must be extracted from the drawing file (*feature extraction*) and converted into a network representation to support shortest path calculations.

6.4 Current Status

Our objective in this section is not to give a detailed description of the current status of the EWS. In the first place, space limitations do not permit such a full description--a crude user's guide [4] runs to 78 pages. Instead, we will present a general status report, and highlight some important observations, or lessons, that should prove helpful to others attempting to develop integrated facility design methods.

As of this writing (December, 1989) the EWS offers a customized AutoCad™ menu structure, organized as shown in Table 2. We have shown the menu structure only to two levels; in the elements, analysis, and dialog submenus, many of the functions have additional options that may be exercised. The file, sketch, edit, display, snap, and utilities submenus contain functions that are used essentially as provided by AutoCad™. The bulk, although not all, of the customization is in the layout, elements, analysis, and

Table 2 EWS menu structure

File	LAYOUT	Sketch	Edit	Display	Elements	ANALYSIS	DIALOG	Snap	Utils
-Save	-Straight Segment	-Line	-Erase	-Zoom In	-Segment, P/D Info	-Update Files	-Update Files	-Snap/Grid	-Shell
-Save as	-Bidirec. Segment	-Circle	-Erase Window	-Zoom Back	-P/D List	-Flowpath Design	-Flow Reqmnts	-Apertrue	
-Close	-Reverse of Straight	-Arc	-UnErase	-Zoom All	-Segment List	--Path Method	-Workarea Info		
-Abandon	-Curved	-Box	-Move	-Zoom to Fit		--Compute Potential	-Part Routes		
-Open	-P/D Station	-Text	-Move Window	-Zoom Dynamic		--Quit	-Vehicle Specs		
-Plot	-Split Straight		-Copy	-Turn Highlight Off		-Vehicle Reqmnts Analysis	-Profile		
-Print	-Join 2 Straight		-Copy Window	-Redraw		--Minimum Travel			
	-Outer Loop		-Rotate	-Modify Layout		--Worst Case			
			-Rotate Window	-3D Rotate		--Factoring			
			-Reverse Straight	-Split Screen		--Round Trip			
			-Reverse Curved	-Plan View		--Summary			
			-Stretch Segment			--Quit			
			-Shrink			--Simulation			
			Segment						
			-Rescale Drawing						
			-Undo						
			-Redo						
			-Cancel						

dialog submenus. The analysis and dialog submenu selections, along with the P/D list and segment list selections from the elements submenu, all initiate external routines that are accessed using the AutoCad™ shell command.

The first important observation is that by far the largest number of functions in the current version of the EWS are devoted to the tasks of creating, manipulating, and managing the design database. After three years of development and two years of field testing, we are convinced that as the EWS continues to evolve, **the proportion of modeling and analysis functions will decrease, not increase!** The reason is that the current version of the EWS does not integrate all the necessary data. For example, there are no files containing financial data and no files containing reference data, such as standards and codes. Furthermore, as additional models are integrated, they will have their own specific data integration requirements, resulting in the need, possibly, for additional functions for creating, manipulating, and managing the additional data.

We conjecture that, in general, an integrated facility design environment will have much more functionality devoted to data integration than to model integration. Especially for the case of analytic models, model integration is very straightforward, given that the necessary data integration already has been accomplished.

It may also be observed from Table 2 that the EWS menu has a particular structure. When there are so many functional options (in the EWS, well over 100, including all the options in the external modules), some structure is absolutely essential, in order for the user to understand them. The menu structure illustrated in Fig. 2 is the result of a process of evolution, with the need for structure driven by the implementation of additional functionality.

The menu structure in a design environment should be consistent with, and supportive of, the way that a designer will actually use the functions. This means that the developers of a design environment should have some concept of the design process. This is especially true for AGVS applications, because the major design decisions are so tightly related. Based on our field tests of an earlier version of the EWS, we proposed a strategy for AGVS application design, which is described in greater detail in [5]. The menu structure has evolved to support this design strategy.

A menu structure is essential for the design environment, especially for the purposes of technology transfer. On the other hand, once the menu structure becomes relatively stable, it may become an impediment to research. There is a tendency for the questions and issues to become "channeled" by the structure of the EWS. Facility design researchers must recognize this potential problem.

Most of the data integration in the current version of the EWS is of the "programming solution" variety, i.e., the ability to access different types of data from what appears to be a single user interface. However, there are some aspects of an intelligent database. For example, the network structure of the flow path design is initially represented in the CAD database as points and lines. A feature extraction process translates the points and lines representation to a nodes and arcs representation suitable for network analysis. The extracted database is checked for errors in the CAD drawing, such as deadend segments (i.e., the vehicle cannot leave a control point), or unreachable segments (i.e., once the vehicle leaves, it can never return). Similarly, consistency in labels is checked, since labels assigned in the CAD system are subsequently used in the editor for defining flow requirements.

The EWS in its current version does provide significant capabilities for integrated modeling of two specific issues: initial flowpath layout, and vehicle requirements analysis (or fleet sizing). In the current literature, there are no operationally feasible models for specifying the initial flowpath layout. Thus, one area of basic research in the EWS project has been models and algorithms for proposing initial designs. Bakkalbasi [6] has proposed three specific heuristic methods, two of which are new and one which is based on a network design heuristic from the transportation literature. At the present, only one of the heuristics is accessible in the EWS, although the current state of data integration would support all three.

The issue of fleet sizing had been addressed previously by a number of authors (an issue of *Material Flow* was devoted to this topic soon after the EWS project was underway), who have tended to focus on vehicle travel and congestion delays. Unfortunately, there was no general framework within which to view all the different models. We proposed a framework for analyzing the vehicle requirements which accommodates all the prior research. Also, we developed new methods for estimating unloaded vehicle travel which permitted the estimation of upper bounds on fleet size as well as lower bounds. In addition, we have developed statistical estimates for unloaded vehicle travel that are appropriate for several of the most popular vehicle dispatching rules. These results all are described in Bakkalbasi [6].

Model integration in the EWS is more than simply automating the translation from the neutral database to the format of the associated solution procedure. For example, the fleet sizing calculations are presented in such a way that the designer can explore the effect of varying the estimated congestion delay, or the estimated vehicle availability. Since the fleet sizing calculations require computing the loaded vehicle travel and estimating the unloaded vehicle travel, the designer has the option of displaying the most heavily and least heavily traveled path segments **on the CAD drawing.** Likewise, the flowpath layout heuristic presents its results, not as a table of path segments, but as graphical elements on the CAD drawing.

Over the past two years, the EWS has been used to analyze over twenty real AGVS application design problems. In 90% of these case studies, the designer has been someone from the corporation with the design problem. While the reaction of these "clients" has been uniformly enthusiastic, there have been some hard lessons learned. In particular, the receivers of this kind of technology transfer have very little patience with software that behaves in unexpected ways, or fails without warning or without explanation. There must be a higher standard of software engineering than one finds in much of the "research domain software" developed by an individual to implement a specific algorithm. Even when the software is thoroughly debugged, the designers often find a way to use it that is outside the intended range of application.

Our experience with field testing the EWS has convinced us that **data integration alone is not enough.** The EWS without the integrated models would be of very little value. The integrated models perform enormous amounts of computing, which would be impractical without the integrated environment. On the other hand, **model integration without data integration is not possible.** In many ways, the data integration provides an *infrastructure* within which we can develop and implement integrated models. A very important lesson for computer aided facility design research is that **the effort required to build and maintain the data integration infrastructure is at least two to three times greater than the direct research effort on modeling and analysis which it supports.**

6.5 Summary

The AGVS EWS demonstrates some data and modeling integration in the context of designing AGV applications. While it incorporates no models for analyzing block layouts, it does permit specifying and modifying the facility layout. Likewise, it permits specifying and editing other aspects of the facility design, such as part routes, material flow requirements, workarea definition, etc. Because the EWS uses AutoCad™ as its primary interface, it has a familiar "look and feel" to most facility designers, resulting in a very short learning curve for first-time users.

The models integrated in the EWS perform extensive computations and provide results that the typical facility designer could obtain in no other way. The results are presented in context, so are easy to interpret. Equally important, design results are accompanied by enough supporting information to permit the designer to conduct an intelligent search of similar alternatives.

The EWS does not guarantee successful technology transfer. In the long run, if the research is to have a lasting impact on the practice of AGVS application design, the methodology embodied in the EWS must be commercialized. The significance of the EWS is the demonstration that commercialization is technically feasible.

7 CONCLUSIONS

The central thesis of this chapter is that facility design research has failed to impact the practice of facility design because researchers in the past have focused on models of design decision problems and have ignored the process in which those models might be used. We have proposed a computational model of facility design, and have argued that the critical issues in facility design are data integration and model integration.

To illustrate the applicability of the concepts of data integration and model integration, we have described an engineering workstation for designing AGVS applications. This EWS provides an integrated environment in which the facility, the part routes, the material flows, the workareas, and the AGVS can be specified and manipulated, along with models for analyzing the AGV system. It has been field tested over the past two years, with over twenty case studies.

We have demonstrated that it is possible for facility design research to address the process of facility design and successfully integrate models in the process. Obviously, we have only scratched the surface. Hopefully, others will be inspired to create their own facility design workstations, further demonstrating the value of data and model integration. **The community of facility design researchers will never see large scale technology transfer of facility design models until it has provided overwhelming proof of the value of the models and the technical feasibility of technology transfer.** And if technology transfer is not the objective, then it is not engineering research, it is simply intellectual game-playing.

8 ACKNOWLEDGEMENTS

The AGVS EWS research has been, and continues to be codirected by Professor Marc Goetschalckx. A number of outstanding students have worked on the EWS project; in chronological order they are: Kathleen Henning, Gene Durrence, Omer Bakkalbasi, Dah-Chuan Gong, Brett Peters and Arlin Johnson. The EWS project has been supported by the Computer Integrated Manufacturing Systems program and the Material Handling Research Center, both at Georgia Tech, and by an equipment grant from the Intel Corporation. Invaluable feedback has been provided by the participants of the Annual AGVS Design Workshop held on the Georgia Tech campus in 1988 and 1989, and by seminar attendees at AGVS 88, AGVS 89, and the 39th Annual Material Handling Short Course.

9 REFERENCES

1. Apple, J.M., *Plant Layout and Material Handling*, 3rd edition, John Wiley and Sons, Inc., New York, NY (1977).
2. Armour, F.C. and E.S. Buffa, "A Heuristic Algorithm and Simulation Approach to Relative Location of Facilities," *Management Science*, Vol. 10, pp. 294-309 (1963).
3. *AUTOCAD User's Manual*, Autodesk, Inc., 2320 Marineship Way, Sausolito, CA 94965 (1985).
4. Bakkalbasi, O. and G. Durrence, "An Engineering Workstation for AGVS Design: User's Manual," Material Handling Research Center, Georgia Institute of Technology, Atlanta, GA (1989).

5. Bakkalbasi, O. and L.F. McGinnis, "ABC's of Preliminary In-House Planning and Analysis of AGVS Applications," *Proc. AGVS 89*, Material Handling Institute of America, Charlotte, NC (1989).
6. Bakkalbasi, Omer, *Flow Path Network Design and Layout Configuration for Material Delivery Systems*, PhD Dissertation, School of Industrial and Systems Engineering, Georgia Institute of Technology, Atlanta, GA (1989).
7. Boegli, P., "A Comparative Evaluation of AGV Navigation Techniques," *International Trends in Manufacturing Technology: Automated Guided Vehicle Systems*, R. H. Hollier (Ed.), IFS (Publications) Ltd., London (1987).
8. Bohlander, R.A. and W. Heider, "Control Considerations When Planning AGVS Installations," *Proc. AGVS 88*, Material Handling Institute of America, Charlotte, NC (1988).
9. Bozer, Y.A. and J.A. White, "Design and Performance Models for End-of-Aisle Order Picking Systems," TR-87-09, Material Handling Research Center, Georgia Institute of Technology, Atlanta, GA (1988).
10. Choe, Kyung-Il and Gunter P. Sharp, "Class-Based Storage with Multi-Command Operation," TR-88-08, Material Handling Research Center, Georgia Institute of Technology, Atlanta, GA (1990).
11. CINEMA [1986], Systems Modeling Co., State College, PA.
12. Davis, Deborah A., "Modeling AGV Systems," *Proceedings of the 1986 Winter Simulation Conference*, pp. 568-574 (1986).
13. Davis, Robert P. and Jorge Haddock, "A Simulation Generator for the Selection and Design of Material Handling Devices," *International journal of Modelling and Simulation*, Vol. 6, No. 3, pp. 90-95 (1986).
14. Foley, R.D. and E.H. Frazelle, "Analytical Results for Miniload Throughput and the Distribution of Dual Command Travel Time," TR-87-05, Material Handling Research Center, Georgia Institute of Technology, Atlanta, GA (1989).
15. Foulds, L.R., "Techniques for Facilities Layout: Deciding Which Pairs of Activities Should Be Adjacent," *Management Science*, Vol. 29, No. 12 (1983).
16. Francis, R.L. and J.A. White, *Facility Layout and Location: An Analytical Approach*, Prentice-Hall, Inc., Englewood Cliffs, NJ (1974).
17. Goetschalckx, M. and K. Henning, "Computer Aided Engineering of Automated Guided Vehicle Systems," *Computers in Industrial Engineering*, Vol. 13, Nos. 1-4, pp. 149-152 (1987).
18. Goetschalckx, M. and L.F. McGinnis, "Designing Design Tools for Material Flow Systems," *Proceedings of the 11th Conference on Computers and Industrial Engineering*, Orlando, FL, March 13 (1989).
19. Hales, H. Lee, "Using Computers in State-of-the-Art Facilities Planning," *Industrial Engineering*, Vol. 16, No. 5, pp. 60-70 (1984).
20. Hales, H. Lee, *Computer-Aided Facilities Planning*, Marcel Dekker, Inc., New York, NY (1984).

21. Koff, G.A., "Automatic Guided Vehicle Systems: Applications, Controls and Planning," *Material Flow*, Vol. 4, pp. 3-16 (1987).
22. Kuhn, A., "Efficient Planning of AGVS by Analytical Methods," *Proc. 2nd International Conference on AGVS*, Stuttgart, West Germany (1983).
23. Lee, R.C. and J.M. Moore, "CORELAP-COmputerized RElationship LAyout Planning," *Journal of Industrial Engineering*, Vol. 18, pp. 195-200 (1967).
24. Liu, F.H., *Analytical Approaches for Designing a Class of Material Flow Systems*, Ph.D. Dissertation, School of Industrial and Systems Engineering, Georgia Institute of Technology, Atlanta, GA (1988).
25. Mathewson, S.C., "The Application of Program Generator Software and Its Extensions to Discrete Event Simulation Modeling," *IIE Transactions*, Vol. 16, No. 1 (1984).
26. Maxwell, W.L. and J.A. Muckstadt, "Design of Automatic Guided Vehicle Systems," *IIE Transactions*, Vol. 14, No. 2, pp. 114-124 (1982).
27. McGinnis, L.F. and M. Goetschalckx, "An Engineering Workstation for Computer Aided Engineering of Material Handling Systems," *Proceedings of Manufacturing International '88*, Atlanta, GA, Vol. III, pp. 137-142 (1988).
28. *Microsoft QuickBASIC*, Microsoft Corporation, (1986).
29. Montreuil, B. and H.D. Ratliff, "Optimizing the Location of Input/Output Stations Within Facilities Layout," *Engineering Costs and Production Economics*, Vol. 14, pp. 177-187 (1988).
30. Montreuil, B., H.D. Ratliff, and M. Goetschalckx, "Matching Based Interactive Layout," *IIE Transactions*, Vol. 19, No. 3, pp. 271-279 (1987).
31. Montreuil, B., U. Venkatadri, and H.D. Ratliff, "Generating a Layout from a Design Skeleton," Working Paper 89-01, Faculte des Sciences de L'Administration, Universite Laval, Quebec, Canada (1989).
32. Montreuil, B., "Integrated Design of Cell Layout, Input/Output Station Configuration, and Flow Network of Manufacturing Systems," in *Intelligent and Integrated Manufacturing Analysis and Synthesis*, edited by C. R. Liu, A. Requicha, and S. Chandrasekan, ASME, PED-Vol. 25, pp. 315-326 (1987).
33. Muther, R., Systematic Layout Planning, Industrial Education Institute, Boston, MA (1961).
34. Pegden, C.D., *Introduction to SIMAN*, Systems Modeling Co., Pittsburgh, PA (1984).
35. Reed, R., *Plant Layout: Factors, Principles, and Techniques*, Richard D. Irwin, Inc., Homewood, IL (1961).
36. Schrage, Linus, *User's Manual for Linear, Integer and Quadratic Programming with LINDO*, Third Edition, The Scientific Press, Redwood City, CA (1987).

37. Solberg, J.J., "A Mathematical Model of a Computerized Manufacturing System," *Proceedings of the 4th International Conference on Production Research*, Tokyo (1977).
38. Toro-Ramos, Zulma R. and Leon F. McGinnis, "Performance Modeling for a Single Material Handling Device with Random Service Requests, Part I: Pure Blocking," working paper, School of Industrial and Systems Engineering, Georgia Institute of Technology, Atlanta, GA (1989).
39. Toro-Ramos, Zulma R. and Leon F. McGinnis, "Performance Modeling for a Single Material Handling Device with Random Service Requests, Part II: Blocking with Recourse," working paper, School of Industrial and Systems Engineering, Georgia Institute of Technology, Atlanta, GA (1989).
40. Toro-Ramos, Zulma R., "Network of Queues and Decomposition Methods for Modeling Manufacturing Systems," PhD Dissertation, School of Industrial and Systems Engineering, Georgia Institute of Technology, Atlanta, GA (1988).
41. Yang, Moonhee, *Analysis and Optimization of Class-Based Dedicated Storage Systems*, PhD Dissertation, School of Industrial and Systems Engineering, Georgia Institute of Technology, Atlanta, GA (1988).
42. Yao, D.D. and J.A. Buzacott, "Modeling the Performance of Flexible Manufacturing Systems," *International Journal of Production Research*, Vol. 23, pp. 945-959 (1985).

A MODELLING FRAMEWORK FOR INTEGRATING LAYOUT DESIGN AND FLOW NETWORK DESIGN

Benoit Montreuil
Laval University

ABSTRACT

This chapter introduces a comprehensive modelling framework for integrating layout design and material flow network design toward the direct generation of net layouts including the design of the physical aisle system. Eight models are developed, each being specialized to deal with various facets of this integrated design. The framework offers an alternative conceptual platform to the Quadratic Assignment Problem for layout design researchers.

1 INTRODUCTION

The aim of this chapter is to provide the first comprehensive modelling framework for integrating layout design and material flow network design toward the direct generation of net layouts.

Most research work in layout design has concentrated on designing gross cell layouts depicting only the cell shapes and their location in the layout, with cell shape requirements boosted to take into account aisle space to be added in subsequent design phases. This research activity has for the most part relied on the quadratic assignment model as its underlying modelling framework. Kusiak and Heragu portray this line of research in their recent survey paper [4].

Starting in the early 1980s, there has been a research trend toward integrating more flow travel representativity into layout design. This has lead toward enriching the major layout design phase to include the location of the Input/Output stations of each cell, as well as the physical layout of the aisle system. This enriched output is what will be termed a net layout. Fig. 1.b presents such a net layout. The pioneers in net layout design have been O'Brien and Abdel Barr [7] and Warnecke and Dangelmaier [8]. Both teams introduced net

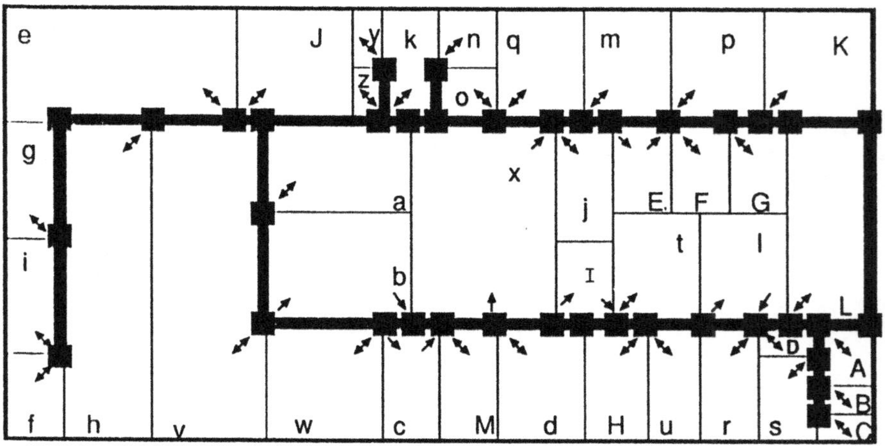

a. A gross layout with overimposed flow network

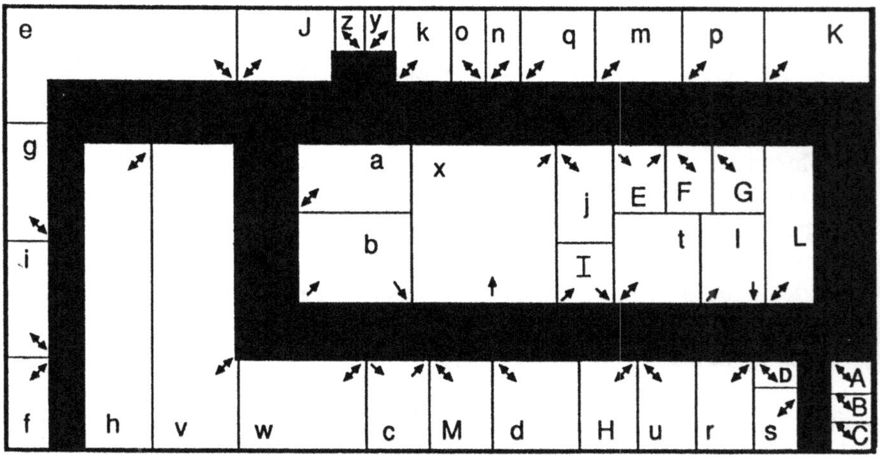

b. A network layout

Fig. 1 An illustrative example

layout design heuristics. The former developed an improvement heuristics while the latter developed a construction heuristics, both hence refining classical lines of research in layout design methodology.

Recently, Montreuil and Venkatadri [5] introduced a linear programming model which transforms optimally a gross cell layout coupled with an overimposed logical flow network into a net layout. A flow network defines the logical network along which flow travel is to circulate from one I/O station to another in the layout. The input to their model is hence a design as illustrated in Fig. 1.a, the output is a net layout, such as shown in Fig. 1.b, which satisfy the relative positioning imbedded in the input design. Their model is capable of sustaining various design objectives and constraints while maintaining its polynomial efficiency.

Their gross-to-net layout model is not based at all upon the traditional discrete quadratic assignment framework. It relies on a continuous definition of space which permits the definition of linear variables and constraints. The actual paper extends far beyond this model in order to formally state the net layout design problem by taking advantage of the tight relationship between flow network design and its physical instanciation into an aisle system.

The paper progresses through a series of models. It starts with limited models in order to show the basics of the modelling framework, and then extends into more refined, flexible and representative models. All models are mixed integer models; most models have linear objective functions and constraints, some have quadratic objective functions and/or quadratic constraints. No solution methodology for these models is offered in this chapter. However, it is our belief that the introduction of these models should foster enthusiasm in the research community. First, the problem is posed. Second, the complexity imposed by added degrees of freedom is explicitly presented. Third, the models are based on multicommodity network flow modelling, coupled with disjunctive and partitioning integer constraints in the layout component of the models. This offers to layout design researchers a wealth of recent research developments in integer and combinatorial optimization, which was not much used for solving the quadratic assignment problem and forced relying on myopic heuristic development.

Before proceeding with the development of the models, let us examine the designs of Fig. 1. This is a real industrial case. The gross cell layout has been generated by the engineers of the company using traditional free flow oriented techniques. If one examines the interstation flow set presented in Table 1, it can be remarked that many intercell proximity decisions have been decided assuming straight travel between any two points. The net layout has also been generated by the company engineers, from their gross layout, while the flow network has been overimposed on the gross layout by the author to basically represent the concept imbedded in their aisle system. What happened through the netting process is that many expected straight paths have been forbidden. Hence, many cells end up on different aisles and a tour of the plant is necessary to go from one to the other, which worsens significantly the expected flow efficiency of the

Table 1 Interstation flow set for the example case

CELL	STATION	DIRECTION	CELL	STATION	INTENSITY
v	2	>	e	1	926
v	2	>	b	1	40
v	2	>	c	1	311
v	2	>	d	1	575
v	2	>	w	1	130
v	2	>	f	1	12
v	2	>	g	1	6
v	2	>	h	1	107
v	2	>	a	1	8
b	2	>	x	1	1002
c	2	>	x	1	184
d	2	>	x	1	1357
w	2	>	x	1	158
x	2	>	c	1	42
x	2	>	E	1	126
x	2	>	d	1	34
x	2	>	l	1	6
x	2	>	o	1	2
x	2	>	p	1	962
x	2	>	j	1	747
x	2	>	e	1	120
x	2	>	u	1	53
p	2	>	w	1	520
D	2	>	w	1	4
c	2	>	w	1	40
d	2	>	w	1	4
A	2	>	x	1	4
w	2	>	l	1	4
w	2	>	s	1	382
w	2	>	r	1	140
b	2	>	w	1	116
h	2	>	w	1	106
f	2	>	w	1	32
j	2	>	w	1	120
w	2	>	t	1	520
w	2	>	p	1	78
w	2	>	u	1	116
w	2	>	I	1	36
g	2	>	w	1	30
w	2	>	A	1	74
w	2	>	B	1	26

Table 1 (continued)

s	2	>	w	1	4
a	2	>	w	1	28
r	2	>	w	1	20
w	2	>	c	1	30
E	2	>	w	1	6
F	2	>	w	1	30
r	2	>	D	1	13
s	2	>	o	1	20
D	2	>	s	1	30
d	2	>	D	1	4
d	2	>	l	1	2
d	2	>	E	1	6
l	2	>	D	1	8
d	2	>	g	1	20
E	2	>	F	1	390
u	2	>	j	1	145
H	2	>	B	1	4
I	2	>	t	1	26
B	2	>	s	1	4
o	2	>	I	1	48
g	2	>	f	1	20
I	2	>	s	1	10
e	2	>	c	1	4
s	2	>	E	1	4
r	2	>	s	1	30

layout. A strong motivation for this research is that by considering simultaneously layout design and flow network design, then the designer may succeed to generate significantly more flow efficient layouts. Also, there are many instances when the logical flow network or the aisle system are constraints for the designer. Then, taking it into consideration while designing the layout may again permit gains in flow efficiency.

2 MODEL 1: GROSS LAYOUT DESIGN GIVEN FREE FLOW TRAVEL

The classical assumption of free flow travel, that is unconstrained movement from any point to any other point in the layout, offers a basic modelling platform from which we will build all further extensions in integrating flow network design and aisle design within layout design.

The model, presented in Fig. 2, uses the variables defined in Fig. 3. It is basically composed of six main components: the objective function (1), the cell configuration constraints (2-4), the cell-within-building constraints (5-6), the station-within-cell constraints (7-8), the interstation distance constraints (9-10), and the intercell relative positioning constraints (11-14). Expressions (1-10) are

$$\text{MIN} \quad \sum_{\forall ijkl \,\in\, F} I_{ijkl}\,(x^{+}_{ijkl} + x^{-}_{ijkl} + y^{+}_{ijkl} + y^{-}_{ijkl}) \qquad (1)$$

$$\text{s.t.} \quad \underline{LX}_i \le \overline{X}_i - \underline{X}_i \le \overline{LX}_i \qquad \forall i \in \{C \cup B\} \qquad (2)$$

$$\underline{LY}_i \le \overline{Y}_i - \underline{Y}_i \le \overline{LY}_i \qquad \forall i \in \{C \cup B\} \qquad (3)$$

$$\underline{P}_i \le 2(\overline{X}_i) - \underline{X}_i + \overline{Y}_i - \underline{Y}_i)) \le \overline{P}_i \qquad \forall i \in \{C \cup B\} \qquad (4)$$

$$\underline{X}_B \le \underline{X}_c \le \overline{X}_c \le \overline{X}_B \qquad \forall c \in C \qquad (5)$$

$$\underline{Y}_B \le \underline{Y}_c \le \overline{Y}_c \le \overline{Y}_B \qquad \forall c \in C \qquad (6)$$

$$\underline{X}_c \le x_{cs} \le \overline{X}_c \qquad \forall c,\ \forall s \in S_c \qquad (7)$$

$$\underline{Y}_c \le y_{cs} \le \overline{Y}_c \qquad \forall c,\ \forall s \in S_c \qquad (8)$$

$$x_{ij} - x_{kl} = x^{+}_{ijkl} - x^{-}_{ijkl} \qquad \forall (ijkl) \in F \qquad (9)$$

$$y_{ij} - y_{kl} = y^{+}_{ijkl} - y^{-}_{ijkl} \qquad \forall (ijkl) \in F \qquad (10)$$

$$R_{i|j} + R_{j|i} + R_{i/j} + R_{j/i} \le 3 \qquad \forall (i < j,\ i, j \in C) \qquad (11)$$

$$R_{i|j},\ R_{j/i} = 0 \text{ or } 1 \qquad \forall (i \ne j \in C) \qquad (12)$$

$$\overline{X}_i + \text{MAX}(\underline{CX}_i, \underline{CX}_j) \le \underline{X}_j + M R_{i|j} \qquad \forall (i \ne j \in C) \qquad (13)$$

$$\overline{Y}_j + \text{MAX}(\underline{CY}_i, \underline{CY}_j) \le \underline{Y}_j + M(R_{j/i}) \qquad \forall (i \ne j \in C) \qquad (14)$$

Fig. 2 Model 1: Gross layout design given free flow travel

directly borrowed from Montreuil et al. [6] and are therefore only broadly explained here.

In constraints (2-4), the physical shape of each rectangular cell, as well as of the building, is controlled by six linear parameters stating the minimal and maximal side length along the X and Y dimensions, as well as the minimal and maximal perimeter. This set of constraints has the net advantage of being linear while permitting satisfactory control of the cell shape. If area constraints are judged absolutely necessary, then the following may be added:

$$\underline{LA}_i \le (\overline{X}_i - \underline{X}_i)(\overline{Y}_i - \underline{Y}_i) \le \overline{LA}_i \qquad \forall \{i \in C \cup B\}$$

However, these constraints are quadratic and therefore complicate drastically the model. If complex non-rectangular shapes are desired, the linear modelling framework introduced by Montreuil and Venkatadri [5] can be applied. Precise shapes can further be imposed through combinations of binary variables and disjunctive constraints.

Constraints (5-8) impose that each cell be laid out within the building, and that each I/O station of each cell be located within its cell. Clearly, more

B:	Building
C:	Set of cells c
$\underline{CX}_i$, $\underline{CY}_i$:	Specified minimal clearance required between cell i and any other cell along the X and Y dimensions
F:	Set of flows with positive intensity I_{ijkl} from I/O station j of cell i to I/O station l of cell k. Each flow is termed either f or (ijkl)
I_{ijkl}:	Intensity of flow from I/O station j of cell i to I/O station l of cell k
($\underline{LX}_i$, $\overline{LX}_i$):	Lower and upper bounds on the length of the X-dimension side of rectangular entity i
($\underline{LY}_i$,$\overline{LY}_i$):	Lower and upper bounds on the length of the Y-dimension side of rectangular entity i
M:	A very large positive number
($\underline{P}_i$, $\overline{P}_i$):	Lower and upper bounds on the perimeter of rectangular entity i
$R_{i\|j}$:	0 if entity i is imposed to be the west of (left of) entity j 1 if the above imposition is relaxed
$R_{i/j}$:	0 if entity i is imposed to be to the north of (above) entity j 1 if the above imposition is relaxed
S_c:	Set of input/output stations s for cell c
($\underline{X}_i$, $\overline{X}_i$):	Variable lower and upper boundaries of rectangular entity i along the X-dimension
($\underline{Y}_i$, $\overline{Y}_i$):	Variable lower and upper boundaries of rectangular entity i along the Y-dimension
(x_{cs},y_{cs}):	Variable coordinates for input/output station s of cell c along the X-dimension and Y-dimension
(x^-_{ijkl},x^+_{ijkl}):	Variable negative and positive components of the X-dimension distance between input/output station j of cell i and input/ output station l of cell k
(y^-_{ijkl},y^+_{ijkl}):	Variable negative and positive components of the Y-dimension distance between input/output station j of cell i and input/ output station l of cell k

Fig. 3 Basic glossary of variables and parameters

specific versions of these constraints are possible, like imposing the I/O stations to be located at the cell centroid or along a specific cell boundary.

The model uses the rectilinear distance metric. Constraints (9-10) take advantage of the classical linear programming trick [1] consisting of decoupling the X-distance and the Y-distance, each into a positive and a negative component, in order to avoid constraints involving nonlinear absolute values. The objective function minimizes the sum, over all interstation flows with positive intensity, of the flow intensity times the rectilinear interstation distance. This distance is expressed as the sum of the negative and positive components of distance along the X and Y dimensions.

The intercell relative positioning constraints (11-14) insure avoidance of spatial interference between cells or, simply stated, that no two cells cross each other. Constraints (13, 14) incorporate the notion of minimal clearance between cells which is particularly helpful in machine layout design when the designer wants to explicitly exhibit machines instead of aggregate cells (e.g. [2]).

Constraints (11-14) use the simple property introduced by Montreuil et al. [6] stating that for any two cells i and j not to cross, at least one of the following must hold: cell i is strictly west of cell j, cell i strictly east of cell j, cell i is strictly north of cell j, or cell i is strictly south of cell j (see Fig. 4). The model defines a 0-1 variable for each of these four conditions ($R_{i|j}$, $R_{j|i}$, $R_{i/j}$, $R_{j/i}$) which takes on the value 0 if the condition is imposed or 1 if it is relaxed.

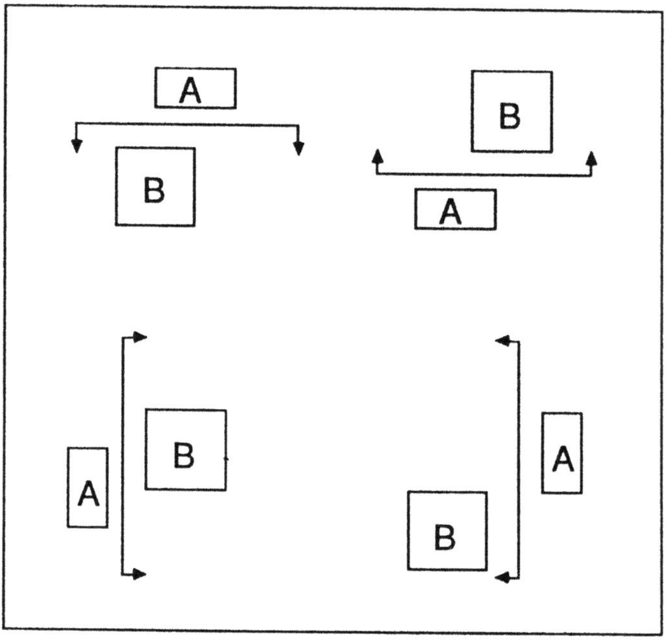

Fig. 4 Possible relative positions of cells A and B

Constraint (13-14) insure imposition of the condition when selected. For example, $R_{i|j} = 0$ states that cell i is to be west of cell j, that is in lower X coordinates than j. This means that $\overline{X}_i + MAX(\underline{CX}_i, \underline{CX}_j) \leq \underline{X}_j$ is to be imposed. Constraint (13) does so by adding $MR_{i|j}$ (where M is a large number) to the right-hand side of the above expression. Therefore, when $R_{i|j} = 0$ the expression strictly holds. When $R_{i|j} = 1$, then $MR_{i|j}$ is large and hence permits i not to be west of j. Constraint set (11) insures that no more than three of the relative positioning conditions are relaxed simultaneously.

3 MODEL 2: GROSS LAYOUT DESIGN GIVEN A SPATIALLY FIXED FLOW NETWORK

As a first step into integrating layout design and flow network design, let us investigate the case where the overall flow network is prespecified, with its nodes located at specific coordinates. Even though appearing very limitative at first glance, this model is adequate in cases where standard module-type cells are desired since the fixed flow network will favor such a modular layout if the nodes are located at "standard coordinates". Furthermore, the model is to support any generic flow network as well as any prespecified length for each edge; this is interesting in multisite multistory applications.

The model (Fig. 5) uses the combined glossaries of variables from Figs. 3 and 6. It uses the same cell configuration constraints (2-4), cell-within-building constraints (5-6), station-within-cell constraints (7-8), and intercell relative positioning constraints (11-16) as the model for the free flow travel assumption (see Fig. 2). To these are added the following: a flow network travel minimization function (15), flow balance constraints (16), edge travel capacity constraints (17), and station-anchoring-to-network constraints (18-22).

The flow travel concept is that the flow departs from the origin I/O station, moves directly to the nearest network node, then travels along the flow network edges up to the node nearest of the destination I/O station, and finally moves directly toward the destination I/O station.

In constraint (19), the 0-1 variable A_{csn} is used to state whether or not station s of cell c is anchored to node n. Constraint (18) impose that each I/O station be anchored to one and only one node. Constraints (20-22) and the first component of the objective function (15) insure that station s of cell c is anchored to the nearest node n. Constraints (20-21) compute the distance between the I/O station and each node. Constraint (22) state that the reaching distance for the I/O station must be greater or equal than the distance to node n if the station is to be anchored to node n. Finally, the first component of the objective function is a dominating goal insuring that the reaching distances are as low as possible, so that the nearest node is selected as anchor.

Constraint (16) are typical multicommodity flow balancing constraints, where each interstation flow is a commodity. They state that flow travel out of a node is equal to flow travel toward the node, except if the node is the anchor to the origin or destination I/O stations. Constraints (17) guarantee that total travel

$$\text{MIN} \sum_{\substack{\forall c \in C \\ \forall s \in S_C}} Mr_{cs} + \sum_{\forall e \in E} l_e \sum_{\forall f \in F} t_{ef} \quad (15)$$

s.t. (2) - (8)
(11) - (14)

$$I_f A_{O(f),n} + \sum_{\substack{\forall m \neq n \in N \\ (m,n) \in E}} t_{mnf} = I_f A_{D(f),n} + \sum_{\substack{\forall m \neq n \in N \\ (n,m) \in E}} t_{nmf} \qquad \forall n \in N, \forall f \in F \quad (16)$$

$$\sum_{\forall f \in F} t_{ef} \leq C_e \qquad \forall e \in E \quad (17)$$

$$\sum_{\forall n \in N} A_{csn} = 1 \qquad \forall c \in C, \forall s \in S_c \quad (18)$$

$$A_{csn} = 0 \text{ or } 1 \qquad \forall c \in C, \forall s \in S_c \quad (19)$$

$$x_{cs} - x_n = x^+_{csn} - x^-_{csn} \qquad \forall c \in C, \forall s \in S_c, \forall n \in N \quad (20)$$

$$y_{cs} - y_n = y^+_{csn} - y^-_{csn} \qquad \forall c \in C, \forall s \in S_c, \forall n \in N \quad (21)$$

$$r_{cs} \geq x^+_{csn} + x^-_{csn} + y^+_{csn} + y^-_{csn} - M(1 - A_{csn}) \qquad \forall c \in C, \forall s \in S_c, \forall n \in N \quad (22)$$

Fig. 5 Model 2: Gross layout design given a spatially fixed flow network

along a network edge never exceeds a prespecified edge capacity in reason of safety, congestion avoidance, etc. Note that similar capacities can be imposed on travel through any node.

The second component of the objective function (15) minimizes the total travelled distance along the flow network. It does so by minimizing the sum over all edges of the specified edge length times the total travel along the edge.

4 MODEL 3: GROSS LAYOUT DESIGN GIVEN AN OPTIONAL SPATIALLY FIXED FLOW NETWORK

This third model differs from the previous one from the fact that each edge in the spatially fixed flow network is now optional, with a fixed cost incurred when it is used for travel. The model presented in Fig. 6 conceptually seeks to balance flow travel efficiency with complexity and overall length of the flow network.

The model uses 0-1 variable U_e (see Fig. 7) to state whether or not edge e is selected for use (constraint 25). Constraints (17) are transformed into constraints

A_{csn}:	1 if node n is the anchor point of station s of cell c to the flow network, 0 otherwise
C_e:	Prespecified travel capacity of edge e
D(f):	Destination of flow f (TO I/O station)
E:	Set of edges e in the flow network. An edge is termed e or (mn) to denote that it travels from node m to node n
l_e:	Prespecified length of edge e in the flow network
N:	Set of nodes n of flow network
O(f):	Origin of flow f (FROM I/O station)
r_{cs}:	Variable reaching distance between station s of cell c and its anchor node in the flow network
t_{ef} or t_{mnf}:	Variable intensity of travel for flow f along edge e from node m to node n
(x_n, y_n):	Fixed coordinates for node n of the flow network along the X-dimension and the Y-dimension
(x^-_{csn}, x^+_{csn}):	Variable negative and positive components of the X-dimension distance between input/output station s of cell c and node n of the flow network
(y^-_{csn}, y^+_{csn}):	Variable negative and positive components of the Y-dimension distance between input/output station s of cell c and node n of the flow network

Fig. 6 Supplementary glossary of variables and parameters

(24) to supply travel capacity to an edge only if it is selected for use. The objective function (23) is obtained by adding the sum of fixed costs for edge usage to objective function (15).

Even though apparently minor, these modifications transform the linear multicommodity network flow component of the problem into a multicommodity network flow design subproblem with fixed charge. This component hence becomes much harder to solve, being NP-Complete by itself.

5 MODEL 4: GROSS LAYOUT DESIGN GIVEN A LOGICAL FLOW NETWORK

The difference between the second model and this fourth model lies in the fact that nodes of the flow network are not located a priori anymore and must now be located by the model. The length of each edge in the flow network is computed rectilinearly from the coordinates of its two end nodes. A simple example of such a case is when the designer wants the flow network to be a cycle, letting the model find the best combination of layout and flow network location.

As with model 3, this fourth model (see Fig. 8) is easily generalized from model 2 by letting (x_n, y_n) and l_e be variables instead of parameters. Constraints (26, 27) compute the X and Y dimension components of distance between the end nodes m and n of each edge e, then constraint (28) simply defines l_e to be sum of these X and Y components of distance.

Minimize $\quad (15) + \sum_{\forall e \in E} FC_e U_e$ (23)

s.t. (2) - (8)
(11) - (14)
(16)
(18) - (22)

$$\sum_{\forall f \in F} t_{ef} \leq C_e U_e \qquad \forall e \in E \tag{24}$$

$$U_e = 0 \text{ or } 1 \qquad \forall e \in E \tag{25}$$

where
FC_e: Fixed cost for using edge e for flow travel

U_e: 1 if edge e is used for flow travel
0 otherwise

Fig. 7 Model 3: Gross layout design for optional spatially fixed flow network

This slight transformation has the impact of modifying the objective function from a linear equation to a quadratic equation multiplying edge length variables by edge travel variables, which increases drastically the model complexity.

6 MODEL 5: NET LAYOUT DESIGN GIVEN A FIXED AISLE SYSTEM

As a first model explicitly integrating physical aisle system design into layout design, let us investigate the case when the orthogonal aisle system along which flows are to travel is prespecified. In this case, the task is to lay out cells in spaces unused by the aisle system and locate I/O stations within each cell such that flow travel distance along the aisles is minimized.

From a modelling standpoint, there are two major elements. First, spatial interference between cells and aisles is to be forbidden. Second, travel along the aisle system must be correctly and efficiently modelled. Fig. 9 presents an appropriate model incorporating these two elements.

The physical aisle system is modelled as a set A of rectangular aisle segments a whose extreme coordinates ($\underline{X}a$, $\underline{Y}a$) and ($\overline{X}_a$, $\overline{Y}_a$) are prespecified parameters. Given an aisle system, the combination of aisle segments is arbitrary. The only requirement is that the union of all aisle segments be spatially identical to the total aisle system. Then each of these aisle segments can be thought of as a "circulation cell". As we have spatial interference constraints between every pair of regular cells, we now need to add such constraints between regular cells and aisle segments. Constraints (29-34) perform this function.

The aisle system being fixed, it is easy to impose over it a spatially fixed flow network (model 2). With undirected aisles, one needs simply to locate a node in the middle of each intersection and a user-selected number of nodes along the

Min	(15)		
s.t.	(2-8)		
	(11-14)		
	(16)		
	(18-22)		
	$x_m - x_n = x^+_{mn} - x^-_{mn}$	$\forall(m, n) \in E$	(26)
	$y_m - y_n = y^+_{mn} - y^-_{mn}$	$\forall(m, n) \in E$	(27)
	$l_{mn} = x^+_{mn} + x^-_{mn} + y^+_{mn} + y^-_{mn}$	$\forall(m, n) \in E$	(28)

where

l_{mn}: Variable length of edge e from node m to node n in the flow network

(x_n,y_n): Variable coordinates for node n of the flow network along the X-dimension and the Y-dimension

(x^-_{mn},x^+_{mn}): Variable negative and positive components of the X-dimension distance between nodes m and n of the flow network

(y^-_{mn},y^+_{mn}): Variable negative and positive components of the Y-dimension distance between nodes m and n of the flow network

Fig. 8 Model 4: Gross layout design given a logical flow network

center of every aisle segment, and to join these nodes through edges according to the topology of the aisle system. Multiple nodes along an aisle segment augment flexibility in anchoring the I/O stations into the aisle system. Unidirectional and bidirectional aisles can be treated by refining the flow network. Aisle travel is hence modelled as in model 2 with objective function (15) and constraints (16-22) applied to the spatially fixed flow network overimposed upon the aisle system.

7 MODEL 6: NET LAYOUT GIVEN A SPATIALLY FIXED AISLE FLOW NETWORK

The task imposed to this sixth model is to generate a net layout including a physical aisle system, given a prespecified spatially fixed logical flow network representing desired aisle travel logic. In the basic case investigated here in which the width of each aisle segment is predetermined, the result of this model is identical to the result of model 5. The difference lies in the fact that here the model has to automatically lay out the aisles, while this was preprocessed by the human layout designer in model 5. This model serves the purpose of demonstrating clearly how to instantiate a flow network into an aisle system. The subsequent models build on this modelling framework.

As can be seen in Figs. 10 and 11, this model requires to be more rigorous in the definition of the set A of rectangular aisle segments. In fact, A is to be

MIN	(15)		
s.t.	(2 - 8)		
	(13 - 14)		
	(16 - 22)		
	$R_{i\|j}+R_{j\|i}+R_{i/j}+R_{j/i} \le 3$	$(\forall i \in C, \forall j \in (A \cup C))$	(29)
	$R_{i\|j}, R_{j\|i}, R_{i/j}, R_{j/i} = 0$ or 1	$(\forall i \in C, \forall j \neq i \in (A \cup C))$	(30)
	$\overline{X}_i + \underline{CX}_i \le \underline{X}_a + M(R_{i\|a})$	$\forall(i \in C, a \in A)$	(31)
	$\overline{X}_a + \underline{CX}_i \le \underline{X}_i + MRR_{a\|i}$	$\forall(i \in C, a \in A)$	(32)
	$\overline{Y}_i + \underline{CY}_i \le \underline{Y}_a + M(R_{i/a})$	$\forall(i \in C, a \in A)$	(33)
	$\overline{Y}_a + \underline{CY}_i \le \underline{Y}_i + M(R_{a/i})$	$\forall(i \in C, a \in A)$	(34)
where			
A:	Set of rectangular aisle segments a		
$\underline{X}_a, \overline{X}_a, \underline{Y}_a, \overline{Y}_a$:	Fixed parameters for $a \in A$		

Fig. 9 Model 5: Net layout design given a fixed aisle system

divided into three subsets A_X, A_Y and A_N. A_X, and A_Y are respectively the sets of aisle segments associated with edges along the X-dimension and the Y-dimension. A_N is a set of aisle segments explicitly defined around a node n of the flow network. These refer to an intersection, a 90° turn, an end-of-aisle, or the zone in which an aisle is changing its width. Each of the aisle segments in A_N is centered on the node location. Fig. 12 shows an illustrative spatially fixed flow network and its underlying set of aisle segments.

Contrary to model 5, $(\underline{X}_a, \underline{Y}_a)$ and $(\overline{X}_a, \overline{Y}_a)$ are now variables. Constraints (35-38) guarantee the integrity of each aisle segment in A_X and A_Y by imposing a non-negative length and a fixed width. Constraints (39, 40) insure the integrity of aisle segments in A_N by forcing their X-side to be as wide as their adjacent A_Y-aisles, and their Y-side to be as wide as their adjacent A_X-aisles.

The model must insure that the aisle segments are physically connected into a valid aisle system according to the flow network topology. Connectivity is achieved by recognizing that all aisle segments in A_X, and A_Y must be adjacent to exactly two other segments, these being in A_N and corresponding to the end nodes of the path of edges instantiated into the aisle segment. This is enforced by constraints (41-44). Alignment of nodes into the center of their aisle segment is achieved by constraints (45-46).

8 MODEL 7: NET LAYOUT GIVEN A LOGICAL AISLE FLOW NETWORK

Transforming model 6 such that the aisle flow system is not spatially fixed anymore is conceptually identical to what we performed when we transformed the

Minimize (15)

s.t. (2 - 8)
(13 - 14)
(16 - 22)
(29 - 34)

$$\underline{X}_a \leq \overline{X}_a \qquad \forall a \in A_x \qquad (35)$$

$$\overline{Y}_a - \underline{Y}_a = W_a \qquad \forall a \in A_x \qquad (36)$$

$$\underline{Y}_a \leq \overline{Y}_a \qquad \forall a \in A_y \qquad (37)$$

$$\overline{X}_a - \underline{X}_a = W_a \qquad \forall a \in A_y \qquad (38)$$

$$\overline{X}_a - \underline{X}_a \geq W_\alpha \qquad \forall a \in A_N \quad \forall \alpha \in (A_{\eta(a)} \cap A_y) \qquad (39)$$

$$\overline{Y}_a - \underline{Y}_a \geq W_\alpha \qquad \forall a \in A_N \quad \forall \alpha \in (A_{\eta(a)} \cap A_x) \qquad (40)$$

$$\underline{X}_a = \overline{X}_{\theta(\underline{n}(a))} \qquad \forall a \in A_X \qquad (41)$$

$$\overline{X}_a = \underline{X}_{\theta(\overline{n}(a))} \qquad \forall a \in A_X \qquad (42)$$

$$\underline{Y}_a = \overline{Y}_{\theta(\underline{n}(a))} \qquad \forall a \in A_Y \qquad (43)$$

$$\overline{Y}_a = \underline{Y}_{\theta(\overline{n}(a))} \qquad \forall a \in A_Y \qquad (44)$$

$$\frac{\underline{X}_a + \overline{X}_a}{2} = x_n \qquad \forall a \in (A_y \cup A_N)\ \forall n \in N_a \qquad (45)$$

$$\frac{\underline{Y}_a + \overline{Y}_a}{2} = y_n \qquad \forall a \in (A_x \cup A_N)\ \forall n \in N_a \qquad (46)$$

Fig. 10 Model 6: Net layout design given a spatially fixed aisle flow network

gross layout design model 2 into model 4. In fact, as can be seen in Fig. 13, we do not need to define any new parameter, variable or constraint. The model is a combination of model 4 and model 6.

This model has a critical underlying hypothesis which is that the logical flow network is orthogonal and planar, and that its physical mapping into relative positions is to be maintained. It still permits tremendous flexibility in node location, and associated edge length shortening and lengthening.

As was brought up in model 4, freedom in locating the network nodes imposes a quadratic objective function with its added complexity.

$A = A_X \cup A_Y$	
A_N:	Set of aisle segments explicitly defined around a node n of the flow network. These refer to an intersection, a 90° turn, an end-of-aisle, or the zone in which an aisle is changing its width. They are rectangles centered on the node location
A_n:	Set of aisle segments supporting flow network edges which are linked to node n
E_a:	Set of flow network edges logically imbedded in aisle segment a
θ_n:	The aisle segment $\in A_N$ anchored to node n
η_n:	Anchor node for an aisle segment $a \in A_N$
$\overline{n}(a)$:	East (north) most node $\in N_a$ for an aisle segment $a \in A_x(A_y)$
$\underline{n}(a)$:	West (south) most node $\in N_a$ for an aisle segment $a \in A_x(A_y)$
N_a:	Set of flow network nodes linked through edges $e \varepsilon Ea$
A_x:	Set of aisle segments whose main travel axis is along the X-dimension
A_y:	Set of aisle segments whose main travel axis is along the Y-dimension
W_a:	Prespecified width of aisle a

Fig. 11 Supplementary glossary for modelling net layout design given a spatially fixed aisle flow network

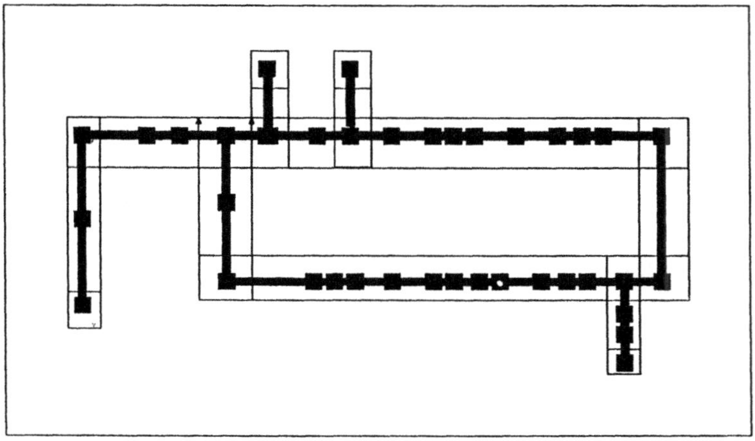

Fig. 12 A spatially fixed flow network and its underlying set of aisle segments

9 MODEL 8: NET LAYOUT GIVEN AN OPTIONAL LOGICAL AISLE FLOW NETWORK

This last model offers the most flexibility to the designer. In fact, for most orthogonally defined layout design cases, the model adequately represents the overall net layout design problem when a sufficiently generic aisle flow network is allowed. Flow travel is modelled through a flow network imposed over the aisle system. This flow network is defined only in logical terms, with flexibility in node location and edge length. Furthermore, aisle segments are instantiated only when their imbedded flow network edge(s) is(are) used, with the possibility of imposing fixed costs for using an aisle segment.

As can be seen in Fig. 14, this added flexibility is obtained at an extensive modelling price, requiring 28 new sets of constraints as well as new binary variables. Basically, new constraint sets are required for controlling space requirements for used aisle segments (47-51), for spatial interference avoidance given optional aisle segments (52-53), for valid instanciation of edges into aisle segments (54), for aisle system connectivity (54-66), and for aisle-to-node alignment (67-74).

The model uses binary variable υ_a to state whether aisle segment a is used or not (constraints 47). When an aisle segment a is used, then it must be spatially configured correctly as was done in constraints (35-38) for instantiating a fixed flow network into an aisle system. Constraints (48-51) perform this function by multiplying ν_a to the right-hand-side, so that the constraints are binding only when the aisle segment is used.

Constraints (52) insure that used aisle segments do not spatially interfere with cells, while constraints (53) insure that pairs of used aisle segments do not spatially interfere with each other. When two entities are known to be laid out, then the maximum number of relative positioning relaxation variables allowed to be 1 is known to be 3. In constraints (52), it needs to be 3 instead of 4 only when the aisle segment a is used. In constraints (53), it similarly needs to be 3 only when both aisle segments i and j are used, and therefore can potentially be overlapping if valid relative positioning is not enforced.

Constraints (54) force an aisle segment a to be used whenever one of the flow network edges e ε E_a, the set of edges logically imbedded in aisle segment a, is used for flow travel.

When an aisle segment a ε A_X is used then it must be adjacent to the pair of aisle segments ε A_N which are centered around $\bar{n}(a)$ and $\underline{n}(a)$, the east and west

Minimize	(15)
s.t.	(2 - 8)
	(13 - 14)
	(16 - 22)
	(26 - 46)

Fig. 13 Model 7: Net layout design given a logical aisle flow network

most nodes attached to the aisle segment. This insures that the aisle system is connected since for each edge e that is used for flow travel, then its aisle segment is connected to its extreme node aisle segments. For the west most node of aisle segment a, this is achieved as follows. Constraint (55) measures the distance between the west side of aisle segment a and the east side of aisle segment $\theta(\underline{n}(a))$. Constraints (56, 57) insure that both positive and negative components of distance are forced to zero whenever aisle segment a is used, and are unconstrained otherwise. Constraints (58-60) repeat a similar process for the east most node. Then constraints (61-66) insure connectivity for aisle segments $a \in A_y$.

A "node" aisle segment $a \in A_N$ is to be used whenever an edge touching its node $\eta_{(a)}$ is used (constraints 54). Then aisle segment a must be centered around the location of node $\eta(a)$. This is achieved through constraints (67, 68). Similarly, whenever an aisle segment $a \in \{A_x \cup A_y\}$ is used, then all nodes $n \in N_a$ must be aligned through the center of the aisle segment a. When $a \in A_y$, then the X coordinate of the nodes must be forced to be at the center of the aisle. To do so, constraints (69) measure the X-distance between the center of aisle segment a and the location of node n. Then constraints (70-71) insure that both positive and negative components of this X-distance are forced to zero whenever aisle segment a is used, and are unconstrained otherwise. Constraints (72-74) similarly insure node alignment for aisle segments $a \in A_x$.

10 CONCLUSION

This chapter has presented eight models dealing with various facets of integrating cell layout design, I/O station location, flow network design, and physical aisle system design. Through these models, a comprehensive modelling framework has been introduced. This framework has the advantage of easily permitting various modelling refinements and extension into multiobjective modelling (see [5] for example).

The chapter also poses a nice challenge to the research community (obviously including the author!) by offering a battery of realistic unsolved models. It is our belief that within a few years, clever decomposition, relaxation, branching, or other combinatorial and integer optimization techniques, will permit to solve medium size cases up to near-optimality. A factor reinforcing this statement is that the generic flow networks necessary to allow satisfactory flexibility to the designer are of quite limited size in proportion with sizes dealt with in many active network flow and network design research projects.

Minimize	(23)		
s.t.	(2 - 8) (11 - 14) (16 - 22) (24 - 28) (30 - 34) (35, 37)		
	$\upsilon_a = 0$ or 1	$\forall a \in A$	(47)
	$\overline{Y}_a - \underline{Y}_a = W_a \upsilon_a$	$\forall a \in A_x$	(48)
	$\overline{X}_a - \underline{X}_a = W_a \upsilon_a$	$\forall a \in A_y$	(49)
	$\overline{X}_a - \underline{X}_a \geq W_\alpha \upsilon_\alpha$	$\forall a \in A_N$ $\forall \alpha \in (A_{\eta(a)} \cap A_y)$	(50)
	$\overline{Y}_a - \underline{Y}_a \geq W_\alpha \upsilon_\alpha$	$\forall a \in A_N$ $\forall \alpha \in (A_{\eta(a)} \cap A_x)$	(51)
$R_{c\|a} + R_{a\|c} + R_{c/a} + R_{a/c} \leq 4 - \upsilon_a$		$\forall c \in C, \forall a \in A$	(52)
$R_{i\|j} + R_{j\|i} + R_{i/j} + R_{j/i} \leq 5 - \upsilon_i - \upsilon_j$		$\forall i, j \in A$	(53)
$\upsilon_a \geq U_e$		$\forall a \in A, \forall e \in E_a$	(54)
$\underline{X}_a - \overline{X}_{\theta(\underline{n}(a))} = \Delta X^+_{a,\theta(\underline{n}(a))} - \Delta X^-_{a,\theta(\underline{n}(a))}$		$\forall a \in A_x$	(55)
$0 \leq \Delta X^+_{a,\theta(\underline{n}(a))} \leq M(1 - \upsilon_a)$		$\forall a \in A_x$	(56)
$0 \leq \Delta X^-_{a,\theta(\underline{n}(a))} \leq M(1 - \upsilon_a)$		$\forall a \in A_x$	(57)
$\overline{X}_a - \underline{X}_{\theta(\overline{n}(a))} = \Delta X^+_{a,\theta(\overline{n}(a))} - \Delta X^-_{a,\theta(\overline{n}(a))}$		$\forall a \in A_x$	(58)
$0 \leq \Delta X^+_{a,\theta(\overline{n}(a))} \leq M(1 - \upsilon_a)$		$\forall a \in A_x$	(59)
$0 \leq \Delta X^-_{a,\theta(\overline{n}(a))} \leq M(1 - \upsilon_a)$		$\forall a \in A_x$	(60)
$\underline{Y}_a - \overline{Y}_{\theta(\underline{n}(a))} = \Delta Y^+_{a,\theta(\underline{n}(a))} - \Delta Y^-_{a,\theta(\underline{n}(a))}$		$\forall a \in A_y$	(61)

Fig. 14 Model 8: Net layout design given an optional aisle flow network

$$0 \le \Delta Y^{+}_{a,\theta(\underline{n}(a))} \le M(1 - \upsilon_a) \qquad \forall a \in A_y \qquad (62)$$

$$0 \le \Delta Y^{-}_{a,\theta(\overline{n}(a))} \le M(1 - \upsilon_a) \qquad \forall a \in A_y \qquad (63)$$

$$\overline{Y}_a - \underline{Y}_{\theta(\overline{n}(a))} = \Delta Y^{+}_{a,\theta(\overline{n}(a))} - \Delta Y^{-}_{a,\theta(\overline{n}(a))} \qquad \forall a \in Ay \qquad (64)$$

$$0 \le \Delta Y^{+}_{a,\theta(\overline{n}(a))} \le M(1 - \upsilon_a) \qquad \forall a \in A_y \qquad (65)$$

$$0 \le \Delta Y^{-}_{a,\theta(\overline{n}(a))} \le M(1 - \upsilon_a) \qquad \forall a \in A_y \qquad (66)$$

$$\frac{\underline{X}_a + \overline{X}_a}{2} = x_{\eta}(a) \qquad \forall a \in A_N \qquad (67)$$

$$\frac{\underline{Y}_a + \overline{Y}_a}{2} = y_{\eta}(a) \qquad \forall a \in A_N \qquad (68)$$

$$\frac{\underline{X}_a + \overline{X}_a}{2} - x_m = \Delta X^{+}_{n} - \Delta X^{-}_{n} \qquad \forall a \in A_y, \forall n \in N_a \qquad (69)$$

$$0 \le \Delta X^{+}_{n} \le M(1 - \upsilon_a) \qquad \forall a \in A_y, \forall n \in N_a \qquad (70)$$

$$0 \le \Delta X^{-}_{n} \le M(1 - \upsilon_a) \qquad \forall a \in A_y, \forall n \in N_a \qquad (71)$$

$$\frac{\underline{Y}_a + \overline{Y}_a}{2} - y_m = \Delta Y^{+} - \Delta Y^{-}_{n} \qquad \forall a \in A_x, \forall n \in N_a \qquad (72)$$

$$0 \le \Delta Y^{+}_{n} \le M(1 - \upsilon_a) \qquad \forall a \in A_x, \forall n \in N_a \qquad (73)$$

$$0 \le \Delta Y^{-}_{n} \le M(1 - \upsilon_a) \qquad \forall a \in A_x, \forall n \in N_a \qquad (74)$$

where

$\upsilon_a =$ 1 if aisle segment a is to be laid out
0 otherwise

$\Delta X^{-}_{ij}, \Delta X^{+}_{ij}$: Negative and positive components of the X-distance between he extremities of aisle segments in $\in A_X$ and $j \in A_N$ that must be adjacent when both segments are laid out

$\Delta Y^{-}_{ij}, \Delta Y^{+}_{ij}$: Negative and positive components of the Y-distance between he extremities of aisle segments in $\in A_y$ and $j \in A_N$ that must be adjacent when both segments are laid out

$\Delta X^{-}_{n}, \Delta X^{+}$: Negative and positive components of the X-distance location of node n and the center of its supporting aisle a when $a \in A_y$

$\Delta Y^{-}_{n}, \Delta Y^{+}_{n}$: Negative and positive components of the Y-distance location of node n and the center of its supporting aisle a when $a \in A_X$

Fig. 14 (continued)

11 ACKNOWLEDGEMENTS

This work was supported in part by the Natural Sciences and Engineering Research Council of Canada under grant OGPIN 020, by the "Fonds pour la Formation de Chercheurs et l'Aide à la Recherche" of Quebec under grant 89-TM-1012, and by the "Centre Francophone de Recherche en Informatisation des Organisations (CEFRIO)".

12 REFERENCES

1. Francis R.L. and J.A. White, *Facility Layout: An Analytical Approach*, Prentice-Hall, Englewood Cliffs, NJ, USA (1974).
2. Heragu, S.S. and A. Kusiak, "Machine Layout: An Optimization and Knowledge Based Approach", to appear in *International Journal of Production Research* (1989).
3. Koopmans, T.C. and Beckmann, "Assignment Problems and the Location of Economics Activities", *Econometrica*, Vol. 25 , pp. 229-251 (1957).
4. Kusiak, A. and S.S. Heragu, "The Facility Layout problem", *European Journal of Operational Research*, Vol. 29 , pp. 229-251 (1987).
5. Montreuil B. and U. Venkatadri, "From Gross to Net Layout: An Efficient Design Model", submitted for journal publication, available as Research Document FSA-88-56, Laval University, Québec, Canada (1988).
6. Montreuil, B., U. Venkatadri and H.D. Ratliff, "Generating a Layout from a Design Skeleton", submitted for journal publication, available as Research Document FSA 89-1, Laval University, Québec, Canada (February 1989).
7. O'Brien, C. and S.E.Z. Abdel Barr, "An Interactive Approach to Computer Aided Facility Layout", *International Journal of Production Research*, Vol. 18, No. 2, pp. 201-211 (1980).
8. Warnecke, H.J. and W. Dangelmaier, "Progress in Computer Aided Plant Layout", Technical Report, Institute of Manufacturing Engineering and Automation - Fraunhofer, Fed. Rep. of Germany (1982).

BACKTRACKING OF JOBS AND MACHINE LOCATION PROBLEMS

Bhaba R. Sarker
Louisiana State University
Wilbert E. Wilhelm
Gary L. Hogg
Texas A&M University
Min-Hong Han
Pohang Institute of Technology

ABSTRACT

In flexible manufacturing systems, production lines are often laid out along a linear track. In these lines, facilities engineers attempt to simplify the workflow of jobs by avoiding a job shop structure in favor of a generalized flow line (GFL) by assigning machines to appropriate locations in the layout. The upstream flow of materials in a production line is called *backtracking*. Determining an optimum assignment of M unique machines to exactly M locations along a track to minimize the total backtracking of jobs is computationally intractable. Therefore, an efficient heuristic is developed in this chapter to overcome these difficulties.

An ideal (GFL) incurs no backtracking. In many cases, achieving a GFL is not feasible because of the routing required by jobs. The relative amount of backtracking which does occur in such a configuration is an indication of the degree to which the ideal (and most productive) case is achieved. A measure of backtracking is developed in this chapter to assess the degree which a configuration achieves a (GFL). A simulation model is used to assess the effect of reducing backtracking on overall system performance in day-to-day dynamic operations.

1 INTRODUCTION

The problem of assigning indivisible facilities, like machines, to mutually exclusive locations has long been of interest to engineers, economists, only to this problem but also to other, allied problems. The interest in this problem is mainly due to its wide range of applications in the general field of location and layout design.

There are two related problems in this context. First, the static problem (designing a layout) is based on a long-range horizon for a given set of jobs (i.e., a known product mix and volume). Second, the dynamic problem is based on a short-range horizon such as day-to-day operations in which the product-mix is uncertain. The static problem is the primary focus of this research.

In a job shop, a job may start production at any machine and may leave the system after processing at any machine, and it may move either backward or forward from one operation to the next, depending upon machine locations along the transporter track. Transporting jobs from one machine to another may impose a substantial workload on the material handling system. The two-directional flow of materials not only incurs added costs but also creates a problem of controlling the jobs and scheduling limited resources such as transporters and machines.

A GFL is a serial line in which a job does not necessarily visit all of the stations in the line [Conway et al. (1967) and Saboo et al. (1989)]. The machines in this line are arranged along the sides of a track (or path) on which a transporter moves, as shown in Fig. 1. According to its (perhaps unique) sequence of operations, a job may begin production at any machine and complete processing at any machine downstream. For convenience, it is assumed that jobs generally flow from left to right.

In an industrial environment, the facilities engineer is concerned about systematically laying out the machines in a line so that the products travel the shortest distance possible. Such an objective may be achieved by reducing the distance traveled in one or both directions.

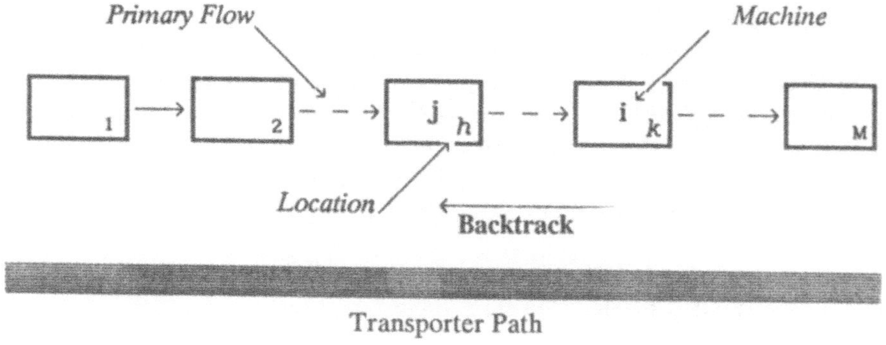

Fig. 1 Backtracking and layout configuration

1.1 The General Problem, Dθ

The general problem is to assign a set of M machines to M locations $[M = \sum_{t=1}^{T} \theta_t \ (1 \le T \le M)$ and θ_t is the number of *identical* machines of type $t]$ along a line to minimize the total distance travelled by a given set of jobs. In general, jobs may flow in either direction (i.e., downstream or upstream) between successive operations. It is assumed that the routing and production volume over the planning period are known for each job, and that the transporter can move only one job at a time.

1.1.1 Special Cases

Problem (D1), *Unique Machines with Two-Directional Flow:* If all machines are unique ($\theta_t = 1$ for all t), the general problem (Dθ) reduces to a more simple problem of assigning M machines to M locations to minimize the *total distance* travelled by the given set of jobs.

Problem (Bθ), *Sets of Identical Machines with Backtracking:* If the downstream flow of material is not considered (as it is inescapable), problem (Dθ) reduces to the *Backtracking* problem, the objective of which is to assign M machines to M locations to minimize the *total backtrack distance* of the set of jobs.

Problem (B1), *Unique Machines with Backtracking:* Problems (Bθ) and (D1) specialize to problem (B1) if $\theta_t = 1$ for all t in problem (Bθ), and forward (inescapable) movement of the products is not considered in problem (D1). The location problem (B1) deals with unique machines and backtracking of materials. These four problems are related as shown in Table 1.

1.1.2 Backtracking Problem

When the sequence of operation for a job does *not* specify a machine located downstream of its current location, the job has to travel to the left (i.e., upstream) in order to complete the required operation. This reverse travel is termed backtracking or upstream flow and changes the (GFL) to a less efficient job shop flow structure. Minimizing the flow of materials in the backward direction is called the backtracking problem. If all backtracking can be

Table 1 Model identification

	Number of Identical Machines of type t, θ_t	
Distance	1	θ_t
Backtracking, B	B1	Bθ
Bi-Directional	D1	Dθ

avoided by a judicious assignment of machines to locations, an efficient (GFL) results instead of a less productive job shop.

The explicit goal of reducing backtracking of materials in a production line serves several implicit goals. The implicit goal of this research is to reduce the backtracking distance that a material handling transporter must travel to support a production line. Such a reduction in backtracking distance is desirable, because, as the backtracking distance is reduced, the throughput of jobs can be expected to increase. In a general sense, an increased throughput rate is a measure of the increased performance of the production line which ultimately is translated into increased profitability. As the load levels of the flow line (i.e., product types and the production volume) increases, the effects of reducing travel distance can be expected to have greater impact on system performance measures such as average machine utilization, transporter utilization, throughput rate and makespan.

1.1.3 Applications and Scope

In addition to machine location problems, other applications in which backtracking should be minimized include assigning tools to locations in magazines which have unidirectional motion (Bozer and Srinivasan, 1989), placing operators at sewing machines in a garment production line, placing dies in bending presses in a (sheet metal) production line, locating (i.e., loading computer programs in) machines in a robotic cell, among others.

It is acknowledged that, even in an ideal GFL, the transporter will have to backtrack to position itself to initiate a transit (i.e., deadheading). The movement of the transporter is not directly treated in this chapter; it is expected that improving material flow will (indirectly) reduce transporter workload. This expectation is supported by simulation analysis reported in this chapter.

In lieu of improving the workflow structure by a judicious layout, a sophisticated control system could improve efficiency even in a job shop configuration by selecting the best transit to prescribe dynamically, considering all pending moves and transporting jobs while backtracking to reduce deadheading. It is, however, expected that developing such a control algorithm would be difficult; and implementing it, costly.

2 LITERATURE REVIEW

One-dimensional machine location problems along a line may be classified into two categories: unidirectional (backtracking) and bidirectional (two-way movement along the line). A vast literature on machine location problems in classical manufacturing systems exists; but, apparently, no research has been directed to study the one-dimensional, one-directional problem of *backtracking*. Most of the literature on location problems deals with either bidirectional or two-dimensional layout. The backtracking problem has recently become important due to the increased use of automated material handling in small-lot manufacturing.

There seems to be no 'standard' set of one-dimensional layout problems for testing solution methods. Moreover, there is, apparently, no prior experience related to solving a backtracking (unidirectional flow) problem. However, several authors have mentioned the existence of certain backtracking problems [Bozer and Srinivasan (1989), Khare et al. (1988)]. Obata (1979, pp. 137-138) compiled a family of eight problems of sizes M = 5, 6, 7, 8, 9, 10, 15, 20 to study a one-dimensional layout involving bidirectional flow.

3 THE PROBLEM

The problem of assigning M machines to M locations, (B1), along a line to minimize the *total backtracking* of all jobs is treated in this chapter. In a realistic problem, it may *not* be possible to prescribe an ideal (GFL) in which products flow in one direction only. Thus, a measure to assess the degree to which a configuration achieves an ideal (GFL) is developed in this chapter. Finally, a simulation model is used to assess the degree to which achieving the *explicit goal* of reducing backtracking attains the *implicit goals* of improving machine utilization and system throughput.

3.1 Formulation of the Problem

Assume that, over the planning horizon, the system must process N jobs in the set $\mathbf{J} = \{J_1, J_2, J_3, ... J_N\}$. Let the N jobs be defined by known measures p_{ij}^n and v_{ij}^n:

$$p_{ij}^n = \begin{cases} 1 & \text{If the } n^{th} \text{ job is processed at machine } j \\ & \text{immediately after machine } i \\ 0 & \text{Otherwise} \end{cases} \quad (1)$$

and

$$v_{ij}^n = \text{the number of moves made from machine } i \text{ to machine } j \text{ by the } n^{th} \text{ job.} \quad (2)$$

The total flow for all N jobs from machine i to machine j, r_{ij}, is given by

$$r_{ij} = \sum_{n=1}^{N} v_{ij}^n \, p_{ij}^n \quad , \quad i,j = 1,2,..,M \quad (3)$$

Considering all possible flows from machine i to machine j, a flow matrix called the *requirement matrix*, **R**, may be defined as

$$\mathbf{R} = [r_{ij}]_{M \times M} \quad (4)$$

3.1.1 Example 1: An Illustration of the Computation of R

Suppose a set of jobs, $\mathbf{J} = \{J_1, J_2, J_3, \ldots, J_6\}$ require the following routing:

$J_1 : 2 - 3 - 1 - 4$
$J_2 : 2 - 1 - 4$
$J_3 : 1 - 4 - 3 - 2$
$J_4 : 3 - 4 - 2$
$J_5 : 4 - 1 - 3$
$J_6 : 1 - 3 - 2 - 4$

To compute the value of r_{ij}, i,j=1,2,3,...,M, one has to count movements of all jobs from machine i to machine j. To illustrate this, consider an example which demonstrates use of Equation (3).

$$r_{32} = v_{32}^1 p_{32}^1 + v_{32}^2 p_{32}^2 + v_{32}^3 p_{32}^3 + v_{32}^4 p_{32}^4 + v_{32}^5 p_{32}^5 + v_{32}^6 p_{32}^6$$

$$= 0(0) + 0(0) + 1(1) + 0(0) + 0(0) + 1(1) = 2$$

Thus, the flow matrix, $\mathbf{R}$, for the set $\mathbf{J} = \{J_1, J_2, J_3, \ldots, J_6\}$ is given by

$$\mathbf{R} = \begin{bmatrix} 0 & 0 & 2 & 3 \\ 1 & 0 & 1 & 1 \\ 1 & 2 & 0 & 1 \\ 1 & 1 & 1 & 0 \end{bmatrix} \quad \begin{matrix} \text{Row Sum} \\ 5 \\ 3 \\ 4 \\ 3 \end{matrix} \tag{4a}$$

Column Sum 3 3 4 5

It is assumed that exactly one machine is to be assigned to each location. It is also assumed that the locations are equally spaced and numbered 1,2,3,...,M sequentially from left to right with 1 being the first location and M being the last location (see Fig. 1). Decision variable that prescribes the layout may be defined as,

$$x_{ik} = \begin{cases} 1 & \text{if machine } i \text{ is assigned to location } k \\ 0 & \text{otherwise} \end{cases} \tag{5}$$

The measure of distance between machines *i* and *j*, a constant, is given by

d_{ikjh} = the total backtracking distance from machine *i* at location *k* to machine *j* at location *h*

$$= r_{ij}\, \delta_{kh}^{b} \tag{6}$$

in which r_{ij} is defined in Equation (3) and

$$\delta^{b}_{kh} = \begin{cases} k-h, & \text{if } h<k \\ 0 & \text{otherwise} \end{cases} \tag{7}$$

The problem of minimizing the backtracking distance incurred by all jobs may be formulated as

$$QAP\text{: Minimize} \quad f(\mathbf{x}) = \sum_{i=1}^{M} \sum_{k=1}^{M} \sum_{j=1}^{M} \sum_{h=1}^{M} d_{ikjh}\ x_{ik}\ x_{jh} \tag{8}$$

$$\text{Subject to} \quad \sum_{i=1}^{M} x_{ik} = 1, \quad k = 1,2,3,\ldots,M \tag{8a}$$

$$\sum_{k=1}^{M} x_{ik} = 1, \quad i = 1,2,3,\ldots,M \tag{8b}$$

$$x_{ik} = 0 \text{ or } 1, \quad \text{for all } i,k \tag{8c}$$

in which Equation (8a) ensures that each machine is assigned to exactly one location and Equation (8b) ensures that each location is assigned exactly one machine. If machine i is assigned to location k and machine j is assigned to location h, then both $x_{ik}$and x_{jh} equal 1, and the cost term d_{ikjh} is included in the total backtracking distance, $f(\mathbf{x})$.

Solving QAP optimally for large M is computationally prohibitive [Murtagh et al. (1982),Bazaraa and Kirca (1983)]. The structure of the backtracking matrix is exploited in this chapter to devise an efficient heuristic and a measure for its solution quality.

3.2 Backtrack Matrix

The backtrack matrix plays an important role in determining the total backtracking for a set of jobs and may be defined using assignment vector,

$$\mathbf{a} = (a_1, a_2,, a_3, \ldots, a_M) \tag{9}$$

in which element a_k denotes the machine i which is assigned to location k. Given an assignment, $\mathbf{a}$, the backtrack matrix, which manifests some unique properties defined in Sarker (1989) and Sarker et al. (1988), may be easily determined. In general, if machine m is assigned to location k, (i.e., $a_k = m$), there are $(k-1)$ preceding locations; and, hence, at most $(k-1)$ steps of backtracking exist for a job leaving this machine and moving to next machine in its operating sequence. The term *step* is used to mean the distance between two adjacent machine locations. Thus, *backtrack matrix*, **B**, is defined as,

$$\mathbf{B} = [b_{ij}]_{M\times M} \tag{10}$$

in which b_{ij} = the number of backtrack steps from machine i at location k, to machine j at location h.

$$= \sum_{k=1}^{M} \sum_{h=1}^{M} \delta_{kh}^{b}\, x_{ik}\, x_{jh} \tag{11}$$

For example, if machine 3 is placed at location 4 and machine 2 is placed at location 1, then b_{32} = 4 - 1 = 3 steps, while , by definition, b_{23} = 0. Therefore, for **a** = (2,4,1,3) the *unique* backtrack matrix, **B**, is,

$$\mathbf{B} = \begin{bmatrix} 0 & 2 & 0 & 1 \\ 0 & 0 & 0 & 0 \\ 1 & 3 & 0 & 2 \\ 0 & 1 & 0 & 0 \end{bmatrix} \tag{12}$$

					Row Sum
	0	2	0	1	3
	0	0	0	0	0
	1	3	0	2	6
	0	1	0	0	1
Column Sum	1	6	0	3	10

Each element in column j of the **B** matrix indicates the backtracking distance a job would travel going to machine j from the machine indicated by the corresponding row i. Clearly, there is always one row corresponding to the machine in the first location, a_1, and one column corresponding to the machine in the last location, a_M, which contain all zeros.

3.3 Total Backtracking Distance

Introducing r_{ij} and b_{ij} into Equation (8), the backtracking distance TB(**a**), for a given assignment, **a**, may be written as

$$f(\mathbf{x}) = \mathrm{TB}(\mathbf{a}) = \sum_{k=1}^{M} \sum_{h=1}^{M} \sum_{i=1}^{M} \sum_{j=1}^{M} r_{ij}\, \delta_{kh}^{b}\, x_{ik}\, x_{jh} = \mathbf{R}\bullet\mathbf{B} \tag{13}$$

The matrix **R** is constant for a given set of jobs. However, the backtrack matrix **B** changes as a function of the assignment vector **a**. The assignment vector is optimum if its corresponding **B** matrix produces the minimum TB(**a**). Therefore, if a **B** matrix that minimizes the total backtracking distance TB(**a**) = **R•B**, can be constructed, the machine assignment problem is solved.

4 SOLUTION METHOD

Before a heuristic is described, several motivating ideas underlying the procedure are discussed. Since **R** represents the machine-to-machine flow and **B** represents the machine-to-machine distance, a new matrix, called the total backtrack matrix, RB(**a**), may be defined as

$$RB(\mathbf{a}) = [r_{ij}\, b_{ij}]_{MxM} \tag{14}$$

Element (i,j) of the RB(**a**) matrix represents the total backtracking distance incurred by all jobs going from machine i to machine j. A row (column) sum of the RB(**a**) matrix indicates the total backtracking distance incurred by all jobs in going from (to) the corresponding machine to (from) all other machines. The total backtracking distance form machine i to all other machines, TB_i, is given by

$$TB_i = \sum_{j=1}^{M} r_{ij}\, b_{ij} \tag{15}$$

Consider the **R** matrix given in Equation (4a). For initial assignment, a = (4,3,2,1), the resulting backtrack matrix

$$\mathbf{B} = \begin{bmatrix} 0 & 1 & 2 & 3 \\ 0 & 0 & 1 & 2 \\ 0 & 0 & 0 & 1 \\ 0 & 0 & 0 & 0 \end{bmatrix} \tag{16}$$

produces machine-to-machine distance matrix

$$RB(\mathbf{a}) = \begin{bmatrix} 0 & 0 & 4 & 9 \\ 0 & 0 & 1 & 2 \\ 0 & 0 & 0 & 1 \\ 0 & 0 & 0 & 0 \end{bmatrix} \quad \begin{matrix} \text{Row Sum} \\ 13 \\ 3 \\ 1 \\ 0 \end{matrix} \tag{17}$$

The value of $TB_1 = 13$ indicates that, because machine 1 is at location M, it incurs 13 total backtracking steps. Likewise, machine 2 is in location 3 and incurs 3 total backtracking steps and so on. The objective of the heuristic is to search for an assignment **a*** that minimizes the total backtracking steps, so the procedure relocates machine(s) to reduce backtracking. For example, consider inserting (instead of interhanging) machine 1, which is initially at location 4, at different locations between other machines without altering the order of other machines. M-1 such alternatives can be generated. Thus, three alternatives for the example case are: (1,4,3,2) (4,1,3,2) and (4,3,1,2). The procedure is to evaluate whether these three new (i.e., M-1) assignments give better solutions than the incumbent. The following heuristic was devised to minimize backtracking.

4.1 Depth-first Insertion Heuristic (DIH)

Step 0: Setup the **R** matrix from the given set of jobs, **J**. Assume an arbitrary assignment, $\mathbf{a}_0$. Generate **B** and RB($\mathbf{a}_0$) matrices, and compute TB($\mathbf{a}_0$). Set **a*** = $\mathbf{a}_0$. and TB(**a**) = TB($\mathbf{a}_0$).

Step 1: Set TB(**a***) = TB(**a**). Call the available assignment **a*** an incumbent solution and TB(**a***) the value of the solution.

Step 2: Obtain all possible assignments by placing a machine corresponding to the largest row sum of the RB matrix, in each of the other M-1 locations, resulting in (M-1) new assignment vectors $\mathbf{a}_1,\mathbf{a}_2,\mathbf{a}_3,\ldots,\mathbf{a}_{M-1}$. These assignments are evaluated by computing TB($\mathbf{a}_\iota$), $\iota = 1,2,3,\ldots,$ M–1.

Step 3: Search for a better solution by comparing each of these M-1 assignments with the incumbent solution [i.e., TB(**a**) ← min {TB($\mathbf{a}_\iota$), ι = 1,2,3,...,M–1}]. If a better solution (i.e., if TB(**a**) < TB(**a***)} is found, a search is continued by going to Step 1. If none of these solutions is better than the incumbent solution, choose the machine corresponding to the next largest row sum of the RB(**a**) matrix under consideration.

Step 4: Stop when all machines corresponding to all row sums of the first RB($\mathbf{a}_0$) matrix are scanned.

A few iterations of this method are shown in the following example.

4.1.1 Example 2: An Illustration of DIH

Step 0: For $M = 4$, let **R** and **B** [for an initial $\mathbf{a}_0$ = (4,3,2,1)] be given as

$$\mathbf{R} = \begin{bmatrix} 0 & 0 & 2 & 3 \\ 1 & 0 & 1 & 1 \\ 1 & 2 & 0 & 1 \\ 1 & 1 & 1 & 0 \end{bmatrix} \quad \text{and } \mathbf{B} = \begin{bmatrix} 0 & 1 & 2 & 3 \\ 0 & 0 & 1 & 2 \\ 0 & 0 & 0 & 1 \\ 0 & 0 & 0 & 0 \end{bmatrix}$$

Thus,

$$\text{RB}(\mathbf{a}_0) = \begin{bmatrix} 0 & 0 & 4 & 9 \\ 0 & 0 & 1 & 2 \\ 0 & 0 & 0 & 1 \\ 0 & 0 & 0 & 0 \end{bmatrix} \quad \begin{matrix} \text{Row sum} \\ 13 \\ 3 \\ 1 \\ 0 \end{matrix}$$

and TB($\mathbf{a}_0$) = 13 + 3 + 1 + 0 = 17.

Step 1: *Incumbent solution,* **a*** = (4,3,2,1) and TB(**a***) = 17

Step 2: Since the largest row sum in RB($\mathbf{a}_0$) matrix is 13 corresponding to machine 1, all possible assignments are obtained by placing machine 1 in each of the other 3 locations in $\mathbf{a}_0$= (4,3,2,1), keeping the order of other machines constant:

$$\mathbf{a}_1 = (1,4,3,2) \Rightarrow \text{TB}(\mathbf{a}_1) = 10$$
$$\mathbf{a}_2 = (4,1,3,2) \Rightarrow \text{TB}(\mathbf{a}_2) = 12$$
$$\mathbf{a}_3 = (4,3,1,2) \Rightarrow \text{TB}(\mathbf{a}_3) = 15$$

Step 3: TB(a) $\leftarrow$ TB($\mathbf{a}_1$) = 10 < TB (**a***) = 17

Step 1: For $\mathbf{a}_1$ = (1,4,3,2), **B** and $RB(\mathbf{a}_1)$ are given as

$$\mathbf{B} = \begin{bmatrix} 0 & 0 & 0 & 0 \\ 3 & 0 & 1 & 2 \\ 2 & 0 & 0 & 1 \\ 1 & 0 & 0 & 0 \end{bmatrix} \quad \text{and } RB(\mathbf{a}_1) = \begin{bmatrix} 0 & 0 & 0 & 0 \\ 3 & 0 & 1 & 2 \\ 2 & 0 & 0 & 1 \\ 1 & 0 & 0 & 0 \end{bmatrix} \begin{matrix} 0 \\ 6 \\ 3 \\ 1 \end{matrix}$$

Incumbent solution, **a*** = (1,4,3,2) and TB(**a***) = 10

Step 2: Since the largest row sum in RB(**a**) matrix is 6 corresponding to machine 2, all possible assignments are obtained by placing machine 2 in each of the other 3 locations in $\mathbf{a}_1$ = (1,4,3,2), keeping the order of the other machines constant:

$$\mathbf{a}_1 = (2,1,4,3) \Rightarrow \text{TB}(\mathbf{a}_1) = 12$$
$$\mathbf{a}_2 = (1,2,4,3) \Rightarrow \text{TB}(\mathbf{a}_2) = 12$$
$$\mathbf{a}_3 = (1,4,2,3) \Rightarrow \text{TB}(\mathbf{a}_3) = 11$$

Step 3: TB(**a**) $\leftarrow$ TB($\mathbf{a}_3$) = 11 > TB(**a***) = 10
that is, none of these solutions is better than the incumbent solution; hence choose the next largest row sum of RB(1,4,3,2).

Step 2: Since the next largest row sum in RB(**a**) matrix is 3 corresponding to machine 3, all possible assignments are obtained by placing machine 3 in each of the other 3 locations in **a** = (1,4,3,2), keeping the order of the other machines constant:

$$\mathbf{a}_1 = (3,1,4,2) \Rightarrow \text{TB}(\mathbf{a}_1) = 11$$
$$\mathbf{a}_2 = (1,3,4,2) \Rightarrow \text{TB}(\mathbf{a}_2) = 10$$
$$\mathbf{a}_3 = (1,4,2,3) \Rightarrow \text{TB}(\mathbf{a}_3) = 11$$

Step 3: TB(**a**) $\leftarrow$ TB($\mathbf{a}_2$) = 10 = TB(**a***) = 10
that is, none of these solutions is better than the incumbent solution; hence choose the next largest row sum of RB(1,4,3,2).

Step 2: Since the next largest row sum in RB(**a**) matrix is 1 corresponding to machine 4, all possible assignments are obtained by placing machine 4 in each of the other 3 location in **a** = (1,4,3,2), keeping the order of the other machines constant:

$$\mathbf{a}_1 = (4,1,3,2) \Rightarrow TB(\mathbf{a}_1) = 12$$
$$\mathbf{a}_2 = (1,3,4,2) \Rightarrow TB(\mathbf{a}_2) = 10$$
$$\mathbf{a}_3 = (1,3,2,4) \Rightarrow TB(\mathbf{a}_3) = 10$$

Step 3: $TB(\mathbf{a}) \leftarrow TB(\mathbf{a}_2) = 10 = TB(\mathbf{a}^*) = 10$
that is, none of these solutions is better than the incumbent solution; hence choose the next largest row sum of RB(1,4,3,2). Since the next largest row sum in RB(**a**) matrix is 0 corresponding to machine 1, there is no chance of improving the value of TB(**a**) by inserting this machine.

Step 4: The *depth-first* search for the largest row sum of RB($\mathbf{a}_0$) matrix is completed. Hence, the machine corresponding to the next row sum of RB($\mathbf{a}_0$) will be chosen for insertion in the $\mathbf{a}_0$ = (4,3,2,1).

Step 2: Since the next largest row sum in RB($\mathbf{a}_0$) matrix is 3 corresponding to machine 2, all possible assignments are obtained by placing machine 2 in each of the other 3 locations in $\mathbf{a}_0$ = (4,3,2,1), keeping the order of the other machines constant:

$$\mathbf{a}_1 = (2,4,3,1) \Rightarrow TB(\mathbf{a}_1) = 14$$
$$\mathbf{a}_2 = (4,2,3,1) \Rightarrow TB(\mathbf{a}_2) = 16$$
$$\mathbf{a}_3 = (4,3,1,2) \Rightarrow TB(\mathbf{a}_3) = 15$$

Step 3: $TB(\mathbf{a}) \leftarrow TB(\mathbf{a}_1) = 14 > TB(\mathbf{a}^*) = 10$
Hence, the incumbent solution persists at **a*** = (1,4,3,2) with TB(**a***) = 10. The next row sum of RB($\mathbf{a}_0$) will be considered.

Step 2: Since the next largest row sum in RB($\mathbf{a}_0$) matrix is 1 corresponding to machine 3, all possible assignments are obtained by placing machine 3 in each of the other 3 locations in $\mathbf{a}_0$ = (4,3,2,1), keeping the order of the other machines constant:

$$\mathbf{a}_1 = (3,4,2,1) \Rightarrow TB(\mathbf{a}_1) = 16$$
$$\mathbf{a}_2 = (4,2,3,1) \Rightarrow TB(\mathbf{a}_2) = 16$$
$$\mathbf{a}_3 = (4,2,1,3) \Rightarrow TB(\mathbf{a}_3) = 15$$

Step 3: $TB(\mathbf{a}) \leftarrow TB(\mathbf{a}_3) = 15 > TB(\mathbf{a}^*) = 10$
Hence, the incumbent solution **a*** = (1,4,3,2) with TB(**a***) = 10 is still the best. The next largest row sum of RB($\mathbf{a}_0$) matrix is 0 corresponding to machine 1; hence, there is no chance of improving the

value of TB(**a**) by inserting this machine. Since all row sums of RB($\mathbf{a}_0$) are scanned, the heuristic is terminated.

Thus the final solution of the example problem produced by DIH is $\mathbf{a}^*$ = (1,4,3,2) with TB($\mathbf{a}^*$)) = 10. Exhaustive enumeration produces a global optimal value of TB_{OPT} = 10 achieved by assignments a*: (2,3,1,4), (1,4,3,2),(1,3,4,2) and (1,3,2,4).

4.1.2 Computational Performance of DIH

Since there are no alternative methods which can be applied to the one-dimensional, *one-directional* layout problems, a program was developed to measure the performance of DIH in comparison with exhaustive enumeration. Table 2 reports the results obtained from a set of test problems. For each value of *M*, 52 to 75 randomly generated problems were solved and DIH solutions were found, on the average, to be about 3.33 percent larger than optimum. Ninety-five percent of the DIH solutions were within 5 percent of the optimum value.

Problems with $M \leq 50$ were solved successfully with the heuristic in reasonably short time. A *Cray Y-MP2/116*, a supercomputer, took about 8.21 CPU-hours to obtain an optimal solution to a 12-machine problem by complete enumeration. It is not possible to comment on the exact computational performance of the heuristic for larger *M* because optimal solutions could not be obtained for problems with $M \geq 13$. However, a lower bound on the optimum solution of a larger problem may be established.

Table 2 Computational performance of the DIH

M[1]	TB_{DIH}/TB_{OPT}		% Time
	Randomly Assigned $\mathbf{a}_0$	Assigned MPH[2] Solution as $\mathbf{a}_0$	$TB_{DIH} = TB_{OPT}$ using MPH solution as $\mathbf{a}_0$
4	1.0078	1.000000	80.00
5	1.0339	1.000000	50.00
6	1.0490	1.000001	30.00
7	1.0235	1.000001	20.00
8	1.0256	1.000012	16.67
9	1.0276	1.000915	13.23
10	1.0334	1.004228	12.02
11	1.0472	1.012105	11.13
12	1.0488	1.027210	10.51

1. Sample sizes varied from 52 to 75
2. Sarker et al. (1988)

4.1.3 Lower Bound on the Optimal Solution

A lower bound on the minimum value of the total backtracking distance for a given **R** was obtained by using the characteristics of **B** matrix, called the *amoebic properties* . The term *amoebic* was coined to represent the characteristics of the **B** matrix that acts much like an *amoeba* to represent different configurations of the linear layout (see Sarker 1989).

The following lemma may be stated before establishing an efficient lower bound using the amoebic properties of the **B** matrix.

4.1.3.1 LEMMA 1:

Let α_j and β_j represent the non-diagonal elements of matrices **R** and **B** respectively. If

$$\mathbf{r} = \left\{(\alpha_j)_{j=1}^{K} : \alpha_k \leq \alpha_{k+1},\ k = 1,2,\ldots,K\text{-}1,\ and\ K=M(M\text{-}1)\right\} \text{ and}$$

$$\mathbf{b} = \left\{(\beta_j)_{j=1}^{K} : \beta_k \geq \beta_{k+1},\ k = 1,2,\ldots,K\text{-}1,\ and\ K=M(M\text{-}1)\right\}$$

then $LB_o = \mathbf{rb'}$

Proof: See Gavett and Plyter (1966).

For a given pair of machines (i and j), the minimum possible number of backtracking *transits* required to process a job set is

$$\zeta_{ij}^{b} = \min\,(r_{ij}, r_{ji}) \tag{18}$$

no matter where machines i and j are located along the track. Now, let $\mathbf{p}_B^0$ be the set of all M(M-1)/2 ζ_{ij}^{b} -values obtained from **R**, and $\mathbf{p}_B$ is the set of all elements of $\mathbf{p}_B^0$ arranged in *non-decreasing* order such that $\mathbf{p}_B = (p_1, p_2, \ldots, p_L)$ in which L = M(M-1)/2. According to an amoebic property of the **B** matrix [see Sarker (1989)], a **B** matrix contains M(M-1)/2 positive elements which are:

$$\mathbf{q}_B^0 = \left\{\pi_{ij} : \pi_{ij} = M\text{-}i\ ,\ i=1,2,3,\ldots,M\text{-}1;\ j=1,2,3,\ldots,i\right\} \tag{19}$$

If $\mathbf{q}_B$ is the set of elements of $\mathbf{q}_B^0$ arranged in *non-increasing* order such that $\mathbf{q}_B = (q_1, q_2, \ldots q_L)$, then, by Lemma 1, the lower bound, LB_B, is given by

$$LB_B = \mathbf{p}_B\, \mathbf{q}_B' = \sum_{m=1}^{L} p_m q_m \tag{20}$$

in which $\mathbf{q}_B'$ is the transpose of $\mathbf{q}_B$. Example 3 illustrates the two bounds.

4.1.3.2 Example 3: Amoebic Lower Bound

From the given **R** in Equation (4a), $\mathbf{r}$ = (0,1,1,1,1,1,1,1,1,2,2,3) and $\mathbf{b}$ = (3,2,2,1,1,1,0,0,0,0,0,0). Thus, by Lemma 1, $LB_0 = \mathbf{rb}' = 7$.

Using the minimal pair value $\zeta^{b}_{ij} = \min(r_{ij}, r_{ji})$ from **R**, we have, $\mathbf{p}^0_B = \{(0,1,1,),(1,1),(1)\}$ and $\mathbf{p}_B = (0,1,1,1,1,1)$. By the amoebic property defined by Equation (19), $\mathbf{q}_B = (3,2,2,1,1,1)$. Thus, $LB_B = \mathbf{p}_B \mathbf{q}_B' = 7$. Similarly for

$$R = \begin{bmatrix} 0 & 2 & 2 & 3 & 0 \\ 1 & 0 & 5 & 2 & 3 \\ 6 & 1 & 0 & 4 & 2 \\ 1 & 2 & 6 & 0 & 3 \\ 1 & 5 & 3 & 3 & 0 \end{bmatrix} \tag{21}$$

$LB_0 = 22$ and $LB_B = 28$. It may be shown, in general, that $LB_0 \leq LB_B$ [see Sarker (1989)].

The relative performance of this bound with respect to DIH and the optimum solution is shown in Table 3. It is observed that LB_B / TB_{DIH} is within 3.20% of LB_B / TB_{OPT}.

4.2 Comparative Performance of DIH

Two scenarios - one with equal workload and the other with unequal workload - were used to evaluate the heuristic using randomly generated data. Both scenarios involved sets of problems. The workload, ρ, indicates the average

Table 3 Performance of the lower bound

M	Number of Problems	$\frac{LB_B}{TB_{DIH}}$	$\frac{LB_B}{TB_{OPT}}$
3	30	0.97112	0.97112
4	280	0.96481	0.96562
5	320	0.90388	0.90698
6	415	0.86472	0.87049
7	425	0.82178	0.83057
8	433	0.78052	0.79213
9	327	0.75018	0.77599
10	198	0.74816	0.76671
11	125	0.73128	0.75245
12	35	0.72264	0.74657

number of transits from one machine to another required to produce all jobs in the set **J**, and it plays an important role in evaluating the performance of the lower bound. The term *equal* (or *unequal*) workload is used to indicate an equal (or unequal) number of average transits of jobs between any pair of machines.

To measure the effect of *workload* on the performance of a heuristic, an average workload on the transporter, ρ, is defined as

$$\rho = \frac{\mathbf{1R1'}}{M(M-1)}$$

in which the dimension of sum vector **1** is 1×M and (′) denotes the transpose. If the number of operations for a job in a problem is a discrete random variable distributed uniformly on the range [2,Θ] and the machine selected for each operation (no two consecutive operations are performed on the same machine) is a discrete random variable distributed uniformly on the range [1,M], then the workload, ρ, may be written as

$$\rho = \frac{\mathbf{1R1'}}{M(M-1)} \approx \frac{N\Theta}{2M(M-1)} \quad (22)$$

The workload, ρ, may be calculated faster by NΘ/2M(M-1) than by **1R1'**/M(M-1).

Eighteen randomly generated problems, *P-m, ∀m=5,6,...,20,25,30* (in which m indicates M, the size of the problem) were tested and results using backtracking flows are reported in Table 4. The assignment obtained from a construction procedure, MPH (Sarker et al. 1988), was used to initiate DIH, an improvement procedure.

It is observed from Table 4 that, while the MPH solutions to backtracking problems are (approximately) within 3 percent of the optimum when M ≤ 12, the DIH solutions are much better (about 0.30 percent on the average). For M ≥ 13, solutions are compared with the lower bound. The ratio LB_B/TB_{DIH} is observed to decrease with the size of the problem, M, even though the workload, ρ, remains relatively constant.

It is observed that the lower bound becomes tighter when the workload, ρ, increases (i.e., as the transporter becomes busier). The computational results for 'S-m' problems with equal workloads (ρ ≈ 8.75)are shown in Table 5. Tests with these problems indicate that the lower bound for a backtracking problem is a function of problem size, M; Sarker (1989) has shown that lower bound is also dependent on the workload, ρ.

The complexity of DIH is $O(M^4)$ [see Sarker (1989)]. It was apparent from Table 6 that the DIH is more efficient than either the rotating-grid or mixed-exchange heuristics [Obata (1979)] which used one-dimensional and *two-directional* flow problems with symmetric distance matrices. Obata reported that the computational time required by his mixed exchange heuristic was 117.11 seconds on an *IBM 370/3033* for M = 34 [see Obata (1979), p. 65]. This result

Table 4 Relative performance of the heuristics and lower bound

Problem P-m	Workload	Backtracking Flow TB(a)		
m	ρ[1]	$\frac{TB_{MPH}}{TB_{OPT}}$	$\frac{TB_{DIH}}{TB_{OPT}}$[2]	$\frac{LB_B}{TB_{DIH}}$
5	30.00	1.00000	1.00000	0.92733
6	23.33	1.02356	1.00000	0.93979
7	19.05	1.00607	1.00000	0.88271
8	16.07	1.00779	1.00000	0.81620
9	13.88	1.04055	1.00915	0.82761
10	12.22	1.02423	1.01184	0.79510
11	10.91	1.04874	1.00455	0.76326
12	9.85	1.05312	1.02617	0.74305
15	7.61	-	-	0.67192
20	5.53	-	-	0.60817
25	4.16	-	-	0.58821
30	3.45	-	-	0.49839

1 ρ is calculated based on approximate formula in Equation (22).

2 The solution obtained by MPH was used as initial assignment to DIH.

Table 5 Relative performance of the heuristics on problems with equal workload

Problem S-m	Backtracking Flow TB(a)		
m	$\frac{TB_{MPH}}{TB_{OPT}}$	$\frac{TB_{DIH}}{TB_{OPT}}$	$\frac{LB_B}{TB_{DIH}}$
5	1.000000	1.000000	0.91515
6	1.014760	1.000000	0.90036
7	1.000000	1.000000	0.86375
8	1.080279	1.00826	0.76166
9	1.089606	1.02867	0.75842
10	1.070898	1.01401	0.74878
11	1.049779	1.02520	0.72403
12	1.048356	1.03145	0.70641
15	-	-	0.69003
20	-	-	0.68297
25	-	-	0.67917

$\rho = 8.75$ was calculated with $M = 25$, $N = 500$, $\Theta = 21$. The solution obtained by MPH was used as initial assignment to DIH. Detailed results are reported in Sarker (1989).

Table 6 Computational time (sec) required by different heuristics

	VAX 6850			IBM 3090-200E		
M	DIH	Rotating Grid	Mixed Exchange	DIH	Rotating Grid	Mixed Exchange
5	0.01	0.29	0.42	0.06	0.24	0.26
10	0.10	5.10	8.13	0.09	4.11	4.66
15	0.56	16.60	20.60	0.22	11.60	12.23
20	1.69	60.74	73.70	0.66	26.74	31.60
25	3.66	-	164.32	1.30	-	74.91
30	10.89	-	212.49	2.39	-	156.52
40	28.08	-	-	7.44	-	-
50	79.62	-	-	12.53	-	-

is consistent with the time reported in Table 6, since the VAX 8650 is a much slower system.

It is indicated in previous sections that there are no 'standard' methods to compare the results of DIH with other methods; hence, a direct measure of backtracking is developed in the following section to assess the performance of DIH.

5 A MEASURE OF BACKTRACKING

Backtracking is zero when an assignment establishes a (GFL). Given an assignment of machines, it is of interest to know how closely the resulting flow agrees with an ideal GFL.

Assume an assignment **a** and its corresponding backtrack matrix **B**. For a given **a** = $(a_1,a_2,a_3,\ldots,a_M)$, let the rows and columns of **R** and **B** matrices be re-arranged in the order $a_1,a_2,a_3,\ldots,a_M$ such that the row k corresponds to machine a_m (m=1,2,3,...,M). Similarly, the column n corresponds to machine a_n (n=1,2,3,...,M). Hence

$$
R^{a}_{\text{re-arranged}} = \begin{array}{c} \\ a_1 \\ a_2 \\ a_3 \\ \cdots \\ a_M \end{array}
\begin{array}{c} \begin{array}{ccccc} a_1 & a_2 & a_3 & \ldots & a_M \end{array} \\
\begin{bmatrix} 0 & r_{a_1,a_2} & r_{a_1,a_3} & \cdots & r_{a_1,a_M} \\ r_{a_2,a_1} & 0 & r_{a_2,a_3} & \cdots & r_{a_2,a_M} \\ r_{a_3,a_1} & r_{a_3,a_2} & 0 & \cdots & r_{a_3,a_M} \\ \cdots & \cdots & \cdots & 0 & \cdots \\ r_{a_M,a_1} & r_{a_M,a_2} & r_{a_M,a_3} & \cdots & 0 \end{bmatrix} \end{array}
\tag{23}
$$

and

$$B^{a}_{\text{re-arranged}} = \begin{array}{c} \\ a_1 \\ a_2 \\ a_3 \\ \ldots \\ a_M \end{array}\begin{array}{c} \begin{array}{ccccc} a_1 & a_2 & a_3 & \ldots & a_M \end{array} \\ \begin{bmatrix} 0 & 0 & 0 & \ldots & 0 \\ b_{a_2,a_1} & 0 & 0 & \ldots & 0 \\ b_{a_3,a_1} & b_{a_3,a_2} & 0 & \ldots & 0 \\ \ldots & \ldots & \ldots & \ldots & 0 \\ b_{a_M,a_1} & b_{a_M,a_2} & b_{a_M,a_3} & \ldots & 0 \end{bmatrix} \end{array} \tag{24}$$

For assignment, **a**, B\s(a,re-arranged) is always a lower triangular matrix, and $\mathbf{R}^{a}_{\text{re-arranged}}$, consists of two distinct parts in which the lower triangle indicates the *backward* flow and the upper triangle indicates the *forward* flow of jobs. For notational simplicity, let $\mathbf{R}^{a}_{b}$ denote the backward flow matrix and $\mathbf{R}^{a}_{f}$ the forward flow matrix of the assignment **a**. Hence, $\mathbf{R}^{a}_{\text{re-arranged}}$ may be written as

$$\mathbf{R}^{a} = \mathbf{R}^{a}_{\text{re-arranged}} = \mathbf{R}^{a}_{b} + \mathbf{R}^{a}_{f} \tag{25}$$

It may be easily show that for any assignment, **a**, in a problem with requirement matrix, **R**,

$$\text{TB}(\mathbf{a}) = \mathbf{R}^{a}_{b} \cdot \mathbf{B}^{a}_{\text{re-arranged}}$$

A ratio of the total backtracking distance, $\mathbf{R}^{a} \cdot \mathbf{B}^{a}_{\text{re-arranged}}$, to the total distance moved by all jobs in set **J** could be developed if the forward travel distance were known. A measure, based on the total number of machine-to-machine transits, is developed in this section.

Let R^{a}_{b} and R^{a}_{f} denote the sum of all elements of $\mathbf{R}^{a}_{b}$ and $\mathbf{R}^{a}_{f}$, respectively, such that

$$R^{a}_{b} = \mathbf{1}\,\mathbf{R}^{a}_{b}\,\mathbf{1}' \tag{26}$$

and

$$R^{a}_{f} = \mathbf{1}\,\mathbf{R}^{a}_{f}\,\mathbf{1}' \tag{27}$$

so that the total number of transits for all jobs, R^{a}, for the assignment, **a**, may be determined by

$$R^{a} = R^{a}_{b} + R^{a}_{f}. \tag{28}$$

Therefore, two relative measures of the upstream and downstream transits are

$$\lambda_b^a = \frac{R_b^a}{R^a} \text{ and } \lambda_f^a = \frac{R_f^a}{R^a} \qquad (29)$$

which are based only on the relative number of machine-to-machine transits, regardless of the total number of job movements. The limiting values of these measures are $0 \le \lambda_b^a \le 1$ and $0 \le \lambda_f^a \le 1$, and $\lambda_b^a + \lambda_f^a = 1$. In a (GFL), there is no backtracking (i.e., $R_b^a = 0$ and $\lambda_b^a = 0$). For a line with assignment **a** will incur some backtracking if $\lambda_b^a > 0$, since the line is not an ideal GFL. The following example illustrates these measures.

5.1. Example 3: Measures of Backtracking

Assume a requirement matrix:

$$\mathbf{R} = \begin{array}{c} \\ 1 \\ 2 \\ 3 \\ 4 \end{array} \begin{array}{c} \begin{array}{cccc} 1 & 2 & 3 & 4 \end{array} \\ \begin{bmatrix} 0 & 0 & 4 & 1 \\ 3 & 0 & 5 & 4 \\ 2 & 0 & 0 & 2 \\ 6 & 0 & 6 & 0 \end{bmatrix} \end{array} \qquad (30)$$

For assignment **a** = (2,4,1,3), according to Equations (23) and (24), $\mathbf{R}^a_{\text{re-arranged}}$ and $\mathbf{B}^a_{\text{re-arranged}}$ matrices may be written as

$$\mathbf{B}^a_{\text{re-arranged}} = \begin{array}{c} \\ 2 \\ 4 \\ 1 \\ 3 \end{array} \begin{array}{c} \begin{array}{cccc} 2 & 4 & 1 & 3 \end{array} \\ \begin{bmatrix} 0 & 0 & 0 & 0 \\ 1 & 0 & 0 & 0 \\ 2 & 1 & 0 & 0 \\ 3 & 2 & 1 & 0 \end{bmatrix} \end{array}$$

and

$$\mathbf{R}^a_{\text{re-arranged}} = \begin{array}{c} \\ 2 \\ 4 \\ 1 \\ 3 \end{array} \begin{array}{c} \begin{array}{cccc} 2 & 4 & 1 & 3 \end{array} \\ \begin{bmatrix} 0 & 4 & 3 & 5 \\ 0 & 0 & 6 & 6 \\ 0 & 1 & 0 & 4 \\ 0 & 2 & 2 & 0 \end{bmatrix} \end{array}$$

from which,

$$\mathbf{R}^a_b = \begin{array}{c} \\ 2 \\ 4 \\ 1 \\ 3 \end{array} \begin{array}{c} \begin{array}{cccc} 2 & 4 & 1 & 3 \end{array} \\ \begin{bmatrix} 0 & 0 & 0 & 0 \\ 0 & 0 & 0 & 0 \\ 0 & 1 & 0 & 0 \\ 0 & 2 & 2 & 0 \end{bmatrix} \end{array} \quad \text{and} \quad \mathbf{R}^a_f = \begin{array}{c} \\ 2 \\ 4 \\ 1 \\ 3 \end{array} \begin{array}{c} \begin{array}{cccc} 2 & 4 & 1 & 3 \end{array} \\ \begin{bmatrix} 0 & 4 & 3 & 5 \\ 0 & 0 & 6 & 6 \\ 0 & 0 & 0 & 4 \\ 0 & 0 & 0 & 0 \end{bmatrix} \end{array}$$

Using $\mathbf{R}_b^a$ and $\mathbf{R}_f^a$, and Equations (26) and (27), $R_b^a = 5$ and $R_f^a = 28$, resulting in $\lambda_b^a = 0.152$ and $\lambda_f^a = 0.848$. On the other hand, for (reverse) assignment, a =

(3,1,4,2), the **R** in Equation (30) yields $\lambda_b^a = 0.848$ and $\lambda_f^a = 0.152$. If the assignment were **a** = (4,2,1,3), the measure of backtracking would be $\lambda_b^a = 0.273$ and $\lambda_f^a = 0.727$, indicating a less desirable assignment. As summarized in Table 7, the λ-measure assesses the relative percentage of total machine-to-machine transits either downstream or upstream and the TB-measure assesses the total number of backtracking transits of jobs only.

The measure of total backtracking provides information as to the total machine-to-machine upstream transit of materials. The value TB(**a**) = 0 indicates that all product flows are in the downstream direction (i.e., no backtracking). For TB(**a**) > 0, the assignment, **a**, does not indicate how good the configuration is -- it simply indicates that there is a total of TB backtracking steps. On the other hand, the λ-measure provides a relative assessment of the backtracking and forward movement; that is, the λ-measure provides an index to assess the degree to which a configuration achieves an ideal (GFL) structure.

6 EVALUATION OF THE GOAL

A simulation study was undertaken to investigate the degree to which a heuristic solution contributes to the implicit goals of increasing machine utilization and overall throughput as opposed to the static analysis of the problem. This investigation compares the relative performance of a heuristic solution with that resulting from a 'base layout' prescribed by a method that might be used by plant engineers. To achieve this purpose, an experiment was designed as follows.

6.1 Model Configuration

Four test scenarios were defined with M values of 5, 10, 15 or 20. In each of these scenarios, material handling systems with 2 to 4 transporters were tested. The input parameters which define the system were (a) the number of jobs, (b) the routing of each job, (c) the time for each operation, and (d) the release rule to launch jobs into the shop. One hundred jobs were generated by defining their routings and the processing time for each operation. The number of operations for each job was a discrete random variable distributed uniformly on the range $[2, M+1]$. Each processing time was distributed exponentially with a mean that was drawn randomly from a discrete uniform distribution with support on [1,20].

Table 7 Ranking of assignment relative to an ideal generalized flow line

Assignment, **a**	λ_b^a	Rank Relative to GFL
(2,4,1,3)	0.152	1
(4,2,1,3)	0.273	2
(3,1,4,2)	0.848	3

Each operation was randomly assigned to a machine with equal likelihood, except one machine was not permitted to perform two successive operations on any job.

The simulation program modeled the four scenarios, each using the 'base layout' as well as the assignment prescribed by DIH. Machine locations in the 'base layout' were configured by the *single-pass heuristic*, *SPH* [see Sarker (1989)], representing a 'simple' rule that might be used to locate machines in an actual plant.

6.2 Simulation Results

Thirty replications were performed; each processed the same set of jobs. The simulation results were statistically significant at a = 0.05, and the results of only 10-, 15- and 20-machine scenarios are summarized in Table 8.

The average utilization of machines increases with the number of transporters, as machines do not have to wait long for a transporter. It is evident from Table 8 that, in most cases, the transporter is less highly utilized in a layout prescribed by DIH than in the base layouts; reducing the backtracking of jobs decreases total transporter movement. Of course, it does not guarantee that deadheading (i.e., unloaded movement) will be minimized. Reducing backtracking reduces job cycle times and consequently enhances the availability of jobs to increase machine utilization. The DIH solution improved average machine utilization in eight out of eleven cases compared with the base layout. In ten of eleven cases, DIH reduced the makespan achieved by the base layout.

These results are confirmed by simulation results reported in Fig. 2 and 3, which indicate how machine utilization and makespan improve as the backtracking of jobs is reduced in this system. These simulation results demonstrate the impact that the explicit goal of minimizing backtracking might have on achieving the implicit goals of increasing machine utilization and job throughput.

7 CONCLUSION

This research was initiated with the intention of using the backtracking concept to minimize the travel distance of jobs along a material handling track. The performance of the DIH was evaluated by comparisons with complete enumeration and with guaranteed lower bound values. The DIH solution often produced optimum results on smaller test problems. A few trials with different initial assignments frequently produced optimal solutions for problems of moderate size. Because of the computational difficulty of determining the optimum solution for a large problem, this study limited comparisons with optimal solutions to problems with 12 or fewer machines. For larger problems, the lower bound was used to evaluate solution quality.

Solutions fell within about 3 percent of the optimum for problems with $M \leq 12$. The simulation study indicates that average machine utilization and system

Table 8 Simulation results

Number of transporters	Existing System			Backtracking System		
	Machine, M			Machine, M		
	10	15	20	10	15	20
	TRANSPORTER UTILIZATION					
2	0.765	0.967	0.887	0.735	0.770	0.810
3	0.510	0.541	0.594	0.491	0.540	0.552
3	0.395	0.429	0.473	0.367	0.393	0.423
	AVERAGE MACHINE UTILIZATION					
2	0.789	0.680	0.707	0.785	0.689	0.707
3	0.799	0.751	0.762	0.817	0.792	0.736
3	0.802	0.762	0.771	0.821	0.803	0.741
	MAKESPAN					
2	808	825	830	813	799	809
3	790	806	811	781	794	778
4	785	804	799	780	791	777

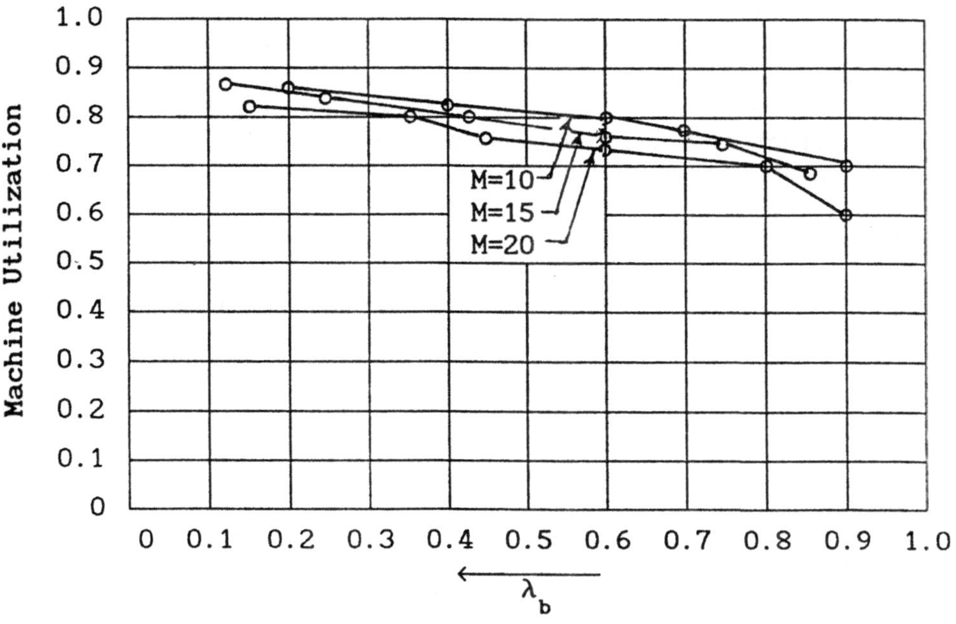

Fig. 2 Effect of reduction of backtracking on machine utilization

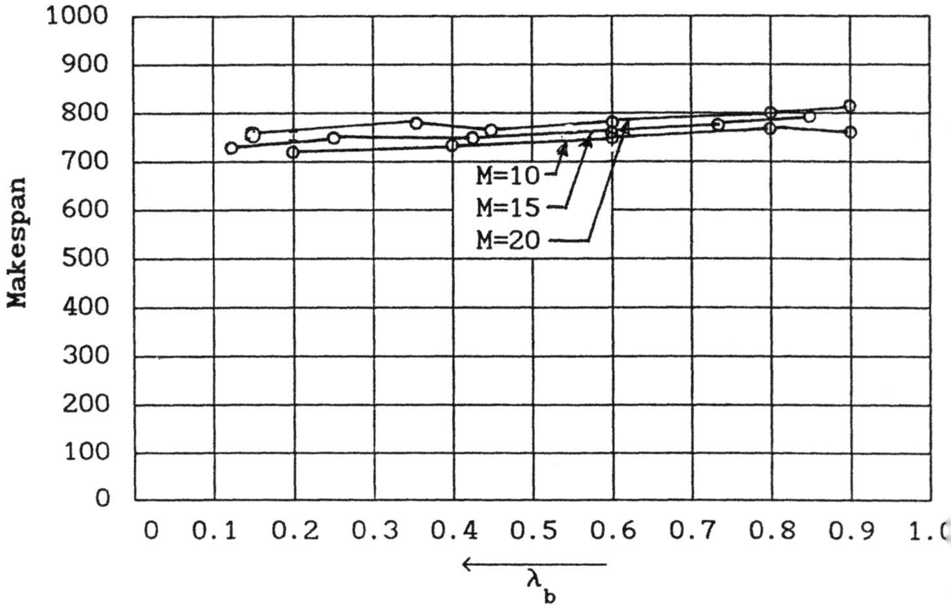

Fig. 3 Effect of reduction of backtracking on makespan

throughput increase and that makespan reduces as the backtracking of jobs is reduced. Hence, backtracking may be treated as a favorable objective to solve machine location problems along a linear track. This method may thus be a useful approach for simplifying the workflow of jobs in a shop where unidirectional flow is of interest to management.

8 ACKNOWLEDGEMENT

We are thankful to Dr. Leon McGinnis of Georgia Tech for first proposing the backtracking problem to one of us (Min-Hong Han).

9 REFERENCES

1. Bazaraa, M.S. and O. Kirca, "An Branch-and-Bound-Based Heuristic for Solving the Quadratic Assignment Problem," *Naval Research Logistic Quarterly*, 30, pp. 287-304 (1983).
2. Bozer, Y.A. and M.M. Srinivasan, "Tandem Configurations for AGV Systems Offer Simplicity and Flexibility," *Industrial Engineering*, 21(2), pp. 23-27 (1989).

3. Conway, R.W., W.L. Maxwell and L.W. Miller, *Theory of Scheduling*, Addison-Wesley, Reading, MA, Chapter 5 (1967).
4. Gavett, J.W. and N.V. Plyter, "The Optimal Assignment of Facilities to Locations by Branch and Bound," *Operations Research*, 14, pp. 210-232 (1966).
5. Hillier, F.S., "Quantitative Tools for Plant Layout Analysis," *Journal of Industrial Engineering*, 14, pp. 33-40 (1963).
6. Murtagh, B.A., T.R. Jefferson and V. Sornprasit,"Heuristic Procedure for Solving the Quadratic Assignment Problem," *European Journal of Operational Research*, 9, pp. 71-76 (1982).
7. Obata, T., "Quadratic Assignment Problem: Evaluation of Exact and Heuristic Algorithms," Unpublished Ph.D. Dissertation, Rensselaer Polytechnic Institute, Troy, NY (1979).
8. Picone, C.J. and W.E. Wilhelm, "A Perturbation Scheme to Improve Hillier's Solution to the Facilities Location Problem," *Management Science*, 30, pp. 1238-1249 (1984).
9. Saboo, S., L. Wang and W.E. Wilhelm, "Recursion Models for Describing and Managing the Transient Flow of Materials in Generalized Flow Lines", *Management Science*, 35, pp. 722-742 (1989).
10. Sarker, B.R., "The Amoebic Matrix and One-Dimensional Machine Location Problems," PhD Dissertation, Department of Industrial Engineering, Texas A&M University, College Station, TX 77843 (December 1989).
11. Sarker, B.R., W.E. Wilhelm and G.L. Hogg, "Amoebic Matrix and Machine Location Problems," Working Paper INEN/MS/WP/07/11-88, Department of Industrial Engineering, Texas A&M University, College Station, TX 77843 (1988).
12. Shapiro, C.W. and H.L.W. Nuttle, "Hoist Scheduling for a PCB Electroplating Facility," *IIE Transactions*, 19, pp. 157-166 (1988).
13. Steinberg, L., "The Backboard Wiring Problem: A Placement Algorithm," *SIAM Review*, 43, pp. 37-50 (1961).

COMPUTER AIDED DESIGN OF UNIT LOADS: A DESIGN AND SELECTION PROCEDURE

Jaime Trevino
North Carolina State University
Juan J. Daboub
Empaques Plasticos, S.A., de C.V.

ABSTRACT

A computer aided approach for unit load design and selection based on product characteristics, pallet/container dimensions, and facility and operational constraints is discussed in detail. The unit load design and selection problem is formulated as a nonlinear integer programming problem and solved by a heuristic approach and by an intelligent enumeration procedure.

1 INTRODUCTION

Traditionally, design of unit loads and selection of material handling, transportation, and storage equipment, and lot sizing and plant layout have been addressed sequentially and without considering the integrated impact of these decisions (see Fig. 1).

Unit loads should be designed considering plant layout, storage system dimensions, lot sizing strategies, product crushability, pallet/container types and sizes, and handling and transportation limitations. Different material handling alternatives are used to move unit loads to the right station, in the right sequence, at the right time, in the right amount, at the right place, and in the most economical and efficient way. Likewise, material handling equipment alternatives are selected according to volumes and characteristics of products being moved, plant layout, type of movement (synchronous vs. non-synchronous), and equipment cost and capabilities (i.e., reaching, speed, and lifting).

The number of parts comprising a unit load is partially determined by the lot sizing policy. Furthermore, lot sizing strategies are developed according to storage cost, inventory carrying cost, demand, ordering and setup practices, transportation, and quality cost. Flow intensity is determined according to the

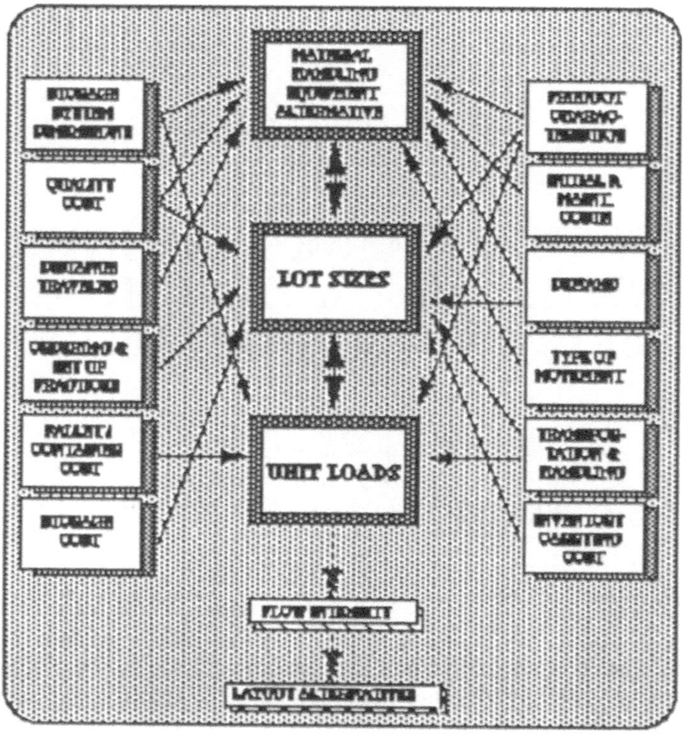

Fig. 1 Relationship among factors affecting unit load size

relationship of the above factors—specifically, material handling equipment, unit load, and lot size. Finally, plant layouts are evaluated and selected based largely on the minimization of flow intensities.

An integrated selection and design procedure of unit loads, material handling, storage, and transportation equipment, and lot sizing and plant layout is proposed in this chapter. The procedure and modeling approach are explained using a warehousing -distribution scenario. However, other scenarios have also been modeled (see [19a]).

Specifically, critical design and selection issues in warehousing-distribution applications are covered in Section 2, the scenario modeled is presented as a warehousing-distribution problem in Section 3, the modeling approach is described in Section 4, an example is included in Section 5, and conclusions and recommendations are provided in Section 6.

2 LITERATURE REVIEW

A review of most relevant literature on the problem of unit load design indicates that the alternatives proposed to solve this problem can be grouped into three approaches.

2.1 Approach One

In this approach, the pallet size and stacking pattern (also called pallet layout or pallet loading) for which the pallet utilization is maximized is determined. This problem is a special case of the cutting stock or bin packaging problem and has been addressed by different authors, such as, Golden [4], Garey and Johnson [3], and Herrera and Sharp [6].

The pallet loading problem falls in the category of problems called NP-Hard [3] and most of the solution procedures developed are approximation algorithms, with specific limitations that most of the time include worst-case bounds on the performance of the packings they produce.

2.2 Approach Two

In this approach for unit load design, costs considerations are included in the analysis. Steudel [14], [15], and Tanchoco [17] considered this case. Steudel [14] developed a model to determine the least-cost pallet deckboard size based on in-plant material handling costs (considering the pallet moves per year) and on space costs (considering storage and aisle space). The pallet employs average values based on representative samples of the parts handled and stored. The unit load size is obtained considering volumetric and weight restrictions and load stability criteria.

Tanchoco [17] included the unit load interactions with various aspects of the material flow function (conveyor, palletizer, lift truck, and storage). Equipment cost (for unit load, lift truck, and storage interface) and cube utilization (for unit load, lift truck, and storage interface) are the performance measures utilized.

2.3 Approach Three

In the third and more recent approach, it is suggested that the unit load design should be integrated with the lot sizing policies, handling and storage equipment, and plant layout. This means, determine the unit load size, that, interacting with the other elements, minimizes the total cost (handling, storage, carrying inventory, etc.). Tanchoco et al. [16], Grasso et al. [5], and Sayin [11] address some of the elements in this integrated process. Tanchoco and Agee [17] present, using an example, the interactive effect that the unit load size and configuration have on the various warehousing system components.

In the third approach, single level, multi-level, single product, multi-product, independent demand, and dependent demand (MRP type) situations are addressed; warehouse area, cube utilization, truck utilization, and minimization of costs are the performance measured considered; and handling, storage, setup, and inventory carrying costs are incorporated in the minimization model and in the evaluation of alternatives. All the contributions present important achievements in the integrated analysis of unit load design.

3 WAREHOUSING-DISTRIBUTION CASE

The Parts Distribution Center (PDC) of a telecommunication systems manufacturer distributes a small number of basic telephone cabinets, each with a variety of models. In storage area #1, the purchased components—chassis, case, fasteners, power supply, and wiring harnesses—and the manufactured subassemblies—printed circuit boards—are received from suppliers in constant lot sizes Following customer orders, the products are packed on boxes and unitized on pallets or in containers. Box dimensions are a function of the product and the number of parts that can be placed in it.

Some of the suppliers are overseas and have to deliver components by ship, using 25, 40, or 45 foot long sea containers, depending on the lot size. The majority of the suppliers are, however, closer to PDC and deliver their products via trailers. Trailer-truck dimensions most commonly arriving to PDC are 40 and 45 feet long. Air transportation has been required in the past to cope with unexpected demand changes.

Upon arrival at the receiving area of PDC, the unit loads are unloaded by a lift truck and transported to the storage area #1. Several types of lift trucks are available at PDC. Each lift-truck alternative possesses different capabilities. These capabilities include reaching, lifting, moving (speed), and depositing/retrieving pallet loads.

The lift trucks carry pallet loads to storage area #1. This storage area consists of selective pallet racks one deep. Each selective pallet rack is designed to accommodate different pallet load sizes and storage policies (i.e., randomized, dedicated, or class-based). Selective pallet racks differ according to their height, opening dimensions, and aisle widths. Currently, the storage area is located next to the receiving area; however, plant layout changes are expected as a result of implementing Just-In-Time. Expected layout changes include number and location of docks and storage racks.

PDC wants to review their facility and distribution activities to determine the unit load design which would minimize operational costs. In order to determine the best unit load design for PDC, the following questions need to be resolved:

1. What is the best box size for each product?
2. What should be the pallet and/or container dimensions?
3. How many boxes should be placed per level and in what pattern?
4. What is the best material handling alternative to move the unit loads?
5. What is the best storage alternative?
6. What is the best facility layout?
7. What is the best transportation mode to use?
8. If the same product can be purchased from several suppliers, which supplier should be selected?
9. If several products can be purchased from several suppliers, which products should be bought from which supplier?
10. What is the best lot sizing strategy?
11. What special considerations would be needed if JIT purchasing is implemented?

A methodology for addressing these questions is presented in this article. This methodology employs as a key element the unit load size which is constrained by physical and operational values. Design and operational combinations (i.e., pallet size, pallet layout, number of parts per pallet, storage equipment, material handling equipment, facility layout, transportation system, and lot sizing policies) that satisfy system constraints are identified through intelligent enumeration analyses and then evaluated using a total cost function to select the preferred combination.

3.1 Scenario Analyzed

In scenario A, PDC is planning to purchase a component from supplier S in constant lot sizes that the vendor will deliver periodically to the receiving area of the company. From receiving, products will be moved to a selective pallet rack where the unit loads will be stored. A counterbalanced lift truck will be used to move the unit loads from receiving to storage and from storage to the next process. Fig. 2 illustrates scenario A.

This scenario can be found in a broad variety of environments and is not exclusive of distribution centers. Manufacturing companies would have similar scenarios for arriving raw materials.

The various interactions involving equipment, facility and production strategies are examined in detail to determine critical factors impacting the unit load design for this particular scenario. A critical factor in these interactions is the specification of the box used and the pallet size. These two factors directly affect the selection of the material handling equipment and the physical configuration of storage area and plant layout (see Fig. 2).

The fundamental design and selection decisions for scenario A are:

1. pallet size,
2. pallet layout,

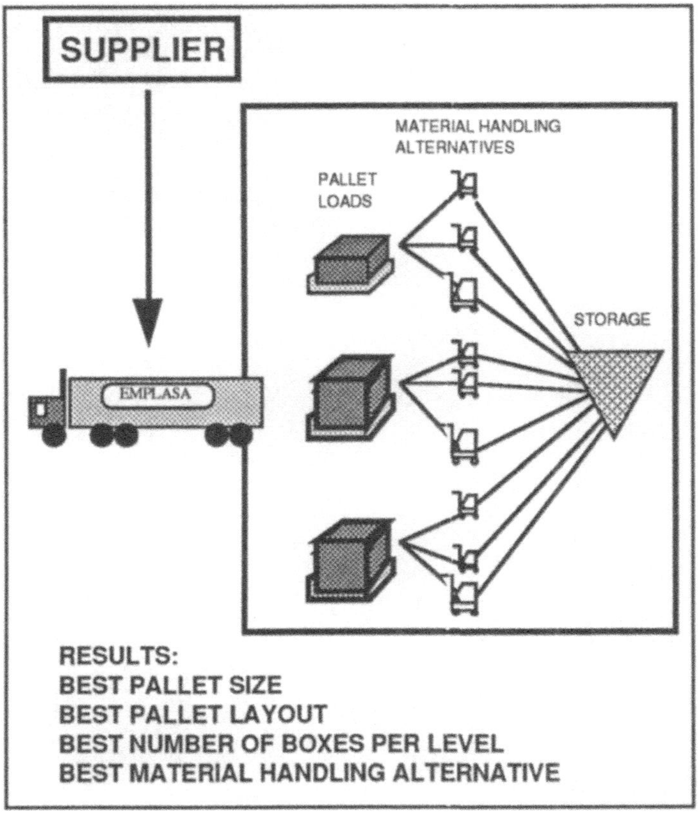

Fig. 2 Scenario A

3. number of levels, and
4. material handling equipment.

System constraints to satisfy in scenario A are:

1. capabilities of the material handling equipment alternatives,
2. dimensions of the storage alternative,
3. facility's layout or product flow,
4. dimensions of the highway transportation alternative,
5. product crushability, and
6. lot size.

The specific operations included in scenario A are described below (see also Fig. 3):

1. Products are packed using closed-top cardboard cartons or containers (with stacking capability) at the supplier's facility.

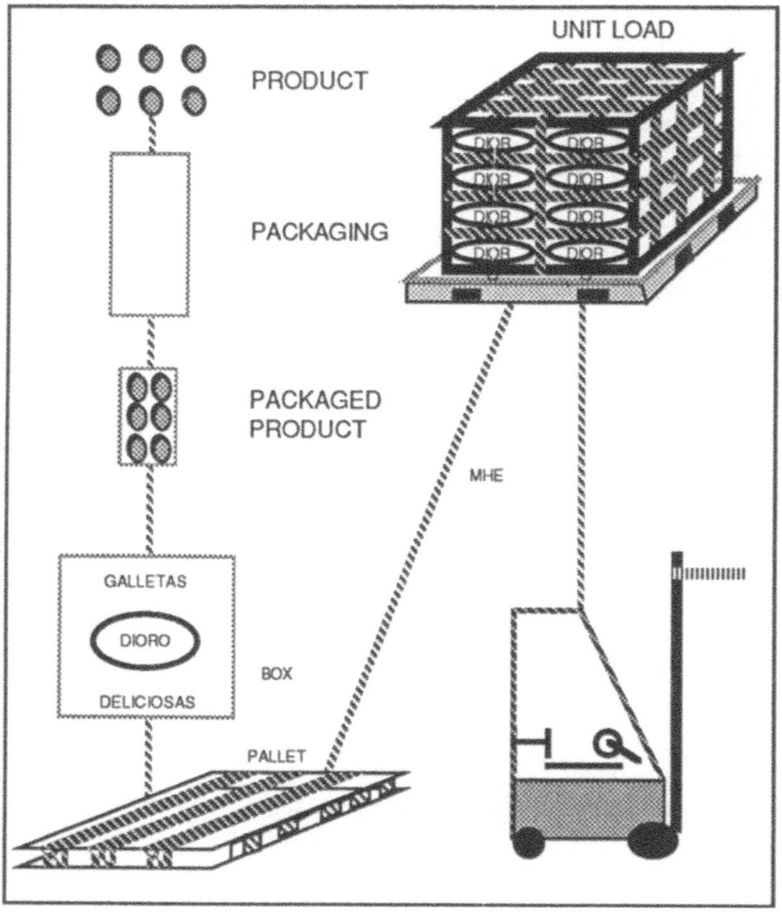

Fig. 3 Product, box size, pallet and material handling relationship

2. The cartons or containers arrive to the plant or distribution center via trailer-truck or any other type of carrier, but in a container form (i.e. sea or air-type).
3. The pallet loads are unloaded from the trailer-truck by a lift truck.
4. The full pallet loads are then transported to the storage area by the same type of lift truck.
5. The pallet loads are stored in a selective pallet rack one deep.
6. The stored pallets are then released upon need.

Only one product category is assumed to move through the system. The intent here is to isolate the interactions among the numerous elements of the system, and to illustrate the importance of the unit load design as it relates to these interactions.

The nomenclature used in this article describe the variables, input data, and performance indicators used in the enumeration and mathematical models for scenario A is included at the end of the chapter. In Table 1 the nomenclature more often used in the rest of the study is classified according to different design and selection decisions.

3.2 System Constraints

Each of the system constraints listed previously can be broken down even further, to include all the variables that they capture and which directly affect the design of the unit load. These variables include:

1. Material handling capabilities. These include the following :
   a. Reaching capacity (R),
   b. Weight capacity (MHwt),
   c. Speed (S), and
   d. Efficiency (Ee).
2. Storage alternative dimensions. The dimensions of interest include:
   a. Opening dimensions:
      i. opening height (OH),
      ii. opening width (OW), and
      iii. opening depth (OD).
   b. Height of selective pallet rack (BUH).
   c. Structural dimensions:
      i. horizontal beam (Rh) and
      ii. vertical beam (Rv).
   d. Rack design clearances:
      i. horizontal clearance (Ch),
      ii. vertical clearance (Cv),
      iii. clearance between floor and first supporting beam (Ca),
      iv. clearance between top of last pallet load and ceiling (Cb), and
      v. flue (f).

In addition, the storage requirements will be affected by the storage policy of each company. The storage policies most often used in industry are: dedicated and random. A dedicated storage policy allows for a higher system throughput, whereas a random storage policy requires less space. In a dedicated storage policy, pallet load storage positions per product are determined based on maximum inventory. Furthermore, products are assigned to these dedicated pallet load positions considering input/output activities, storage requirements, and distance or time traveled.

On the other hand, in a random storage policy, pallet load storage positions per product are determined based on average inventory. Furthermore, products are assigned based on pallet load position availability. Consequently, less storage positions are needed if a random policy is used.

Storage policy decisions are left as a variable which can be adapted to the design procedure recommended in this study according to the particular situation of each company.

3. Facilities layout or product flow. Flow intensity is a function of the distance traveled and the number of products to be moved. This in turn, is used to determine plant layouts along with the relationships between departments. The variables included in here which affect the unit load design are:
    a. aisle width (AW),
    b. distance to be traveled (L), and
    c. location of shipping and receiving docks, storage area, and pick up and deposit stations.
4. Trailer-truck dimensions. The dimensions needed here are:
    a. trailer-truck length (TTL),
    b. trailer-truck height (TTH), and
    c. trailer-truck width (TTW).
5. Characteristics of the product. One of the most important criteria when packaging, moving, storing, stacking, or transporting products is their fragility. Excessive handling can contribute to product damage if the product is not properly protected. This product fragility is referred to as product crushability. The main impact of product crushability restrictions in the unit load design problem is the physical limitations for determining the maximum number of boxes (constrained by weight of the box with the product), that can be placed per level without damaging the product. This information is usually available and, thus, in this study is assumed to be known.
6. Lot sizes. The variables which play a major role in determining lot sizes are: product cost; supplier location, quality, and delivery; demand pattern; and inventory carrying and ordering costs. Different formulations exist to compute the economic lot size. In scenario A this is assumed given. However, in Chapter IV variations are considered where the lot size is not given and the design procedure proposed can be used for selection of lot sizing strategies as a function of the unit load design.

The relationship of the fundamental design decisions on different performance indicators and cost components is presented Fig. 4.

Pallet dimensions will have an impact and will be impacted by material handling, storage, product flow, trailer-truck, crushability, and lot sizing decisions. The larger the pallet, the more parts that can be placed on it. As more parts are placed on the pallet, the heavier it becomes, and the more difficult to move through the plant. Thus, the weight capacity of the material handling equipment could become a limiting factor. Similarly, the more parts on a pallet,

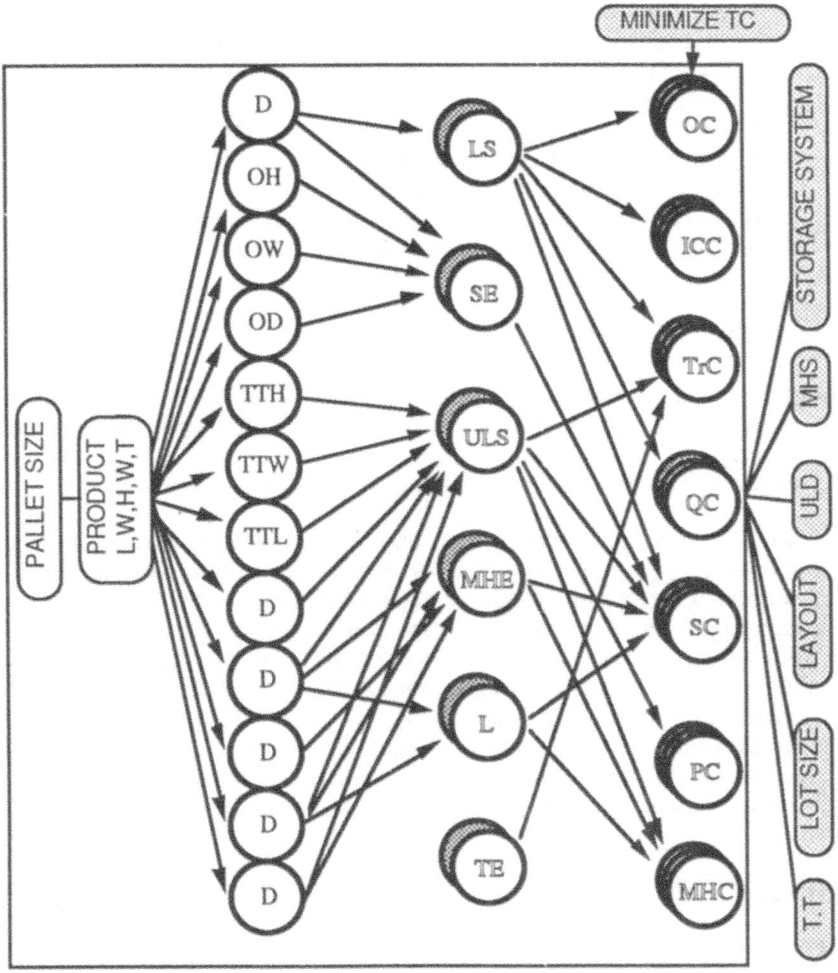

Fig. 4 Relationship between system constraints and design decisions

the higher the probability of damaging the product, thus, crushability has to be considered.

On the other hand, the larger the number of parts placed on a pallet the fewer the number of trips that will be needed for delivering the product. Thus, product flow intensity is minimized, as well as, transportation and handling costs.

As follows, the assumptions used in the model are listed and then the model is explained.

3.3 Assumptions

The assumptions made in developing the unit load design procedure of scenario A are:

1. Products are moved in one size box or container with stacking capability limited by crushability restrictions.
2. The carton or container size is given.

3. Auxiliary equipment such as transfer mechanisms, sorting stations, and palletizers have enough capacity to handle the operations of the system.
4. Lot sizes are known.
5. The product is received from one supplier only.
6. The products are transported, handled ,and stored on pallets.
7. Pallet loads are transported on trailer-trucks,
8. A selective pallet rack (one deep) is used for storage, and
9. Product crushability is defined as the maximum weight a given product would support per level.

4 MODELING APPROACH

In order to obtain the best pallet size, pallet layout, number of parts per pallet, and material handling alternative for scenario A, the following conceptual model is used (Fig. 5) .

The model reflects the following equations and logic flow:

1. Verify if the pallet width is less than or equal to the storage opening dimension (Equation (1)). This is part of the information needed to determine if the existing storage alternative can handle the pallet sizes being evaluated. If this is not satisfied, then other pallet sizes need to be selected from among a given set. Note that for the case in which the storage alternative is also under design, other constraints need to be included. Instead of only one unknown variable (X), pallet width, being evaluated against a known variable opening width (OW), now clearances also will become variables to be evaluated. Equation (1) includes the clearance between pallet loads and between pallet load and vertical rack member

---

MINIMIZE TOTAL COST = STORAGE COST + TRANSPORTATION COST+HANDLING COST+PALLET COST

SUBJECT TO:

- PALLET UTILIZATION
  - PALLET SIZE
  - BOX SIZE
- STORAGE ALTERNATIVE DIMENSIONS
  - PALLET SIZE
  - NUMBER OF LEVELS
- MATERIAL HANDLING CAPABILITIES
  - PALLET SIZE
  - PALLET LAYOUT
  - NUMBER OF LEVELS
- PLANT LAYOUT
  - PALLET SIZE
  - NUMBER OF LEVELS
  - PRODUCT FLOW

Fig. 5 Unit load design model

(beam), Cv. All clearances related to the selective pallet rack which are used throughout this chapter are shown in Fig. 6. Selective pallet racks store 2 pallet loads per opening, thus, OW and Cv are divided by 2.

$$X \leq \frac{OW}{2} - \frac{3C_V}{2} \tag{1}$$

2. Evaluate if the width of the pallet is less than or equal to the trailer-truck width. This is used to calculate the number of pallet loads which can be transported by the trailer-truck. A variation of this situation would be one in which the trailer-truck selection is also under evaluation. Note that in Equation (2) the TTW is divided by 2 because it is assumed that 2 pallet loads should be able to fit widthwise in the trailer-truck container. In addition, four inches are allowed for clearance between pallet loads and between pallet loads and trailer-truck container walls.

$$X \leq \frac{TTW}{2} - 6" \tag{2}$$

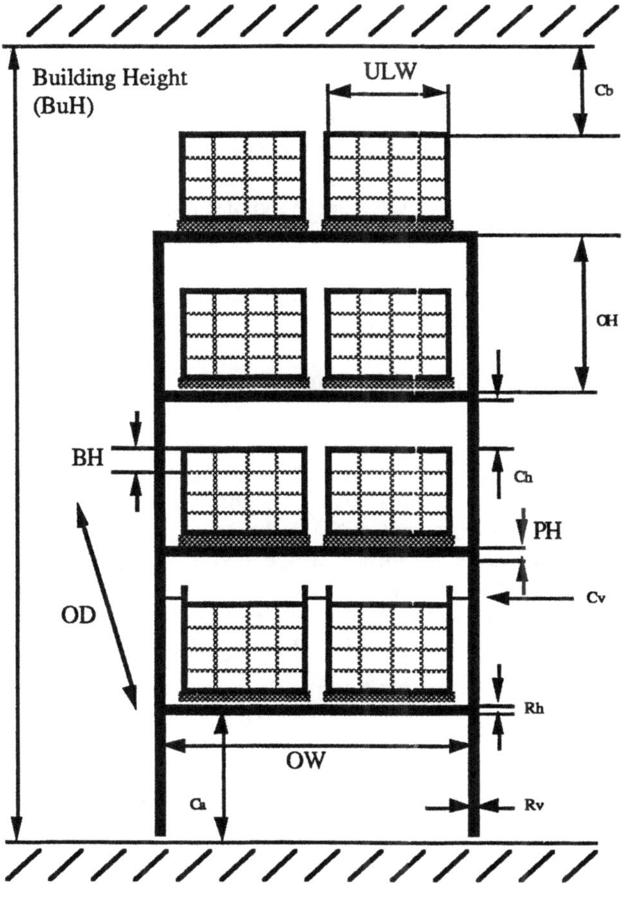

Fig. 6 Typical selective pallet rack and storage dimensions

3. Finally, the pallet width must also be compared against the width of the aisle through which it is to be moved. Aisle width restrictions might exist in the storage area and in any other area through which the unit load will flow. Aisle width can also be used as a variable for plant layout decisions. Note that in Equation (3) a six inch clearance is included as a safety factor.

$$X \leq AW - 6'' \tag{3}$$

4. The depth of the pallet (Y) is then compared to the depth of the opening of the storage alternative (OD). Four inches are added to opening depth since pallet overhang is allowed. Again, this approach can be used to determine feasible storage alternatives.

$$Y \leq OD + 4'' \tag{4}$$

5. If different material handling equipment alternatives are being evaluated, the turning radius (MHtr) and length (MHL) of each alternative has to be considered since the lift truck requires maneuverability space coming in and out of aisles, as well as, for storing and retrieving operations.

$$MHtr + MHL \leq AW - 6'' \tag{5}$$

6. The reaching capability of the lift truck (R) should be enough to pick up and deposit a pallet load from the upper level of the selective pallet rack. The reaching capability of the lift truck has to be at least as high as the upper supporting beam of the selective pallet rack. The height of the upper opening of the selective pallet rack is equal to the height of the storage alternative (BuH) minus rack beam thickness (Ch), minus height of the unit load (NL * BH), and minus the height of the pallet (PH). Equation (6) includes all of these considerations.

$$R \geq BuH - Ch - (NL*BH) - PH \tag{6}$$

7. In order to compute the number of levels (NL), four constraints need to be considered. These four constraints are: opening height (OH), weight capacity of the lift truck (MHEwtc), height of the trailer-truck (TTH), and crushability of the product (CC). Equations (7) thru (10) take these constraints into consideration. Since product crushability (CC) is the weight that a product or level of products can support, then a one must be added to Equation (9) to account for the fact that the number of levels obtained are in addition to an original level. Note that Equation (11) is used to select the minimum of the above.

$$NLoh = [(OH-Ch-PH)/BH] \tag{7}$$
$$NLwt = [MHEwtc/(NBPL*Bwt)] \tag{8}$$
$$NLCC = [CC/(Bwt*NBPL)] + 1 \tag{9}$$
$$NLTTH = [(TTH-2PH-4'')/(2*BH)] \tag{10}$$
$$NL = MIN\{NLoh, NLwt, NLCC, NLTTH\} \tag{11}$$

8. The unit load size is obtained using the number of levels (NL) and the number of boxes per level (NBPL). NBPL is obtained from the interlocking

pattern generator described in the section of the intelligent enumeration approach. NL is obtained from Equation (11).

$$ULS = NL * NBPL \quad (12)$$

9. The number of pallet loads to move per year and per period depends on the demand, unit load size, and lot size as given in Equations (13) and (14). With Equations (13) and (14) lot sizing decisions can be evaluated. For example, if manufacturing follows a JIT approach, smaller lot sizes are used. This decision influences the number of trips that are to be performed by the lift truck in order to deliver smaller, more frequent pallet loads. Similarly, decisions regarding the number of Kanban for the system will be affected by the variables in these two equations.

$$NPLPY = [D/ULS] \quad (13)$$
$$NPLPP = [LS/ULS] \quad (14)$$

10. An important factor to be included in the cost function of the unit load design problem is the handling time per round trip (HPRT). HPRT is equal to the average distance from receiving to storage (L)(round trip), divided by the speed of the vehicle (S), plus the time to store and retrieve a load in the rack, Ts/r. In order to minimize the number of variables used in the model, Ts/r is assumed to be the same for all vehicles. HPRT is used to compute the number of vehicles needed and handling cost. Equation (15) is used to obtain HPRT.

$$HPRT = 2L/S + Ts/r \quad (15)$$

11. The equivalent number of vehicles (NOV) needed for moving the unit load can be obtained from Equation (16). NOV is a function of the number of pallet loads to be moved per period, the distance to be traveled by the lift truck and its speed (given in terms of handling time per round trip from receiving to storage --HPRT and handling time per round trip from storage to the next operation—HPRTsf), and the efficiency of both the operator and the equipment. In addition to equipment efficiency (Ee), operator efficiency (Eo) is considered. This equation implies that the number of vehicles obtained are dedicated to the handling of a particular product arriving to PDC. In other words, a full trailer-truck will be unloaded per shift (480 minutes). If Equation (16) was left without considering the total number of products that can be handled by the number of vehicles, it would imply that once pallet loads of a particular product have been unloaded, the equipment would not be used (stay idle), until the time when the next trailer-truck of that particular product arrives again. Since this is inefficient, it would be assumed that NOVr (number of vehicles needed to move r number of products) are needed per period. Furthermore, in order to reduce the number of variables, it would be assumed that NOVr is equal to 10. That is, the vehicles are shared by 10 products.

$$NOV = \frac{\left\{ \frac{NPLPP*(HPRT + HPRTSF)}{(LS/D)*NOS*480} \right\}}{(NOVr = 10)(Ee*Eo)} \tag{16}$$

12. Handling cost per year (HC) is also another important factor that needs to be minimized in this design procedure. HC is a function of the pallet loads per year that will be handled by the system, the number of vehicles, labor cost (HCL), and each vehicle's annualized fixed cost (K) and annual operating and maintenance costs (MCPVPY). Equation (17) is used to obtain the annual handling cost.

$$HC=\{(NPLPY)(HPRT+HPRTSF)\left(\frac{HCL}{60}\right)\} + \{(NOV*K) +MCPVPY*NOV\} \tag{17}$$

13. Transportation cost is obtained as a function of the number of trips per period, demand, and transportation cost per round trip. Equation (18) summarizes the transportation cost. Transportation cost per trip (TCPT) is assumed given.

$$TrC = \left[\frac{NPLPP}{NPLPTT}\right] \left[\frac{NPLPY}{NPLPP}\right] * TCPT \tag{18}$$

NPLPTT, the number of pallet loads per trailer-truck, can be obtained using Equation (19). The length of the trailer- truck (TTL) , and the depth of the pallet to be placed in the trailer are the important factors. NPLPTT assumes 4 pallet loads per position. That is, 2 pallet loads per level.

$$NPLPTT = \left[\frac{TTL}{PD}\right] * 4 \tag{19}$$

14. The storage cost (SC) can be obtained by multiplying the number of pallet loads per year that need to be stored times the rack cost per pallet load per year (see Equation (20)). These are a function of the opening dimensions, the cost per square foot of storage area (CPSQFT), the rack cost per pallet load position (RCPPLP), and the building space available for storage (or warehouse sizing), WS. For a dedicated storage policy, SC can calculated as follows:

$$SC = \left[\frac{LS}{ULS}\right] \left[\frac{OW*OD*CSPSQFT}{WS} + RCPPLP\right] \tag{20}$$

Where,

$$OW = 2PW + 3Cv + Rv \tag{21}$$

$$OD = 1/2AW + PL +1/2f \tag{22}$$

$$WS = 2\left[\frac{(BuH*12-Ca-Cb)}{(BH*NL+Ch+Rh+PH)}\right] \tag{23}$$

15. Pallet cost (PC) is computed based on the cost of the pallet per year (PCY) and the number of pallets needed per year , which is given by the ratio of lot size to unit load size. The pallet cost equation is as follows:

$$PC = PCY \left[\frac{LS}{ULS}\right] \tag{24}$$

16. The total cost function (TC) which needs to be minimized in order to select the best unit load design includes: handling cost (HC), transportation cost (TrC), pallet cost (PC), and storage cost (SC). Equation (25) is used to obtain the total cost. Equations (17) thru (24) are used to obtain the cost elements PC, TrC, and SC.

$$TC = MHC + TrC + PC + SC \tag{25}$$

4.1 Design Procedure

The objective of this procedure is to determine the best pallet size, pallet layout, number of parts per pallet, and material handling equipment alternative that satisfy a set of system constraints and minimize storage, handling, transportation, and pallet costs.

The unit load design problem is formulated as a mathematical programming problem which can be solved using heuristics or intelligent enumeration. A heuristic approach is useful because if it can be solved mathematically, then results are obtained faster, and are more precise.

An enumeration procedure is useful because it provides insight into the trade-offs between the different design decisions. Furthermore, it supports the convergency of the nonlinear programming algorithm by providing the feasible searching region and an adequate initial searching point. The enumeration procedure and nonlinear integer programming algorithm are explained in the following sections.

4.2 Problem Formulation

In order to incorporate all system constraints and design decisions, the unit load design procedure can be formulated as a nonlinear integer programming problem. In the following sections, the nonlinear programming formulation is presented along with a heuristic solution. At the end, advantages and disadvantages of using the analytical approach are included along with some conclusions.

4.2.1 Nonlinear Programming Formulation

In this model, the relationship between the different factors summarized in Table 3-1 is taken in an integrated manner. In one general form, the nonlinear programming problem is to find $x = (x1,x2,\dots,xn)$ so as to

maximize or minimize $f(x)$
subject to

$$gi \le bi, \text{ for } i = 1,2,\dots,m$$

and

$$x \ge 0,$$

where

$f(x)$ and the $g_i(x)$ are given functions of the n decision variables.

No algorithm that will solve every specific problem fitting this format is available [35]. However, substantial progress has been made for some important special cases of this problem by making various assumptions about these functions, and research is continuing very actively.

In this formulation, constraint (1) is used to ensure that pallet utilization is at least å. Constraint (2) is used to determine the maximum number of levels that can be placed on the pallet considering dimensions of the storage alternative, weight capacity of material handling equipment, height of the trailer-truck, and crushability of the product. Constraints (3), (4), and (5) are used to eliminate pallets that do not satisfy storage alternative dimensions, and layout designs. Constraint (6) is used to ensure that the material handling alternative can handle the weight of the unit load. The complete nonlinear integer programming problem is formulated as shown in Fig. 7.

Note that constraint (6) can be eliminated since constraint (2), in its second factor, MHEwtc/[[X/w][Y/l] * Bwt], is already accounting for the weight capacity of the material handling equipment. In order to make it more convenient for the reader to follow the mathematical models presented throughout this chapter, the conventions shown in Fig. 7 (at the end of the chapter) are used. The modified mathematical formulation is presented in Fig. 8.

All factors included in the model (Fig. 8) can be computed relatively easily, except for the unit load size, ULS. The unit load size is a function of the number of levels, Z, and the layout of the pallet PLAY or number of boxes per level, NBPL, Equation (12). The number of boxes per level, in turn, depends on the dimensions of the box and pallet, and the desired or expected pallet utilization. Specifically, the product of the number of levels and number of boxes per level, (Z * NBPL), is usually referred to as unit load size.

Determining the number of levels is a function of the dimensions of the storage alternative, transportation alternative, crushability of the product, and weight capacity of material handling equipment. On the other hand, determining an optimum pallet layout is an NP-Hard problem which has been studied in the field of operations research. Results from several studies indicate that a closed form expression for obtaining a pallet layout, similarly to Equation 26, gives acceptable values. However, algorithms, similar to that of Smith and DeCani [67] which provide near optimum results are recommended. In this article, both closed form expressions and algorithms will be used in the two design procedures explained in this section.

A closed form expression to compute the number of boxes per level can be obtained by dividing the area of the pallet by the area of the boxes to be placed on top of the pallet. The assumption being that boxes will be placed lengthwise (l) to the length of the pallet (Y) and, widthwise (w) to the width of the pallet (X). Fig. 9, illustrates this procedure. The analytical approach uses Equation (40) to compute the unit load size.

$$NBPL = [X/w][Y/l] \qquad (26)$$

Thus, the unit load size, in number of boxes,becomes,

MINIMIZE TOTAL COST = STORAGE COST + TRANSPORTATION COST + HANDLING COST + PALLET COST

MIN TC = [LS/ULS]{[(2PW+3Cv+Rv)(1/2AW+PL+1/2f)(CSPSQFT)/
{2(144)[(BuH*12+Ca+Cb)/(BH*NL+Ch+Rh+PH)]]+RCPPLP}+
[[LS/ULS]/[TTL/PL]*4][D/LS]*TCPT +[[D/ULS] (HPRT+
HPRTSF)(HC/60)]+{[[LS/ULS]*(HPRT+ HPRTSF)*Ee*Eo]/
[[LS/D](250*480)(NOS)} * (K+MCPVPY) + PCY [LS/ULS]] (27)

SUBJECT TO:

(1) {[X/w][Y/l]w*l}/XY ≥ å (28)
(2) Z ≤ MIN {[(OH-Ch-Ph)/BH],MHEwtc/[X/w][Y/l]
*Bwt],[CC/[X/w][Y/l]*Bwt],
[(TTH-2PH-4")/(2*BH)]-} (29)
(3) PWlb ≤ X ≤ [(OW)/2] - 3Cv - Rv (30)
(4) PLlb ≤ Y ≤ OD - Cod (31)
(5) PWlb ≤ X≤AW - 2Caw (32)
(6) Z[X/w][Y/l] ≤ MHEwtc/Bwt (33)
FOR ALL X,Y,Z > 0

Where,

å = Expected pallet utilization, usually between 80% - 100%.
PWlb = Pallet width lower bound, assumed to be 30 inches.
PLlb = Pallet length lower bound, assumed to be 30 inches.
Ch = clearance between unit loads and between unit load and horizontal beam, assumed to be 4 inches.
Caw = aisle clearance assumed to be 4 inches.
MCPVPY = Maintenance cost per vehicle per year.

Fig. 7 Nonlinear integer programming formulation

MIN TC = [LS/ULS]{[[(2X+µ)(Y + ç)(ø)]/ {œ[(¥)/(rZ + ß)]]+∂}+
[[LS/ULS] /([TTL/Y]*4][D/LS]+(d))]+[[D/ULS]+(t)(h)]
+{[[LS/ULS](t)Ë]÷[LS/D](a)(g)}(K+M) + e [LS/ULS] (34)

SUBJECT TO:

(1) {[X/w][Y/l]w*l}/XY ≥ å (35)
(2) Z ≤ MIN {Z', Z", Z"',Z""} (36)
(3) PWlb ≤ X ≤ (OW)/2 - µ (37)
(4) PLlb ≤ Y ≤ OD - Cod (38)
(5) PWlb ≤ X≤AW - 2Caw (39)
FOR ALL X,Y,Z > 0
for all X,Y,Z integers

Fig. 8 Mathematical model for unit load design using conventions in Fig. 7

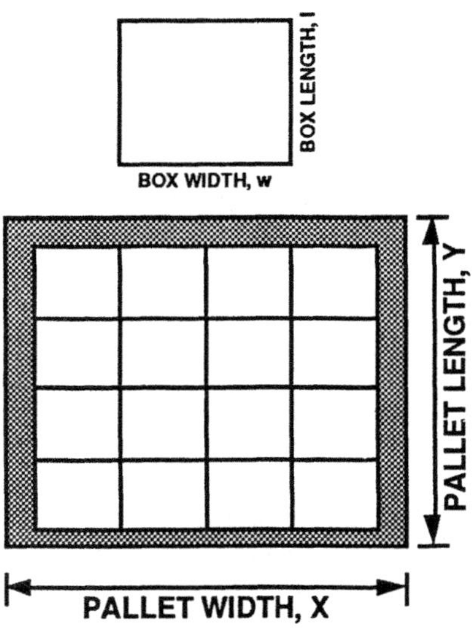

Fig. 9 Typical layout for determining number of boxes per level

$$ULS = Z * ([X/w][Y/l]) \tag{40}$$

Where Z= the number of levels

When Equation (40) is incorporated to Equation (34), the unit load design problem can completely be formulated as a nonlinear integer programming problem, as shown in Fig. 10. For scenario A, X, Y, and Z, are the variables to be determined.

4.2.2 Heuristic Solution

One could start searching for values using the known variables (i.e., Z'). The highest possible values for X and Y can be obtained according to the maximum space available in the storage area (i.e., OW and AW). Then, according to those selected X and Y values, the rest of the Z's can be computed. These results are then used in the total cost function, and the combination which minimizes it is selected.

The heuristic procedure recommended for solving the nonlinear integer programming problem is stated as follows:

HEURISTIC SOLUTION TO UNIT LOAD DESIGN PROBLEM

1. COMPUTE Z' FROM FIRST FACTOR IN CONSTRAINT (2)

   $$Z' = [(OH-Ch-Ph)/BH]$$

2. SELECT HIGHEST POSSIBLE X AND Y FROM CONSTRAINTS (3) AND (4) , SUCH THAT,

   $$[X/w][Y/l] \leq (CC)/[(Z'-1)*Bwt]$$

3. IF STEP 2 IS INFEASIBLE, REDUCE X AND Y UNTIL FEASIBLE, BUT WITHOUT VIOLATING CONSTRAINTS (3) AND (4), ELSE GO TO NEXT STEP
4. COMPUTE Z" ,Z"', AND Z"" FROM CONSTRAINT (2)
5. SELECT HIGHEST POSSIBLE Z FROM CONSTRAINT (2)
6. IF THE Z SELECTED IN 5 IS EQUAL TO Z', THEN THE BEST X,Y, AND Z HAVE BEEN FOUND. GOTO 7 ELSE REDUCE X,Y UNTIL Z=Z', THEN GOTO 7.
7. PLUG X,Y, AND Z INTO EQUATION (34) AND COMPUTE TOTAL COST

4.2.3 Intelligent Enumeration Approach

Consider the problem to minimize $f(x)$ subject to $x \in S$ where f is the objective function and S is the feasible region. A solution procedure or an algorithm for solving this problem can be viewed as an iterative or intelligent enumeration process that generates a sequence of points according to a prescribed set of instructions, together with a termination criterion.

The second approach for unit load design consists of an intelligent enumeration procedure. An enumeration procedure is feasible since the number of possible alternatives to be evaluated is relatively small. In general , there are about 20 different pallet sizes which are used in industry. Among those 20, there are 9 that are the most popular. This implies about 81 combinations of pallet width and pallet length. The other decision being incorporated into the design procedure is the pallet layout. The pallet layout can be computed with Equation (40). However, the closed form expressions presented in Equation (40) exclude combinations of length and width dimensions for both pallet and boxes. This weakness has a negative impact on the pallet utilization. Thus, an algorithm is recommended.

A desirable property of an algorithm for solving the above problem is that it generates a sequence of points converging to a global optimal solution. In many cases, however, less favorable outcomes have to be accepted. Fig 11. is a diagram of the logic behind the pallet layout generator algorithm developed.

MIN TC = [LS/ Z ([X/w][Y/l])]{[[(2X+µ)(Y + ç)(ø)]/
œ[(¥)/(rZ + ß)]]+∂}+[[LS/ Z ([X/w][Y/l])]/([LT/Y]*4]
[D/LS](d)) + [[D/ Z ([X/w][Y/l])](t)(h)]+ {[[LS/ Z ([X/w][Y/l])]
(t)Ë]/[LS/D](a)(g)}(K+M) + e [LS/ Z ([X/w][Y/l])] (41)

SUBJECT TO:

(1)	{[X/w][Y/l]w*l}/XY ≥ å	(42)
(2)	Z ≤ MIN {Z', Z", Z"',Z""}	(43)
(3)	PWlb ≤ X ≤ (OW)/2 - µ	(44)
(4)	PLlb ≤ Y ≤ OD - Cod	(45)
(5)	PWlb ≤ X≤AW - 2Caw	(46)
	FOR ALL X,Y,Z > 0	
	for all X,Y,Z integer values	

Fig. 10 Mathematical model with for determining ULS

The algorithm consists of the following steps:

Step 1. Pick any corner of the pallet as a start point 01, and define the other three corners as 02, 03, and 04, following a counter-clockwise direction.

Step 2. Start from corner 01, fix B value (starting from B=0, to Bmax), starting, try A=0 to Amax.

Step 3. Based on A, C will take the rest of the space available from the Ath box to 02.

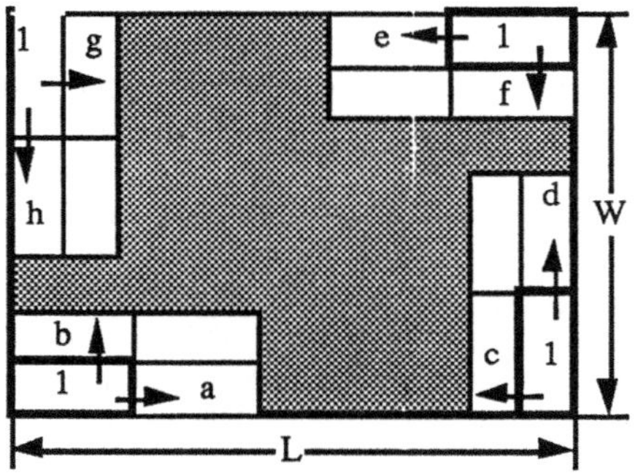

Description of variables

L: Length of the pallet.
W: Width of the pallet.
l: Length of the box.
w: Width of the box.

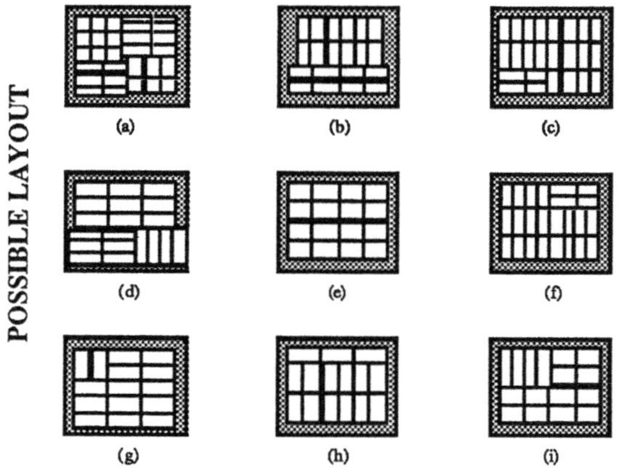

Fig. 11 Pallet layout generator subroutine logic

Step 4. Starting from corner 02, try D=0 to Dmax.

Step 5. Based on D, F will take the rest of the space available from the Dth box to 03.

Step 6. Starting from corner 03, try E=0 to Emax.

Step 7. Based on the value E, G will take the rest of the space available from the Eth box to the corner 04.

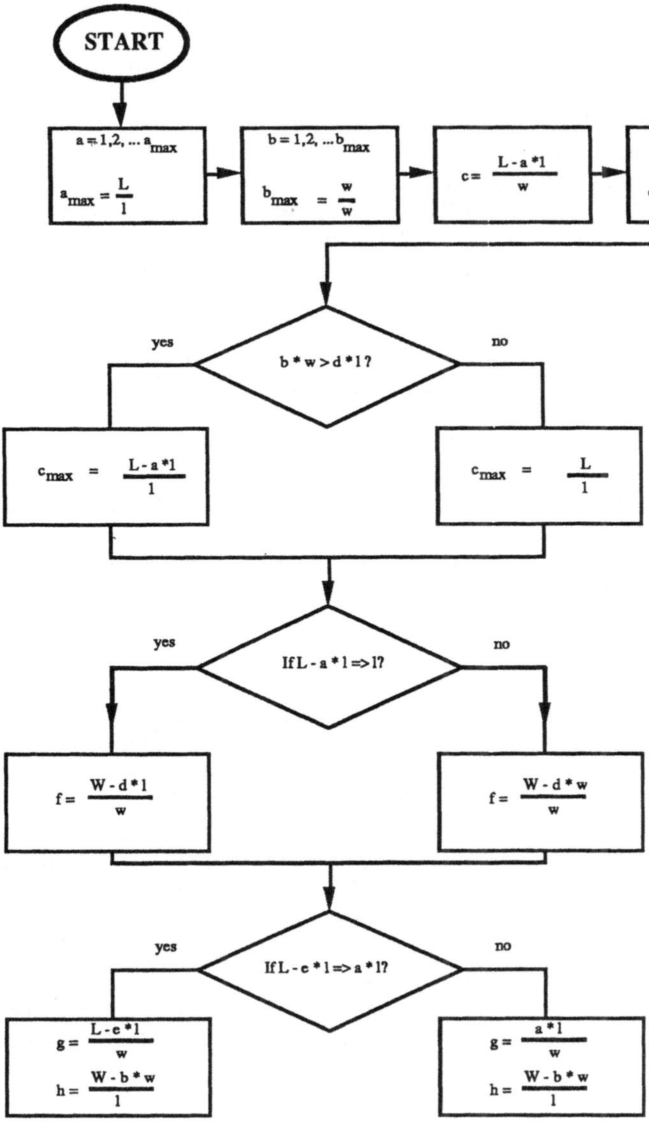

Fig. 11 (continued)

Step 8. H will take the space available from the Bth box to the corner 04.

Step 9. Return to step 2, add 1 to B and repeat step 2-8 until B=Bmax + 1.

The procedure is fast and allows for evaluating several box sizes at the same time. In addition, common pallet sizes are used in the model as the feasible set of pallet sizes from which the system starts generating alternatives for evaluation.

4.2.3.1 Description of Intelligent Enumeration Approach

The logic followed in the intelligent enumeration procedure is illustrated in Fig. 12 (at the end of the chapter). This procedure will be referred to as either the enumeration procedure or CADUL (Computer Aided Design of Unit Loads), which is the name given to the computer program used for implementing the unit load design problem. The steps followed in the enumeration procedure are:

1. Determine if the dimensions of the pallet are feasible subject to the constraints provided by the system. These constraints include: aisle width (AW), trailer-truck container width, storage alternative opening width (OW), and opening depth (OD). Constraints 1, 2, and 3 are used here. These constraints are represented by Equations (1) thru (6).
2. Determine that the material handling equipment alternatives to be evaluated do not violate system constraints such as aisle width. Likewise, lift trucks have to be capable of sufficient speed to satisfy the throughput of the system and reaching capabilities require to store the product into the selective pallet rack. Constraints 3, 4, 5, and 6 must be satisfied. The equations applicable at this stage of the design procedure are 5, 6, 15, and 16.
3. With the information generated above and using the interlocking pattern generator described previously (Fig. 11), the pallet layout and the number of boxes per level can be obtained.
4. Then the number of levels has to be obtained based on four system constraints: height of the opening, weight capacity of the lift truck, height of the trailer-truck, and crushability of the product. Equations (8), (11), (12), (16), and (17) will be used.
5. The next step in the logic of the model consists of determining the number of vehicles required to deliver the products. This is a function of the lot size, the distance to be traveled, the efficiency of the handler and of the equipment. In addition, vehicle maintenance cost is considered. Equations (15) thru (20) will be used in this step.
6. The next step is to determine the utilization and costs related to the trailer-truck. Variables to be considered include the length and height of the trailer-truck as well as the policy followed to load and unload the pallets from the trailer-truck. Equations (19) and (20) are used.
7. The last step is to determine the cost for each of the alternative pallet size and material handling combinations generated. This cost is a function of the pallet cost, the transportation cost, the storage cost, and the material handling cost. The total cost function and its elements are described in Equations (17) thru (25).

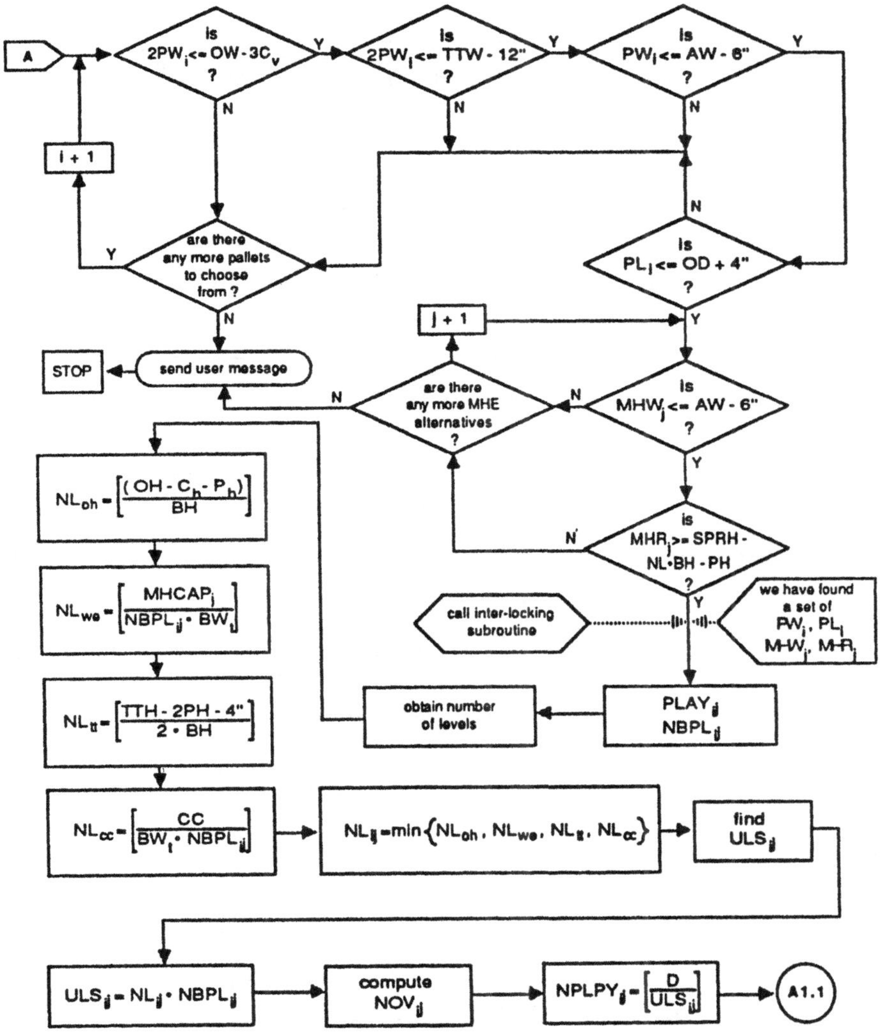

Fig. 12 Enumeration procedure for unit load design problem of scenario A

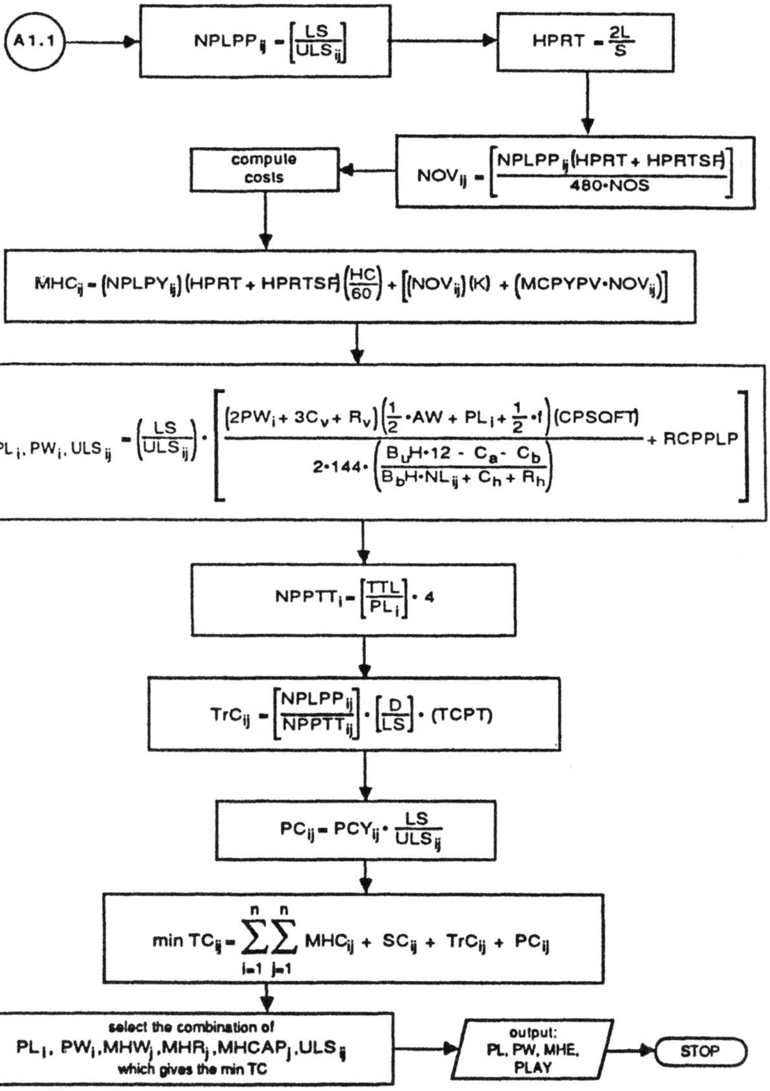

Fig. 12 (continued)

The results of this procedure are the determination of the best pallet size, best pallet layout, best material handling alternative to handle pallet loads, and the number of parts per pallet that can be moved.

In summary, the unit load design which satisfies system constraints is obtained. The procedure finds sets of feasible pallet sizes, material handling alternatives, and pallet layouts. These sets of feasible alternatives are measured against a total cost function to select which of the alternatives minimizes total cost.

An example is presented below to illustrate how the enumeration procedure works and to show how the procedure can be used in the following cases:

1. when the pallet sizes and storage alternative are known, and
2. when sensitivity analysis of the results obtained is necessary.

5 EXAMPLE

PDC is receiving product r in a box of the following dimensions:

BOX LENGTH= 12 in
BOX WIDTH=12 in
BOX HEIGHT=8 in
The weight of the product and box is 25 lbs.
The following three pallet sizes are available for evaluation:

Alternative #1	42 x 42
Alternative #2	46 x 40
Alternative #3	48 x 48

and three material handling alternatives are under consideration, as such:

ALTERNATIVE #	SPEED	LIFT	REACH	COST	WIDTH
	(ft/sec)	(lbs)	(ft)	$/yr)	(in.)
1	3	2500	25	6000	45
2	5	1500	22	4500	40
3	7	4500	30	9500	50

In addition, assume that the information in Table 1 is available to the facilities planner. (Most of these values are already in the computer program and the user has the option to use the default values or to input new values.)

The following information is known:

Given the information given in Table 1 and the pallet alternatives available, PDC wants to determine,

1. the number of parts per pallet,
2. the best pallet layout,
3. the best lift truck to utilize,
4. the pallet size which maximizes utilization, and
5. total cost of the design.

5.1 Solution to Example 1

Example 1 was solved using CADUL. The screens of the computer program and the results obtained are shown in Fig. 13. It can be seen in Screen 32 that the

Table 1 Standard data used for unit load design

LS = 10,000	OH = 60 in	μ = 10 in
ç = 72 in	Ph = 4 in	Ch = 4 in
ø = $50	MHEwtc = 2500 lbs	r = 8 in
S = 3 ft/sec	CC = 2000 lbs/level	ß = 8 in
l = 12 in	TTW = 144 in	MHEw = 45"
∂ = $34	TTH = 120 in	R = 25 ft.
¥ = 272 in	OD = 50 in	Cod = 4 in
œ = 288	OW = 108 in	w = 12 in
D = 120,000	AW = 144 in	e = $150
d = $600	TTL = 540 in	M = $1000
Bwt = 15 lbs	Product Cost=$35	K = $5000
t = 5 min	BUH = 20 feet	a = 48,000
h = $0.25/min	Ë = 0.89	g = 2
E[PALLET UTIL] = 85%		

best (lowest total cost) alternative combination is #2. The characteristics of that best alternative combination are further defined in Screens #33 and #34.

In order to evaluate the results obtained from the intelligent enumeration approach, a sensitivity analysis was performed. The variables believed to be more uncertain and which could impact the unit load design and selection procedure were:

1. Demand (D)
2. Product crushability (CC)
3. Transportation cost per round trip (TCPT)
4. Pallet cost per year (PCY)

A total of 16 runs were performed using CADUL to observe the effect of changing the standard values for the variables listed above. The ranges considered were ±90%, and ±45% of the standard value. The standard values for the four variables listed above are:

VARIABLE	STANDARD VALUE
DEMAND	120,000 units
CRUSHABILITY	2,000 lbs/level
TRANS. COST	600 $/round trip
PALLET COST	20 $/yr.

The results of the sensitivity analysis are summarized in Table 2.

The results obtained from the sensitivity analysis show that the unit load design and the material handling alternative selected are not affected by the transportation cost per trip nor by the cost per pallet per year. Similar conclusions can be drawn for demand and crushability, except at -90% of their standard values. At -90% of the standard (base case) values (demand=12,000 and crushability = 200), the unit load design is affected. The following section discusses the results obtained by the sensitivity analysis in detail.

SCREEN #32

COSTS OF FEASIBLE ALTERNATIVES

ALT.	HANDLING	TRANSP	PALLET	STORAGE	HOLDING	ORDER	TOTAL
1	83457	7211	2080	62400	25500	1200	181848
2	80-62	7205	2040	61200	25500	1200	178096
3	99696	7210	2220	66600	25500	1200	191315
4	91878	14423	4160	124800	25500	1200	261961
5	79069	14447	3560	106800	25500	1200	230576
6	74160	14457	3320	99600	25500	1200	218237
7	247489	36000	12500	375000	25500	1200	697689
8	282557	43220	14280	428400	25500	1200	795157
9	263291	36036	13320	399600	25500	1200	738947

SCREEN #33

THE BEST ALTERNATIVE IS...

PALLET ALTERNATIVE #3: 48X48
MATERIAL HANDLING ALTERNATIVE #1
SPEED: 3 ft/sec
WEIGHT CAPACITY: 2500 lbs
REACHING CAPACITY: 25 ft
WIDTH: 45 IN
COST: $6000
UNIT LOAD SIZE: 98
NUMBER OF BOXES PER LEVEL: 14
BOX SIZE: 12X12X8
PALLET UTILIZATION: 87.5%
TOTAL COST: $178096

SCREEN #34

THE PALLET LAYOUT FOR BEST ALTERNATIVE IS

PALLET UTILIZATION: 87.5%

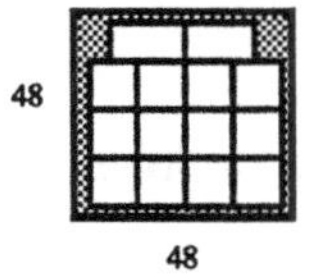

Fig. 13 Screens of CADUL

5.2 Discussion of Sensitivity Results

Example 1 illustrated how the enumeration procedure works using CADUL. A sensitivity analysis was made for several key parameters in the example. The main findings of the sensitivity analysis are discussed as follows:

When demand is reduced by 90% from 120,000 to 12,000 units per year, the unit load size changes from 98 units (boxes per pallet) to 60 units and the material handling alternative selected is changed from #1 to #2. Likewise, the number of boxes per level changes from 14 for the standard values to 15 when the demand is reduced 90%. Furthermore, the pallet utilization increases from 87.5% to 93.8%. Finally, the total cost is reduced from $178, 096 to $64,900.

Table 2 Results of sensitivity analysis (intelligent enumeration approach)

DESIGN/ SELECTION RESULTS	DEMAND (BASE:120K)				TCPT (BASE:$600)			
% DEV.FROM BASE	-90	-45	+45	+90	-90	-45	+45	+90
PALLET ALT.	3	3	3	3	3	3	3	3
MHEA	2	1	1	1	1	1	1	1
ULS	60	98	98	98	98	98	98	98
BOXES/LEVEL	15	14	14	14	14	14	14	14
PALLET UTIL.	94	88	88	88	88	88	88	88
TC(x100000)	67	119	232	286	172	175	181	185

DESIGN/ SELECTION	CC (BASE:2000 LBS/LEVEL)				PCY (BASE:$20) RESULTS			
% DEV.FROM BASE	-90	-45	+45	+90	-90	-45	+45	+90
PALLET ALT.	3	3	3	3	3	3	3	3
MHEA	2	1	1	1	1	1	1	1
ULS	16	98	98	98	98	98	98	98
BOXES/LEVEL	16	14	14	14	14	14	14	14
PALLET UTIL.	100	88	88	88	88	88	88	88
TC(x1000)	836	165	165	165	176	177	179	181

These results can be interpreted as follows:

1. As the yearly demand decreases (from 120,000 to 12,000 units), the number of trips per period is reduced (from 10 to 1). This is because the lot size used was 12000, therefore, there is only one trip per year. Thus, transportation and handling costs will be reduced. Transportation and handling costs are reduced simply because there is only one lot size to move and handle.
2. Although box size and system physical constraints (i.e., rack opening, trailer-truck, and layout dimensions) remain unchanged, the unit load size (a function of these physical constraints and the box size) to be selected would be affected due to changes in the total cost.
3. Since the unit load size selected is lighter (from 98 to 60 boxes per pallet) as a direct result of 1 and 2, then, the material handling alternative would also change. In this case alternative #2 (with lower weight capacity and cost) is selected.

When the crushability was reduced by 90% (from 2000 lbs/level to 200 lbs/level), the unit load size, the material handling alternative selected, the number of boxes per level, the pallet utilization, and the total cost were affected. Specifically, the unit load size was drastically reduced from 98 boxes per pallet to 16 boxes per pallet (one level). The pallet utilization went from 87.5% to 100%, and the total cost increased drastically from $178,096 for the base case to $835,200.

These results can be interpreted as follows:

1. Since the crushability factor limits the number of boxes that can be placed on top of the first level of boxes, it happens that in this example, the number of levels becomes equal to one.
2. As the number of levels decreases, the weight of the pallet load also decreases, and therefore, the weight capacity of the material handling equipment can also be reduced. In this example, alternative #2 is selected instead of alternative #1, (alternative #2 is less expensive and has less weight capacity).
3. The total cost will tend to increase when crushability is drastically reduced because handling, storage, pallet and transportation costs will increase. Since less products are moved now per pallet, the number of trips (transportation and handling) will increase. Furthermore, since pallet loads are smaller, the storage cube utilization will not be maximized, thus, increasing storage cost, and the number of pallets needed.

In addition to the above specific conclusions, the following general statements can be made from the results obtained from the sensitivity analysis:

1. Pallet size (alternative #3 was selected for all cases), will not change as transportation and pallet costs or crushability and demand change. This is due to the fact that the pallet size is mainly constrained by the system (i.e., opening dimensions, aisle width, trailer-truck dimensions, etc.).
2. As demand increases, the total cost will also increase proportionally. However, the unit load design remains the same.
3. As transportation cost increases, the total cost will also increase, but not as much as when changes in demand occur.
4. As the pallet cost per year increases, the total cost will also increase.

The results obtained from the sensitivity analysis begin to demonstrate that the design and selection procedure (intelligent enumeration approach) presented in this chapter is robust. For the values used, the results obtained indicate that only in extreme cases (i.e., demand reduced by 90% or crushability reduced by 90%) would there be any change in the unit load design. It also shows that no physical or operational changes will be required within ±90% changes in transportation and pallet costs.

5.3 Comparison of Heuristic and Enumeration Approaches

Two approaches are proposed for solving the unit load design problem of scenario A, a heuristic and an intelligent enumeration procedure. Both of the approaches are formulated as optimization problems. In this section, their advantages and disadvantages are summarized. The heuristic approach gives acceptable results for pallet size and number of levels. In this approach, the number of boxes per level is obtained through an approximation which excludes several alternative pallet layouts. This approximation does not assure maximization of pallet utilization. Pallet utilization is an important criterion to consider since it affects total cost. Lower pallet utilization implies more number of trips to move pallets and more pallet load storage locations. Likewise, if pallet utilization is low, more transportation trips will be required to deliver the product and more pallets will be needed.

The decisions that can be obtained from the heuristic are limited. For example, it would be extremely complicated to evaluate material handling alternatives . Since one of the objectives of this study is to develop an efficient unit load design procedure which would be applicable to several situations, the heuristic is not recommended.

On the other hand, when intelligent enumeration is used, an optimum solution to the unit load design problem is obtained. In addition, material handling alternatives, as well as other system decisions can be readily incorporated. Optimum results are obtained through intelligent enumeration because, basically, all feasible alternatives are evaluated for a set of constraints. Thus, the approach searches for as many combinations as possible, subject to model limitations and assumptions, and selects the alternative which minimizes the total cost function.

6 CONCLUSIONS

The unit load design and selection problem has been formulated as a nonlinear integer programming problem and solved by a heuristic and an intelligent enumeration procedure (see Fig. 11). The enumeration procedure is recommended over the heuristic.

From the analysis conducted in this chapter, the following conclusions have been reached:

1. The heuristic approach provides a quick and dirty approximation to the unit load design and selection problem. Unfortunately, it cannot be used to evaluate material handling alternatives or any other element of the facility. In addition, the assumption made with regard to the unit load size (closed form expressions were used), does not ensure maximization of pallet utilization.
2. The intelligent enumeration procedure evaluates all possible pallet size, pallet layout, unit load size, and material handling equipment alternatives in an integrated way. Therefore, all system constraints are considered and total cost is minimized. In addition, this procedure allows for real time decision making. Finally, a sensitivity analysis can easily be performed to validate the results obtained.
3. From the example presented for the intelligent enumeration procedure, the following conclusions are obtained:
   a. The maximization of pallet utilization does not necessarily imply minimum total cost. This means that there are pallet layouts that do not maximize pallet utilization, but maximize the number of levels per pallet, thereby reducing total cost.
   b. The maximization of "number of boxes per pallet" or unit load size, considering only the weight capacity of the material handling

equipment alternative, does not necessarily imply that total cost is minimized. This means that there may be combinations of "number of boxes per pallet" that do not maximize unit load cubic utilization, but maximize storage or trailer-truck cubic utilization reducing total cost.

c. Independent maximization of trailer-truck, storage alternative, or unit load cubic utilization, does not necessarily imply the minimization of total cost.
d. The selection of the material handling alternatives to use should not be based exclusively on costs. The interaction with other system components should be considered--specifically, speed, reaching and weight capacities, and dimensions.
e. The following results were obtained after the sensitivity analysis was performed:
   i. location of suppliers only impacts the transportation cost and since the transportation cost is not very high as compared with the rest of the other cost elements, the unit load design is not affected.
   ii. abrupt downward changes in expected demand might change the unit load design, but in general, using the approach proposed in this chapter, such changes will be minor.
   iii. product crushability will impact the unit load design and the total cost function only at extreme values, i.e., -90%. When crushability resistance is low the number of levels is reduced, thereby reducing the number of boxes per pallet. Thus, the number of trips and moves increase. Therefore, total cost increases.
   iv. pallet cost will have a minor impact in the total cost function, unless it becomes extremely large.

7 APPENDIX

NOMENCLATURE

AW =	Aisle width
BH =	Box height
Bwt =	Box and Product weight
CC =	Crushability of the product
Ch =	Clearance between unit load and horizontal beam of the selective pallet rack
CPPY =	Pallet cost per year
CSPSQFT =	Cost per square feet
Cv =	Clearance between pallet loads and between pallet loads and vertical rack members
D =	Demand
Ee =	Efficiency of the equipment
Eo =	Efficiency of the operator
f =	Flue
HC =	Handling cost
HCL =	Handling cost per operator

HPRT =	Handling time per round trip
HPRTSF =	Handling time per round trip from storage area to subsequent process
ICC =	Inventory carrying cost
K =	Annualized fixed cost per vehicle (including depreciation, taxes, etc.)
l =	Box length
L =	Distance traveled from receiving to storage area
LS =	Lot size
MCPVPY =	Maintenance cost per vehicle per year
MHE =	Material handling equipment (lift truck)
MHr =	Turning radius of lift truck
MHwtc =	Weight capacity of lift truck
NBPL =	Number of boxes per level
NL = Z =	Number of levels
NLCC =	Number of levels subject to product crushability
NLoh =	Number of levels subject to opening height
NLwt =	Number of levels subject to weight capacity of material handling equipment
NLTTH =	Number of levels subject to trailer-truck height
NOS =	Number of shifts
NOV =	Equivalent number of vehicles
NOVr =	Number of vehicles needed to move r number of product per period
NPLPP =	Number of pallet loads per period
NPLPTT =	Number of pallet loads per trailer-truck
NPLPY =	Number of pallet load per year
OC =	Ordering or setup cost
OD =	Opening depth of selective pallet rack
OH =	Opening height of selective pallet rack
OW =	Opening width of selective pallet rack
PC =	Total pallet cost
PCY =	Pallet cost per year
PH =	Pallet height
PLAY =	Pallet layout
r =	Box size
R =	Reaching capacity of lift truck
RCPPLP =	Rack cost per pallet load position
Rv =	Width of upright beam of selective pallet rack
S =	Speed of lift truck
SA =	Storage alternative
SAd =	Storage alternative dimensions
SC =	Storage cost
SCPO =	Cost per opening
SPR =	Selective pallet rack
BUH =	SPR height
TC =	Total cost
TCPT =	Transportation cost per trip
Ts/r =	Vehicle time to store or retrieve a pallet load from the selective pallet rack
tP/D =	Time needed to palletize/depalletize a box or carton in a pallet

TrC = Transportation cost
TT = Trailer-truck dimensions
TTH = Trailer-truck height
TTL = Trailer-truck width
ULS = Unit load size
ULW = Unit load width
w = Box width
X = Pallet length
PW = X
PL = Y
NL = Z
3Cv+Rv = µ
1/2(AW+f) = ç
CPSQFT = ø
RCPPLP = ∂
2(144) = two pallet loads per opening
(BuH*12-Cb-Ca) = ¥
BH = r
Ch + Rh + PH = ß
TCPT = d
HCL/60 = h
Eo*Ee = 1/Ë
(250*480) = a
NOS = g
MCPVPY = M
PCY = e
[(OH-Ch-Ph)/BH]- = Z'
[CC/([X/w]-[Y/l]-*Bwt)]- +1 = Z"
[(TTH-2PH-4")/(2*BH)]- = Z'"
[MHEwtc/([X/w]-[Y/l]-*Bwt)]- = Z""
MHEwtc/Bwt.= s
(inches conversion factor) = œ
(HPRT + HPRTSF) = t

8 REFERENCES

1. Albert, Mark, "Pallets Keep Work Within Robot's Reach," *Modern Machine Shop*, V 59, Nov., 1986, pp. 50-55.
2. Apple, James M., Jr., "Designing In-Process Handling," *30th Annual Material Handling Management Course*, Airlie Va., June 12-24, 1983.
3. Apple, James, *Material Handling Systems Design*, John Wiley & Sons, New York,N. Y., 1972.
4. Assad, Arjang and Golden, Druce, "A Categorized Bibliography of Survey Articles in Management Science and Operations Research," *Management Science*, Vol 28, No. 4, April 1982, pp. 425-438.
5. Bacon, F. C., "Design of Efficient Unit Loads," an unpublished project report at Georgia Institute of Technology.
6. Bahl, H., Ritzman, L.and Jatinder, G., "Determining Lot Sizes and Resource Requirements: A Review," *Operations Research*, V 35, May-June, 1987, pp. 329-346.

7. Bashall, Michael, "Returnable Container Systems Boast Handling Economy," *Modern Material Handling*, V 41, Oct. 1986, pp. 74-76.
8. Bashall, Michael, "Returnable Containers Save Automakers Over $20 Million," Modern Material Handling, V 41, Oct. 1986, pp. 76-79.
9. Bazaraa, M., and Shetty, C, *Nonlinear Programming*, John Wiley & Sons, 1979.
10. Beck, Larry, "Cut Handling Costs with the Right Container," *Modern Material Handling*, V 43, April, 1988, pp 60-65.
11. Bohman, Ray, "A Look at Loading Allowances," *Traffic Management*, V 26, Nov. 1987, pp. 23-25.
12. Bolz, H.A. and G. G. Hageman (eds.), *Material Handling Handbook*, Ronald Press, 1958.
13. Botnick, Diane, "Stop Thefts in Loading Docks and Warehouses by Using Rules for Receiving, Delivering Goods," *Merchandising*, V 10, June, 1985, p.102.
14. Brown, Evelyn, "Robotic Pallet Loading at Reynolds Saves Money and Muscle," *Industrial Engineering*, V 19, Feb. 1987, pp. 24-26.
15. Callari, James, "Pallets and Shipping Containers: Containers Provide New Wrinkle in Dan River's Storage Operations," *Traffic Management*, V 25, Feb., 1986, p. 95.
16. Chancellor, Andrea, "Warehousemen Encouraged to Tighten Rein on Pallets," *Journal of Commerce and Commercial*, V 373, Sept. 17, 1987, p 2B.
17. Chiromis, Nicholas, "Computer-Control Systems Speeds Accurate Loading of Unit Trains," *Coal Age*, V 92, April, 1987, pp. 62-65.
18. Chung, F. R., Garey, M. R., and Johnson, D. S., "On Packing Two-Dimensional Bins," *Society of Applied Mathematics*, V 3, No. 1, March 1982.
19. Clugston, Don, "Plotter Package Saves $750,000 a Year," *Packaging*, V 32, Oct. 1987, pp. 80-82.
19a. Daboub, Juan J., "Computer Aided Design of Unit Loads: An Integrated Approach," Ph.D. Dissertation, Department of Industrial Engineering, North Carolina State University, Raleigh, NC 27695, July 7, 1989.
20. Daboub, J., Trevino, J., Liao, H., Wang, J., "Computer Aided Design of Unit Loads," a paper presented at the 11th Conference on Computers and Industrial Engineering, Spring 1989, Orlando, Florida.
21. Dunlop, Craig, "Views Differ on Unloading," *Journal of Commerce and Commercial*, V 376, June 28, 1988, p. 1B.
22. Dunn, Richard, "Custom-Designed Pallets Reduce Costs," V 40, April 10, 1986, p. 85.
23. Ebeling, C. W., "The Ingredient for Shipping Ingredients," *Shipping Management*, V 26, Sept. 1985, pp. 50-54.
24. Fisher, E. L., Farber, J. B., and Kay, M., "MATHES; An Expert System for Material Handling Equipment Selection, " an unpublished paper of the Department of Industrial Engineering, North Carolina State University, Raleigh, NC 27695-7906.
25. Fisher, Edward, "An AI-Based Methodology for Factory Design," *AI Magazine*, Fall 1986.
26. Fitzgerald, Mark, "Palletizing Cuts Paper's Labor Costs in Half: Fort Worth Paper Prefers It Over Building System," *Fort Worth Star-Telegram*, V 121, July 9, 1988, p. 32.

27. Garey, M.R. and Johnson, D.S., *Computers and Intractability: A Guide to the Theory of NP-Completeness*, W. H., Freeman, San Francisco, 1979.
28. Golden, B. L., "Approaches to the Cutting-Stock Problem," *IIE Transactions*, Vol. 8, No. 2, June 1976, pp. 265-272.
29. Gould, Les, "Pallet Trucks : Workhorses of Unit Load Handling," Modern *Materials Handling*, V 42, April, 1987, pp. 73-77.
30. Grasso, E. T. and Tanchoco, J. M. A., "Unit Load and Material Handling Considerations in Material Requirements Planning," *Material Flow*, Vol. 1, 1983, pp. 78-87.
31. Hartman, R. T., "Palletization," *Higher Productivity*, May-June 1958, 24-30.
32. Herrera, C. A. and Sharp, G. P., "Modular Drawer Storage System: A Design Procedure," Material Handling Research Center, RP-83-08, Dec. 20, 1983, Georgia Institute of Technology, Atlanta, Ga.
33. Herrod, Richard, "Industrial Applications for Artificial Intelligence," *IEEE Western Conference on Expert Systems*, June 1987.
34. Hillenius, Corneus, "Packaging and Palletization; The Pallet Pool and the Modulus System," an unpublished paper.
35. Hiller and Liberman, *Introduction to Operations Research*, Fourth Edition, Holden-Day, Oakland, California, 1986.
36. Hinxman, A. I., "The Trim-Loss and Assortment Problems: A Survey," *European Journal of Operations Research*, Vol. 5, 1980, pp. 8-18.
37. Hodgson, T. J. and Lowe, T. J., "Production Lot Sizing with Material Handling Cost Considerations," *IIE Transactions*, March 1982, pp. 237-244.
38. Hodgson, Thom J., "A Combined Approach to the Pallet Loading Problem," *IIE Transaction*, Vol. 14, No. 3, September 1982, pp 22-24.
39. Hodgson, Thom J., "IPLS: Interactive Pallet Loading Systems," Research Report No. 81-9, Industrial and Systems Engineering Department, University of Florida, Gainsville, Fla., 1981.
40. Hodgson, Thom J., "Production and Inventory Control, IE 523 Notes," Course Pack, Department of Industrial Engineering, North Carolina State University, Raleigh, North Carolina, 1988.
41. Hodgson, Thom J., Swift, H. D., and Martin-Vega, L. A., "A Note on a Combined Approach to the Pallet Loading Problem," Vol. 15, No. 3, September 1982, pp. 268-271.
42. Jabbonsky, Larry, "Multipacking is a Many Splendored Thing: Just Ask Your Secondary Packaging Suppliers," *Beverage World*, V 106, June, 1987, pp. 90-94.
43. Johnson, Bruce, "Containers Loaded to 60,000 lbs.," *American Shipper*, V 30, Jan., 1988, pp 26-30.
44. Johnson, Glen R., "Unit Load Theory and Practice," Before U.S. Department of Commerce Trade Center, Cargo and Material Handling Seminar, Bangok,Thailand, 1948.
45. Johnson, L. and D. Montgomery, *Operations Research in Production Planning, Scheduling, and Inventory Control*, John Wiley & Sons, New York, 1974.
46. Knee, Richard, "Overloading Boxes is Risky Business," *American Shipper*, V 30, Feb. 1988, pp 38-39.
47. Kulick, A., "Interlocking Pattern Simulation Program," *Industrial Engineering*, September 1982, pp. 22-24.

48. Mangier, Mark, "Return to Basics, Box Lessons Told," *Journal of Commerce and Commercial*, V 369, Sept. 22, 1986, p. 144.
49. McCarty, Lyle, "Containers Flipped 90º - The Easy Way: Scalloped Ramp Rotates Rectangular Boxes: Other Curves Handle Different Shapes," *Design News*, V 44, April 18, 1988, p. 78.
50. Miller, J. S., "Some Practical Pointers on Pallets and Containers for Automated Conveyors," *Plant Engineering*, Vol. 30, No. 24, pp. 88-90, Nov. 25, 1976.
51. Minart, Scott, "Keeping Parts and Costs Under Containment," *Automotive News*, Sept., 1986, p.12.
52. Mitchel, Joseph, "98% Effective Lot-Sizing for One Warehouse, Multi-Retailer Inventory System With Backlogging," *Operations Research*, V 35, May-June, 1987, pp. 399-405.
53. Moinzadeh, K., Lee, H., "Batch Size and Stocking Levels in Multi-Echelon Respirable Systems," *Management Science*, V 32, Dec., 1986, pp. 1567-1583.
54. Monden, Yasuhiro, "Adaptable Kanban System Helps TOYOTA Maintain Just-In-Time Production," *Industrial Engineering*, May 1981, pp. 29-46.
55. Monden, Yasuhiro, "Kanban and TOYOTA Production System," *Proceeding AIIE Spring Annual Conference*, May 1981, pp. 460- .
56. Morgan, Donald K., "De-Palletizing Unit Loads," *Automation*, June 1966, p.71.
57. Morison, Robert, "Burnley Cracks Down on Overloaded Boxes," *Journal of Commerce and Commercial*, V 377, July 14, 1988, p. 108.
58. Morison, Robert, "Japan Cube Curbs Still Under Review: Japan has Cited Roadway Limitations and Safety Concerns as Reasons for Refusing to Allow Steamship Companies to Use So-Called High-Cube Containers in the Country," *Journal of Commerce and Commercial*, V 3, June 1986, pp 1-10.
59. Ohno, Taiichi and Kumagai, Tomonori, "Toyota Production System," *Proceedings of the International Conference on Industrial Engineering and Management Developing Countries*, Asian Institute of Technology, Bangkok, Thailand, Nov. 3-6, 1980, pp. 349-386.
60. Patterson, Patt, "Containers in the Grocery Distribution Industry," *Grocery Distribution*, Sept./Oct., 1984.
61. Peleg, K, and Peleg E., "Container Dimensions for Optimal Utilization of Storage and Transportation Space," *Computer Aided Design*, Vol 8, No. 3, 1976, pp. 175-178.
62. Ponteous, Evan, "Optimal Lot Sizing, Process Quality Improvement and Setup Cost Reduction," *Operations Research*, V 34, Jan.-Feb., 1986, pp. 137-145.
63. Quinlan, Joe, "The Revolution at the Truck Docks," *Handling & Shipping Management*, V 26, April 1985, pp. 80-85.
64. Ratliff, H. and T. Tran, "Optimum Four Module Pallet Packing," Report No. 82-03, Production and Distribution Research Center, Georgia Institute of Technology, School of Industrial and Systems Engineering, Atlanta, Ga 1982.
65. Reed, R., *Plant Layout: Factors, Principles, and Techniques*, R. D. Irwin, 1961.
66. Sayin, Bulent, "Unit Load Selection in Facilities Design," Technical Report, Middle East Technical University, Ankara.

67. Smith, A., and P. DeCani, "An Algorithm to Optimize the Layout of Boxes in Pallets, "*Journal of Operational Research Society*, Vol. 31, 1980.
68. Somer, T. J., M. Ringrose, *Pallet Pattern Guide*, a private publication on N.V. Philips Gloeilampenfabriken, Eindhoven, Netherlands.
69. Steudel, H. J., "Determination of Economic Pallet Size for In-Plant Handling," *Material Flow*, Vol. 1, 1982, pp. 71-76.
70. Steudel, H. J., "The Right Pallet Cuts Costs," *Industrial Engineering*, Vol. 9, 1977, pp. 14-18.
71. Steudel, Harold J., "Generating Pallet Loading Patterns: A Special Case of the Two-Dimensional Cutting Stock Problem," *Management Science*, Vol. 25, No. 10, Oct. 1979, pp. 997-1004.
72. Stove, Vincent, "Container-Buying Concept Catches on in Australia," *Journal of Commerce and Commercial*, V 366, Nov. 27, 1985, pp. 18-29.
73. Tanchoco, J. M. A., and Agee, Marvin H., "Plan Unit Loads to Interact with All Components of Warehouse System," *Industrial Engineering*, Vol. 13, No.6, June 1981, 36-48.
74. Tanchoco, J. M. A., Davis, R. P., and Wysk, R. A., "Economic Order Quantities Based on Unit Loads and Material Handling Considerations," *Decision Sciences*, Vol. 11, No. 3, pp. 514-521.
75. Tanchoco, J. M. A., Davis, R. P., Egbelu, P. J., and Wysk, R. A., "Economic Unit Loads for Multi-Product Inventory Systems with Limited Storage Space," *Material Flow*, Vol. 1, 1983, pp. 141-148.
76. Tanchoco, Jose, M., "Unit Load Design - A Key Element in Integrating Material Management and Storage,"
77. Thomas, Gordon C., "What is really Happening in Unit Load Shipping," *Modern Material Handling*, March 1965.
78. Thomas, Gordon C., "What to Consider When You Switch to a Pallet Pool," *Modern Material Handling*, April, 1965, p. 62.
79. Tompkins, J. A., 1985. *Personal Consultation*, Columbia, S C.
80. Tompkins, J. A., and John White, *Facilities Planning*, John Wiley & Sons, 1984.
81. Trevino, J., Clincy, V., Jang, E., Daboub, J., "SITSES; An Expert System for Selection of Industrial Trucks and Storage Alternatives," an unpublished paper of the Department of Industrial Engineering, North Carolina State University, Raleigh, NC 27695-7906.
82. Trevino, Jaime, "Collection of Notes of Advanced Material Handling Systems," North Carolina State University, Raleigh, NC, 1988.
83. Trevino, Jaime, "Collection of Notes of Facilities Design," North Carolina State University, Raleigh, NC, 1987.
84. Trevino, Jaime, "Design Procedures for 'Pull' Production Systems," Ph.D Dissertation, Georgia Institute of Technology, Georgia, 1985.
85. Watson III, Ripley, "Truckers Face Major Intermodal Challenges," *Journal of Commerce and Commercial*, V 375, May 30, 1988, pp. 1A-1B.
86. Wilson, Linda, "Plastics Move Into the Warehouse," *Chemical Week*, V 142, May 4, 1986, pp. 46-48.
87. Wood, Charles W., "Planning for Unit Loads," *Transportation and Distribution Management*, Vol. III, No. 3, p.11.
88. Zein, Abudi, and Agoos, Alice, "Potential Outruns Profits in Plastics Packaging," *Chemical Week*, V 139, Oct. 15, 1986, pp. 38-41.

89. "Automakers Leading Way in Inventory Control: The Automotive Industry is Becoming the Leader in Using Bar Codes and Reusable Containers to Control Inventories," *Modern Materials Handling*, V 40, Oct. 1985, p. 9
90. "Container," Packaging, V 32, April, 1987, pp. 118-128.
91. "Equipment Selection Guide," *Modern Materials Handling*, V 41, Fall 1987, pp. 161-167.
92. "Exchange Pallets for Profits," in Material Handling Engineering, *Presidential Issue*, October 1966, p. 43.
93. "Guidelines for Selecting Bulk Handling Containers," *Modern Materials Handling*, V 42, August 1985, pp. 62-65.
94. "Japan Creates Multi-Layer Can," *Packaging*, V 32, Oct., 1987, p 85.
95. "Making Pallets In-Plant Saves $1 per Pallet," *Packaging*, V 32, Feb. 1987, p. 90.
96. "Material Handling: Specialized Handling Systems," *Plant Engineering*, Reprints series 45-1.
97. "First Complete Study Shows a Pallet Can Save $500," *Modern Material Handling*, August 1969, "
98. "Material Handling: Unit Load Conveyors," Plant Engineering, Reprints series 45-2.
99. "Open the Door to Safety: Shipping and Receiving Materials Properly," *National Safety and Health News*, V 131, Sept., 1985, pp. 63-67.
100. "Palletizing Helps Nabisco Wrap Up $4 Million Savings," *Packaging*, V 30, Jan., 1985, pp. 88-92.
101. "Pallets vs. Slipsheets; The Debate Continues," *Traffic Management*, V 25, April, 1988, pp. 83-86.
102. "Pallets: Playing a Vital Role in PD Systems," *Retail & Distribution Management*, V 13, Nov.-Dec., 1985, pp. 28-31.
103. "Pallets: The Difficulty of Learning How to Delegate Responsibility," *Business Atlanta*, V 16, Nov., 1987, pp. 46-48.
104. "Plastic Pallets Free Floor Space, Reduce Fork Truck Handling," *Modern Materials Handling*, V 42, June, 1987, pp. 97-99.
105. "Plastic Pallets Withstand Auto Assembly Abuse," *Design News*, V 44, Feb., 1988, p. 37.
106. "Roll Pallets in Multi-Level Order Picking," *Retail & Distribution Management*, V 15, Jan. - Feb., 1987, pp. 56-58.
107. "Selecting the Right Pallet for Your Needs," *Modern Materials Handling*, V 42, March, 1987, pp. 80-82.
108. "Unit Loads," *Production Engineering*, May 1979, pp. 30-34.
109. "Warehouse Modernization and Layout Planning Guide," NAVSUP Publication 529, Naval Supply Systems Command, Department of the Navy, Washington, D. C., December 1978.

A THREE DIMENSIONAL DYNAMIC PALLETIZING HEURISTIC

Russell D. Tsai
Digital Equipment Corporation
Eric M. Malstrom
University of Arkansas
Way Kuo
Iowa State University

ABSTRACT

This chapter extends earlier work on three dimensional palletization completed by the authors. Earlier work focused on the development of a mixed 0-1 integer programming model. While effective, the model proved to be computationally too complicated to permit its use in real time palletizing applications. This chapter describes the development of a three dimensional dynamic programming heuristic. The heuristic's computational requirements lend themselves more readily to real time palletizing applications.

1 INTRODUCTION

Material handling pallets are a common tool in warehousing industries. Traditionally, palletizing methods load only boxes of one size on the same pallet. In retail distribution operations, a wide mix of different box sizes must be loaded onto the same pallet and shipped to different retail destinations.

This chapter extends previous work by Tsai, Malstrom and Meeks [37,38] and Tsai, Malstrom and Kuo [39]. References [37] and [38] describe the development of a two dimensional palletizing model. This work has been extended in reference [39] which describes a dynamic three dimensional palletizing methodology. The three dimensional model employs mixed 0-1 integer programming. The model has robust computational requirements which impede its use in real time palletizing applications.

This chapter describes the development of a three dimensional palletizing heuristic that employs the use of dynamic programming. The performance of the developed heuristic is comparable to the integer programming model. The advantage of the developed heuristic lies in its reduced computational requirements. This will permit its use in real time palletizing applications.

2 REVIEW OF RELEVANT LITERATURE

The two-dimensional pallet loading problem is generally addressed by attempting to maximize the number of small rectangles that can be placed orthogonally within a large containing rectangle. This problem has been described in a variety of operations research literature on two-dimensional stock cutting and bin packing problems.

The pallet loading problem has been classified as NP-Hard [15]. No polynomial algorithms exist for this class of problems. They are generally considered to be intractable. Many algorithms and approaches found in the literature rely heavily on dynamic programming and heuristic techniques. As a result, suboptimal solutions are often obtained.

A review of the literature in the area of loading and cutting problems has been reported by many authors. Tsai [35] has classified this type of problem into plane tiling, pallet stacking, stock cutting, bin packing and container packing. This particular problem has also been the subject of a number of survey papers including those of Brown [6], Dowsland [12], Golden [19], Hinxman [23] and Israni and Sanders [26].

In recent years, new developments on two-dimensional pallet loading related problems have included the linear programming approach of Tsai, Malstrom and Meeks [37,38]; the goal programming model of Fleming, Malstrom and Meeks [15]; dynamic programming approaches of Beasley [2], Steudel [34], Hodgson [24,25], Roberts [32], and Israni and Sanders [26]; the integer programming model of Beasley [3,4]; the combinatorial approach of Wang [40]; the network flow algorithm of Biro and Boros [5]; and the bin packing approaches of Coffman et al. [9], Golan [18] and Baker [1]. Microcomputer based palletization has been described by Carlo et al. [7].

Most of the existing literature on the subject addresses only two dimensional problems. A mixed 0-1 integer programming model for three-dimensional pallet packing has been developed by Tsai, Malstrom and Kuo [39]. This chapter addresses the development of a more computationally efficient three dimensional palletizing model. This development is described in the sections that follow.

3 BACKGROUND

The heuristic approach of the chapter has been inspired from the early developmental work of Haims and Freeman [20]. Their algorithm is only applicable for two-dimensional pallet packing problems in which no restriction is placed on the number of small pieces of each size.

The developed heuristic procedure is carried out with a dynamic programming (DP) algorithm to solve three-dimensional pallet packing problems. The algorithm involves a sequence of iterations, each of which allows a higher degree of packing. Each stage in the dynamic programming algorithm allocates one type of box.

Unlike traditional dynamic programming, there are two goals to be achieved in the heuristic DP algorithm. First, it is desired to place as many boxes as possible onto the pallet so as to maximize pallet space utilization. Secondly, for each box type, the number of boxes packed on a pallet (or box proportion) should not exceed some user-specified number. These two goals usually contradict each other. One has to compromise between these two goals to determine a "rational" solution. The recursive dynamic programming procedure and the criteria to determine a solution from these two conflicting goals are described in the sections that follow.

4 NOTATION AND ASSUMPTIONS

To apply the dynamic programming procedure, the dimensions of the pallet and boxes of all types are restricted to integer values. It is assumed that the orientations of boxes and pallet are predetermined and remain fixed. The respective lengths, widths, and heights of the boxes and pallet are assumed to be parallel to each other. No interchange of orientations is allowed. Boxes of different orientations are considered to be a unique type even though they may have exactly the same dimensions.

Each box type is considered to be a stage in the dynamic programming algorithm. Also, the allocated box is placed in an index pallet and is represented by (L_x,W_x,H_x). The dimensions of the index pallet are then increased by integer values of 1 from a unit cube (1,1,1) to the pallet size (L,W,H).

Denote (L,W,H) = dimensions of the pallet;
= length, width and height, respectively

(l_i,w_i,h_i) = dimensions of a box of type i;
= length, width and height, respectively

v_i = volume of a box of type i
= $(l_i \cdot w_i \cdot h_i)$;

r = total number of box types.

Define the following allocation vector

$$\underline{a}_1(L_x, W_x, H_x) = (a_1)$$
$$\underline{a}_2(L_x, W_x, H_x) = (a_1, a_2)$$
$$\vdots$$

$$\underline{a}_r (L_x, W_t, H_x) = (a_1, a_2, ..., a_r)$$
$$\underline{a}_k (L_x, W_t, H_x) = \underline{0}, \text{ if any of the three arguments equals zero, } k=1,2,...,r.$$

The term $\underline{a}_i (L_x, W_x, H_x)$ is an i-dimensional vector at stage i. The jth element of $\underline{a}_i$, a_j , represents the number of type j boxes allocated in the index pallet (L_x, W_x, H_x), where i = 1,2,...,r and j = 1,2,...,i.

Also, define the following return function:

$$F_1 (L_x, W_x, H_x) = \max (v_1 a_1)$$
$$F_2 (L_x, W_x, H_x) = \max (v_1 a_1 + v_2 a_2)$$
$$\vdots$$
$$F_r (L_x, W_x, H_x) = \max (v_1 a_1 + ... + v_r a_r)$$
$$F_k (L_x, W_x, H_x) = 0, \text{ if any of the three arguments equals zero, } k=1,2,...,r.$$

$F_i (L_x, W_x, H_x)$ is the maximum return function at stage i given the index pallet (L_x, W_x, H_x).

The index pallet (L_x, W_x, H_x) varies from (1,1,1) to (L,W,H) and increases by integer values of 1. Therefore, F_r (L,W,H) is the final optimal function value. The term $\underline{a}_r$ (L,W,H) gives the required number of boxes of each type.

5 DEVELOPMENT OF A TWO DIMENSIONAL HEURISTIC

The following two-dimensional dynamic programming procedure is adapted from the work of Haims and Freeman [20]. For all values of L_x , W_x such that $L_x > l_i$ and $W_x > w_i$, a box i can be allocated as shown in Fig. 1.

Eight subrectangles, A, B, C, D, E, F, G and H, as defined in Fig. 1 are created by allocating a box of type i at location (x,y). This generates sixteen combinations of the subrectangles. One of the combinations is

$$A+B+C = L_x \cdot (W_x - w_i - y)$$
$$D+E = (L_x - l_i - x) \cdot (w_i + y)$$
$$F+G = (l_i + x) \cdot y$$
$$H = (x) \cdot (w_i)$$

This means four smaller index pallets A+B+C, D+E, F+G and H are generated from the original index pallet of size L_x by W_x when a box i is allocated at coordinate (x,y). The subindex pallet A+B+C has the dimensions of (L_x) by $(W_x - w_i - y)$, D+E is $(L_x - l_i - x)$ by $(w_i + y)$, etc. The combination of these four subindex pallets is shown in Fig. 2.

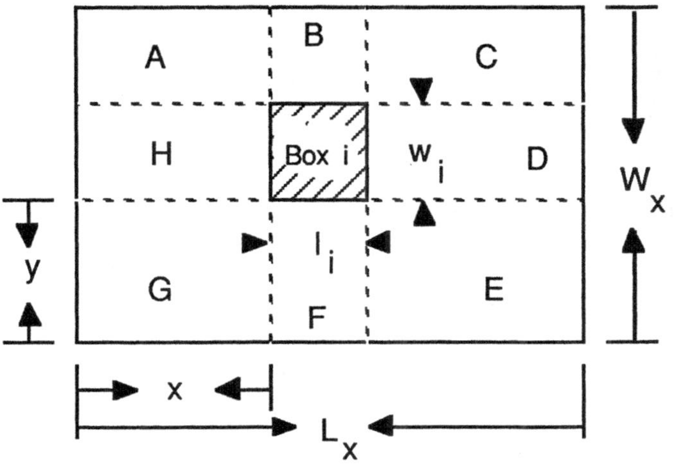

Fig. 1 Allocation of a box i at (x,y)

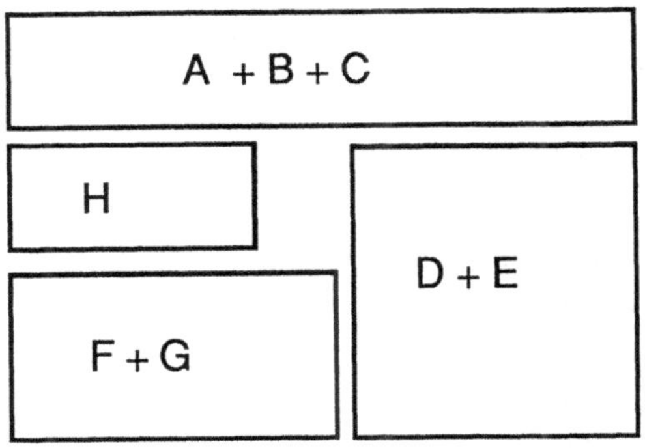

Fig. 2 A combination of four subindex pallets

5.1 Subindex Pallet Combinations

Consider the subindex pallet A+B+C and two smaller index pallets A+B and C. The sum of areas covered on both A+B and C by pieces is at most as good as the subindex pallet A+B+C. The subindex pallet A+B+C contains all possible solutions of the two smaller index pallets A+B and C. Therefore, combinations like A+B and C are discarded from further consideration. Based on this logic, the sixteen possible combinations of subindex pallets exist and are listed below:

1) A+B+C D+E F+G H	5) A+B C+D E+F H+G	9) A+H G+F E+D B+C	13) A+H+G B+C D E+F
2) A+B+C D+E F H+G	6) A+B C+D+E F+G H	10) A+H G+F+E D+E B	14) A+H+G B+C D+E F
3) A+B+C D E+F G+H	7) A+B C+D+E F G+H	11) A+H G+F+E D B+C	15) A+H+G B C+D E+F
4) A+B+C D E+F+G H	8) A+B C+D E+F+G H	12) A+H G+F C+D+E B	16) A+H+G B C+D+E F

Each subindex pallet defined in terms of the dimensions is listed below. (See Fig. 1.)

$$B = (l_i) \cdot (W_x - w_i - y)$$
$$D = (L_x - l_i - x) \cdot (w_i)$$
$$F = (l_i) \cdot (y)$$
$$H = (x) \cdot (w_i)$$
$$A+B = (l_i + x) \cdot (W_x - w_i - y)$$
$$A+H = (x) \cdot (W_x - y)$$
$$B+C = (L_x - x) \cdot (W_x - w_i - y)$$
$$C+D = (L_x - l_i - x) \cdot (W_x - y)$$
$$D+E = (L_x - l_i - x) \cdot (w_i + y)$$
$$E+F = (L_x - x) \cdot (y)$$
$$F+G = (l_i + x) \cdot (y)$$
$$H+G = (x) \cdot (w_i + y)$$

$$A+B+C = (L_x) \cdot (W_x - w_i - y)$$
$$A+H+G = (x) \cdot (W_x)$$
$$C+D+E = (L_x - l_i - x) \cdot (W_x)$$
$$E+F+G = (L_x) \cdot (y)$$

5.2 Allocation Procedure

For every index pallet (L_x, W_x) currently considered, the maximum return function values of the above subindex pallets have been determined in the previous procedure since all these subindex pallets have smaller dimensions than those of the index pallet.

Since the pallet is symmetrical, it is only necessary to evaluate the discrete placement of a box of type i in one-fourth of the rectangle (L_x, W_x). Observe the allocations of box i in Fig. 3a and 3b.

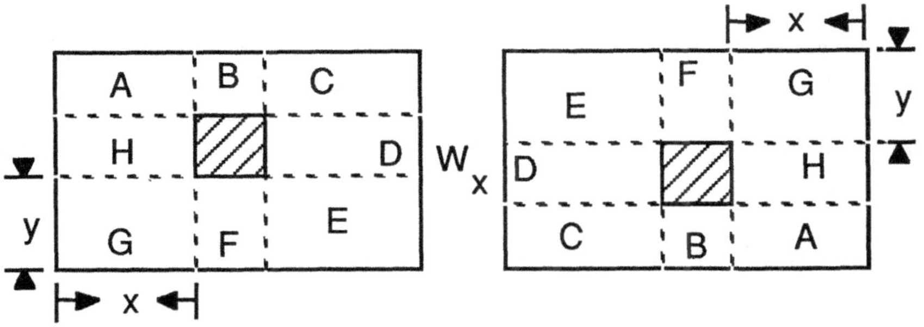

Fig. 3 Symmetrical allocations

Because Fig. 3b is equivalent to Fig. 3a by rotating 180°, both patterns result in the same subrectangles. Therefore, it is only necessary to evaluate (x, y), the allocation of a box i in (L_x,W_x), from (1,1) to ([L_x / 2],[W_x / 2]) increasing by integer values of 1.

For the two-dimensional case, there are sixteen combinations to be considered for each given value of (x, y). The objective is to select the combination with the maximum F_i (L_x, W_x). The recursive function has the form

$$F_i(L_x, W_x) = \text{Max} \begin{cases} \text{1) Combination 1:} \\ l_i \bullet w_i \text{, the area of a box of type i} \\ + F_i(L_x, W_x - w_i - y)\text{, the subindex pallet A+B+C} \\ + F_i(L_x - l_i - x, w_i + y)\text{, the subindex pallet D+E} \\ + F_i(l_i + x, y)\text{, the subindex pallet F+G} \\ + F_i(x, w_i)\text{, the subindex pallet H} \\ \text{2) Combination 2:} \\ l_i \bullet w_i + F_i(L_x, W_x - w_i - y) \\ + F_i(L_x - l_i - x, w_i + y) \\ + F_i(l_i, y) \\ + F_i(x, w_i + y) \\ \bullet \\ \bullet \\ \bullet \\ \text{16) Combination 16:} \\ l_i \bullet w_i + F_i(x, W_x) \\ + F_i(l_i, W_x - w_i - y) \\ + F_i(L_x - l_i - x, W_x) + F_i(l_i, y) \\ \text{17) } F_i - 1(L_x, W_x) \end{cases}$$

The maximum return function value of F_i (L_x, W_x) is selected from one of the sixteen combinations of subindex pallets. In the instance that all return function values of these sixteen combinations are less than the maximum function value at the previous stage, the best previous function value is used for the current stage.

6 THE THREE DIMENSIONAL HEURISTIC

Twenty-six subcubes are created when a box i is allocated at coordinate (x,y,z) of a three-dimensional index pallet of size (L_x, W_x, H_x). There may be hundreds of combinations of subindex pallets with these twenty-six subcubes. To reduce computation time, only six combinations of subindex pallets are taken into consideration. Since not every possible combination of subindex pallets is considered, only "suboptimal" solutions may be obtained using this heuristic dynamic programming procedure. The allocation of a box i in the index pallet (L_x, W_x, H_x) at location (x,y,z) is shown in Fig. 4. The six combinations of subindex pallets are illustrated in Fig. 5a, b, c, d, e, and f.

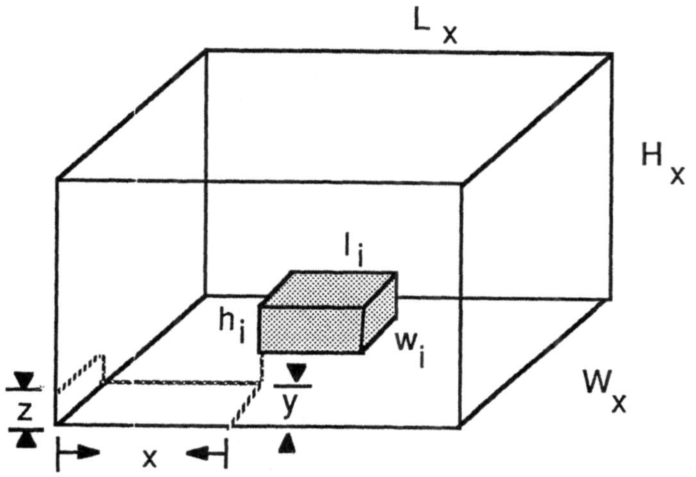

Fig. 4 Allocation of a box in the index pallet

6.1 Allocation Procedure

With an index pallet (L_x , W_x , H_x),
where L_x = 1,2, ...,L;
W_x = 1,2,...,W;
H_x = 1,2,...,H,

allocate a box i at coordinate (x,y,z) in terms of the front-bottom-left corner of the box. For every given index pallet (L_x , W_x , H_x) such that ($L_x > l_i$, $W_x > w_i$ and $H_x > h_i$), the allocated coordinate (x,y,z) is revised as follows:

$x = 1,2,\ldots, \min (L_x - l_i , [L_x / 2])$;
$y = 1,2,\ldots, \min (W_x - w_i , [W_x / 2])$;
$z = 1,2,\ldots, \min (H_x - h_i , [H_x / 2])$.

The return function and allocation vector are then determined as shown below.
If $L_x < l_i$ or $W_x < w_i$ or $H_x < h_i$, then

$F_i (L_x , W_x, H_x) = F_{i-1}(L_x , W_x, H_x)$, and

$\underline{a}_i (L_x , W_x, H_x) = \underline{a}_{i-1} (L_x, W_x, H_x)$.

Otherwise,

$$
F_i(L_x, W_x, H_x) = \max \left\{ \begin{array}{ll}
1) & v_i + F_i(L_x, W_x, z) \\
 & + F_i(L_x, W_x, H_x - h_i - z) \\
 & + F_i(x, W_x, h_i) + F_i(L_x - l_i - x, W_x, h_i) \\
 & + F_i(l_i, y, h_i) \\
 & + F_i(l_i, W_x - w_i - y, h_i) \text{ (see Fig. 5a)} \\
2) & v_i + F_i(L_x, W_x, z) \\
 & + F_i(L_x, W_x, H_x - h_i - z) \\
 & + F_i(x, w_i, h_i) + F_i(L_x, y, h_i) \\
 & + F_i(L_x - l_i - x, w_i, h_i) \\
 & + F_i(L_x, W_x - w_i - y, h_i) \text{ (see Fig. 5b)} \\
3) & v_i + F_i(x, W_x, H_x) \\
 & + F_i(L_x - l_i - x, W_x, H_x) \\
 & + F_i(l_i, y, H_x) \\
 & + F_i(l_i, W_x - w_i - y, H_x) \\
 & + F_i(l_i, w_i, H_x - h_i - z) \\
 & + F_i(l_i, w_i, z) \quad \text{(see Fig. 5c)} \\
4) & v_i + F_i(x, W_x, H_x) \\
 & + F_i(L_x - l_i - x, W_x, H_x) \\
 & + F_i(l_i, y, h_i) \\
 & + F_i(l_i, W_x - w_i - y, h_i) \\
 & + F_i(l_i, W_x, H_x - h_i - z) \\
 & + F_i(l_i, W_x, z) \quad \text{(see Fig 5d)} \\
5) & v_i + F_i(L_x, y, H_x) \\
 & + F_i(L_x, W_x - w_i - y, H_x) \\
 & + F_i(x, w_i, H_x) \\
 & + F_i(L_x - l_i - x, w_i, H_x) \\
 & + F_i(l_i, w_i, H_x - h_i - z) \\
 & + F_i(l_i, w_i, z) \quad \text{(see Fig. 5e)} \\
6) & v_i + F_i(L_x, y, H_x) \\
 & + F_i(L_x, W_x - w_i - y, H_x) \\
 & + F_i(x, w_i, h_i) \\
 & + F_i(L_x - l_i - x, w_i, h_i) \\
 & + F_i(L_x, w_i, H_x - h_i - z) \\
 & + F_i(L_x, w_i, z) \quad \text{(see Fig 5f)} \\
7) & F_{i-1}(L_x, W_x, H_x)
\end{array} \right. \tag{1}
$$

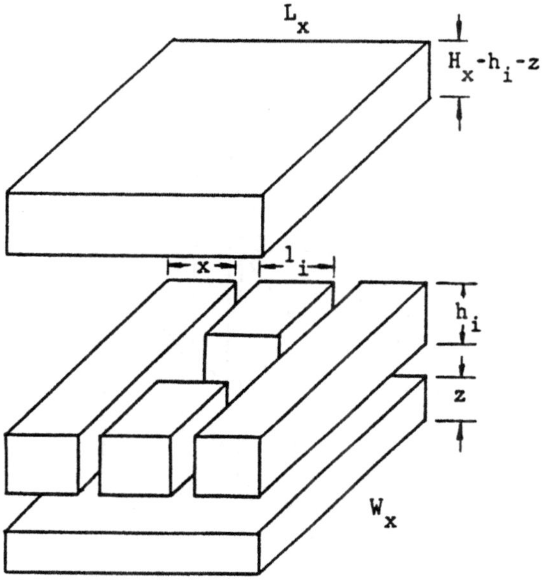

(a)

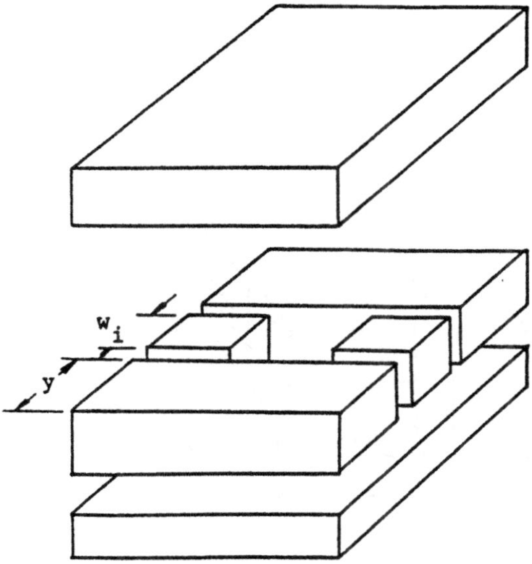

(b)

Fig. 5 Six combinations of subindex pallets

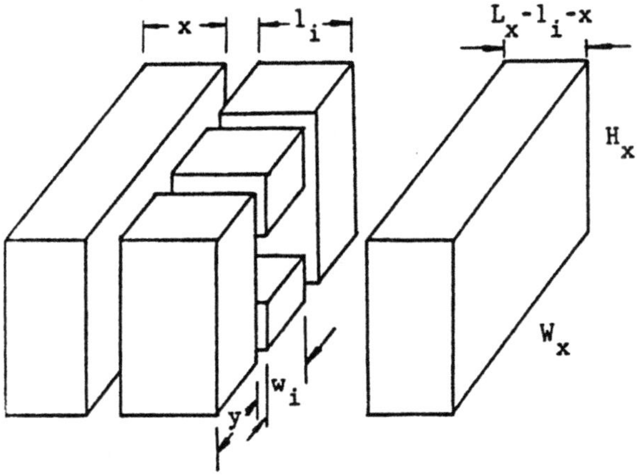

(c)

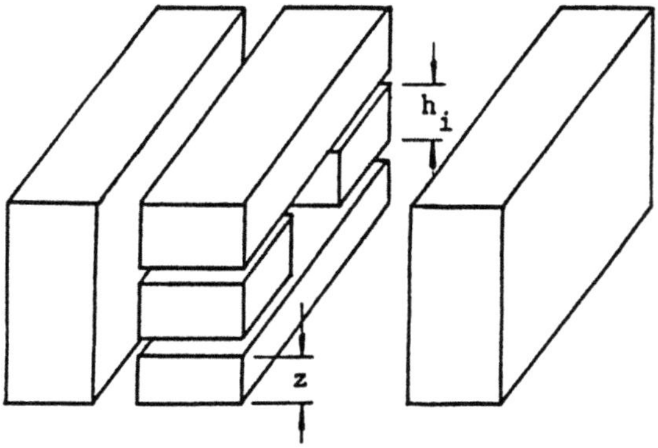

(d)

Fig. 5 (continued)

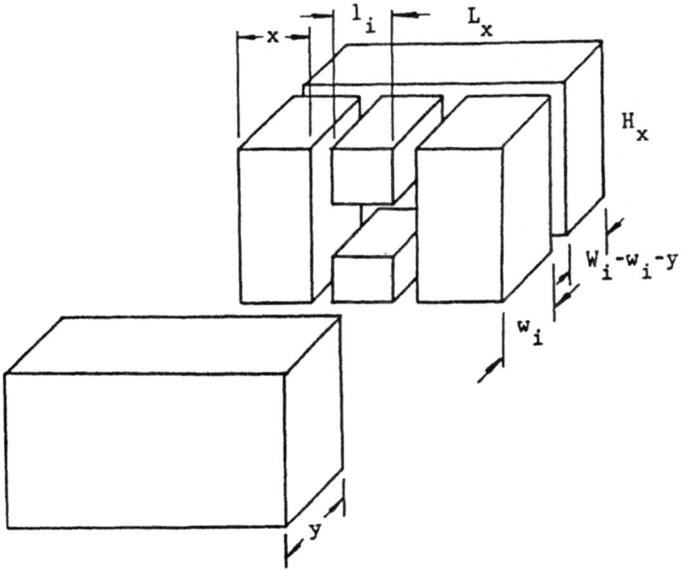

(e)

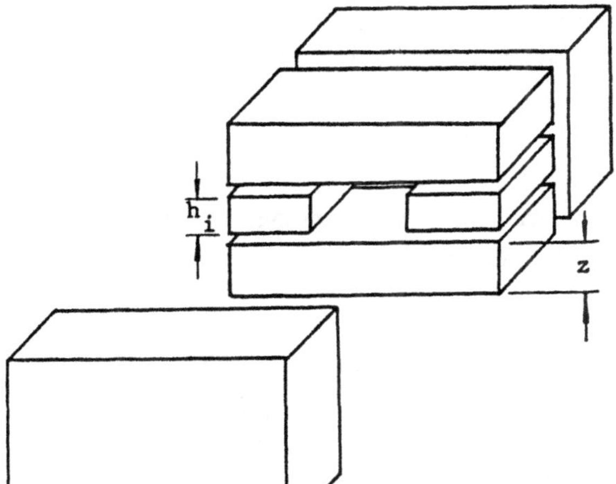

(f)

Fig. 5 (continued)

Let $\underline{e}_i$ be the unit vector. The ith element of $\underline{e}_i$ is set to one, and all other elements are set to zero. The corresponding allocation vectors $\underline{a}_i\,(L_x, W_x, H_x)$ for above seven possible function values are:

1) $\underline{e}_i + \underline{a}_i\,(L_x, W_x, z) + \underline{a}_i\,(L_x, W_x, H_x - h_i - z)$
$+ \underline{a}_i\,(x, W_x, h_i) + \underline{a}_i\,(L_x - l_i - x, W_x, h_i)$
$+ \underline{a}_i\,(l_i, y, h_i) + \underline{a}_i\,(l_i, W_x - w_i - y, h_i)$
2) through 6) can be obtained using a calculation similar to that shown in 1)
7) $\underline{a}_{i-1}\,(L_x, W_x, H_x)$

6.2 The DP Procedure

Based on the three-dimensional allocation algorithm described in the preceding section, the following heuristic dynamic programming procedure applies.

STEP 1: Reorder the box types $v_1, v_2, ..., v_r$ such that

$$v_1 \leq v_2 \leq ..... \leq v_r.$$

STEP 2: Select box type v_1 to be allocated at the first stage. Let [t] represent the largest integer number no greater than t. Compute

$\underline{a}_1\,(L_x, W_x, H_x) = [\,L_x / l_1\,] \bullet [\,W_x / w_1\,] \bullet [\,H_x / h_1\,]$,
$F_1\,(L_x, W_x, H_x) = \underline{a}_1\,(L_x, W_x, H_x) \bullet v_1$, for all
L_x, W_x, H_x. Let $i = 2$.

STEP 3: Compute the best return function value and allocation vector for every index pallet (L_x, W_x, H_x).

(a) For $L_x = 1,2,....,L;$
$W_x = 1,2,...,W;$
$H_x = 1,2,...,H;$
allocate a box v_i at coordinate (x,y,z) in the index pallet (L_x, W_x, H_x).

Let MAX = 0, and
$AL = \min(L_x - l_i, [\,L_x / 2\,]);$
$AW = \min(W_x - w_i, [\,W_x / 2\,]);$
$AH = \min(H_x - h_i, [\,H_x / 2\,]).$

(b) For $x = 1,2,...,AL;$
$y = 1,2,...,AW;$
$z = 1,2,...,AH;$
calculate the function value $F_i\,(L_x, W_x, H_x)$ using Equation (1) and the associated allocation vector $\underline{a}_i\,(L_x, W_x, H_x)$.

(c) If MAX < F_i (L_x, W_x, H_x), then

MAX = F_i (L_x, W_x, H_x) and

$\underline{a}^*$ = $\underline{a}_i$ (L_x, W_x, H_x).

If (x,y,z) is not increased up to (AL,AW,AH), go to (b). Otherwise, go to (d).

(d) Let F_i (L_x, W_x, H_x) = MAX, and

$\underline{a}_i$ (L_x, W_x, H_x) = $\underline{a}^*$

If (L_x, W_x, H_x) is not increased up to (L,W,H), go to (a).

Otherwise, go to STEP 4.

STEP 4: Let i = i + 1

Let $i \leq r$ (total number of box types), go to STEP 3. Otherwise, go to STEP 5.

STEP 5: Deliver the solutions F_r (L,W,H) and $\underline{a}_r$ (L,W,H).

6.3 Restriction on the Number of Boxes

The first goal of the dynamic programming procedure tends only to maximize the utilization of a pallet cube without placing any restriction on the number of boxes of each type to be loaded. To make the box proportion of each type satisfy some user specified number (the second goal), criteria have been developed to select a rational solution from these two conflicting goals. The criteria are determined based on the number of boxes that have been allocated in the index pallet currently considered.

The procedure for the second goal selects a return function whose associated allocation vector best fits the criteria. In the case that no allocation vectors satisfy the criteria, the dynamic programming procedure then selects the return function that maximizes the allocated space of a given index pallet as though no restriction is placed on the number of boxes of each type.

The selection criteria for a best allocation vector are based on the value of box ratio which is defined as follows:

Denote r = total number of box types to be allocated

R_g = the desired box proportion of type g,

g = 1,2,...,r

n_g = the desired number of boxes of type g.

$$n_g = R_g \cdot V / \left(\sum_{i=1}^{r} R_i \cdot v_i\right)$$

where V is the pallet's volume;

v_i is the box's volume of type i.

a_{ij} = the jth element (number of type j boxes allocated) of the allocation vector $\underline{a}_i$ (L_x , W_x , H_x), i = 1,2,...,r; j = 1,2,...,i.

BR = box ratio

$$= \sum_{j=1}^{r} (a_{ij} / n_j), \text{ for stage } 1,2,\ldots,r-1 \quad (2)$$

$$= \sum_{j=1}^{r} \text{ABS} \left[\left(a_{ij} / \sum_{k=1}^{r} a_{ik} \right) - R_j \right] \text{ for stage } r \quad (3)$$

$a_{ij} / \sum_{k=1}^{r} a_{ik}$ = box proportion of type j in the allocation vector $\underline{a}_i$ (L_x , W_x , H_x).

In each stage of the dynamic programming procedure, the box ratio should be as small as possible. The box ratio of Equation (2) calculates the cumulative proportions of the allocated number to the desired number of boxes. The allocation vector associated with the minimum box ratio, Equation (2), tends to have an allocated number far smaller than the desired number for each type of box. This allows the subsequent stages to have more box allocation alternatives. The box ratio of Equation (2) is calculated for all allocation vectors in all stages except the last stage (stage r) of the dynamic programming procedure.

The box ratio of Equation (3) calculates the cumulative difference between the box proportion of an allocation vector and the desired box proportion specified by the user. The allocation vector associated with the minimum box ratio, Equation (3), has the allocated box proportion closest to the desired box proportion. This box ratio is only calculated for the allocation vectors in the last stage (the last box size to be allocated). Recall that each stage of the DP procedure considers only one box size of a unique orientation. In the case that the box types to be allocated in the last two stages have the same dimensions with different orientations (l-by-w and w-by-l), the box ratio of Equation (3) should be applied for the last two stages of the DP procedure.

6.4 Decision Criteria Rules

The following four rules describe the decision criteria used for selecting a best function value and allocation vector between the first and second goals of the model.

Rule 1: For a given index pallet (L_x , W_x , H_x) and the placement location (x,y,z), compute the box ratio for each allocation vector of the six combinations of subindex pallets (see Fig. 5 a-f), and the allocation vector at the previous stage. In case that two or more combinations have the same best function values, select the allocation vector that has the minimum box ratio. In this way, the subsequent process will have more box allocation alternatives so as to obtain a maximum return function value.

Rule 2: For a given index pallet (L_x ,W_x , H_x), suppose that the maximum function values are the same for some different box locations (x,y,z), where

$$x = 1,2,\dots, \min (L_x - l_i , [L_x / 2]),$$
$$y = 1,2,\dots, \min (W_x - w_i , [W_x / 2]),$$
$$z = 1,2,\dots, \min (H_x - h_i , [H_x / 2]),$$

then select the allocation vector such that its box ratio is as small as possible.

Rule 3: For a given index pallet (L_x ,W_x ,H_x), suppose no allocation vector $\underline{a}_i$ (L_x ,W_x , H_x) exists such that

$$a_{ij} \le n_j, \text{ for all } j,$$

then let F_i (L_x ,W_x , H_x) = maximum function value without considering the restriction on the number of boxes. Since no solution exists for goal 2, the return function that has the maximum value is selected. Rules 1 and 2 are still applied to select the maximum value.

Rule 4: For a given index pallet, suppose that at least one allocation vector $\underline{a}_i$ (L_x ,W_x , H_x) exists such that

$$a_{ij} \le n_j, \text{ for all } j,$$

then let F_i (L_x ,W_x , H_x) = the best function value selected from those that their corresponding allocation vectors satisfy $a_i \le n_j$, for all j. In this case, the maximum function value and its corresponding allocation vector are selected among the candidates which have all their allocated number less than the desired number of boxes of every type.

The above four selection rules for determining a best return function value and allocation vector are implemented in STEP 3 of the dynamic programming procedure for goal 1.

6.5 Post-Solution Adjustment

Because of the design nature of the heuristic dynamic programming procedure, goal 1 (maximizing the utilization of a pallet cube) has a higher priority than goal 2 (satisfying desired box proportion). The final solution generated by the dynamic procedure may give a value that only approximates the desired box proportion.

However, a larger box can always be substituted with one or more smaller boxes and still maintain the feasibility of the pallet pattern. A feasible pallet pattern also holds by deleting some number of boxes from the heuristic DP solution if the stability of a pallet load is not the main consideration. Many alternatives with different box proportions and pallet space utilizations can be generated by substituting or reducing number of boxes. It is up to the user to weigh the importance between goals 1 and 2, and select the best pallet pattern from these alternatives.

7 SUMMARY AND CONCLUSIONS

The developed model has been evaluated by the authors with the use of a physical simulator. A RHINO XR2 miniature robot was used to perform the palletizing process. Four different box sizes were used. The box dimensions were structured so as to be divisible as integers into the pallet cube dimensions. This ensures that the sides of the load cube will be smooth and that no spatial voids will exist in the pallet load.

The simulator employed a turntable to permit the simultaneous loading of up to four different pallets. In all, twenty different distributions of the box sizes were simulated. Boxes that could not be directly palletized from the conveyor were temporarily stored in off-line storage areas. Boxes stored off-line were then placed on the pallet as soon as a space for them was available on the pallets being loaded.

The developed heuristic was used to dynamically determine the pallet pattern based on the contents of the palletizing cell's look-ahead queue. The obtained results were comparable to the integer programming model described in reference [39]. None of the off-line areas for box storage overflowed. This indicates the dynamic viability of the developed model. Generation of a separate manuscript to completely describe the simulation results is a task currently being addressed by the authors.

8 REFERENCES

1. Baker, B.S., D.J. Brown, and H.P. Katseff, "A 5/4 Algorithm for Two-Dimensional Packing," *Journal of Algorithms*, 2, No. 4, pp. 348-368 (1981).
2. Beasley, J.E., "Algorithms for Unconstrained Two-Dimensional Guillotine Cutting," *Journal of the Operational Research Society*, 36, No. 4, pp. 297-306 (1985).
3. Beasley, J.E., "An Exact Two-Dimensional Non-Guillotine Cutting Tree Search Procedure," *Operations Research*, 33, No. 1, pp. 49-64 (1985).

4. Beasley, J.E., "Bounds for Two-Dimensional Cutting," *Journal of the Operations Research Society*, 36, No. 1, pp. 71-74 (1985).
5. Biro, M. and E. Boros, "Network Flows and Non-Guillotine Cutting Patterns," *European Journal of Operational Research*, 16, No. 2, pp. 215-221 (1984).
6. Brown, A.R., *Optimum Packing and Depletion*, New York: American Elsevier (1971).
7. Carlo, H., T.J. Hodgson, L.A. Martin-Vega, and E. Stern, "Pallet Loading on a Microcomputer," *Journal of Computers and Industrial Engineering*, Vol. 9, No. 1 (1985).
8. Cheng, M.H.R., and A.W. Pila, "Maximized Utilization of Surface Areas with Defects by the Dynamic Programming Approach," *ISA Transactions*, 17, No. 4, pp. 61-69 (1978).
9. Coffman, E.G., Jr., M.R. Garey, D.S. Johnson, and R.E. Targan, "Performance Bounds for Level-Oriented Two-Dimensional Packing Algorithms," *SIAM Journal on Computing*, 19, No. 4, pp. 808-826 (1980).
10. Cooper, M.W. and K. Farhangian, "A Dynamic Programming Algorithm for Multiple-Choice Constraints," *Computers and Mathematics with Applications*, 10, No. 3, pp. 279-282 (1984).
11. Davis, R.E., D.A. Kendrick and M. Weitzman, "A Branch-and-Bound Algorithm for Zero-One Mixed integer Programming Problems," *Operations Research*, 19, No. 4, pp. 1036-1044 (1971).
12. Dowsland, W.B., "Two and Three Dimensional Packing Problems and Solution Methods," *New Zealand Operational Research*, 13, No. 1, pp. 1-18 (1985).
13. Farley, A., "Trim-Loss Pattern Rearrangement and its Relevance to the Flat-Glass Industry," *European Journal of Operational Research*, 14, No. 4, pp. 386-392 (1983).
14. Farley, A., "A Note on Modifying a Two-Dimensional Trim-Loss Algorithm to Deal with Cutting Restrictions," *European Journal of Operational Research*, 14, No. 4, pp. 393-395 (1983).
15. Fleming, J.R., E.M. Malstrom, and H.D. Meeks, "A Robotic Pallet-Loading Algorithm for Dynamic Carton Distributions," *Proceedings of the 7th International Conference on Automation in Warehousing*, Institute of Industrial Engineers, San Francisco, CA (December 1986).
16. Garey, M.R. and D.S. Johnson, *Computer and Intractability*, New York: W.H. Freeman (1979).
17. George, J.A. and D.F. Robinson, "A Heuristic for Packing Boxes into a Container," *Computers and Operations Research*, 7, No. 3, pp. 147-156 (1980).

18. Golan, I., "Performance Bounds for Orthogonal Oriented Two-Dimensional Packing Algorithms," *SIAM Journal on Computing*, 10, No. 3, pp. 571-582 (1981).
19. Golden, B.L., "Approaches to the Cutting Stock Problem," *AIIE Transactions*, 8, No. 2, pp. 265-274 (1976).
20. Haims, M.S. and H. Freeman, "A Multi-Stage Solution of the Template-Layout Problem," *IEEE Transactions on Systems, Science and Cybernetics*, SSC-6, No. 2 (1970).
21. Healy, V.C., Jr., "Multiple Choice Programming," *Operations Research*, 12, No. 1, pp. 122-138 (1964).
22. Hillier, F.S. and G.J. Lieberman, *Operations Research*, San Francisco: Holden-Day Inc. (1974).
23. Hinxman, A.I., "The Trim-Loss and Assortment Problems: A Survey," *European Journal of Operational Research*, 5, No. 1, pp. 8-18 (1980).
24. Hodgson, T.J., "A Combined Approach to the Pallet Loading Problem," *IIE Transactions*, 14, No. 3 (1982), pp. 175-182.
25. Hodgson, T.J., D.S. Hughes, and L.A. Martin-Vega, "A Note on a Combined Approach to the Pallet Loading Problem," *IIE Transactions*, 15, No. 3, pp. 268-271 (1983).
26. Israni, S. and J. Sanders, "Two-Dimensional Cutting Stock Problem Research: A Review and a New Rectangular Layout Algorithm," *Journal of Manufacturing Systems*, 1, No. 2, pp. 169-182 (1982).
27. Jeroslow, R.G. and J.K. Lowe, "Experimental Results on the New Techniques for Integer Programming Formulations," *Journal of Operations Research Society*, 36, No. 5, pp. 393-403 (1985).
28. Kochenberger, G.A. and V.H. Richard, "A Simple, All Primal Branch and Bound Approach to Pure and Mixed Integer Binary Problems," *Operations Research Letters*, 1, No. 5, pp. 182-185 (1982).
29. Martin, R.K. and L. Schrage, "Subset Coefficient Reduction Cuts for O/I Mixed-integer Programming," *Operations Research*, 33, No. 3, pp. 505-526 (1985).
30. Martin, R.K., D.J. Sweeney, and M.E. Doherty, "The Reduced Cost Branch-and-Bound Algorithm for Mixed Integer Programming," *Computers and Operations Research*, 12, No. 2, pp. 139-149 (1985).
31. Ohtake, Y. and N. Nishida, "A Branch-and-Bound Algorithm for 0-1 Parametric Mixed Integer Programming," *Operations Research Letters*, 4, No. 1, pp. 41-45 (1985).
32. Roberts, S.A., "Application of Heuristic Techniques to the Cutting-Stock Program for Worktops," *Journal of the Operational Research Society*, 35, No. 5, pp. 369-377 (1984).
33. Ruthedge, R.W., "A Simplex Method for Zero-One Mixed Integer Linear Programs," *Journal of Mathematical Analysis and Applications*, 18, No. 3, pp. 377-390 (1967).

34. Steudel, H.J., "Generating Pallet Loading Patterns with Considerations of Item Stocking on End and Side Surfaces," *Journal of Manufacturing Systems*, 3, No. 2, pp. 135-143 (1984).
35. Tsai, D.M., *The Use of a Robotic Manipulator to Unitize Pallet Loads in a Warehouse Distribution System*, M.S. Thesis, Department of Industrial Engineering, Iowa State University, Ames, Iowa (1984).
36. Tsai, D.M., *A Three Dimensional Robotic Palletizing Algorithm for Mixed Box Sizes*, Ph.D. Dissertation, Department of Industrial Engineering, Iowa State University, Ames, Iowa (August 1987).
37. Tsai, D.M., E.M. Malstrom, and H.D. Meeks, "Robotic Unitization of Pallet Loads," *Proceedings of the Fall Annual Conference*, Institute of Industrial Engineers, Chicago IL (December 1985).
38. Tsai, D.M., E.M. Malstrom, and H.D. Meeks, "Modeling and Analysis of Robotic Palletizing Cell," *IIE Transactions*, Vol. 20, No. 4 (December 1988).
39. Tsai, R.D., E.M. Malstrom, and W. Kuo, "Three Dimensional Palletization of Mixed Box Sizes," Working Paper #89-057, Department of Industrial Engineering, University of Arkansas, Under review by *IIE Transactions*.
40. Wang, P.Y., "Two Algorithms for Constrained Two-Dimensional Cutting Stock Problems," *Operations Research*, 31, No. 3, pp. 573-586 (1983).

SECTION THREE
STORAGE SYSTEMS AND LOGISTICS

The use of the distribution function as a competitive tool and the accelerating geo-political changes require an ever increasing attention to the design and operation of global industrial logistics systems. The scope of industrial logistics systems goes all the way from the supplier, through the manufacturing and distribution organizations, to the final customer. At the same time, strategic decisions have to be based on the correct understanding of the operational principles and constraints.

The significant fraction of both the Gross National Product and individual product cost attributable to industrial logistics, have stimulated active research in these areas. In this section, we see results which range from the modeling and optimization of a single storage system, through the framework for a complete warehousing system for small parts order picking and the optimal selection of transportation between ware-houses and customer, to the computer aided design of strategic industrial logistics systems.

Warehousing is one of the most visible areas of industrial logistics. Bozer and White and Elsayed have addressed different aspects of models to determine the throughput of End-of-Aisle storage systems such as Automated Storage/Retrieval Systems (AS/RS). Bozer and White focus on the interaction between the number of human pickers and equipment configuration. Elsayed computes the optimal retrieval sequence when the retrieval request have a due date. Medeiros and Emamizadeh show that the throughput also can be improved based on the location of goods in the storage system.

These and other results for single equipment systems are then combined and integrated by Sharp, Choe, and Yoon into a framework which allows the design of complete small parts order picking systems. Hall and Racer then address the problem on a larger scale by optimizing the transportation function between the warehouse and the customer. Finally, Goetschalckx investigates the problem on a strategic level. Here the question is if and where the goods should be warehoused and which transportation channel is to be used.

Marc Goetschalckx

A GENERALIZED DESIGN AND PERFORMANCE ANALYSIS MODEL FOR END-OF-AISLE ORDER PICKING SYSTEMS

Yavuz A. Bozer
The University of Michigan
John A. White
Georgia Institute of Technology

ABSTRACT

In earlier work we presented a design algorithm to determine the near-minimum number of pickers required in an end-of-aisle order picking operation (which is assumed to be based on a miniload automated storage/retrieval system). In this chapter we present an analytical model to determine the approximate values of the expected picker utilization and storage/retrieval machine utilization for general system configurations where two or more pick positions per aisle are allowed and/or two or more aisles are assigned to each picker. Using the estimated picker utilization, we extend the design algorithm presented in the previous work to general system configurations. We also investigate the possibility of improving the picker utilization by sequencing container retrievals within each order.

1 INTRODUCTION

As remarked in [6], order picking is a fundamental component of the retrieval function performed in warehousing. The main purpose of an order picking operation is to fill customer orders by selecting the appropriate amount of material from a predesignated storage medium known as the picking or forward area. Orders typically consist of a list of stock keeping units (SKUs) or "line items" which specify the type and amount of material to be picked.

Although order picking represents only a subset of the material handling operations performed in a warehouse, it is "one of the most costly and time consuming functions of warehousing. In many warehouses, the difference between profit and loss depends on how well the order picking operation is run" [15]. Furthermore, to some extent, order picking takes place in manufacturing systems as well since a specific set of parts and tools required to process a job (or set of jobs) must be retrieved from storage.

Two fundamental approaches to order picking are based on how the part containers are presented to the picker. If the picker travels to the container, then the system is generally defined as a "picker-to-part" system. In-the-aisle order picking (based on person-on-board automated storage/retrieval machines or order picking trucks) and walk-and-pick systems are well-known examples of picker-to-part systems.

On the other hand, if the container is brought to the picker, the system is generally defined as a "part-to-picker" system. End-of-aisle order picking (based on unit load or miniload automated storage/retrieval machines) and some carousel-based systems are well-known examples of part-to-picker systems. (The reader may refer to Tompkins and White [Ch.6, 14] to review a wide array of material handling equipment including the ones named above.) Note that, in a part-to-picker system, once the picker picks the required parts from a container, it must be placed back in storage unless it is empty.

The following study is concerned with the design and performance (i.e., throughput capacity) analysis of end-of-aisle order picking systems. Although we use a miniload automated storage/retrieval system (AS/RS) as an example throughout the chapter, the results remain valid for other applications as long as a storage/retrieval (S/R) machine brings the containers to the end of the aisle.

2 DEFINITIONS AND ASSUMPTIONS[1]

The part containers are delivered automatically (by the S/R machine) to the picker who performs the pick at one of the pick positions located at the end of the aisle. In general, the pick positions represent the "input/output" (I/O) point. As illustrated in Fig. 1, each aisle contains a storage rack located on either side of the aisle and a S/R machine which stores and retrieves the containers.

The S/R machine performs *dual command* cycles, i.e. immediately after storing a container, it travels directly to a retrieval location and retrieves the next container. That is, between I/O departure and arrival *both* a storage and a retrieval are performed by the S/R machine. Furthermore, the movement of the S/R machine is characterized by the l_∞-norm (or the Chebyshev metric) since it travels horizontally and vertically simultaneously. That is, the travel time between two points is equal to the maximum of the horizontal travel time and the vertical travel time. The system depicted in Fig. 1 is a 3-aisle miniload AS/RS with *two pick positions* per aisle.

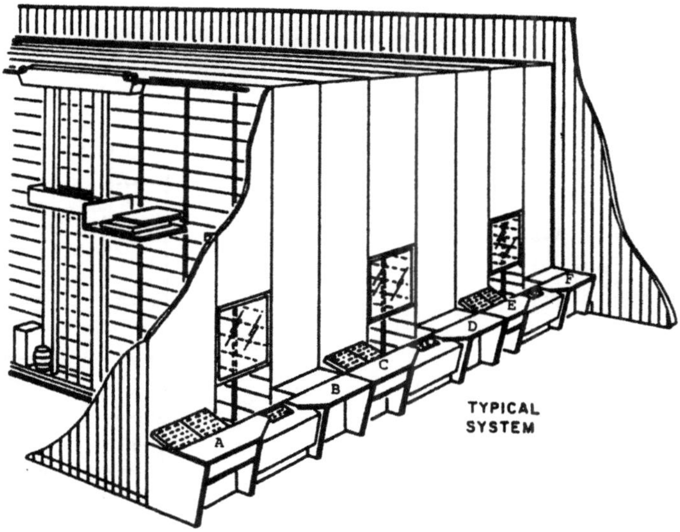

Fig. 1 Three-aisle miniload AS/R system (reprinted from (16))

A miniload AS/R system is typically used for small parts picking. The throughput of the system is expressed in number of picks/hour, where one "pick" is performed from each container. The definition of a "pick" allows multiple pieces of the same part to be picked from a container, as well as one or more pieces of multiple parts.

The miniload system with two pick positions per aisle was analyzed in [6] assuming that exactly one aisle is assigned to each picker. (Refer to section 7 for a brief description of system operation.) Although the two-pick-position system is used in industry quite often, two alternative system configurations may have to be considered in order to obtain the most efficient or the most economical solution. The first alternative is to increase the number of pick positions per aisle. The second alternative is to assign more than one aisle to each picker.

The number of pick positions per aisle may be increased by providing a U-shaped (or "horseshoe") conveyor at the end of the aisle as shown in Figs. 2(a) and 2(b)[2]. With three or more pick positions, the I/O point resembles a U-shaped conveyor (or transfer device) with the two "open ends" interfaced with the S/R machine. The containers delivered by the S/R machine are deposited in position 1. Those that are returned to the rack are picked up at position 5 (see Fig. 2(a)). The picker stands at position 3 where the picking operation is performed. Containers are automatically indexed through the pick positions and automatic 90-degree transfer devices are provided between positions 2 and 3 as well as positions 3 and 4.

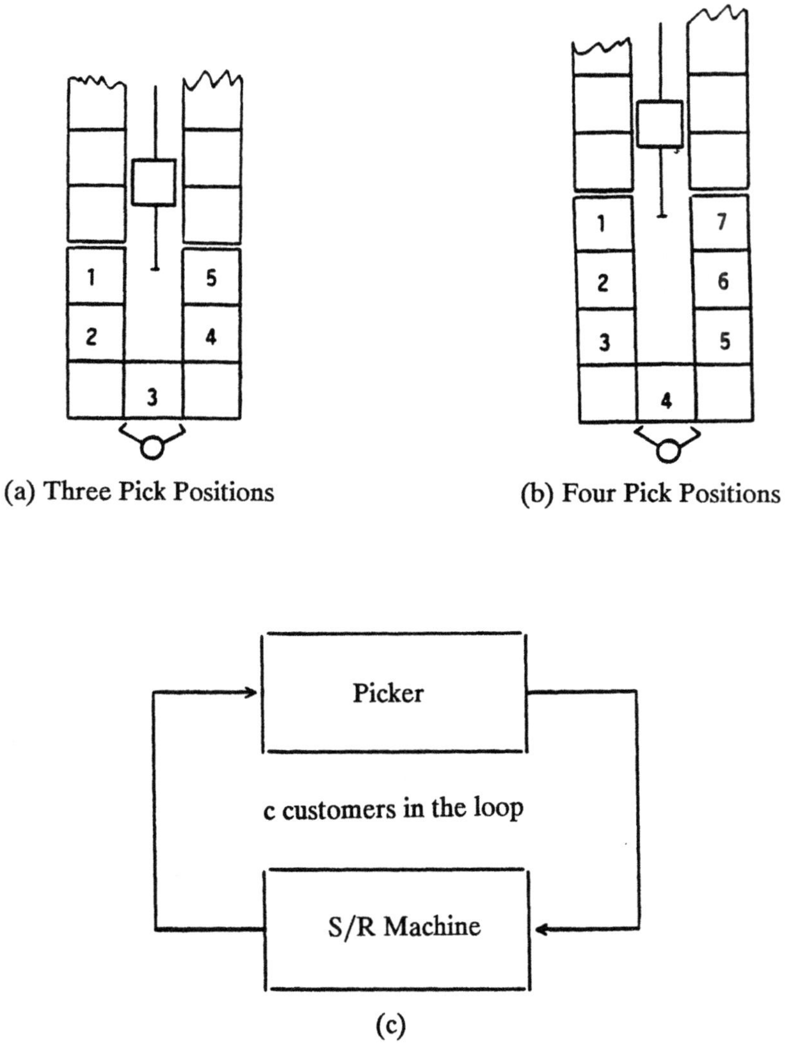

(a) Three Pick Positions

(b) Four Pick Positions

(c)

Fig. 2 End-of-aisle order picking systems with multiple pick positions per aisle

Suppose when the system starts operating, the S/R machine is idle at the I/O point with positions 1, 2 and 3 full and positions 4 and 5 empty. Since the S/R machine operates on a dual command basis, it removes a container from the I/O point (position 5) before delivering a new one (at position 1). Consequently, the maximum number of containers at the I/O point will never exceed three. With positions 1, 2 and 3 full, the S/R machine will remain idle at the I/O point while the picker is busy.

On the other hand, if positions 4 and 5 are full and the S/R machine is busy retrieving the next container, then the picker will idle. Note that positions 3, 4 and 5 cannot all be full while the S/R machine is busy because it implies that four containers were present at the I/O point immediately before the one in position 5 was picked up by the S/R machine.

Assuming that the container transfer time between each position is negligible, the system shown in Fig. 2(a) can be modeled as a closed queueing cyclic server system (see Fig. 2(c)) where the first server is the picker, the second server is the S/R machine, and the fixed number of "customers" (or pick positions) is equal to three. In reference to Fig. 2(a), positions 1 & 2 and 4 & 5 represent the "queue space" for the picker and the S/R machine, respectively. A system with four pick positions, on the other hand, is shown in Fig. 2(b) where picking is performed at position 4.

The second alternative in miniload design is to assign multiple aisles to one picker or vice versa. Unless the mean pick time is considerably long (compared to the mean S/R machine cycle time), assigning multiple pickers to a single aisle is quite uncommon in end-of-aisle order picking. Furthermore, the number and location of the I/O points and the order in which they are served would require further analysis in order to avoid an excessive number of empty S/R machine trips.

Systems where multiple aisles are assigned to each picker are more frequently encountered in practice. Suppose three aisles are assigned to a picker. Further suppose that the pick positions are labeled (A,B), (C,D) and (E,F) as shown in Fig. 1. It is assumed that the picker always follows the sequence: A, C, E, B, D, F, A, C, E, B, ... and so on. Although strictly following such a sequence may somewhat increase picker idle time, it allows each S/R machine the maximum time between two consecutive picks from the same aisle. It will also avoid excessive picker walk times from one aisle to another.

With, say, N aisles/picker and two pick positions/aisle, a single picker system can again be modeled as a closed queueing network with N closed loops where the number of "customers" in each loop remains equal to two and the number of servers is equal to $N+1$ as shown in Fig. 3. In those cases where more than two pick positions are provided in each aisle, the number of customers in each loop in Fig. 3 would be increased appropriately. Also, the picker would still follow a fixed sequence and would visit each aisle exactly once before returning to the first aisle for his/her next pick.

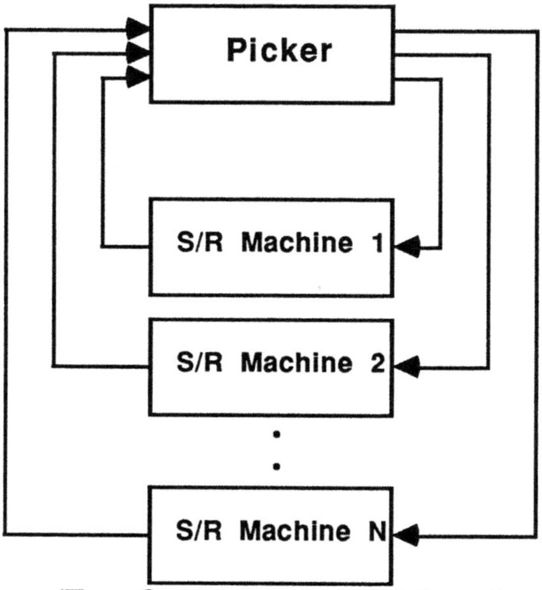

Fig. 3 End-of-aisle order picking systems with two pick positions per aisle and multiple aisles per picker

The remaining assumptions are concerned with the operation of the S/R machine. Storage/retrieval locations are assumed to be continuously and uniformly distributed over the rack (see [5] and [11]). Also, initially in the study, containers are assumed to be retrieved on a first-come, first-served (FCFS) basis, i.e. containers are not sequenced in any manner. Later in the study we relax this assumption and investigate the impact on system throughput of several container sequencing heuristics developed for this purpose. For the case where the containers are retrieved on a FCFS basis, the results obtained in [5] and [11] can be used to compute the travel time for the S/R machine. Following the approach used in [5], the rack is "normalized" as follows. Let:

v_h = horizontal velocity of the S/R machine (in fpm),
v_v = vertical velocity of the S/R machine (in fpm),
L = rack length (in feet),
H = rack height (in feet),
t_h = time required to travel horizontally a distance of L (in minutes), and
t_v = time required to travel vertically a distance of H (in minutes).

The scaling factor, T, and the shape factor, b, are defined as follows:

$$T = max\,[L/v_h\,,\, H/v_v] = max\,[t_h\,,\, t_v], \tag{1}$$

and

$$b = min\,[t_h/T\,,\, t_v/T], \tag{2}$$

where $0 \leq b \leq 1$. Hence, the normalized rack is b time units long in one direction and one time unit long in the other. If $b = 1$, then the rack is said to be "square-in-time."

The principal assumptions for the study can be summarized as follows:

1. The I/O point is located at a corner of a normalized rectangular storage rack defined by the scaling factor, T, and the shape factor, b.
2. The S/R machine follows the Chebyshev metric and travels at constant speed, i.e. acceleration and deceleration are not considered.
3. The S/R machine performs only dual command cycles, i.e. each storage operation is coupled with a retrieval operation; each S/R trip originates and terminates at the I/O point.
4. Every trip performed by the S/R machine involves two containers handled twice each. That is, in addition to travel time, each trip involves two pick-up operations and two deposit operations. Assuming the container handling times for the two are equal, each trip contains four handling operations.
5. The two pick positions (labeled A and B in Fig. 1) are identical in terms of container handling and S/R machine travel time. (Systems with more than two pick positions have only one position where the actual picking is performed.)
6. In retrieving a container, the S/R machine is equally likely to visit any point in the (continuous) rack. That is, randomized storage is used instead of turnover based storage.
7. The containers for each order are retrieved consecutively in a random sequence. In systems with two pick positions, the last two containers (one for each position) of a given order are interleaved with the first two containers of the next order. That is, when the S/R machine stores the last container from, say, position A, it retrieves the first container of the next order for position A. Later in the study we relax the random sequence assumption and study container sequencing heuristics. Note that no container from the next order may be retrieved before all the containers of the current order have been retrieved.
8. The throughput of the system is expressed in expected number of picks per hour. The pick time is independent of the S/R machine cycle time. (The pick time distribution is assumed to be given.)
9. If more than one aisle is assigned to the picker, he/she is assumed to visit each aisle exactly once in a predetermined sequence. All the pickers in the system follow the same sequence and they are all assigned the same

number of aisles. No aisle is served by two or more pickers; that is, each picker has his/her own set of dedicated aisles. Walk times between aisles are assumed to be negligible. The same pick time distribution is used at each aisle.

10. If more than one aisle is required, then each aisle is identical in size and average activity.

3 LITERATURE SEARCH

Not including numerous descriptive articles that mostly appear in trade journals, very few technical studies have been reported on end-of-aisle order picking. Goetschalckx [9] studied the throughput performance of end-of-aisle systems and container retrieval sequencing. However, the analysis was relatively brief and the scope was limited only to the S/R machine; that is, the study did not include the picker at the end of the aisle.

Another study concerned with end-of-aisle order picking is reported by Bozer and White [6]. The authors present a design algorithm to analytically determine the near-minimum number of pickers required given that the storage space and throughput requirements are specified by the user. During the execution of the algorithm, for an arbitrary rack, they determine the expected picker utilization by assuming that only one aisle is assigned to each picker and that two pick positions are provided in each aisle. (Using the special L^2R^2 configuration, they also study the impact of increasing the number of pick positions from two to four per aisle; however, their results with four pick positions are only empirical results based on simulation and they are not part of the design algorithm.)

The approach presented by Bozer and White is an exact one; however their results are approximate since they approximate the S/R machine cycle time with a uniform distribution. For the same type of system (i.e., one aisle/picker and two pick positions/aisle), Foley and Frazelle [7] determine the expected picker utilization by deriving the exact distribution of the S/R machine cycle. However, their results apply only if the rack is square-in-time (that is, $b=1$).

A paper concerned specifically with miniload AS/R systems is presented by Bengtson and Gomez [1]. The paper does not contain specific quantitative results; however, the authors discuss several issues that complicate the design of a single-aisle miniload order picking system. They also describe a simulation model in reasonable detail. Although simulation is a powerful tool for analysis purposes, the model developed by the authors is perhaps most suitable in conducting a detailed analysis of system performance assuming that the user has already selected a particular miniload configuration.

The results presented in this paper represent an extension of those presented by Bozer and White [6]. Since the analytical results presented in [6] are valid only for systems with two pick positions per aisle and one aisle per picker, generalizing them to other possible configurations considerably extends the scope of the resulting design algorithm. Coupled with an appropriate cost model (which is assumed to be supplied by the user), the design algorithm presented here can be used to determine the near-optimum configuration for a wide variety of end-of-aisle order picking systems. The design algorithm is not, however, viewed as a substitute for simulation; rather, it provides benchmark solutions which can be further evaluated by simulation.

In addition to the above generalized design algorithm, we present two types of heuristics for sequencing container retrievals in systems with two pick positions. Han et al. [12] consider a similar sequencing problem for unit load AS/R systems. However, since the authors focus on storage and retrieval rather than order picking, they are concerned with reducing the cycle time of the S/R machine as opposed to improving the utilization of the picker. Furthermore, a container may be stored in any open location in the rack; that is, containers do not have dedicated locations.

4 THE MATHEMATICAL MODEL

The mathematical model and the corresponding algorithm are generalized versions of those developed in [6]. In addition to v_h and v_v, values for the following parameters are assumed to be known:

- S = total storage space required, including clearances and rack members (in total square feet of rack facing required);
- R = total throughput required (in number of picks per hour);
- p = average pick time (in minutes per container);
- W = the total container handling time (in minutes) for a dual command cycle.

As motivated in [6], estimating the cost of a miniload AS/RS is not straightforward since it "is highly dependent on market conditions at the time of quotation by suppliers ... (and) it is not unusual to find extreme variations in prices quoted on such systems. ... Although we are unable to estimate accurately the cost of the system, it appears that the S/R machines and the pickers are the major contributors to the overall cost of the system". Consequently, the design algorithm will attempt to minimize the number of pickers given that a certain number of aisles will be assigned to each picker.

Let M denote the number of pickers and N denote the number of aisles to be assigned to each picker. (Although N itself is a decision variable, it will be treated as a parameter since the algorithm can be executed over several N values.) The model can be presented as follows:

Minimize	M	(3)
Subject to:	*System throughput* $\geq R$	(4)
	System storage capacity $= S$	(5)
	M positive integer	(6)

The throughput constraint is expressed as an inequality since the number of pickers is an integer variable. However, since the rack is assumed to be continuous, the storage space constraint is expressed as an equality.

If we let $E(PU)$ denote the expected picker utilization, the above model can be mathematically expressed as follows:

Minimize	M	(7)
Subject to:	$E(PU)M \geq Rp/60$	(8)
	$(2LH)MN = S$	(9)
	M positive integer	(10)

where the right-hand side of constraint (8) is the required pick rate and the left-hand side is the expected throughput *capacity* of the system. The left-hand side of constraint (9) is based on having (LH) square feet of rack space on each side of an aisle and a total of MN aisles.

5 EXPECTED PICKER AND S/R MACHINE UTILIZATION[3]

In [6], a single aisle with two pick positions is modeled as a two-server closed queueing network with the number of pick positions representing the fixed number of "customers" in the system and the picker and the S/R machine representing the two "servers." The service time for the S/R machine consists of the time required to store the old container and retrieve a new one.

Since there are only two pick positions, the authors model the system as a renewal process where an event occurs when the picker and the S/R machine begin service. (Note that neither server can begin serving a new customer if the other one is still busy.) Hence, the time between two events, say, the cycle time, is given by the maximum of two independent random variables: the pick time and the S/R machine cycle time. Approximating the S/R machine cycle time by a uniform distribution, the authors derive the distribution of the above cycle time and its expected value. Once the expected cycle time is determined, obtaining the expected picker utilization, $E(PU)$, and the expected S/R machine utilization, say, $E(SRU)$, is straightforward.

Unfortunately, the above approach does not extend to three or more pick positions. However, one may consider using the diffusion approximation developed by Gelenbe [8] for two-server cyclic queues. Recall that the mean pick time is given by p. Then the mean pick rate is given by $\mu = 1/p$ and the pick time variance will be denoted by V_p. Furthermore, given the rack dimensions, the S/R machine travel velocities, and the container handling time, suppose the mean S/R machine cycle time, say, $E(SCT)$, has been determined. By definition, $E(SCT) = E(DC)+W$ where $E(DC)$ is the mean *travel time* and W is the total container *handling time* (which consists of two pick-ups and two deposits). Then, the mean S/R machine service rate, λ, is equal to $1/E(SCT)$. Lastly, let V_m denote the variance of the S/R machine cycle time.

Using the approach presented in [8], $E(PU)$ and $E(SRU)$ may be obtained as follows. Let $\alpha = \mu^3 V_p + \lambda^3 V_m$ and $\gamma = [2\lambda\text{-}\mu)]/\alpha$. Then

$$\Omega = \frac{\rho}{1-\rho^2 e^{\gamma(c-1)}} \tag{11}$$

$$E(PU) = 1-\Omega\,[\frac{1-\rho}{\rho}], \tag{12}$$

and

$$E(SRU) = 1 - \Omega\,(1-\rho)e^{\gamma(c-1)}, \tag{13}$$

where $\rho = \lambda/\mu$ and c denotes the number of "customers" in the loop (i.e., the number of pick positions).

The mean and variance of the pick time are assumed to be supplied by the user. The mean and variance of the S/R machine cycle time, on the other hand, can be obtained from the following expressions. The expected travel time and cycle time for dual command cycles are given in [5] as follows:

$$E(DC) = T\,[\frac{4}{3} + \frac{b^2}{2} - \frac{b^3}{30}], \tag{14a}$$

$$E(SCT) = E(DC) + W, \tag{14b}$$

where b and T are as defined in (1) and (2). An approximate expression for the variance of dual command cycles is given in [2] as follows:

$$V_m \approx [(0.3588\text{-}0.1321b)E(DC)]^2. \tag{15}$$

Assuming that only one aisle is assigned to a picker, the performance of the above approximation was tested via simulation for systems with three pick positions. The results are shown in Table 1 where p denotes the mean pick time (in minutes), b denotes the shape factor, and the handling time, W, is assumed to be equal to 0.30 minutes. (It is implicitly assumed that $T=1$ for *both rack shapes tested.*) Given a 95% confidence interval shown for $E(PU)$ and $E(SRU)$ in Table 1, one may observe that the approximation performs reasonably well for both exponential and deterministic pick times tested over

Table 1 Expected picker and S/R machine utilization (one aisle per picker, three pick positions per aisle, W=0.30)

	1	2	3	4
p	95% CI on E(PU)	E(PU)	95% CI on E (SRU)	E(SRU)
Deterministic Pick Time, Shape Factor = 1.00				
0.50	23.66 - 23.92	23.81	100.00 -100.00	100.00
1.00	47.47 - 47.88	47.62	100.00 -100.00	100.00
1.50	71.13 - 71.80	71.43	100.00 -100.00	100.00
1.75	82.82 - 83.72	83.33	99.99 - 99.99	100.00
2.00	94.37 - 95.23	95.22	99.49 - 99.75	99.98
2.50	100.00 -100.00	100.00	83.55 - 84.23	84.00
Deterministic Pick Time, Shape Factor = 0.25				
0.50	29.80 - 30.13	30.05	100.00 -100.00	100.00
1.00	59.79 - 60.66	60.09	99.96 -100.03	100.00
1.50	89.15 - 89.86	90.12	99.38 - 99.60	99.98
1.75	99.12 - 99.39	99.68	93.46 - 94.90	94.79
2.00	100.00 -100.00	100.00	82.96 - 83.92	83.20
2.50	100.00 -100.00	100.00	65.95 - 66.87	66.56
Exponential Pick Time, Shape Factor = 1.00				
0.50	23.25 - 24.12	23.60	99.96 -100.00	99.11
1.00	46.50 - 47.86	46.06	98.64 - 99.26	96.72
1.50	65.54 - 68.48	65.67	92.30 - 93.77	91.94
1.75	72.83 - 75.02	73.80	88.15 - 90.16	88.56
2.00	78.58 - 80.71	80.59	83.09 - 85.10	84.62
2.50	86.82 - 88.68	90.17	72.43 - 75.70	75.75
3.00	91.98 - 92.80	95.43	63.83 - 65.41	66.80
Exponential Pick Time, Shape Factor = 0.25				
0.50	29.38 - 30.36	29.64	99.84 - 99.98	98.63
1.00	57.27 - 58.79	56.71	96.10 - 97.09	94.37
1.50	76.19 - 78.89	77.54	84.67 - 86.68	86.03
1.75	82.82 - 84.68	84.82	77.46 - 80.40	80.66
2.00	86.84 - 88.42	90.10	72.33 - 74.56	74.97
2.50	92.19 - 93.53	96.09	61.02 - 63.77	63.96
3.00	95.58 - 96.21	98.54	52.43 - 54.27	54.66

Cols. 1 and 3: 95% confidence interval based on simulation (10 batches, 1000 picks/batch).
Cols. 2 and 4: Analytical estimates obtained from the diffusion approximation.

a wide range of $E(PU)$ and $E(SRU)$ values. (Note: although we do not need $E(SRU)$ in the mathematical model presented earlier, it is provided as additional information since it may affect the final decision.)

The results of a similar experiment with four pick positions are shown in Table 2. The approximation still appears to generate satisfactory results. (No statistical tests will be performed, however, since we already know that the results are approximate ones.) Note that, with deterministic pick times, the additional pick position does not appear to be justified. With exponential pick times, however, a slight improvement in $E(PU)$ and $E(SRU)$ is obtained when c is increased from three to four.

The trade-off between $E(PU)$ and $E(SRU)$ can be clearly observed in both Tables 1 and 2. That is, as one increases, the other must decrease. If the S/R machine is too slow relative to the picker, a very low picker utilization is obtained. (Note that, when the opposite is true, $E(SRU)$ is not as low.) In fact, both servers are reasonably well utilized only if the system is "balanced"; that is, if the mean S/R machine cycle time, $E(SCT)$, is approximately equal to the mean pick time. (Note that, for $b=1.00$ and $b=0.25$, $E(SCT)$ is equal to 2.10 and 1.66 minutes, respectively.)

The above results (combined with others not shown here) indicate that the diffusion approximation by Gelenbe [8] yields acceptable estimates on $E(PU)$ and $E(SRU)$ for design purposes. However, the above results hold only if one aisle is assigned to each picker, i.e., $N=1$. Analyzing the closed queueing system with $N \geq 2$ is not straightforward. However, as shown below, one may use the model developed for $N=1$, i.e., the 2-server cyclic model, to obtain approximate values for $E(PU)$ and $E(SRU)$ with $N \geq 2$.

Consider first the case where two aisles are assigned to each picker, i.e. $N=2$. From the viewpoint of, say, the first S/R machine, the picker makes a pick from the first aisle and he/she is "unavailable", on the average, for at least p minutes (while he/she performs a pick from the second aisle -- after potentially waiting on the second S/R machine to deliver the next container). Hence, considering a two server cyclic system where one of the servers is the picker and the other one is the *first* S/R machine (as shown earlier in Fig. 2(c)), the "effective" mean service time, say, p', for the picker is equal to $2p + r$ where r represents the expected time the picker waits on the second S/R machine every time he/she visits that aisle.

In other words, from the viewpoint of S/R machine 1, the picker cannot serve the next "customer" in the loop until, on the average, p' minutes has elapsed. Note that, in the actual system, at the end of p minutes -- as opposed to p' minutes -- the "customer" will be ready to be served by the first S/R machine. However, our empirical results indicate that shifting the time at which the "customer" is ready to be served by the first S/R machine by $p' - p$ (i.e., $p+r$) minutes does not affect the approximate value of $E(PU)$ or $E(SRU)$.

Given that the picker has to wait on the second S/R machine, his/her wait time is equal to the residual S/R machine cycle time, say, t'. The

Table 2 Expected picker and S/R machine utilization (one aisle per picker, four pick positions per aisle, W=0.30)

	1	2	3	4
p	95% CI on E(PU)	E(PU)	95% CI on E (SRU)	E(SRU)
Deterministic Pick Time, Shape Factor = 1.00				
0.50	23.71 - 23.93	23.81	100.00 -100.00	100.00
1.00	47.39 - 47.84	47.62	100.00 -100.00	100.00
1.50	70.99 - 71.77	71.43	100.00 -100.00	100.00
1.75	83.10 - 83.79	83.33	100.00 -100.00	100.00
2.00	94.81 - 95.67	95.24	99.93 - 99.97	100.00
2.50	100.00 -100.00	100.00	83.61 - 84.40	84.00
Deterministic Pick Time, Shape Factor = 0.25				
0.50	29.85 - 30.13	30.05	100.00 -100.00	100.00
1.00	59.71 - 60.75	60.09	100.00 -100.00	100.00
1.50	89.37 - 90.37	90.14	99.96 - 99.98	100.00
1.75	99.81 - 99.93	99.92	94.12 - 95.40	95.02
2.00	100.00 -100.00	100.00	83.00 - 83.75	83.20
2.50	100.00 -100.00	100.00	65.88 - 66.97	66.56
Exponential Pick Time, Shape Factor = 1.00				
0.50	23.42 - 24.24	23.76	100.00 -100.00	99.80
1.00	46.76 - 49.30	47.07	99.65 - 99.84	98.85
1.50	67.62 - 70.01	68.39	96.45 - 97.60	95.75
1.75	76.33 - 79.50	77.37	91.98 - 94.50	92.85
2.00	82.69 - 85.39	84.72	88.32 - 90.13	88.95
2.50	92.25 - 93.61	94.10	75.26 - 78.86	79.05
3.00	96.16 - 96.97	98.09	66.04 - 67.93	68.66
Exponential Pick Time, Shape Factor = 0.25				
0.50	29.61 - 30.55	29.94	99.97 -100.01	99.65
1.00	57.94 - 61.00	58.59	98.37 - 99.01	97.49
1.50	80.23 - 82.44	81.53	90.08 - 91.74	90.45
1.75	87.30 - 89.71	89.14	82.21 - 84.65	84.76
2.00	91.21 - 93.31	94.08	76.52 - 79.07	78.27
2.50	96.59 - 97.48	98.49	63.01 - 65.92	65.56
3.00	98.38 - 98.85	99.65	53.51 - 55.28	55.28

Cols. 1 and 3: 95% confidence interval based on simulation (10 batches, 1000 picks/batch).
Cols. 2 and 4: Analytical estimates obtained from the diffusion approximation.

expected value of t' may be approximated (see Tijms [p. 8, 13], among others) by the expression $E(t') = E(SCT^2)/[2E(SCT)]$ where $E(SCT^2)$ is the second moment of the S/R machine cycle time -- which can be easily obtained from Equations (14) and (15).

Hence, in a 2-server cyclic queue (where the picker and the first S/R machine represent the two servers), the effective mean pick time is assumed to be equal to p' minutes. The variance of the effective pick time, say, V_p', on the other hand, is difficult to determine since it depends on the (original) pick time distribution as well as the distribution of the residual S/R machine cycle time. However, since the variance of the complete S/R machine cycle time is relatively small -- its largest value is equal to approximately 0.167 -- we will assume that V_p' depends solely on the variance of the (original) pick time. For example, if the pick time is exponentially distributed, then the effective pick time will follow the Gamma distribution (since it is the sum of two exponential random variables). Hence, for exponentially distributed pick times, V_p' is equal to $2/\mu^2$ or $2p^2$ since $\mu = 1/p$. For deterministic pick times, we will assume that V_p' is equal to zero. For other pick time distributions, the user must determine the appropriate value for V_p'.

Thus, to estimate the values of $E(PU)$ and $E(SRU)$ for non-zero r values, we propose the following iterative scheme:

ALGORITHM I:

0. Given the values of b and T, compute the values of $E(SCT)$ and V_m from Equations (14a), (14b) and (15). Based on the pick time distribution, determine the appropriate value for V_p'. Set $r=0$.
1. Set $p' = 2p + r$ and determine the values of $E(PU)$ and $E(SRU)$ from Equations (12) and (13) by using p' in place of p, V_p' in place of V_p and the appropriate number of pick positions, c.
2. Compare the value of $E(PU)$ with that obtained in the previous iteration. If the difference is sufficiently small go to step 5. Otherwise, go to step 3.
3. Assume the probability that the picker becomes idle when he/she switches to the second aisle is given by $1\text{-}E(PU)$.
4. Compute the new value of r from the expression $[1\text{-}E(PU)]E(t')$ where $E(t')$ is approximated by the expression $E(SCT^2)/[2E(SCT)]$. Go to step 1.
5. $E(PU) \leftarrow E(PU)[2p/p']$. Stop.

Note that, when the algorithm terminates in step 5, the utilization of the picker has to be adjusted since he/she is actually busy, on the average, only $2p$ minutes during his/her assumed service time of $p' = 2p + r$ minutes.

In order to test the above approximation we simulated various configurations. The results for a system with $c=2$ (i.e., two pick positions per aisle) and $N=2$ (i.e., two aisles per picker) are presented in Table 3 where a 95% confidence interval is shown for $E(PU)$ and $E(SRU)$ along with their

Table 3 Expected picker and S/R machine utilization (two aisles per picker, two pick positions per aisle, W=0.30)

	1	2	3	4
p	95% CI on E(PU)	E(PU)	95% CI on E (SRU)	E(SRU)
Deterministic Pick Time, Shape Factor = 1.00				
0.25	23.41 - 23.65	23.81	98.11 - 99.33	100.00
0.50	46.66 - 47.10	47.62	97.49 - 98.88	100.00
0.75	69.74 - 70.41	71.43	97.71 - 98.64	100.00
0.90	83.19 - 84.04	85.68	96.79 - 97.58	99.96
1.00	91.96 - 92.59	94.55	95.60 - 96.91	99.28
1.25	99.97 -100.01	100.00	82.95 - 84.04	84.00
Deterministic Pick Time, Shape Factor = 0.25				
0.25	29.13 - 29.64	30.05	97.19 - 98.31	100.00
0.50	57.65 - 58.18	60.09	96.21 - 97.74	100.00
0.75	85.28 - 86.43	89.16	94.18 - 95.98	98.91
0.90	97.72 - 98.16	98.71	89.82 - 91.61	91.25
1.00	99.95 -100.01	99.82	82.44 - 84.04	83.05
Exponential Pick Time, Shape Factor = 1.00				
0.25	22.68 - 23.87	23.81	97.92 - 98.84	99.99
0.50	45.62 - 47.74	46.59	96.05 - 97.27	97.85
0.75	65.16 - 67.35	65.26	90.88 - 92.56	91.36
0.90	72.88 - 75.29	74.11	86.24 - 88.65	86.46
1.00	78.33 - 80.68	79.07	81.23 - 84.67	83.02
1.25	87.47 - 88.59	88.41	73.01 - 75.33	74.27
1.50	91.91 - 93.28	94.09	64.35 - 65.97	65.86
Exponential Pick Time, Shape Factor = 0.25				
0.25	28.82 - 29.74	29.98	96.04 - 97.54	99.79
0.50	54.89 - 57.65	56.60	92.49 - 94.29	94.18
0.75	75.10 - 77.23	75.73	82.59 - 84.97	84.01
0.90	82.16 - 84.30	83.74	77.28 - 79.18	77.41
1.00	87.23 - 88.57	87.81	70.42 - 72.48	73.06
1.25	92.38 - 93.41	94.43	61.71 - 63.09	62.86
1.50	95.72 - 96.45	97.59	52.42 - 53.52	54.13

Cols. 1 and 3: 95% confidence interval based on simulation (10 batches, 1000 picks/batch).
Cols. 2 and 4: Analytical estimates obtained from the diffusion approximation and Algorithm I.

approximate values obtained from Algorithm I. As before, it is assumed that $W = 0.30$ and $T=1.0$ for both b values tested. With deterministic pick times, $E(PU)$ and $E(SRU)$ are slightly overestimated since the variance of the effective service time for the picker was assumed to be equal to zero. Overall, the approximation appears to perform quite well over a wide range of $E(PU)$ and $E(SRU)$ values.

By setting $p' = Np + r$ and setting V_p' equal to Np^2 (for exponential pick times) or zero (for deterministic pick times), the above approximation can also be extended to those cases where $N \geq 3$. However, estimating the value of r is a complicating factor. As before, we will assume that $r \approx [1\text{-}E(PU)]E(t')$ where $E(t') \approx E(SCT^2)/[2E(SCT)]$. That is, from the viewpoint of the first S/R machine, the picker will become idle with probability 1-$E(PU)$ while serving the other aisles. Obviously, the picker may become idle more than once during this time. However, given that the picker follows a fixed sequence in visiting the aisles, if the picker becomes idle at, say, the ith aisle ($i \geq 2$), it decreases the probability that he/she will become idle at the $i+1$st, $i+2$nd, ... aisles since the S/R machines in these aisles will have more time to retrieve the next container. With the same token, even if the picker becomes idle (for the second time during a cycle) at, say, the $i+1$st aisle, his/her idle time is likely to be quite small.

In order to evaluate the performance of the above approximation we made simulation runs with $N=3$, $c=2$, $W=0.30$ and $T=1$. The results are shown in Table 4 where 95% confidence intervals are given for $E(PU)$ and $E(SRU)$. The model performs reasonably well and yields acceptable estimates for $E(PU)$ and $E(SRU)$. To further test the model we made additional simulation runs with arbitrary b and p values. The results are presented in Table 5 where odd rows have deterministic pick times while even rows have exponential pick times. Based on the range of values assumed for the experiment ($1 \leq N \leq 4$ and $2 \leq c \leq 5$), the model yields reasonably accurate estimates for $E(PU)$ and $E(SRU)$ for evaluation and design purposes.

In the above experiments, we assumed that both container locations -- one for storage and one for retrieval -- on each dual command cycle are selected randomly. In most real-life systems, however, containers have dedicated locations in the rack. This would create dependence between alternate S/R machine cycles. However, we modified the simulation model to "remember" the location of each container when it was stored back in the rack. Our empirical results indicate that the above dependence does not noticeably affect the values of $E(PU)$ and $E(SRU)$.

6 THE DESIGN ALGORITHM

Using the approximate $E(PU)$ and $E(SRU)$ values derived in the previous section, we will present a generalized version of the design algorithm developed by Bozer and White [6] for $N=1$ and $c=2$. Except for determining

Table 4 Expected picker and S/R machine utilization (three aisles per picker, two pick positions per aisle, W=0.30)

	1	2	3	4
p	95% CI on E(PU)	E(PU)	95% CI on E (SRU)	E(SRU)
Deterministic Pick Time, Shape Factor = 1.00				
0.15	20.86 - 21.11	21.43	97.15 - 99.13	100.00
0.25	34.81 - 35.09	35.71	97.58 - 98.73	100.00
0.50	69.27 - 69.97	71.43	96.62 - 97.55	100.00
0.75	99.39 - 99.67	99.73	91.54 - 93.22	93.08
0.90	100.00 -100.00	100.00	77.16 - 78.23	77.78
1.00	100.00 -100.00	100.00	69.71 - 70.83	70.00
Deterministic Pick Time, Shape Factor = 0.25				
0.15	26.01 - 26.48	27.04	96.74 - 97.97	100.00
0.25	43.02 - 43.63	45.07	95.77 - 97.82	100.00
0.50	84.50 - 85.64	89.16	94.06 - 96.03	98.91
0.75	100.00 -100.00	99.98	73.16 - 75.37	73.94
0.90	100.00 -100.00	100.00	61.20 - 62.59	61.63
1.00	100.00 -100.00	100.00	55.25 - 56.43	55.47
Exponential Pick Time, Shape Factor = 1.00				
0.15	20.48 - 21.76	21.43	96.39 - 98.22	100.00
0.25	34.43 - 35.92	35.68	96.62 - 98.36	99.91
0.50	66.36 - 69.23	67.73	93.43 - 94.65	94.82
0.75	87.90 - 89.32	88.22	81.12 - 83.38	82.34
0.90	92.78 - 94.58	94.60	72.30 - 75.37	73.57
1.00	95.83 - 96.87	96.98	65.46 - 68.51	67.89
1.25	98.03 - 98.90	99.38	53.54 - 56.49	55.65
Exponential Pick Time, Shape Factor = 0.25				
0.15	25.72 - 26.71	27.04	95.43 - 96.95	99.99
0.25	42.38 - 44.99	44.72	94.80 - 96.00	99.23
0.50	78.12 - 80.99	79.15	87.09 - 89.14	87.81
0.75	93.65 - 94.93	94.88	67.63 - 70.54	70.17
0.90	96.89 - 97.62	98.09	59.96 - 61.36	60.46
1.00	98.33 - 98.74	99.03	52.29 - 55.65	54.93
1.25	99.41 - 99.67	99.81	43.19 - 44.81	44.29

Cols. 1 and 3: 95% confidence interval based on simulation (10 batches, 1000 picks/batch).
Cols. 2 and 4: Analytical estimates obtained from the diffusion approximation and Algorithm I

Table 5 Expected picker and S/R machine utilization for arbitrary N, c, W, b, and p values

1	2	3	4	5	6	7	8	9
N	c	W	b	P	95% CI E(PU)	E(PU)	95% CI E(SRU)	E(SRU
1	3	0.25	0.50	1.48	86.28 - 86.92	86.85	99.83 - 99.93	100.00
1	5	0.20	0.62	1.80	90.74 - 92.60	91.67	84.45 - 87.37	87.47
2	2	0.35	0.80	0.82	80.02 - 80.79	82.55	96.97 - 98.12	99.97
2	3	0.24	0.44	0.44	50.82 - 53.56	52.49	96.62 - 98.23	99.44
2	4	0.28	0.73	1.15	100.00 -100.00	100.00	81.13 - 82.02	81.17
3	2	0.20	0.90	0.52	74.48 - 76.51	74.63	90.79 - 92.04	91.57
3	3	0.25	0.29	0.91	100.00 -100.00	100.00	59.29 - 60.28	59.51
3	4	0.24	0.35	0.58	95.23 - 96.95	95.87	87.71 - 90.61	89.98
4	2	0.30	0.85	0.22	42.86 - 43.60	44.58	95.45 - 97.89	100.00
4	4	0.22	0.57	0.59	99.93 -100.00	99.97	71.25 - 73.99	72.42
4	4	0.20	0.68	0.10	22.23 - 22.49	22.80	97.65 - 99.24	100.00
4	2	0.27	0.40	1.19	99.98 - 99.98	99.99	34.48 - 36.47	35.32

Col.1: Number of aisles per picker.
Col. 2: Number of pick positions per aisle.
Col. 3: Total S/R machine handling time per dual command cycle.
Col. 4: The shape factor.
Col. 5: Mean pick time (even rows: exponential; odd rows: deterministic.
Cols. 6 & 8: 95% confidence interval based on simulation (10 batches, 1000 picks/batch).
Cols. 7 & 9: Analytical estimation obtained from the diffusion approximation and Algorithm I.

the values of $E(PU)$ and $E(SRU)$, extending the algorithm presented in [6] to other N and c values is straightforward. We assume that the user supplies the required storage space, S, the required throughput, R, the pick time distribution (with mean, p, and variance, V_p), the S/R machine handling time, W, the S/R machine travel velocities, v_h and v_v, the desired shape factor, b, the number of pick positions per aisle, $c(\geq 2)$, and the number of aisles to be assigned to each picker, $N(\geq 2)$. Alternative configurations can be quickly generated and evaluated by executing the algorithm with different b, c, and N values.

The design algorithm is based on searching over the number of pickers, M. (The reader is referred to [6] for the mathematical derivation of the algorithm.) Rather than starting at $M=1$, one may derive a simple lower bound as follows: if the pickers were 100% utilized, then each picker would complete $60/p$ picks per hour. Substituting $E(PU)=1.0$ in the throughput constraint given by (8) and rearranging yields the expression $M \geq \lceil Rp/60 \rceil$, where $\lceil j \rceil$ denotes the smallest integer number greater than or equal to j.

ALGORITHM II:

0. Set $\hat{E}(DC) = [(4/3)+(b^2/2)-(b^3/30)]$ and $\hat{V}_m = [(0.3588-0.1321b)\hat{E}(DC)]^2$.
1. Set $M = \lceil Rp/60 \rceil$ and $K = [S/2bv_v v_h]^{1/2}$.
2. Set $T = K/\sqrt{MN}$, $E(SCT) = \hat{E}(DC)T + W$, and $\hat{V}_m = V_m T^2$. Also, let $E(SCT^2) = V_m + E(SCT)^2$ and $E(t') = E(SCT^2)/[2E(SCT)]$. If the pick time is deterministic, set $V_p = 0$. If the pick time is exponentially distributed, set $V_p = p^2$. Lastly, set $r=0$, $E(PU) = 1.0$, $E(SRU) = 1.0$ and $\epsilon = 0.001$.
3. Set $p' = Np + r$. If the pick time is deterministic, set $V_p' = 0$. If the pick time is exponential, set $V_p' = Np^2$.
4. Let $\mu = 1/p'$ and $\lambda = 1/E(SCT)$. Let $\alpha = \mu^3 V_p' + \lambda^3 V_m$, $\gamma = [2(\lambda - \mu)]/\alpha$, and $\rho = \lambda/\mu$. Then compute the following values:

$$\Omega = \frac{\rho}{1-\rho^2 e^{\gamma(c-1)}},$$

$$E'(PU) = 1 - \Omega \left[\frac{1-\rho}{\rho}\right],$$

and

$$E'(SRU) = 1\,\Omega-(1-\rho)e^{\gamma(c-1)}.$$

5. If $|E'(PU) - E(PU)| \leq$, go to 7. Otherwise, go to 6.
6. $E(PU) \leftarrow E'(PU)$ and $E(SRU) \leftarrow E'(SRU)$. Set $r=[1-E(PU)]E(t')$ and go to 3.
7. $E(PU) \leftarrow E(PU)[Np/p']$. If $E(PU)(60/p)M \geq R$ then go to 9. Otherwise, go to 8.
8. $M \leftarrow M + 1$, go to 2.
9. STOP; M is the near-minimum number of pickers.

Recall that in the above algorithm $E(SRU)$ is provided only as additional information. Also, each aisle in the MN-aisle system will be L feet long and H feet high, where

$$L = [(Sv_h)/(2bMNv_v)]^{1/2} \text{ and } H = bLv_v/v_h.$$

Since the expected throughput capacity may exceed R, the resulting throughput capacity utilization, in percent, will be $[Rp/(60ME(PU))](100)$.

As remarked in [6], the above expressions for L and H are valid only if the *vertical* dimension of the rack yields the value of b. Otherwise, the expressions will be $L = bHv_h/v_v$ and $H = [(Sv_v)/(2bMNv_h)]^{1/2}$. Regardless of which is the dominant dimension, the $E(PU)$ and $E(SRU)$ values will not change as long as the I/O point is at one corner of the rack. Thus, one can select the longer (in time) side of the rack to stand horizontally or vertically, whichever is appropriate in terms of available space and/or cost.

7 SEQUENCING THE RETRIEVALS

For container sequencing we will only consider systems with two pick positions since one is most likely to observe the largest improvement in $E(PU)$ with such a system. In a system with two pick positions (labeled A and B), the picker alternately picks from positions A and B while the S/R machine alternately replaces the container in each position after the picker is done with it. (Note that the picker picks from one position while the S/R machine retrieves the next container for the other position.) Consequently, whenever the picker is done with picking from the container in, say, position A, he/she will become idle if the S/R machine has not retrieved the new container for position B.

Hence, an alternative approach to retrieving the containers on a FCFS basis is to rearrange the sequence of retrieval in order to improve the throughput performance of the system. In determining the retrieval sequence, it is assumed that all the containers required to fill an order must be retrieved consecutively. That is, interleaving the containers of one order with another (except for the first and last containers in the sequence) is not allowed.

A sequence of retrievals implies that each container must be matched with two other containers. More specifically, any given container, say, container k, must have two corresponding distinct containers, say, j and l. After the S/R machine stores container j, it will retrieve container k. Subsequently, when it later stores container k, the S/R machine will retrieve container l and so on. (In most order picking systems each container has a dedicated location in the rack.)

Consider a system with only two pick positions, namely, A and B as described earlier. Disregarding the containers of the previous and subsequent orders, the S/R machine travel pattern to handle four containers is graphically illustrated in Fig. 4 where the two pick positions are assumed to be initially empty. The retrieval sequence is indicated by the number assigned to each container. Note that each container must be retrieved and stored exactly once (as shown in solid lines in Fig. 4). Furthermore, the travel time required to retrieve and store a given container is independent of the retrieval sequence. However, the total S/R machine cycle time, defined by a dual command cycle, is a function of the retrieval sequence.

As shown in Fig. 4, the retrieval sequence divides the entire set of nodes (or containers) into two subsets where each subset represents the containers assigned to one of the pick positions. As implied by the operation of the system, for an even number of containers, the cardinality of the two subsets must be equal. If an odd number of containers are involved, however, the cardinality of the two subsets will differ exactly by one.

Omitting the "out" and "back" leg of each S/R cycle, that is, taking only the dashed lines in Fig. 4 into account, the above retrieval sequence resembles the Traveling Salesman Problem with two salesmen or the 2-TSP in short. An additional constraint is the cardinality constraint described above. Solving the 2-TSP with the cardinality constraint will indeed minimize the total time required for the S/R machine to handle (that is, retrieve and later store) all the containers within a given order. Hence, by definition, it will minimize the average S/R cycle time to perform a dual command cycle. However, based on the arguments presented in the following paragraphs, minimizing the average S/R machine cycle time is not necessarily the appropriate objective.

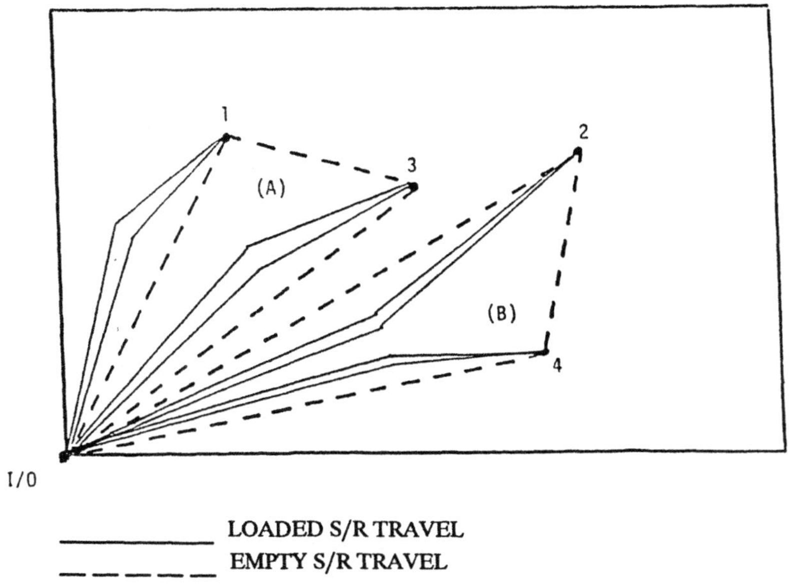

Fig. 4 The S/R machine travel for retrieving and storing containers of a single order

Consider first the list of orders with the corresponding containers for each order. In the general case, sequencing the retrievals implies sequencing the orders as well as the containers within each order. Depending on the complexity of the sequencing rules used and the potentially large size of the order list, solving such a container sequencing problem may prove to be a formidable task. However, the problem may be simplified to a certain extent if an exact solution is not sought. For example, the sequence in which the orders are picked may not have a significant impact on throughput if the number of containers per order is sufficiently large. On the contrary, sequencing the retrievals within each order may not significantly improve throughput if the number of containers per order is relatively small.

Above all, a critical (operational) issue is the definition of the optimum sequencing rule (or criteria). Given a fixed number of pickers and aisles, the throughput of the system is a function of $E(PU)$. The expected picker utilization, however, is a function of the pick time distribution as well as the distribution of the S/R machine cycle time. Determining the probability density function of the S/R machine cycle time in closed form for a specific retrieval sequencing rule is likely to be analytically intractable. Consequently, the following sequencing rules will be evaluated via simulation.

7.1 Heuristic Type A:

Assuming that the pick time distribution is a parameter, $E(PU)$ may be improved by reducing the mean and/or variance of the S/R machine cycle time. Consider first the case where the pick time p is constant. Given a list of containers, if the retrievals are sequenced such that no S/R machine cycle time exceeds p, the picker will be 100% utilized. Note that such a retrieval sequence does not necessarily minimize the average S/R machine cycle time as described earlier for the 2-TSP.

The above approach would lead to a model where the objective function is to minimize the total amount by which the S/R machine cycle time exceeds p. That is, given a list of n containers and letting t_{gh} denote the total S/R machine cycle time to store container g and retrieve container h, the objective function can be written as follows:

$$\text{Minimize} \quad \sum_{g,h} C_{gh}$$

where

$$C_{gh} = \begin{cases} 0 & \text{if } t_{gh} < p \\ t_{gh} - p & \text{if } t_{gh} \geq p \end{cases}$$

The constraints are similar to those of the 2-TSP. That is, every node must be assigned to two other distinct nodes and no container should be moved more than once (retrieval followed by subsequent storage). An additional constraint is the cardinality constraint described earlier. Formulating the above problem appears to be a difficult task. Expressing the 2-TSP constraints along with the additional cardinality constraints in closed form is likely to lead to a fairly complex model. Moreover, obtaining an exact solution appears to be even more difficult especially in an order picking environment where the problem must be solved with significant frequency. Consequently, the following heuristic is proposed for solving the problem:

HEURISTIC A.1

0. Let $S = \{1,2,...,n\}$ and t_{gh} and C_{gh} be defined as above $(g \in S, h \in S)$.
1. Select any two nodes, say, i and j, from S. Let $P = S - \{i,j\}, A = \{i\}$, and $B = \{j\}$.

2a. For node i, determine node $k \in P$ such that C_{ik} is minimized.

2b. $P \leftarrow P - \{k\}$ and $A \leftarrow A + \{k\}$.

2c. For node j, determine node $l \in P$ such that C_{jl} is minimized.

2d. $P \leftarrow P - \{l\}$ and $B \leftarrow B + \{l\}$.

2e. $i \leftarrow k$ and $j \leftarrow l$.

3. Repeat steps 2a through 2e until P is either empty or it contains only one node. If P is empty (i.e., n is even), go to step 5. Otherwise, go to step 4.
4. Let the remaining node in P be labeled m. If $C_{im} < C_{jm}$ then $A \leftarrow A + \{m\}$. Otherwise, $B \leftarrow B + \{m\}$.
5. Stop. The containers assigned to pick positions A and B are in sets {A} and {B}, respectively.

In other words, heuristic A.1 simultaneously builds two subsets (one for each pick position) by adding unassigned containers one at a time to each subset such that the resulting S/R machine cycle time exceeds p as little as possible. The performance of the heuristic generally depends on the starting nodes i and j. Hence, in implementing the heuristic, a solution was obtained for all possible pairs i and j and the best one was selected. That is, the solution which minimizes the sum of all the C_{gh} values.

A slightly different version of heuristic A.1 can be obtained by changing the definition of C_{gh} as follows:

$$C_{gh} = |t_{gh} - p| \tag{16}$$

In other words, the modified heuristic, say, heuristic A.2, will attempt to minimize the total deviation from p by matching containers that yield a cycle time as close as possible to p. Hence, heuristic A.2 is identical to heuristic

A.1 except for the definition of C_{gh} used in steps 2a, 2c and 4 in the above algorithm.

The containers within an order may be sequenced by one of the above heuristics. However, a remaining issue is concerned with order sequencing where the first and last containers of each order are matched with containers from the previous and subsequent orders, respectively. For the purposes of this research, the order sequencing problem will not be considered. That is, the orders will be assumed to be served on a FCFS basis.

In picking the orders on a FCFS basis, one must still match the containers of successive orders (in order to eliminate empty travel to and from the I/O point). For example, a list of orders with seven containers each is shown in Fig. 5. The two subsets can be identified by applying either heuristic A.1 or heuristic A.2 described earlier. Assuming that the two subsets are not interchangeable, the containers of adjacent orders can be matched either by the "backtracking" rule or the "alternating" rule as shown in Fig. 5. Using "backtracking", the last container of the current order is matched with the first container of the next order. With the "alternating" rule, however, the last container of the current order is matched with the last container of the next order (because the containers within an order may be picked in either direction).

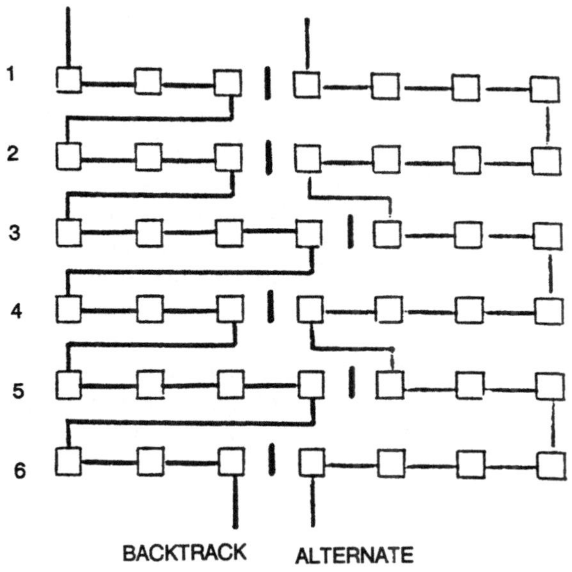

Fig. 5 Alternative strategies for connecting containers of adjacent orders

Other rules may also be developed for matching the containers of successive orders. For example, one alternative is to choose the "best" between "backtracking" and "alternating" at each decision point. Preliminary computational results have shown that the picker utilization is fairly insensitive to the decision rule adopted (especially as the number of containers per order is increased).

Hence, the following computational experiment is based on selecting the "best" match at each decision point. Recall that the orders are assumed to be served on a FCFS basis and that the definition of "best" is based on the type of retrieval sequencing heuristic used. The purpose of the experiment is to compare the two heuristics developed earlier, namely, heuristic A.1 and heuristic A.2 using simulation. Since the two heuristics of interest are identical from an algorithmic standpoint, the basis of comparison will be restricted to solution quality. That is, execution time and related issues will not be considered.

The two heuristics will be compared for $n = 5$ and $n = 16$ containers per order. The results of the numerical experiment are presented in Table 6 where the performance of each heuristic is measured by the mean effective pick time q given by the following expression:

$$q = p + E(IT) \tag{17}$$

where $E(IT)$ is the expected value of the picker idle time following a pick. The column labeled *VAR* in Table 6 designates the sample variance obtained for the corresponding q. Assuming a zero handling time and *using the same list of containers for both heuristics*, the mean effective pick time was estimated over a range of p values as shown in Table 6. Applying the z-test on the q values obtained, it can be observed from Table 7 that there are only three instances where the difference between the q values of interest is statistically significant at 5%. In all of the above three cases, the picker utilization is "practically equal" since the statistical difference can be mainly attributed to very small sample variances obtained from deterministic pick times. Hence, based on the above results, it can be concluded that heuristics A.1 and A.2 do not differ significantly in their performance.

Furthermore, the retrieval sequence generated by the two heuristics from the same original list was observed to be quite different. Hence, there is strong indication that the picker utilization may be relatively insensitive to the retrieval sequence.

Table 6 Mean effective pick time observed for heuristic A.1 and heuristic a.2

		HEURISTIC A.1							
		DETERMINISTIC p				EXPONENTIAL p			
		VAR	q	PU	S/RU	VAR	q	PU	S/RU
n=5									
p	OBS.								
0.50	500	0.1458	1.7126	29.9448	99.9416	0.6094	1.7294	28.7502	98.9908
1.00	450	0.1281	1.7307	57.7789	99.3800	1.3428	1.8960	51.1352	90.7245
1.50	425	0.0634	1.7880	83.8943	96.3548	2.4952	2.1629	67.7827	79.5663
2.00	425	0.0045	2.0198	99.0191	86.4682	3.9654	2.5396	76.9722	68.6750
2.50	425	0.0000	2.5000	100.0000	71.1423	6.1069	2.9550	82.6890	60.1185
3.00	425	0.0000	3.0000	100.0000	59.7598	8.5196	3.3849	86.6237	52.9444
3.50	425	–	–	–	–	11.3554	3.8235	89.4684	46.8859
3.50	425	–	–	–	–	14.6492	4.2731	91.4930	41.9557
n=16									
p	OBS.								
0.50	425	0.2271	1.5643	31.9633	99.8218	0.5894	1.5918	30.7013	98.0604
1.00	425	0.1624	1.5847	63.1051	98.3432	1.3224	1.7711	55.1843	87.9198
1.50	425	0.0634	1.7112	87.6568	93.9027	2.5118	2.1006	69.7935	76.4094
2.50	425	0.0026	2.0100	99.5038	85.1127	4.0280	2.5105	77.8641	68.0563
2.00	425	0.0000	2.5000	100.0000	70.6278	6.0321	2.9387	83.1479	60.0187
3.00	425	0.0000	3.0000	100.0000	59.1223	8.4507	3.3746	86.8884	52.5029
3.50	425	–	–	–	–	11.2929	3.8094	89.7996	46.5251
4.00	425	–	–	–	–	14.5920	4.2576	91.8258	41.6412

Table 6 (continued)

		HEURISTIC A.2							
		DETERMINISTIC p				EXPONENTIAL p			
		VAR	q	PU	S/RU	VAR	q	PU	S/RU
n=5									
p	OBS.								
0.50	500	0.1458	1.7126	29.1948	99.9416	0.6094	1.7294	28.7502	98.9908
1.00	500	0.1249	1.7165	58.2578	99.5773	1.4471	1.9027	52.2632	89.8657
1.50	450	0.0547	1.7814	84.2049	98.3667	2.4118	2.1954	66.2408	79.8227
2.00	500	0.0060	2.0335	98.3533	91.2546	4.3731	2.6364	75.4374	70.3861
2.50	450	0.0004	2.5033	99.8681	75.7591	5.9790	3.0003	80.7850	63.2094
3.00	450	0.0000	3.0000	100.0000	63.1230	8.1194	3.3998	85.5519	55.7007
3.50	450	--	--	--	--	10.8726	3.8289	88.6233	49.4575
4.00	450	--	--	--	--	14.0665	4.2719	90.7811	44.3290
n=16									
p	OBS.								
0.50	500	0.2173	1.5753	31.7403	99.9033	0.6098	1.6089	30.9027	97.8328
1.00	500	0.1533	1.6006	62.4781	99.5265	1.3639	1.8032	55.1474	88.3516
1.50	500	0.0561	1.7243	86.9898	98.8719	2.6342	2.1769	68.5127	78.3319
2.00	500	0.0010	2.0126	99.3743	92.0056	4.2972	2.6328	75.5391	70.3309
2.50	500	0.0004	2.5048	199.8096	76.6026	6.5368	3.0751	80.8439	62.3977
3.00	500	0.0000	3.0000	100.0000	63.8880	9.0373	3.4941	85.3787	54.8533
3.50	500	--	--	--	--	12.0238	3.9291	88.5817	48.7810
4.00	500	--	--	--	--	15.4923	4.3788	90.8394	43.7713

Table 7 Statistical comparison of mean effective pick time (heuristic A.2 versus heuristic A.2)

	HEURISTIC A.1 vs HEURISTIC A.2	
n=5	DETERMINISTIC p	EXPONENTIAL p
p=0.50	E*	E
1.00	E	E
1.50	E	E
2.00	NE*	E
2.50	NE	E
3.00	E	E
3.50	--	E
4.00	--	E
n=16	DETERMINISTIC p	EXPONENTIAL p
p=0.50	E	E
1.00	E	E
1.50	E	E
2.00	E	E
2.50	NE	E
3.00	E	E
3.50	--	E
4.00	--	E

(*) H_0: $q_{A1} = q_{A2}$ (E)
H_1: $q_{A1} \neq q_{A2}$ (NE)
Tested at 5% significance

7.2 Heuristic Type B:

Although heuristic type A may seem quite suitable for a deterministic pick time, it is not clear that minimizing variance will produce the best results. In fact, within the framework of the queueing model shown in Fig. 2(c), the ideal S/R machine cycle time will be the one with minimum expected value *and* variance.

However, identifying such a retrieval sequence is far from trivial. Hence, an alternative to heuristic type A is one where only the average S/R machine cycle time is minimized. As noted earlier, the average cycle time can be minimized by solving the 2-TSP with the cardinality constraint. Since an exact solution is difficult to obtain, we propose the following heuristic, namely, heuristic B.1.

Consider a 1-TSP tour obtained by using either the convex hull heuristic (Goetschalckx and Ratliff [10]) or the 2-band insertion heuristic (Bozer et al.

[4]). The above heuristics have been shown to generate optimum or near-optimum tours for all the order picking problems examined by the authors. Since the total travel time of the S/R machine can be minimized by minimizing the travel between element, the 1-TSP tour obtained above is a potentially good starting point. However, retrieving the containers in the exact same sequence dictated by the 1-TSP tour may not generate a good solution, as the following example shows.

Consider the container locations given in Fig. 6(a). The optimum 1-TSP tour is shown in dashed line where each container is numbered. Given two pick positions, retrieving the containers in the above sequence will generate the two subtours (or subsets) depicted in Fig. 6(b). Note that in traveling from one container to the next, the S/R machine does not follow the dashed line dictated by the 1-TSP tour. Hence, generally speaking, the above solution cannot be a good one.

Instead, consider the solution given by Fig. 6(c). It can be obtained by "evenly dividing" the above 1-TSP tour into two subtours. The containers will now be retrieved in the following sequence: 1-4-2-5-3-6. (Alternative sequences are: 1-6-2-5-3-4, 3-4-2-5-1-6 and 3-6-2-5-1-4). Since the S/R machine generally follows the sequence dictated by the 1-TSP tour, the above solution may be quite close to the optimum. (In the remainder of the chapter, the 2-TSP tour is assumed to follow the general pattern implied by the sequence 1-6-2-5-3-4). Thus, heuristic B.1 can be presented as follows:

HEURISTIC B.1

0. Using the container locations and the convex hull heuristic (or the 2-band insertion heuristic), determine the 1-TSP tour.
1. If the number of containers, n, is even, then "evenly divide" the 1-TSP tour obtained in step 1 and form two subtours; go to step 4. If n is odd, then go to step 2.
2. Let point g denote the $(n+1)/2$th point in the 1-TSP tour found in step 1. Also let f and h denote the points immediately before and after point g, respectively. Lastly, let $d(i,j)$ denote the travel time between points i and j. Go to step 3.
3. Compute $C(f,g)$ and $C(g,h)$ as follows:

$$C(f,g) = d(I/O,f) + d(I/O,g) - d(f,g),$$

and

$$C(g,h) = d(I/O,g) + d(I/O,h) - d(g,h).$$

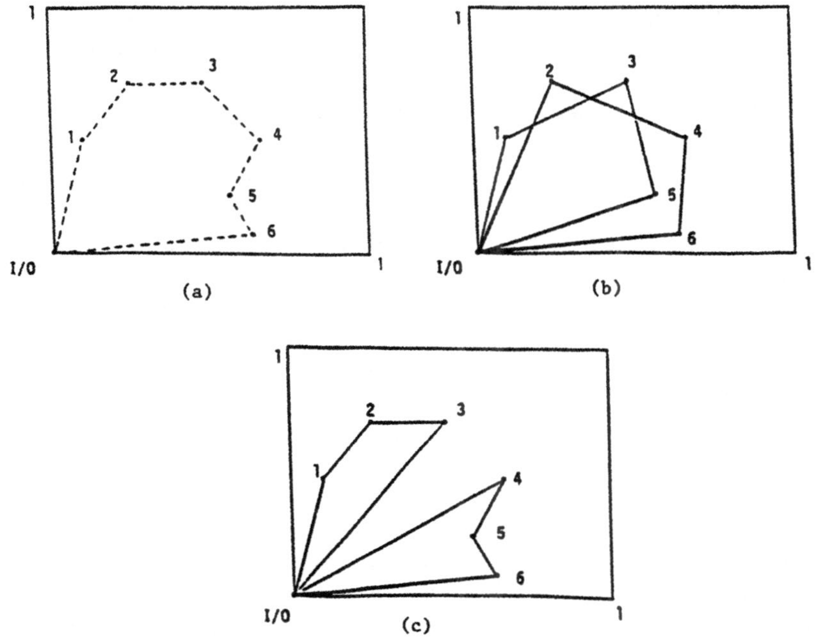

Fig. 6 Retrieval sequencing for heuristic type B

If $C(f,g) < C(g,h)$ then construct the two subtours by deleting the edge (f,g). Otherwise, delete the edge (g,h). Go to step 4.

4. Stop.

Bozer [3] evaluated the performance of heuristic B.1 by comparing the resulting S/R machine cycle times to those obtained from the optimum solution to the 2-TSP problem with the cardinality constraint. Considering the entire dual command cycle length, he empirically showed that heuristic B.1 is no more than 0.93% above optimum for $7 \leq n \leq 13$. He also empirically showed that the reduction in S/R machine cycle time compared to FCFS retrievals (i.e., no container sequencing) is nearly 7% for $n=7$ and 14% for $n=13$.

Note that the above heuristic is used for sequencing the containers within a given order. That is, the containers corresponding to the previous and subsequent orders are not considered. Assuming that the orders are picked on a FCFS basis, the remaining task is to match the first and last containers of adjacent orders as discussed earlier for heuristic type A. Recall that possible strategies are to use "backtracking", "alternating" or the "best" of the two.

Due to the general pattern assumed for the 2-TSP tour (see the sequence 1-6-2-5-3-4 in Fig. 6(c)), the last container of the first order has the tendency to be farther from the I/O point. Hence, using the "alternating" rule and matching the above container with the last container of the second order generally produces better results than using "backtracking." Likewise, the last retrieved container of the second order will be generally closer to the I/O point. Therefore, it will be matched with the first container of the third order and so on.

Using the "alternating" rule, heuristic B.1 was evaluated over a range of pick times. The mean effective pick time q (given by (17)) as well as $E(PU)$ and $E(SRU)$ are presented in Table 8 where the column labeled *VAR* designates the variance of the mean effective pick time. The results presented in Table 8 are based on a zero handling time. Furthermore, they are obtained by using the same data (that is, the same container locations) that were used in numerically evaluating heuristics A.1 and A.2. Since no significant difference was observed between heuristics A.1 and A.2, heuristic B.1 will be compared with heuristic A.2 only.

The results of the above comparison are presented in Table 9. With only one exception, the difference between effective pick times obtained under the two heuristics is not statistically significant at 5%. As before, the single case where the difference is statistically significant is due to a very small sample variance obtained with deterministic pick times.

Hence, based on the computational results presented above, it can be concluded that both heuristics type A and B yield similar results in throughput. From an implementation standpoint, the two heuristics appear to have different characteristics despite the fact that both could be classified as "tour construction" procedures. Perhaps, the most important drawback of heuristic type A is the necessity to consider alternative starting points. Also, note that the exact solution was not obtained for heuristic type A. Heuristic type B, however, was numerically shown to generate near-optimum results. Consequently, the above conclusion may change if the performance of heuristic type A can be improved after comparing its performance with the corresponding exact solution.

A remaining critical issue to be studied is concerned with the amount of improvement gained in mean effective pick time q by sequencing the retrievals. In order to make the above comparison, the data used in preparing Table 8 was also used for determining q under FCFS retrievals. The results obtained from the simulation model are shown in Table 10 for a range of p values. Table 11 presents the comparison between heuristic B.1 and FCFS retrievals based on the same container locations. For deterministic pick times, it is observed that the q value obtained under heuristic B.1 is statistically less than or equal to the q value obtained from FCFS retrievals (at 5%). The only exception occurs at $p = 3.0$ time units.

Table 8 Mean effective pick time observed for heuristic B.1

		HEURISTIC B.1							
		DETERMINISTIC p				EXPONENTIAL p			
		VAR	q	PU	S/RU	VAR	q	PU	S/RU
n=5									
p	OBS.								
0.50	500	0.1842	1.6885	29.6118	99.9190	0.6001	1.7072	29.1249	98.8465
1.00	500	0.1530	1.7068	58.5901	98.9087	1.4806	1.8826	52.8210	89.6947
1.50	500	0.0795	1.7889	83.8506	94.3932	2.7356	2.1817	68.3686	77.3972
2.00	500	0.0140	2.0450	97.7988	82.5714	4.3976	2.5537	77.8804	66.1238
2.50	500	0.0004	2.5025	99.8999	67.4763	6.5146	2.9613	83.9518	57.0230
3.00	500	0.0000	3.0000	100.0000	56.2865	9.0946	3.3923	87.9418	49.7776
3.50	500	–	–	–	–	12.1410	3.8427	90.5726	43.9429
4.00	500	–	–	–	–	15.6537	4.3047	92.4018	39.2266
n=16									
p	OBS.								
0.50	500	0.2173	1.5310	32.6588	99.7535	0.6408	1.5620	31.8315	97.7927
1.00	500	0.1521	1.5752	63.4853	97.0186	1.4614	1.7701	56.1769	86.3675
1.50	500	0.0526	1.7143	87.5014	89.2051	2.6600	2.0901	71.3644	73.2072
2.00	500	0.0035	2.0153	99.2419	75.9266	4.2573	2.4805	80.1782	61.6864
2.50	500	0.0000	2.5000	100.0000	61.2053	6.3465	2.9017	85.6743	52.7319
3.00	500	0.0000	3.0000	100.0000	51.0044	8.9107	3.3422	89.2591	45.7820
3.50	500	–	–	–	–	11.9358	3.7982	91.6339	40.2857
4.00	500	–	–	–	–	15.4461	4.2627	93.3128	35.8959

Table 9 Statistical comparison of mean effective pick time (heuristic B.1 versus heuristic A.2)

	HEURISTIC B.1 vs HEURISTIC A.2	
n=5	DETERMINISTIC p	EXPONENTIAL p
p=0.50	E*	E
1.00	E	E
1.50	E	E
2.00	E	E
2.50	E	E
3.00	E	E
3.50	--	E
4.00	--	E
n=16	DETERMINISTIC p	EXPONENTIAL p
p=0.50	E	E
1.00	E	E
1.50	E	E
2.00	E	E
2.50	NE*	E
3.00	E	E
3.50	--	E
4.00	--	E

(*) H_0: $q_{B1} = q_{A2}$ (E)
H_1: $q_{B1} \neq q_{A2}$ (NE)
Tested at 5% significance

In Table 11, the percent reduction obtained in q by using heuristic B.1 is shown in parentheses. Note that it is generally higher for $n=16$ containers and that it decreases as p is increased (an intuitive result). The maximum reduction obtained is 14.90% while the minimum is 0.14%. For exponential pick times, however, with the exception of very small p values, a majority of the improvements are not statistically significant (at 5%). Recall that, for n = 13 containers, the average reduction gained in the S/R machine cycle time (for a single set of containers) was found to be equal to 14%. Given the relatively large variance imposed by an exponentially distributed pick time, it appears that such reductions in the S/R machine cycle time are not adequate for significantly improving the picker utilization.

Table 10 Mean effective pick time observed for FCFS retrievals (two pick positions)

		FCFS RETRIEVALS							
		DETERMINISTIC p				EXPONENTIAL p			
		VAR	q	PU	S/RU	VAR	q	PU	S/RU
n=5									
p	OBS.								
0.50	500	0.1791	1.8092	27.6362	99.9447	0.6245	1.8245	27.2513	99.1239
1.00	500	0.1647	1.8167	55.0460	99.5906	1.4879	1.9910	49.9443	90.8994
1.50	500	0.1079	1.8703	80.2014	96.7883	2.7063	2.2739	65.5982	79.6494
2.00	500	0.0284	2.0849	95.9258	86.8666	4.3264	2.6317	75.5710	68.8187
2.50	500	0.0012	2.5068	99.7292	72.2486	6.4093	3.0272	82.1240	59.8290
3.00	500	0.0000	3.0000	100.0000	60.3707	8.9574	3.4502	86.4651	52.4929
3.50	500	--	--	--	--	11.9862	3.8902	89.4658	46.5555
4.00	500	--	--	--	--	15.4859	4.3446	91.5528	41.6863
n=5									
p	OBS.								
0.50	500	0.1878	1.7991	27.7914	99.9056	0.6047	1.8133	27.4204	99.1432
1.00	500	0.1622	1.8119	55.1894	97.2538	1.4695	1.9737	50.3843	91.1530
1.50	500	0.0968	1.8728	80.0935	96.0814	2.7186	2.2486	66.3344	80.0590
2.00	500	0.0238	2.0775	96.2711	86.6552	4.3843	2.6085	76.2436	69.0138
2.50	500	0.0006	2.5036	99.8545	71.9045	6.4877	3.0122	82.5330	59.7654
3.00	500	0.0000	3.0000	100.0000	60.0077	9.0532	3.4371	86.7957	52.3769
3.50	500	--	--	--	--	12.1001	3.8834	89.6238	46.3573
4.00	500	--	--	--	--	15.6213	4.3435	91.5764	41.4464

Table 11 Statistical comparison of mean effective pick time (heuristic B.1 versus FCFS retrievals)

	HEURISTIC B.1 vs	FCFS RETRIEVALS
n=5	DETERMINISTIC p	EXPONENTIAL p
p=0.50	LT* (6.67)	LT (6.43)
1.00	LT (6.05)	NLT
1.50	LT (4.35)	NLT
2.00	LT (1.91)	NLT
2.50	LT (0.17)	NLT
3.00	NLT*	NLT
3.50	--	NLT
4.00	--	NLT
n=16	DETERMINISTIC p	EXPONENTIAL p
p=0.50	LT (14.90)	LT (13.86)
1.00	LT (13.06)	LT (10.32)
1.50	LT (8.46)	NLT
2.00	LT (2.99)	NLT
2.50	LT (0.14)	NLT
3.00	NLT	NLT
3.50	--	NLT
4.00	--	NLT

(*) H_0: $q_{B1} \geq q_{FCFS}$ (NLT = Not less than)
H_1: $q_{B1} < q_{FCFS}$ (LT = Less than)
Tested at 5% significance

8 CONCLUSIONS AND EXTENSIONS

In this study we generalized the algorithm developed earlier in [6]. Using the generalized algorithm, one can design and evaluate the throughput performance of end-of-aisle order picking systems with multiple pick positions per aisle and multiple aisles per picker. We believe such an extension enhances the scope and usefulness of the algorithm. Although we did not perform extensive numerical experiments to compare the performance of alternative configurations, our results indicate that a good strategy is to design "balanced" systems where the mean pick time at each aisle is approximately equal to the mean S/R machine cycle time. The final selection, however, must be made on an economic basis. Coupled with an appropriate economic analysis model, the algorithm enables the analyst to rapidly evaluate -- from a *cost and throughput standpoint* -- a wide variety of design alternatives for end-of-aisle order picking.

Our empirical results on retrieval sequencing in systems with two pick positions and one aisle per picker indicate that if the variance of the pick time is large, no or very little improvement in throughput is obtained by sequencing container retrievals within each order. However, if the pick time is constant (or nearly constant), retrieval sequencing can reduce the mean effective pick time up to 15%, which implies that the throughput of the system will increase by 17.65%. The amount of reduction in the mean effective pick time, however, also depends on the number of containers per order. Obviously, smaller improvements in throughput is obtained as the number of containers per order decreases.

A number of extensions can be made to our analysis. First, one may consider using turnover-based dedicated storage in order to reduce the S/R machine cycle time. (Our results indicate that, if the system is not "balanced," the picker may become the bottleneck. In such cases, reducing the S/R machine cycle time will not improve throughput.) Second, one may use remote picking stations where each picking station is interfaced to the storage/retrieval system via a conveyor loop. With such a system, each picking station has access to all the aisles and workload balancing among the pickers becomes less complicated. Third, by inspecting the work content of each pick and developing standard time data for performing specific types of picks, one may be able to estimate the pick time in advance and use this information in retrieval sequencing. Fourth, one may relax the assumption that each container has a dedicated location in the rack. Although in most real-life systems containers have fixed locations in the rack, one may reduce the S/R machine cycle time by embedding additional empty openings in the rack and storing each container in any one of the empty openings. With such a strategy one can also use the sequencing algorithms developed in [12].

9 ACKNOWLEDGEMENT

This study was partially supported by the Material Handling Research Center at Georgia Tech and the National Science Foundation under Grants ISI 8300965 and DMC 8858562.

10 FOOTNOTES

[1]Except for the extensions treated in this study, the assumptions and definitions presented here are the same as those given in [6].

[2]A special configuration, namely, the L^2R^2 configuration provides two pick positions on the left and two pick positions on the right side of the aisle. See [6] for simulation results with this special configuration.

[3]This section can be omitted by the person interested only in the application of the research.

11 REFERENCES

1. Bengtson, N.M. and R.J. Gomez, "Why a Single-Aisle Miniload System is Not Simple to Model," *Proceedings of the 1988 Winter Simulation Conference*, Abrams, Haigh, and Comfort (eds.), pp. 603-608 (1988).
2. Bozer, Y.A., "A Minimum Cost Design for An Automated Warehouse," Unpublished Master's Thesis, Georgia Institute of Technology, Atlanta, GA (1978).
3. Bozer, Y.A., "Optimizing Throughput Performance in Designing Order Picking Systems," Unpublished Ph.D. Dissertation, Georgia Institute of Technology, Atlanta, GA (1985).
4. Bozer, Y.A., E.C. Schorn, and G.P. Sharp, "Geometric Approaches to Solve the Chebyshev Traveling Salesman Problem," to appear in *IIE Transactions*.
5. Bozer, Y.A. and J.A. White, "Travel-Time Models for Automated Storage/Retrieval Systems," *IIE Transactions*, Vol. 16, No. 4, pp. 329-338 (1984).
6. Bozer, Y.A. and J.A. White, "Design and Performance Models for End-of-Aisle Order Picking Systems," to appear in *Management Science*.
7. Foley R.D. and E.H. Frazelle, "Analytical Results for Miniload Throughput and the Distribution of Dual Command Travel Time," to appear in *IIE Transactions*.
8. Gelenbe, E., "On Approximate Computer System Models," *Journal of the ACM*, Vol. 22, No. 2, pp. 261-269 (1975).
9. Goetschalckx, M.P., "Storage and Retrieval Policies for Efficient Picking Operations," Unpublished Ph.D. Dissertation, Georgia Institute of Technology, Atlanta, GA (1983).
10. Goetschalckx, M.P. and H.D. Ratliff, "Sequencing Picking Operations in a Man-Aboard Order Picking System," *Material Flow*, Vol. 4, No. 4, pp. 255-264 (1988).
11. Graves, S.C., W.H. Hausman and L.B. Schwarz, "Storage Retrieval Interleaving in Automatic Warehousing Systems," *Management Science*, Vol. 23, No. 9, pp. 935-945 (1975).
12. Han, M.H., L.F. McGinnis, J.S. Shieh, and J.A. White, "Sequencing Retrievals from Automated Storage/Retrieval Systems," *IIE Transactions*, Vol. 19. No. 1, pp. 56-66.
13. Tijms, H.C., *Stochastic Modeling and Analysis: A Computational Approach*, John Wiley (1986).

14. Tompkins, J.A. and J.A. White, *Facilities Planning*, John Wiley, 1984.
15. "Split-Case Order Picking -- How to Tackle the Problem," *Modern Materials Handling*, Vol. 36, No. 2, pp. 63-71 (1981).
16. *Warehouse Modernization and Layout Planning Guide*, Department of the Navy, Naval Supply Systems Command, NAVSUP Publication 529 (1978).

ORDER SEQUENCING IN AUTOMATED STORAGE/RETRIEVAL SYSTEMS WITH DUE DATES

E. A. Elsayed
Rutgers, The State University of New Jersey

ABSTRACT

In this chapter we investigate the multiple address AS/R (Automated Storage/Retrieval) interleaving systems with due date constraints. Orders for retrieval arrive to the AS/RS according to a static arrival pattern. Algorithms are developed to batch and sequence orders to tours such that the total travel time and the total lateness of retrievals per group of orders are minimized.

1 INTRODUCTION

The automated storage and retrieval systems (AS/RS) are major material handling support systems which are commonly used in the automated factories, distribution centers, warehousing, and nonmanufacturing environments. Their applications vary widely from a simple storage and retrieval system for small parts to central systems where production, assembly, and manufacturing operations are concentrically located around them. Applications of AS/RS (Automated Storage/Retrieval Systems) exist in the assemblies of small electronic components where assembly work stations are installed in the openings of the storage racks. The AS/R (Automated Storage and Retrieval) machine delivers materials directly from the storage location to the work station when needed. This significantly reduces the transportation time between the AS/RS and the work stations. AS/RS have also been used in the clean room manufacturing environment. This is due to the fact that automatic handling of devices such as disk drives and semiconductor devices reduces the contamination of the products from manual handling.

Other applications of the use of AS/RS are found in health-care distribution centers where pallet loads of medical products, ranging from IV solutions to heart valves are temporarily stored for distribution at later times. Frozen food processing environments where temperature is always kept at -20° F making it extremely hostile to human operators represent other implementations of the AS/RS. Wine producers are also employing AS/RS in wine processing, storage, and aging. The aging of wine requires storing the bottled wine in a dark place for a specified period which can be accomplished by constructing underground AS/RS where the AS/R vehicle can perform the storage of the bottled wine into specified openings in the storage structure and the retrieval of those bottles that aged for a specified time.

A recent application of the AS/RS is in the auto industry [15]. After car bodies are painted, they are moved into storage in an AS/RS to coordinate the production schedule with the number of bodies painted a specific color. The selected bodies are then retrieved and returned to production.

Mini-load automated storage and retrieval systems have been used in assembly operations to make storage on the factory floor more flexible and responsive to the needs of production. They allow work-in-process to be stored adjacent to production, thereby minimizing inventory levels and allowing manufacturers to institute just-in-time deliveries of materials.

A typical AS/RS consists of four basic components: (1) a storage structure, (2) S/R machines, (3) conveying devices such as conveyors or AGVS (Automated Guided Vehicle Systems) to transport material to areas outside the reach of the AS/RS, and (4) computers to control the storage and retrieval operations within the AS/RS [5].

The performance of AS/RS varies by the definition of the measure criteria and the operating policies adopted. Several measure criteria of performance are often used, for instance, the travel time per storage/retrieval request, the number of storage/retrieval requests carried out per unit time (i.e. the system throughput), the total time required to store/retrieve a batch of orders, the average cost per storage or retrieval operation, the average waiting time for a storage/retrieval request, and the average or maximum queue length for requests. Many parameters affect the performance of the AS/RS. Although some of the parameters are inter-related, we shall divide them into three main groups: *demand requirements*, *physical design*, and *operating policies*.

Demand requirements represent the orders that need to be stored or retrieved to meet the required production (distribution) schedule. The demand may be defined by several parameters such as: (i) Number of orders received per unit time, (ii) The arrival pattern of orders to the picking and delivery station of AS/RS may be static or dynamic. A static arrival pattern implies that the orders are grouped into a group and the storage and retrieval processes are performed on the group until all orders are completed. New arrivals, while a group of storage and retrieval is being processed, form a different group that can be processed after the completion of the current group. A dynamic arrival pattern implies that a new arrival during the processing of a group is added to the group and rescheduling of storage and retrieval processes is made to accommodate the new arrival(s), (iii) Number of items to be stored or retrieved per order, orders with large number of items will require much longer time to process than orders with smaller number of items, (iv) Weights and sizes of items to be processed have a direct effect on the storage and retrieval times. The accelerations and speeds of the S/R machine are proportional to the weights being stored or retrieved, and (v) the due date constraints affect the retrieval and storage operating policies, since orders with early due dates may require early processing which in turn will affect the performance of the AS/RS.

The second group of parameters which affect the performance of AS/RS relates to its *physical design*. Some of these parameters are: size of storage bins, length and height of storage structure (building the aisle too long may cause the S/R machines to operate at a too high percentage of their capability), single or double deep rack, and capacity and number of S/R machines.

The third group of parameters which affect the performance of AS/RS are the *operating policies* of the system which involve rules for storage and retrieval (storage cycle, retrieval cycle, storage and retrieval in the same cycle) of materials, turnover time and item popularity, order retrieval policies (FCFS, LCFS, priority ...), order storage policies, and routing of the S/R machine.

2 LITERATURE SURVEY

Designing an efficient AS/RS has been a topic of investigation by many researchers. The research areas are divided into (i) assignment of products to storage locations in the storage structure, (ii) configurations of the storage structure (length to height ratio), and (iii) operating policies for order storage and retrieval. Elsayed [4] studied the effect of demand on the configuration of the storage structure. He concluded the study by stating that as the variance of the quantity per order increases, the effect of rack configuration on the AS/RS performance becomes insignificant and the differences in performances in the operating policies are reduced.

The physical design, configurations or layout of an automated warehouse and an AS/RS have been studied since the early '60s. This early research is focused on basic design concept of a warehouse; including such factors as length, height and the number of racks. In the beginning of the 1970's, German researchers and engineers publish research papers which cover design and application of high-rise parallel rack configuration as well as carousel conveyors. Schwarz [18] studies a rotary racking system for storing and removing goods which is a space saving system, and requires short handling times. Bozer and White [2], and Rosenblatt and Roll [17] investigate the optimization of a warehouse design considering the storage policy. Bozer and White [2] derive analytical expressions for the travel time of the S/R machine in an AS/RS along a single rack. By changing the I/O

(Input/Output) location, rack shape and dwell-point strategy, they evaluate the trip time of the S/R machine. The results show that their method gives accurate results as the number of locations increases. A square-in-time rack configuration (in square-in-time configurations the S/R machine travels along the two axis of the rack, till the end, at the same time duration), first mentioned in Hausman et al. [9], is verified to be optimal. In addition, they show a travel time reduction can be obtained when the I/O point is located in the middle of the rack, or moving only the output point to a higher place.

A considerable amount of research effort has been invested in assigning items to locations on a rack in order to maximize picking efficiency. One of the earlier studies, is the cube-per-order (CPO) index rule. The cube-per-order index is the ratio of the amount of cubic footage space for each item to the average number of orders per shipping day. (Heskett [10,11], Harmatuck [8], Kallina and Lynn [12]). It establishes a quantitative tradeoff between the dual objectives of placing closest to the order shipping area those items which take up the least space, and also those items which are most popular.

Kanet and Ramirez [13] formulate the stock retrieval in batches as an integer programming problem. Another paper on batch sizing is Barrett [1]. He concludes that the batch size was the singularly most important factor in reducing both the number of trips and the S/R machine travel time. On the other hand, the zero-one programming solutions for increasing batch size show that both the travel time and the number of trips reach an upper bound beyond a certain batch size. Elsayed [3, 4] investigates the problem of order batching (grouping) to minimize the picking time for AS/RS with a vehicle capacity constraint. A combinatorial algorithm is developed to process the orders by grouping some of the orders according to a criterion. Han et al. [7] develop sequencing retrieval policies for dual command cycles in AS/RS. Equations for approximating mean dual command cycle time using nearest-neighbor sequencing rule and for a lower bound on the mean dual command cycle time for any block sequencing rule are given. Recently, Elsayed and Unal [6] develop algorithms to batch and schedule orders in AS/RS. They also develop expressions for the travel time needed to retrieve a set of orders as a function of the locations to be visited by the S/R machine and the ratio of the storage structure (time ratio between the height and length of the structure). It should be noted that no research work has been reported which investigates order batching and sequencing when orders have due date requirements.

3 PROBLEM STATEMENT

One of the important problems in just-in-time production systems is the assurance of availability of parts, materials, tools,...etc. at the work station at specified times. In using AS/RS as a source for supplying an assembly line with the needed parts and tool, one must ensure the delivery of these items when needed. In this chapter we investigate approaches for batching and sequencing orders with due date requirements in such AS/RS. The AS/RS under investigation, consists of two storage racks separated by one aisle. Each rack consists of 40 locations (bins). We refer to the front end of the rack as an I/O (input/output) point. The I/O point is where receiving or delivery of orders is performed. In addition to the 80 locations in the storage racks, an overflow zone is reserved to accommodate the overflow from other bins in the racks when their capacities are exceeded due to order replenishments.

We classify orders that arrive to the I/O point into two types: those orders that have items needed to be stored in the storage racks, we refer to them as s-orders (storage orders), and those orders that require retrievals from the storage racks, we refer to them as r-orders (retrieval orders).

Processing the above orders can be performed in three different types of cycles: (i) storage-only cycle (the S/R machine can perform storage functions only), (ii) retrieval-only cycle (the S/R machine can perform retrieval functions only), and (iii) interleaving cycle (the S/R machine can store and retrieve items in the same cycle).

A static order arrival (whether it is s-order or r-order) pattern is considered. Under this pattern of arrivals, orders are accumulated into one group to be processed during a time period. During that period no new orders are added to the group. New orders form a new group to be processed at a later time period. Orders in a group that cannot be processed during the time period due to the unavailability of some items are processed at the next time period as part of the next group.

The r-orders are defined by four characteristics: (i) the group size which is the number of r-orders in one group, (ii) the number of locations per order, (iii) the quantity per location per order, and (iv) the due date for processing (retrieving) the order. Only r-orders have due dates associated with them.

The inventory policy for replenishing items in bins is a periodic review policy under which items are replenished when their inventory levels are at or below a specified reorder level at the end of the period (time needed to process a group of orders). Two inventory policies are considered in our investigation: (R,r) policy under which a replenishment quantity is ordered at the end of the period to bring the inventory level up to the target value R once the inventory level drops below the reorder level r, and (Q,r) policy under which an order of fixed quantity Q is always placed at the end of the period once the inventory level drops below the reorder level r.

The objective of this chapter is to develop algorithms for order batching such that the total lateness of retrieving and the total travel time per group of orders are minimized.

3.1 Assumptions

1. The S/R machine travels simultaneously horizontally and vertically as it moves along the storage aisle. The horizontal and vertical accelerations are equal.
2. Each storage location accommodates one type of product, and each type of product is stored in one location. All items have the same size and the lead time for order replenishment is negligible.
3. The r-order group size is a normally distributed random variable, while the number of locations per r-order follows a Poisson distribution.
4. Each product has equal probability to be retrieved and the quantity to be retrieved from any location is normally distributed. The total quantity of any r-order does not exceed the capacity of the S/R machine.
5. The due date of retrieval of each r-order follows a uniformly distributed random variable and the s-orders do not have due date constraints.
6. Each r-order must be retrieved in one S/R cycle; however, each S/R machine cycle can process multiple number of r-orders provided that the capacity of the S/R machine is not exceeded.
7. The pick-up and deposit (P/D) time of items along the rack and at the I/O point is ignored. Since, the P/D time is generally independent of the warehouse configuration, and S/R machine velocity and is usually deterministic.

3.2 Notations:

The following notations will be used throughout the chapter:

c_i completion time of order i (time at which order i is available, after processing, at the I/O point)

$C_{(i)}$ completion time of the tour which processes one or more orders with r-order i as the seed r-order

d_i due date of order i (date at which order i is needed, after processing, at the I/O point)

p_i processing time of order i (time required to retrieve/store order i in a single tour by itself)

s_i starting time of order i (start time of processing order i)

T_i lateness of order i, $T_i = \mathrm{Max}(c_i - d_i, 0)$

T_{ij} total lateness of orders i and j, $T_{ij} = T_i + T_j$

TTIME total travel time per group

4 ALGORITHMS FOR ORDER BATCHING

There are three steps to form the interleaving tours, (i) schedule the r-orders with the due date constraints and the local objective of minimizing the total lateness of retrieving the orders, (ii) batch the r-orders to tours based on the schedule generated from previous step with the objective of minimizing both the total lateness and the total travel time per group of orders, (iii) include the s-orders in the r-order tours without significantly increasing the total lateness of the tours and with the objective of minimizing the total travel time per group of orders. If a tour processes both r-orders and s-orders, it is an interleaving tour. If a tour processes r-orders only, it is a retrieval-only tour. If there are some s-orders that cannot be added to any tour containing r-orders due to the due time constraints or the S/R machine capacity, we form the storage-only tours which process s-orders only using the *MaxLTT* (Maximum Longest Travel Time) rule described below.

4.1 Sequencing s-Orders Using the Maximum Longest Travel Time (MaxLTT) Rule

Consider the following rule: s-orders are batched to tours by first adding to one tour the s-order having the maximum LTT (the s-order containing the farthest location from the S/R machine). Since every single s-order does not exceed the S/R machine capacity by itself; therefore, we do not consider the S/R machine capacity as a constraint at this point of time. Then from the remaining s-orders, we select those orders which will not violate the S/R machine capacity (if added to the current tour) to form the candidate list. Repeatedly add s-orders to the current tour until the tour cannot accommodate any more s-orders due to the S/R machine capacity. If there is no s-order in the candidate list, then no more s-orders can be added to the current tour. To start batching s-orders to a new tour, we first select from the remaining s-orders the one having the maximum LTT and add it to the tour. Then update the candidate list from which we select the next s-order having the maximum LTT and add it to the tour. Repeat the above process to batch s-orders until all of them are batched to tours.

4.2 Sequencing r-orders

In this section, we describe the details of sequencing the r-orders. The usual approach of sequencing jobs (orders) in a process is to seek a sequence compatible with the constraints and satisfies an overall objective. In step (i), our constraints are the due dates of order retrieval, and the objective is to minimize the total lateness of r-orders. Before presenting the algorithm, we first examine the effect of the due date constraints on sequencing the r-orders.

4.2.1 Effect of the Due Date Constraint

The objective of sequencing orders for retrievals is to minimize the total lateness of r-orders. Consider two arbitrary r-orders i and j with $d_i \leq d_j$. There are two possible sequences for them "i-j" and "j-i" (sequence "i-j" means that order i is processed before order j). The two sequences are shown in Fig. 1.

In the following description, we use c_i, c_j, p_i, p_j, s_i, and s_j, for sequence "i-j" and c'_i, c'_j, p'_i, p'_j, s'_i, and s'_j, for sequence "j-i" to represent the completion time, the processing time and the starting time of orders i and j respectively. T_{ij} and T'_{ij} are used to represent the total delay of r-orders i and j for sequence "i-j" and sequence "j-i" respectively.

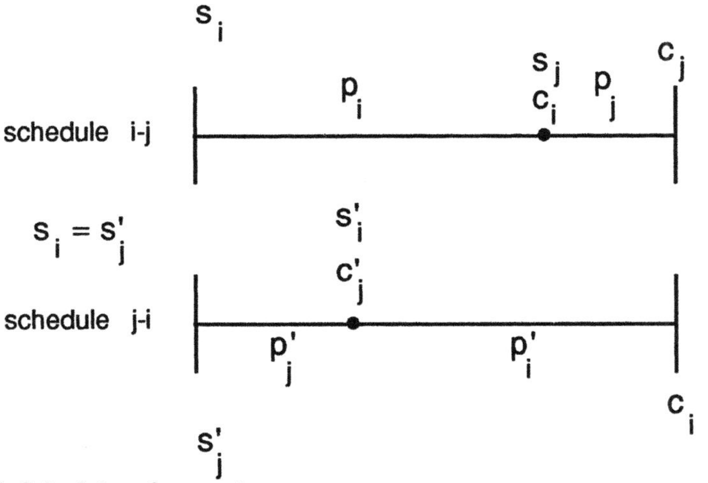

Fig. 1 Schedules of two orders

The total lateness of r-orders i and j is

for i - j, $T_{ij} = \max(c_i - d_i, 0) + \max(c_j - d_j, 0)$
$= \max(s_i + p_i - d_i, 0) + \max(s_i + p_i + p_j - d_j, 0)$;

for j - i, $T'_{ij} = \max(c'_j - d_j, 0) + \max(c'_i - d_i, 0)$
$= \max(s'_j + p_j - d_j, 0) + \max(s'_j + p_i + p_j - d_i, 0)$;

Let $s = s_i = s'_j$ be the starting time.

When $s + p_i + p_j \leq d_i$, then

$T_{ij} = 0$ sequence "i-j" and sequence

$T'_{ij} = 0$ "i-j" are equivalent.

When $d_i \leq s + p_i + p_j \leq d_j$, then

if $s + p_i \leq d_i$, then

$T_{ij} = 0$ sequence "i-j" is preferred

$T'_{ij} \geq s + p_i + p_j - d_i$

if $s + p_i > d_i$, then

$T_{ij} = s + p_i - d_i$ sequence "i-j" is preferred

$T'_{ij} \geq s + p_i + p_j - d_i$

When $s + p_i + p_j > d_j$, then

if $s + p_i \leq d_i$ and $s + p_j \leq d_j$, then

$T_{ij} = s + p_i + p_j - d_j$ sequence "i-j" is preferred

$T'_{ij} = s + p_i + p_j - d_i$

if $s + p_i > d_i$ and $s + p_j \leq d_j$, then

$T_{ij} = 2s + 2p_i + p_j - d_j - d_i \rightarrow$ if $s + p_i \leq d_j$ sequence "i-j" is preferred

$T'_{ij} = s + p_i + p_j - d_i \rightarrow$ if $s + p_i > d_j$ sequence "j-i" is preferred

if $s + p_i \leq d_i$ and $s + p_j > d_j$, then

$T_{ij} = s + p_i + p_j - d_j$ sequence "i-j" is preferred

$T'_{ij} = 2s + p_i + 2p_j - d_i$

if $s + p_i > d_i$ and $s + p_j > d_j$, then

$T_{ij} = 2s + 2p_i + p_j - d_j - d_i \rightarrow$ if $p_i < p_j$ sequence "i-j" is preferred

$T'_{ij} = 2s + p_i + 2p_j - d_i - d_j \rightarrow$ if $p_i > p_j$ sequence "j-i" is preferred

In summary:

If $(s + p_i + p_j > d_j, s + p_i > d_i, s + p_j \leq d_j, s + p_i > d_j)$ or

$(s + p_i + p_j > d_j, s + p_i > d_i, s + p_j > d_j, p_i > p_j)$

sequence "j-i" is preferred; otherwise, sequence "i-j" is preferred.

More compactly, if $(s + p_i > d_i, s + p_j \leq d_j)$ or $(s + p_i > d_i, s + p_j > d_j, p_i > p_j)$, sequence "j-i" is preferred; otherwise, sequence "i-j" is preferred.

Lemma 1: *For two orders i and j with due date d_i and d_j,*

if $(s + p_i > d_j, s + p_j \leq d_j)$ or
$(s + p_i > d_i, s + p_j > d_j, p_i > p_j)$,

sequence "j-i" is preferred;
otherwise, sequence "i-j" is preferred.

4.2.2 Earliest Due Date (EDD) Rule

We develop an algorithm, the Earliest Due Date **EDD** rule, to sequence the r-orders with due dates with the objective of minimizing the lateness of retrievals. Consider this rule, r-orders are first sequenced in non-decreasing d_i's. Then the sequence of r-orders is revised two orders at a time, from the first sequenced r-order to the last sequenced r-order, using Lemma 1 (this is referred to as the forward checking). If reverse of the sequence of two adjacent r-orders i and j occurs, we go backward from r-order i to r-order 1 two orders at a time to check if any exchanges of sequences of r-orders are possible and should be performed (this is referred to as the backward checking), then come back to r-order j to continue the forward checking. (Note: Indices i and j are ordinal indices and do not refer to any specific r-orders.)

The following notations are used in the procedure of the **EDD** rule:

- i — ordinal index (used in forward checking), r-order i is the i-th r-order in the sequence
- j — ordinal index (used in forward checking), r-order j immediately follows r-order i in the sequence, i.e. r-order j is the (i+1)th r-order in the sequence
- s — starting time of r-order i
- k — ordinal index (used in backward checking), r-order k is the k-th r-order in the sequence

t — ordinal index (used in backward checking), r-order t immediately follows r-order k in the sequence, i.e. r-order t is the (k+1)-th r-order in the sequence

The summary of the **EDD** rule is listed below:

1. Rank r-orders by non-decreasing d_i's. This forms a preliminary sequence.
2. Let i=1, j=2.
3. (Forward checking)
   Let s be the starting time of r-order i in the current sequence.
   If $(s + p_i > d_i, s + p_j \leq d_j)$ or
   $(s + p_i > d_j, s + p_j > d_j, p_i > p_j)$ then
   reverse the sequence of r-orders i and j.
   Let k = i - 1, t = i
   Goto step 4
   If no exchange between r-orders i and j, then
   let i = i + 1, j = j + 1,
   repeat step 3 until all r-orders are checked for scheduling.
4. (Backward checking)
   If exchange between r-orders k and t is needed,
   reverse the sequence of r-orders k and t.
   Let k = k - 1, t = t - 1
   Repeat step 4 until t = 1
   Otherwise, continue to step 5.
5. Let i = i + 1, j = j+1 and goto step 3.

4.3 Batching r-orders

In this section, we describe the details of step (ii) in which r-orders are batched to tours (batching means selection of one or more orders to form a single S/R tour) based on the sequence generated in step (i), described in the above section, with the objective of minimizing both the total lateness and the total travel time per group of orders. Three rules are developed, the Nearest Schedule **NS** rule, the Shortest Processing Time **SPT** rule and the Most Common Location **MCL** rule.

Before presenting the procedures and detailed descriptions of the algorithms, we first define the following notations:

$C_{(i)}$ — completion time of the current tour with r-order i as the seed rorder (time the tour is finished, after retrieval/storage, at the I/O point). The seed order is the first order included in a tour.

$CC_{(i)}$ — completion time of the temporary batch which contains the current tour (with r-order i as the seed) and r-order k ($CC_{(i)}$ is used to check if r-order k can be added to the current tour) i ordinal index for the seed r-order of the current tour in the sequence obtained in step (i) by the EDD rule

k ordinal index for the r-order being checked for inclusion in the current tour

4.3.1 Nearest Schedule (NS) Rule

Consider this rule: r-orders are batched to tours by first adding to a new tour the first sequenced r-order, based on the ranking obtained in step (i) in the group (referenced as r-order i, and is referred to as the seed r-order of the current tour). Now the completion time, $C_{(i)}$, of the current tour is the completion time of the seed r-order we just selected. Then we select the first sequenced r-order (referenced as r-order k), which will not violate the S/R machine capacity if added to the current tour, from the r-orders which have not been batched yet. For r-order k, first we perform a loose check: if $C_{(i)} + p_k \leq d_i$, we add order k to the current tour. Since $C_{(i)} + p_k \leq d_i$, is greater than or equal to the completion time of the batch including the current tour and r-order k, this condition insures that the addition of r-order k will not result in lateness of retrieving the seed r-order, and thus will not result in lateness of any other r-orders in the current tour because the seed r-order has the earliest due date in the tour. If the loose check cannot guarantee that the r-order k can be added to the current tour, we perform the tight check which calculates the completion time, $CC_{(i)}$, of the batch including the current tour and r-order k using the TSP (Traveling Salesman Problem) procedure. (The reason for performing the loose check is that the TSP is a time consuming procedure and thus the utilization of the loose check saves a great deal of the CPU time.) If $CC_{(i)} > d_i$, r-order k cannot be added to the current tour, and this is the end of batching r-orders to the current tour. Start batching r-orders to a new tour. If $CC_{(i)} = d_i$ add rorder k to the current tour. And then, since the current tour cannot accommodate any more r-orders due to the due date constraint of the seed r-order, start to batch r-orders to a new tour. If $CC_{(i)} > d_i$, add r-order k to the current tour then select the first sequenced r-order (referenced as r-order k) from the r-orders which are not batched to tours and will not violate the S/R machine capacity if added to the tour. Then repeat the loose and tight checkings to find out whether r-order k can be added to the current tour. Repeatedly add r-orders to the current tour in this way until the tour cannot accommodate any more r-orders due to the due date constraints or the S/R machine capacity. To start batching r-orders to a new tour, we first select a seed r-order (referenced as r-order i) which is the first scheduled r-order among the r-orders that have not been included in any tour. Repeat the selection and checking processes described above. Batches of r-orders are formed in this way until all r-orders are assigned to tours.

The steps of the **NS** rule are:

1. Let i=1
2. Include r-order i in a new tour, r-order i is called the seed r-order.
3. Compute the completion time, $C_{(i)}$, of the current tour.
   If $C_{(i)} \geq d_i$ this is the end of forming the current tour.
   Goto step 7
   Otherwise, continue to step 4.
4. From r-order j's, which have not been included in any tour
   and will not violate the S/R machine capacity if added to the current tour,
   $j \in (i+1, i+2, ..., n)$,
   find the r-order k which is 1st sequenced in r-order j's.
   If no r-order j exists, then the formation of the current tour is ended.
   Goto step 7.
5. (Loose check)
   Let $CC_{(i)} = C_{(i)}$
   $CC_{(i)} = CC_{(i)} + p_k$
   If $CC_{(i)} \leq d_i$, add r-order k to the tour
   Goto step 4
   If $CC_{(i)} > d_i$, continue to step 6.
6. (Tight check)
   Compute the completion time, $CC_{(i)}$, of the tour with r-order k
   If $CC_{(i)} > d_i$, this is the end of the current tour formation
   Goto step 7
   If $CC_{(i)} = d_i$, add r-order k to the current tour
   $C_{(i)} = CC_{(i)}$
   This is the end of the current tour formation, goto step 7
   If $CC_{(i)} < d_i$, add r-order k to the current tour
   $C_{(i)} = CC_{(i)}$
   Goto step 4
7. Let the 1st scheduled r-order which has not been included in any tour be the seed r-order i
   If no r-order i exists, stop; otherwise, goto step 2.

4.3.2 Shortest Processing Time (SPT) Rule

In this rule the r-orders are batched to tours by first adding to a new tour the first sequenced r-order, based on the ranking obtained in step (i) in the group (referenced as r-order i and is referred to as the seed r-order, similar to in the NS rule). Now the completion time, $C_{(i)}$, of the current tour is the completion time of the seed r-order we just selected. Then from the r-orders which are not assigned to tours yet, we select the r-order k which has the shortest processing time and will not violate the S/R machine capacity if added to the current tour. First we perform the loose check as described in the NS rule to determine whether the r-order k can be added: if $C_{(i)}$ + $p_k \leq d_i$ add r-order k to the tour. If the loose check cannot guarantee the addition of r-order k to the tour, we perform the tight check as described in the NS rule which calculates the completion time, $CC_{(i)}$, for the batch containing the current tour and r-order k using the TSP procedure described later in this chapter. If $CC_{(i)} > d_k$, r-order k cannot be added to the current tour, and this is the end of batching r-orders to the current tour. Start a new tour. If $CC_{(i)} = d_i$, add r-order k to the current tour. Since the current tour cannot accommodate any more r-orders due to the due date constraint of the seed r-order, then start batching r-orders to a new tour. If $CC_{(i)} < d_i$, add r-order k to the current tour. Then, from the r-orders which are not assigned to tours, we select the r-order (referenced as r-order k) which has the shortest processing time and will not violate the S/R machine capacity if added to the current tour. Perform the loose and tight checks to determine if the r-order k can be added to the current tour. Repeatedly add r-orders to the current tour until the tour cannot accommodate any more r-orders due to the due date constraints or the S/R machine capacity. To start batching r-orders to a new tour, we first select a seed r-order (referenced as r-order i) which is the first sequenced r-order among the r-orders that have not been included in any tour. We then repeat the selection and checking processes described above to batch r-orders, which are not included in any tour, to the new tour. Batches of r-orders are formed in this way until all r-orders are assigned to tours. As evident, the difference between the SPT rule and the NS rule is the criterion used to select r-orders (excluding the seed r-order). However, the criterion used to select the seed r-orders under both the SPT rule and the NS rule are the same.

The steps of the **SPT** rule are:

1. Let i=1
2. Include r-order i in the tour, r-order i is called the seed r-order.
3. Calculate the completion time of the current tour, $C_{(i)}$
   If $C_{(i)} \geq d_i$, this is the end of the current tour formation
   Goto step 7
   Otherwise, continue to step 4.
4. From r-order j's which have not been included in any tour,
   $j \in (i+1, i+2, \ldots, n)$,
   Find the r-order k which has the shortest processing time and will not result in the violation of the S/R machine capacity.
   If no r-order k exists, this is the end of the current tour formation
   Goto step 7.

5. (Loose check)
   If $C_{(i)} + p_k \leq d_i$, add r-order k to the tour
   Goto step 4
   If $C_{(i)} + p_k > d_i$, continue to step 6
6. (Tight check)
   Calculate the completion time, $CC_{(i)}$, of the
   tour containing the current tour and r-order k using the TSP procedure
   If $CC_{(i)} > d_i$, this is the end of the current tour formation
   Goto step 7
   If $CC_{(i)} = d_i$, add r-order k to the current tour
   $C_{(i)} = CC_{(i)}$
   This is the end of the current tour formation, goto step 7
   If $CC_{(i)} < d_i$, add r-order k to the current tour
   $C_{(i)} = CC_{(i)}$
   Goto step 4.
7. Let the 1st scheduled r-order which has not been included in any tour be the seed r-order i
   If no r-order i exists, stop; otherwise, goto step 2.

4.3.3 Most Common Location (MCL) Rule

Under this rule the criterion used to select the seed r-orders is the same as those used in the NS rule and the SPT rule. The difference between the MCL rule and the two rules described above is the criterion used to select the r-orders other than the seed r-orders. The following description presents the details of the MCL rule. The r-orders are batched by adding to a new tour the first sequenced r-order, based on the ranking obtained in step (i) in the group (referenced as r-order i and is referred to as the seed r-order of the current tour). Then from the r-orders that have not been yet assigned to tours, select the r-order k which will not violate the S/R machine capacity if it is added to the tour and has the most common locations with the current tour. If there is no r-order k, then the batching r-orders to the current tour is completed. Start a new tour. For r-order k, first we perform the loose check as described in the NS rule to determine if the r-order k can be added: if $C_{(i)} + p_k \leq d_i$, add r-order k to the tour. If the loose check cannot guarantee the addition of r-order k to the tour, we perform the tight check as described in the NS rule, which calculates the completion time, $CC_{(i)}$, for the tour containing the current tour and r-order k using the TSP procedure described later in this chapter. If $CC_{(i)} > d_k$, r-order k cannot be added to the current tour, and this completes the batching of r-orders to the current tour. Start a new tour. If $CC_{(i)} = d_i$, add r-order k to the current tour. Since the current tour cannot accommodate any more r-orders due to the due date constraint of the seed r-order, start a new tour. If $CC_{(i)} < d_i$, add r-order k to the current tour. From the r-orders that have not been assigned to tours, select the r-order k which will not violate the S/R machine capacity if it is added to the current tour and has the most common locations with the current tour. Repeat the loose and tight checks to determine whether an r-order k can be added to the current tour. Repeatedly add r-orders to the current tour until the tour cannot accommodate any more r-orders due to the due date constraints or the S/R machine capacity. Batching r-orders to other tours involves the selection of a seed r-order (referenced as r-order i) which is the first scheduled r-order among the r-orders not included in previous tour. Then repeat the selection and checking processes described above to batch r-orders, which are not included in any tour, to the new tour. Batches of r-orders are formed in this way until all r-orders are assigned to tours.

The steps of the MCL rule are:

1. Let $i=1$
2. Include r-order i in the tour, r-order i is called the seed r-order.
   Calculate the completion time of the current tour, $C_{(i)}$
   If $C_{(i)} \geq d_i$, this is the end of the current tour formation
   Goto step 7
   Otherwise, continue to step 4.
3. From r-order j's which have not been included in any tour $j \in (i+1, i+2, \ldots, n)$, Find the r-order k which has the most common locations with the current tour and does not violate the S/R machine capacity If no r-order k exists, this completes the current tour formation Goto step 7
4. (Loose check)
   If $C_{(i)} + p_k \leq d_i$, add r-order k to the tour
   Goto step 4
   If $C_{(i)} + p_k > d_i$, continue to step 6
5. (Tight check)
   Calculate the completion time, $CC_{(i)}$, of the
   tour containing the current tour and r-order k using the TSP procedure
   If $CC_{(i)} > d_i$, this is the end of the current tour formation
   Goto step 7
   If $CC_{(i)} = d_i$, add r-order k to the current tour
   $C_{(i)} = CC_{(i)}$
   This is the end of the current tour formation, goto step 7
   If $CC_{(i)} < d_i$, add r-order k to the current tour
   $C_{(i)} = CC_{(i)}$
   Goto step 4.

6. Let the 1st scheduled r-order which has not been included in any tour be the seed r-order i
   If no r-order i exists, stop; otherwise, goto step 2.

4.4 Inclusion of s-orders in Tours

With r-orders being batched to specific tours, the last step (step (iii)) is to add s-orders to each tour to form the interleaving tours without significantly increasing the lateness of the r-orders and keeping the travel time as short as possible. The inclusion of the s-orders in the tours under the Nearest Location (NL) rule is described below.

4.4.1 Nearest Location (NL) Rule

Suppose after batching the r-orders we have m tours scheduled as $\{t_1,t_2,t_3,...,t_m\}$. The NL (Nearest Location) rule of adding s-orders to tours containing r-orders involves the selection of an s-order which has location nearest to one of the locations in the tour. This is done such that the total lateness is not significantly affected.

The following notations are used in the procedure and description of the NL rule:

e_i	earliness of retrieving r-order i, $e_i = \text{Max}\,(d_i - C_{(i)}, 0)$
g	ordinal index for the last scheduled tour which has the lateness of retrievals
h	index for the s-order having the location nearest to one of the locations in the current tour
k	ordinal index for current tour
m	total number of tours formed in step (ii)

We use the NL rule to add s-orders to the tours formed in step (ii). From the 1st to the last tour, we first check if the lateness of retrievals exists in any tour. If the last tour which has the lateness of retrieval is tour g, then no s-orders can be added to any tour scheduled before tour g, since the addition of s-orders will result in more lateness of retrievals. Then for each tour k scheduled after tour g, we check for the smallest earliness of retrieval, $e_i = d_i - C_{(i)}$, from the r-order i's processed in tour j's which are scheduled after and including tour k. ($C_{(i)}$ is the completion time of a tour which includes the r-order i.) In other words, we check for the smallest earliness of retrieval e_i from r-orders in those tours which can accommodate s-orders. Then with e_i on hand, we find the s-order h which has not been added to any tour and has location nearest to one of the locations in tour k. Route the tour containing the current tour k and s-order h using the TSP procedure. If the addition of s-order h to tour k does not result in extra travel time greater than e_i, we add the s-order h to tour k; otherwise, s-order h is rejected by tour k. If the S/R machine can still accommodate more items for storage, we update the current e_i and find the s-order (referenced as s-order h) which has not been included in any tour or rejected by tour k and has location nearest to one of the locations in tour k, then check if s-order h can be added to tour k. Repeatedly add s-orders to tour k until the tour cannot accommodate any more s-orders due to the due date constraints or the S/R machine capacity. The addition of s-orders to the next scheduled tour is performed using the process described above. If there are some s-orders that cannot be added in any one of the tours 1 through m, we form the storage-only tours using the **MaxLTT** rule described earlier in this chapter.

The algorithm is explained below:

1. Let k=1
2. If no s-order need to be processed, then stop; otherwise continue to step 3.
3. Calculate the processing and completion times for tour j's, j=k,k+1,...,m
4. Check for tour j's, j=k,k+1,...,m
   If delay of retrieval occurs in tour g, $k \leq g \leq m$,
   no more s-orders can be added to
   any tours scheduled before tour g+1
   Let k=g+1
   Goto step 5
   If no delay in any tour results, continue to step 5.
5. From the r-order i's which are batched in tour j's, j=k,k+1,...,m
   find the smallest earliness of retrieving, $e_i = d_i - C_{(j)}$
6. For tour k, if the S/R machine is at its maximum capacity then conclude the tour
   Goto step 9
   Otherwise, find the s-order h which has not been added to any tour or rejected by tour k and has location nearest to one of the locations in tour k.
7. If there is no s-order h, this completes the addition of s-orders to tour k
   Goto step 9
   Otherwise, continue to step 8
8. Compute the completion time, $CC_{(k)}$, of the tour with s-order h
   If $CC_{(k)} - C_{(k)} \leq e_i$, then
   add s-order h to the tour, and
   $C_{(i)} = CC_{(i)}$,
   Goto step 2

Otherwise, reject s-order h from being added to the tour
Goto step 6.

9. Let k = k + 1
   If k ≤ m, goto step 2
   If k > m, continue to step 10
10. If all s-orders have been processed, then stop; otherwise, form storage-only tours to store the remaining s-orders.

5 RESULTS

We develop a simulation program written in FORTRAN which calls IMSL routines to simulate AS/R interleaving systems. The simulation program contains three major components: (i) the order generator, (ii) the order batching procedure, and (iii) the TSP procedure. The order generator calls the IMSL routines to generate random numbers used to form r-orders and check the inventory levels at the end of retrieving/storing each group of orders against the inventory policy, (R,r) or (Q,r). The order batching procedure applies the order batching algorithms presented earlier in this chapter to batch orders in tours. The TSP procedure produces a route with minimum travel time for each tour. The TSP procedure uses the network software program for the Traveling Salesman Problem in [16] with some changes. The simulation program applies the Branch-and-Bound algorithm described in [14] to find the traveling salesman route which is a closed Hamilton cycle. In our TSP procedure, the S/R machine capacity is considered to produce both the capacitated closed Hamilton cycle and the capacitated open Hamilton cycle.

The procedure of the simulation program is listed below:

1. The AS/RS starts with loaded storage bins.
2. Generate r-orders for a new group using the order generator.
3. Form storage/retrieval tours for the r-orders and s-orders using the order batching procedure which applies the order batching rules. If there are both r-orders and s-orders in a tour, we have an interleaving tour. If there are only r-orders or s-orders in a tour, we have a retrieval-only or storage-only tour.
4. Route each tour using the TSP procedure and accumulate the total travel time of this group as well as the total lateness.
5. Generate s-orders for next group according to the inventory policy and the inventory levels at the end of processing the current group.
6. Goto step 2 and continue processing next group.

Simulations were run and comparisons were made for the three combinations of the order batching rules. The combinations, CD(i), are listed in Table 1 with CD(1) for the combination of EDD schedule rule, NS r-order batching rule and NL s-order batching rule, CD(2) for the combination of EDD schedule rule, SPT r-order batching rule and NL sorder batching rule, and CD(3) for the combination of EDD schedule rule, MCL r-order batching rule and NL s-order batching rule.

Table 1 Combinations of order batching rules

Combination	Schedule Rule	Batch r-order Rule	Batch s-order Rule
1. CD(1)	EDD	NS	NL
2. CD(2)	EDD	SPT	NL
3. CD(3)	EDD	MCL	NL

The combined retrieval and storage locations are visited in the same tour. This is performed under the condition that the capacity of the S/R machine cannot be exceeded during any time of the tour.

The simulation results for six examples (Examples 1 through 6) are listed in Tables 2 through 4. Each result in Table 2 is the mean of 200 groups of orders (200 experiments). The group size (number of r-orders per group) has been generated using a normal distribution with μ = 8 and 10, σ = 2 and 3, and the number of locations visited per r-order has been generated using a Poisson distribution with λ = 5.8 and 7; while, the quantity retrieved from each location follows a normal distribution with μ = 40 and 50, σ = 12 and 15. The parameters used in each example are listed on the top of the corresponding figure. Examples 1, 2 and 3 are run under the (R,r) inventory policy, and examples 4, 5 and 6 are run under the (Q,r) inventory policy.

Table 2 E(TTIME) and expected lateness

	E(TTIME)			Lateness		
Example	Combination			Combination		
	CD(1)	CD(2)	CD(3)	CD(1)	CD(2)	CD(3)
1	218.70	233.70	226.74	1.15	10.21	12.62
2	189.00	202.20	194.83	6.70	19.00	15.40
3	232.48	251.35	235.34	11.56	33.60	18.42
4	238.27	247.43	238.76	4.42	13.27	11.39
5	202.14	212.52	206.01	2.91	11.76	8.73
6	247.15	258.54	249.65	5.62	20.78	13.85

*Table 3 Number of results with minimum travel time or minimum lateness of retrievals for examples 1 through 6**

Example	No. of Results with Min. Travel Time			No. of Results with Min. Lateness		
	Combination			Combination		
	CD(1)	CD(2)	CD(3)	CD(1)	CD(2)	CD(3)
1	90	41	97	76	7	32
2	106	55	111	52	14	19
3	80	41	102	64	19	27
4	61	35	122	67	16	31
5	101	50	113	50	12	15
6	66	45	110	63	15	27

*More than one combination may give the result with minimum travel time for the experiment.

*Table 4 Number of times each combination of batching rules produced a minimum travel time or a minimum lateness in 1200 experiments for examples 1 through 6**

Combination	Number of Results with Minimum Travel Time	Minimum Lateness
CD(1)	504	372
CD(2)	267	83
CD(3)	655	151

* More than one combination may give the result with minimum travel time for the experiment.

From Table 2, the simulation results show that combination 1, i.e. CD(1), always gives the shortest expected travel time and the smallest expected lateness. This suggests that the Nearest Schedule (NS) r-order batching rule is better than the other two rules (the Shortest Processing Time (SPT) rule and the Most Common Location (MCL) rule). From Tables 3 and 4, the simulation results also indicate that combination CD(1) most frequently gives the minimum lateness and the frequency of CD(1) giving the minimum travel time is slightly less than that of combination CD(3). So we conclude that the NS r-order batching rule is superior to the other two rules considered in this research.

6 CONCLUSION

In this chapter we present the first investigation of sequencing orders in automated storage/retrieval systems subject to due date constraints. We develop one scheduling rule, three r-order batching rules and one s-order batching rule. The results of the research show that the NS (Nearest Schedule) r-order batching rule results in the smallest lateness of retrieval and the shortest travel time. This rule is superior to other rules investigated in this research.

7 REFERENCES

1. Barrett, B.G., "A Further Digression on the Over-Automated Warehouse: Some Evidence," *Interfaces*, 1, pp. 46-49 (1977).
2. Bozer, A.Y. and J.A. White, "Travel-Time Models for Automated Storage/Retrieval Systems," *IIE Transactions*, 16, 4, pp. 329-338 (1984).
3. Elsayed, E.A., "Algorithms for Optimal Material Handling in Automatic Warehousing Systems," *International Journal of Production Research*, 19, 5, pp. 525-535 (1981).
4. Elsayed, E.A., "Design and Scheduling Rules for Automated Storage/Retrieval Systems," Tech. Report 85-112, Department of Industrial Engineering, Rutgers University (1985).
5. Elsayed, E.A., "Automated Storage Systems," Chapter in *International Handbook of Production and Operations Management*, Editor: Ray Wild, Cassell Educational Ltd., pp. 174-193 (1989).
6. Elsayed, E.A. and O.I. Unal, "Order Batching Algorithms and Travel-Time Estimation for Automated Storage/Retrieval Systems," *International Journal of Production Research*, 27, 7, pp. 1097-1114 (1989).
7. Han, M.H., L.F. McGinnis, J.S. Shieh, and J.A. White, "On Sequencing Retrievals in an Automated Storage/Retrieval System," *IIE Transactions*, 19, 3, pp. 56-66 (1987).
8. Harmatuck, D.J., "A Comparison of Two Approaches to Stock Location," *Logistics and Transportation Review*, 12, 4, pp. 282-284 (1976).
9. Hausman, W.H., L.B. Schwarz, and S.C. Graves, "Optimal Storage Assignments in Automated Warehousing Systems," *Management Science*, 22, 6, pp. 629-638 (1976).
10. Heskett, J.L., "Cube-per-Order Index - A Key to Warehouse Stock Location," *Transportation and Distribution Management*, 3, pp. 27-31 (1963).
11. Heskett, J.L., "Putting The Cube-per-Order Index to Work in Warehouse Layout," *Transportation and Distribution Management*, 4, pp. 23-30 (1964).

12. Kallina, C. and L. Lynn, "Application of the Cube-per -Order Index Rule for Stock Location in Distribution Warehouse," *Interfaces*, 7, 1, pp. 37-46 (1976).
13. Kanet, J.J. and R.G. Ramirez, "Optimal Stock Picking Decisions in Automatic Storage and Retrieval System," *International Journal of Management Science (OMEGA)*, 14, 3, pp. 239-244 (1986).
14. Little, J.D.C., K.G. Murty, D.W. Sweeney, and C. Karel, "An Algorithm for the Traveling Salesman Problem," *Operations Research*, 22, 5, pp. 809-821 (1963).
15. Manufacturing Engineering, "ETVs Automate Storage/Retrieval," pp. 28-29 (September 1989).
16. Phillips, D.T. and A. Garcia-Diaz, *Fundamentals of Network Analysis*, Prentice-Hall, New Jersey, 1981.
17. Rosenblatt, M.J. and Y. Roll, "Warehouse Design with Storage Policy Considerations," *International Journal of Production Research*, 22, 5, pp. 809-821 (1984).
18. Schwarz, J., "Rotary Racking System According to a Simple Principle," *Foerdern und Heben*, 19, 12, p. 761 (1969).

COMPUTER AIDED DESIGN OF INDUSTRIAL LOGISTICS SYSTEMS

Marc Goetschalckx
Georgia Institute of Technology

ABSTRACT

The efficient design of complex and effective industrial logistics systems will require the synergy between design engineer(s) and computers. The design engineers must be provided with a powerful, flexible, and intuitive modeling environment for all phases and aspects of the design. If such design environments are to be used successfully in practice, they must address and integrate the data base design and management, user interface, algorithmic, and technology transfer functions.

This chapter reports on the development of an interdisciplinary methodology for building engineering workstations for the integrated design of industrial logistics systems. System requirements and design principles are derived, and implementation aspects are illustrated based on two prototype implementations for the modeling of manufacturing transportation and distribution systems.

1 INTRODUCTION

The efficient design of complex and effective industrial logistics systems will require the synergy between design engineer(s) and computers. The design engineers must be provided with a powerful, flexible, and intuitive modeling environment for all phases and aspects of the design.

In this chapter, a fundamental design methodology for material flow systems is presented. Based on this methodology, the goals and functional requirements for Computer Aided Engineering (CAE) of industrial logistic systems are developed. The requirements concern data base design and management, user interface, solution procedures and algorithms, and technology transfer. To illustrate the design principles, the internal structure and capabilities of two prototype engineering workstations are then presented.

1.1 Definition and Importance of Industrial Logistics Systems

The definition of Logistics as given by the Council of Logistics Management (COLM) is:

"Logistics is the process of planning, implementing and controlling the efficient, cost-effective flow and storage of raw materials, in-process inventory, finished goods, and related information from point of origin to point of consumption for the purpose of conforming to customer requirements".

As such, logistics is concerned with the movement, storage, and control of material from the suppliers, through the facility, all the way to the customer. Logistics is one of the central functions of any production or manufacturing organization. Logistics encompasses many other disciplines such as facilities design, production and inventory control, and distribution and transportation. Because of the many interrelated subfunctions it is hard to measure the true cost of logistics and to design effective logistics systems.

Kearney (1984), in a report to the National Council of Physical Distribution Management (NCPDM) estimated that logistics in industrial organizations accounted for about 21% of the Gross National Product and, along with manufacturing, formed one of the largest value added cost components for most companies. On the average, about 22.5% of the retail price of a product can be attributed to logistics expenses. In 1982, approximately $650 billion was devoted to logistics in the United States. Warner (1988) reports that 500 billion dollars are spent each year on transportation alone.

In many organizations the different subfunctions are controlled and optimized independently, e.g. transportation costs are reduced by selecting a larger transportation shipment quantity, causing larger inventory cost both at the origin and destination points. Such suboptimizing behavior is encouraged because there does not exist a methodology or a set of tools to address the problem of designing the overall logistics system. Implementation of such a design typically would require the establishment of the function of vice president for logistics.

The logistics function has been undergoing dramatic changes in the last few years primarily due to three factors. First, most of the industrialized nations have achieved a mature replacement economy, where increased competition and marketing pressures dictate shorter product life cycles, larger variety of products, and manufacturing in smaller quantities. Inventory reduction goals and the Just-In-Time (JIT) manufacturing and distribution philosophy have reduced the size and increased the frequency of material movements and stores. This makes logistics and service goals more difficult to attain.

Second, the deregulation of the transportation industry has created many new transportation alternatives and opportunities. For example, the overnight and second day delivery services have virtually appeared out of nowhere. In addition, companies compete more and more on a global level, where markets and production and distribution facilities are defined without regard for national boundaries. This makes it possible and necessary to use radically different distribution strategies. Examples of which are direct distribution, public warehousing, and even third party distribution.

Third, the capabilities of computer based information systems used in logistics systems have exploded, making it possible to track and control materials in real time. Large data bases are now available interactively to support production and distribution decisions. Real time communications can be established with vendors and customers through Electronic Data Interchange (EDI). Trucks and other vehicles can be tracked in real time by satellite. Containers and packages can be identified by bar codes and radio frequency tags. Further information on the importance of logistics can be found in Fox [4].

The result of all these changes is that management has recognized distribution and logistics as a competitive tool to increase service and decrease cost. The above factors have created the need for a continuing, integrated review and redesign of the logistics systems used by the corporation to adapt to the constantly changing business environment.

1.2 Logistics and Distribution Systems Research

In the research literature, the problem of designing these complex logistics systems is largely sidestepped by focusing on some small aspect or dimension of the overall problem. Currently, there exist a large number and variety of analytical models for specific areas of logistics planning. The more specific and specialized the model, the more effective it is in solving its problem. Some of the more prominent examples include vehicle routing programs, which are used to find minimum cost routes to execute the delivery of goods to geographically dispersed customers. A recent survey is given in Golden and Assad [10]. Another example are the systems to determine the best location of distribution centers and the allocation of customers to them. A survey is provided in Aikens [1]. Robeson [15] gives a survey of transportation and distribution systems. Tompkins and White [17] have collected research on material handling and facilities design. Silver and Peterson [16] focus on inventory modeling.

Notwithstanding this large quantity and variety of research on isolated logistics systems, the literature on how to efficiently analyze and design ***integrated*** logistics systems is virtually nonexistent. This observation is even more valid for multi-disciplinary models which combine elements of economic analysis, project management, and traditional operations analysis. There exists a need to integrate operational models with aggregate business models in order to achieve a consistent design environment and to achieve a feasible implementation. This need for integration has been advocated strongly by Perl and Sirisponsilp [14] for distribution problems and by McGinnis [13] for facilities design.

The gains in productivity derived from integrated logistics systems can be substantial. Kearney [11] reports that industrial corporations saved more than 12% when they went to an integrated approach of design and operation of distribution systems. The strategic goal of this research is thus to develop a basic understanding, design methodology, and decision making theory for integrated logistics systems.

2 INTEGRATED, MULTI-DISCIPLINARY LOGISTICS SYSTEMS

2.1 Strategic Steps in Distribution System Design

An integrated approach to industrial logistics planning requires a sequence of four steps.

In the first step, realistic customer service goals are developed to support the organization's business goals and strategy and marketing plans. A typical example might be to "meet customer demand from a single distribution center 98% of the times within 24 hours of order placement". Or in the manufacturing scenario the service goal could be "let no machine wait longer than 5 minutes for its next job".

Based on these goals, a materials deployment strategy is developed which makes the tradeoffs between raw materials procurement, production, and finished goods management. A typical decision at this level might be to "make-to-stock of individual components with a two week safety stock and assemble-to-order for a customer order". In the manufacturing environment one might select to operate under a Materials-Requirements-Planning (MRP) or a pull oriented Just-In-Time (JIT) philosophy.

Using this materials deployment strategy a facility and network strategy is then developed. Typical decisions at this level are the number and location of distribution echelons and centers and the allocation of customers to distribution centers. In manufacturing, possible decisions are to use "point-of-use" or central storage.

A transportation strategy is then developed to support the movement of the goods. A typical example is to "ship by air overnight for customers farther than 300 miles away and to ship by truck within 300 miles". In manufacturing, examples of decisions are to operate the material handling system in a dynamic response mode or to operate a fixed set of schedules.

Finally, the operational decisions are made, such as which truck and which driver will serve a set of customers and in what order. A schematic view of the strategic steps is given in Fig. 1.

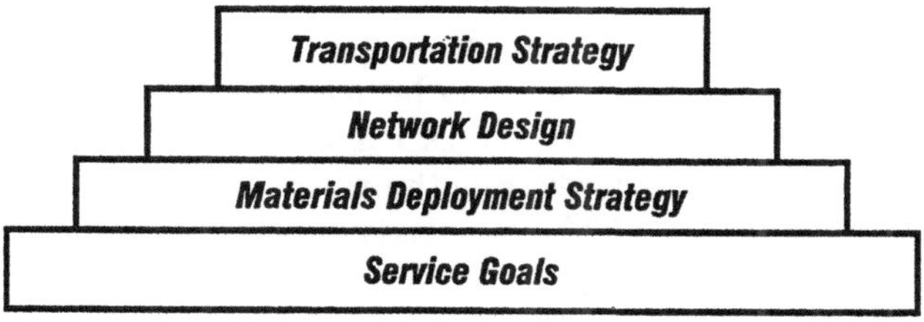

Fig. 1 Strategic steps in integrated logistics systems design

The research presented here focuses on the development of models for the last two of the strategic steps and the identification of the parameters and decisions which connect the different steps. Specific questions to be answered are: 1) what are the objectives, constraints, and decisions that must be made at each step, 2) how are previous step decisions passed as constraints to following step models, and 3) how are the results of later steps aggregated and passed back as parameters to previous steps.

2.2 Support for All Stages of Design

The design of any industrial logistics system involves three major phases. Initially, a great variety of different alternatives are evaluated through rapid prototyping, which is also called rough cut design. At the end of this phase a limited number of alternatives are retained. In the second phase, these alternatives are designed and optimized in detail. At the end of the second phase, a single alternative is selected for implementation. In the third phase, this alternative is then validated through simulation, and sensitivity analysis is performed. The design process might cycle through these different stages, if alternatives have been rejected at a later phase.

The models and modeling techniques in the different phases are quite different with respect to realism and resource requirements. In the preliminary design, the objective is to evaluate as large a variety as possible in a short amount of time. In the validation phase a highly realistic simulation requires large amounts of computer time. To reduce the computational burden on the simulation, the rapid prototyping phase will reduce the range of feasible parameter values. Similarly, the simulation will provide more accurate estimates of parameters which in turn can be used in the preliminary design.

To support these activities for the different phases of the design, the design tool should provide at least the following three capabilities in an integrated way.

First, it must allow the easy, interactive and *graphical definition and modification* of the system by providing specialized drawing symbols and specialized operations on those symbols. Many existing commercial drawing packages can be extended to allow this graphical definition. For example, a user should be able to move a customer by just dragging the customer symbol to a different location. Similarly, the user must be able to execute all of the solution procedures from anywhere in the program by selecting a command from the menu structure. The commands should be named following the problem terminology.

Second, the design tool must support direct, *interactive analysis* of the design from within the design package to allow for rough cut design and evaluation. This requires the extension and modification of the menu and function structure of the drawing package. The design tool must also be able to generate the appropriate reports based on the results of the preliminary design.

Third, the design tool must provide an *easy interface with off-line simulation and optimization* programs, which are used for detailed design and verification. These detailed design and verification packages are often industry standards and their input and output cannot be modified. The design tool should provide means to integrate these packages.

These capabilities require different types of highly specialized software. The drawing and user interaction functions should be based on interactive colorgraphics and human factors principles. The interactive analysis requires knowledge of topology to translate a geometrical design into a logical design on which the interactive analysis algorithms can then operate. Finally, off-line optimization and simulation are well known software types, with their own specialized requirements and techniques.

The developed methodology supports this natural flow of the design process by providing different models for different phases, but by collecting data at all levels of detail in a single data base. This assures data consistency and easy aggregation and disaggregation depending on the desired level of detail and realism. A schematic illustration of the different components and modules and the information flow in CIMPEL and AGVS-EWS implementations is given in Fig. 2.

Many industrial logistics problems have a natural geometrical and graphical representation. For example the location of customers and plants on a map and the location of aisles and machines on facility blueprint. At the same time, a large variety of alphanumeric data is required. Examples are the average and variance of customer demands and the speed of the automated guided vehicle.

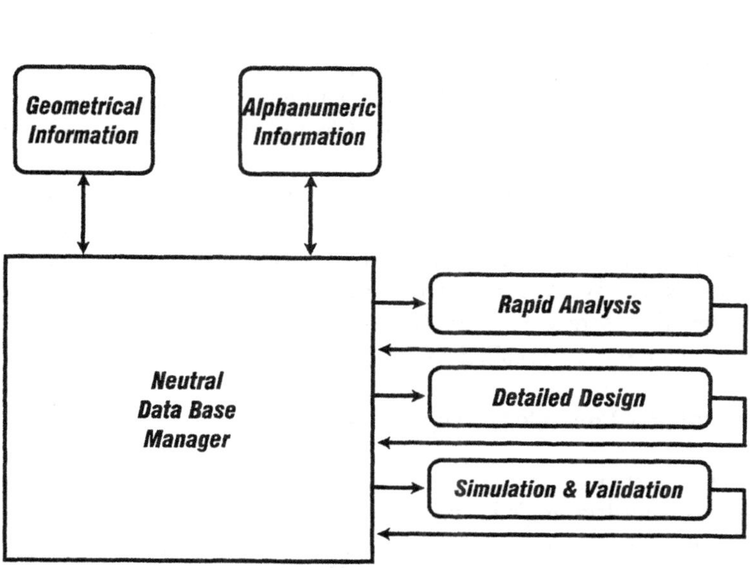

Fig. 2 Modules and inforamtion flow

Examples of rapid analysis models are the continuous location-allocation model to locate distribution centers in the plane and the creation of a planar relationship diagram for facilities design. Detailed design might involve the solution of discrete location-allocation models with site dependent costs and individual transportation channel costs. Detailed facilities design creates a layout with aisles and machine locations and transportation network.

Specific questions to be addressed here are: 1) what are the required level of detail and realism for parameters and variables at each stage, 2) how to achieve automatic aggregation and disaggregation when moving between stages, 3) which data must be collected in the field and how can other data be derived from it, and 4) which kind of data base to use and how to structure it so as to support the modeling effort most efficiently.

3 COMPUTER INTEGRATED DESIGN METHODOLOGY

The research reported on in this chapter extends the state of the art in understanding, analyzing, and designing industrial logistics systems. It should be emphasized again that industrial logistics is concerned with transportation, inventory, and production location and allocation between different facilities and within a single facility. The same methodologies and tools can be used for the design of material flow systems in the plant and between plants, e.g. order picking and routing delivery vehicles have the same fundamental structure. Similarly, the location of departments in plant layout shows similarities with the location of distribution centers in a distribution network.

The main research main focus and primary research tool is to develop the methodology and a prototype of an integrated, multi disciplinary, modeling environment implemented on an engineering workstation. This environment also functions as the primary vehicle for technology transfer to industry and constitutes at the same time a learning laboratory for university education on industrial logistics and material flow. Most of the examples and illustrations will be drawn form two scenarios.

The first scenario focuses on the movement of goods between plants. The goal for this scenario is to create a Computer Integrated Modeling and Planning Environment for Logistics or **CIMPEL**. The second scenario builds an Engineering Workstation (EWS) for the design of in-plant, carrier-based transportation systems such as Automated Guided Vehicle Systems (AGVS). This implementation is called **AGVS-EWS.**

3.1 Computer Integrated Design Tools

In many instances, there are two primary difficulties associated with designing industrial logistics systems. First, it is difficult to understand and model the interrelationships between service, inventory, facility, production, and transportation costs. Second, it is difficult to collect accurate data for the different models. As a result logistics systems are often designed, based on previous experience and with a limited use of algorithms. In addition, analysis and design algorithms focus on components without taking an integrated look at the whole problem. Finally, the use of simulation as a design and prototyping tool either increases the cost and duration of the design process or reduces the number of alternatives that are investigated.

The approach in developing a prototype CAE tool to assist the designers has been based on the principle of "providing the designer with the right tool at the right time in an integrated environment". The **integration of existing, commercially available, and specialized software** provides all the different tools. The main mechanism to achieve the integration is the concept of a **neutral data base**.

Hence, the design tool structure consists of two main segments: the central neutral data base and the specific function modules interfacing with it. The design engineer or user directs the flow of control between the different modules, interprets analysis or simulation results, and adjusts the design accordingly.

This integration approach has the following advantages when compared with the development of a stand-alone tool. The development process of the CAE tool was substantially shortened because most of the software was available and debugged. It was felt that these commercial packages could provide a higher level of quality and functionality than software designed and developed specifically for this project. The use of commercially available software has the additional advantages of reducing the amount of training required, increasing the level of proficiency of the users, improving the ease of transfer of CAE technology, and increasing the portability of the CAE

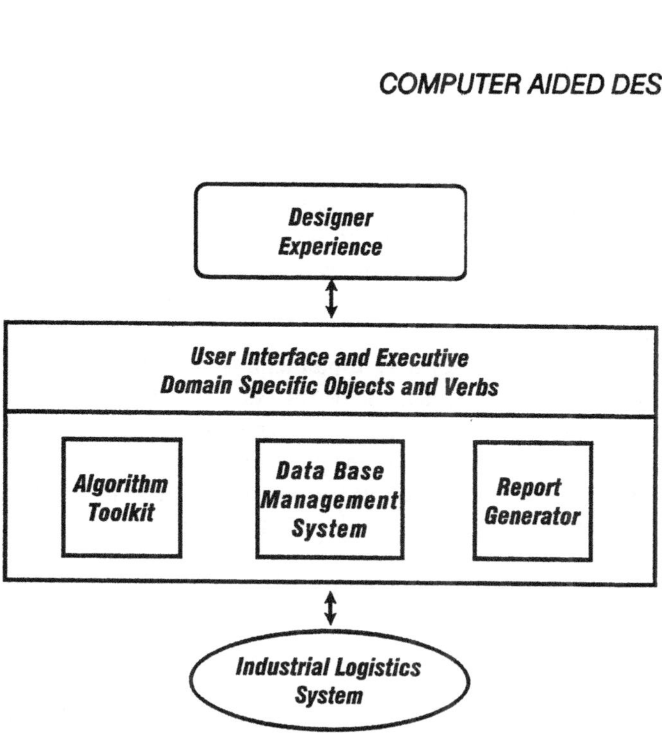

Fig. 3 CIMPEL computer integrated design environment

tool. Finally, detailed design and acceptance testing can be done with industry standard software.

The disadvantage is that the interaction between the different components is not as smooth and instantaneous as with a stand-alone program developed by a single programming team. Most of the commercial packages have their own specialized input and output formats and requirements. The integration required a translation of the output of one package so that it can be used as the input of another package.

The concept of a *neutral data base* was chosen over the development of custom one-to-one interfaces between the different software modules to facilitate this translation. A neutral data base is the repository of all the data for a particular design project *in a format independent of any module*. The output of one module is translated and stored into the neutral data base and then the next module retrieves its data from the data base and translate it to correspond to its input requirements. This requires two translation steps between the execution of any two modules.

In the case of one-to-one interfaces, the number of the interfaces required would grow quadratic with the number of modules and, any time a new module is added, interfaces for the existing modules would have to be created. With the neutral data base concept, the number of interfaces grows linearly with the number of modules and interfaces for existing modules never have to be changed or added. In addition, the neutral data base allows all the modules in the system to share the same set of data, insuring data and design consistency. This data consistency and function modularity far outweigh the need for additional processing time.

Results of programs which solve the same design question but at different levels of detail will be saved in the same format. For example, a continuous location-allocation rough cut solution can be directly compared with the solution of a mixed integer programming model regardless of their native input and output formats.

The developed methodology and implementation provide a powerful and flexible toolkit to design industrial logistics systems. In the future, it will provide intuitive use of the data base manager, report generator and algorithm execution through a colorgraphics interface. One can think of CIMPEL or AGVS-EWS as the Excel (an integrated graphical spreadsheet) for logistics. An illustration of the developed structure is given in Fig. 3.

4 PROTOTYPE IMPLEMENTATIONS

4.1 AGVS EWS

Currently, the above methodology for building design tools has been implemented and tested in two prototypes. The first prototype is an "Engineering Workstation for the Design of Carrier Based Material Flow Systems". This system allows the integrated design of carrier based material handling systems such as Automated Guided Vehicle Systems (AGVS), Automated Electrified Monorail Systems (AEMS), and traditional fork lift truck systems. The design system has been developed by a team of faculty and graduate students over the last four years led by McGinnis and the author. It uses AutoCAD as its interactive executive and colorgraphics interface. At the current time the conversion of the neutral data base from a flat file system to the Oracle relational data base management system has been undertaken.

The workstation has been a crucial tool in developing new guidepath layout algorithms which extend current methodology. It has also been successfully tested in two industrial workshops where participants out of industry bring their problem to the workshop and after three days have a preliminary design of their system. As such the workstation functioned as a technology transfer tool. Further information on the workstation can be found in McGinnis and Goetschalckx [12] and in McGinnis [13]. A schematic illustration of a AGVS network design is given in Fig. 4.

Other research based on or related to the Engineering Workstation are Computer Aided Layout integrated with material handling, reported on by McGinnis [13], and the direct interaction between system design and hardware control, reported on by Bohlander et al. [2].

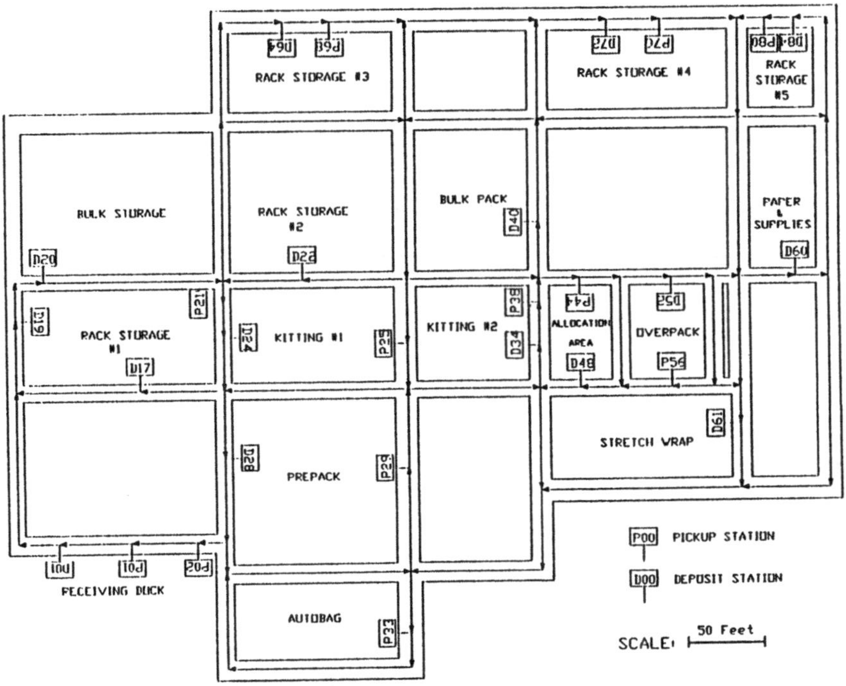

Fig. 4 AGVS network illustration

4.2 CIMPEL

The second prototype implementation is an engineering workstation for the design of strategic distribution systems or CIMPEL (Computer Integrated Modeling and Planning Environment for Logistics). This system allows the integrated design of distribution networks. The design system has been developed by a team of faculty and graduate students over the last two years. The CIMPEL project has been conducted within the framework of the Material Handling Research Center (MHRC). The MHRC is industry supported and the member companies have had significant impact on the direction of the research. At the same time technology transfer has been a major concern from the start. The CIMPEL environment has been tested on several case studies with the member companies.

Geoffrion and Graves [6] and Geoffrion et al. [7] have developed a successful tool for solving the one echelon transportation and location-allocation question, but they ignore inventory and service level considerations. Perl and Sirisponsilp [14] developed a model which includes inventory considerations, but no local delivery costs. No solution algorithm is specified. Bookbinder and Reece [3] include local delivery but no inventory costs and their solution algorithm is only suitable for small problems.

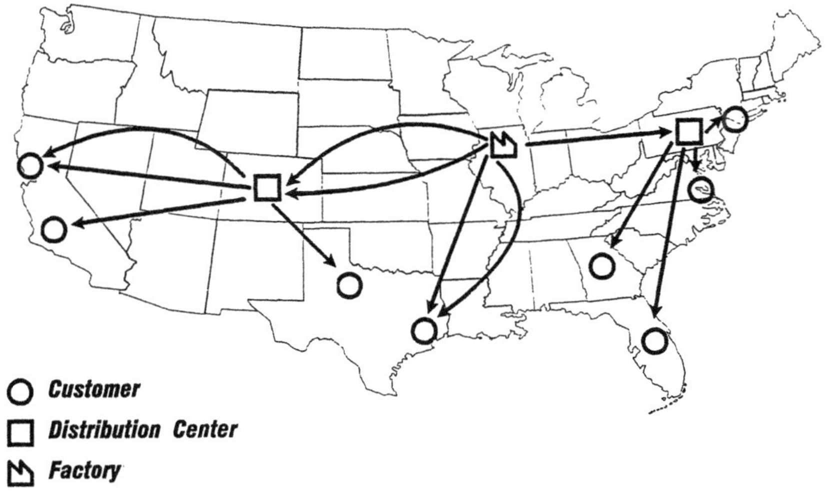

Fig. 5 Network location-allocation illustration

CIMPEL is based on the Microsoft Windows environment for its interactive executive and colorgraphics interface. It is currently being used to develop new algorithms to solve Multi-Echelon, Multi-Channel Location-Allocation problems. A schematic illustration of such a network location-allocation problem is given in Fig. 5. Questions arising in the strategic design of distribution networks are the number and locations of distribution centers and the material flows and transportation channels to the customers.

The data elements in CIMPEL are essentially organized into six groups. The first group contains the project definition data. The data input screen for an example is shown in Fig. 6. It essentially defines the project, data and modeler, basic units and names of more detailed data groups such as customer data, distribution center data, supplier data, and transportation channel data.

The customer data block contains the following items for each customer: name, label, x and y coordinate, average demand and standard deviation of demand. The supplier data contains the following items for each plant: name, label, x and y coordinate, and capacity. The distribution center data block contains the following items for each center: names, label, x and y coordinate, lower and upper bound on throughput capacity, fixed site cost and variable throughput cost. The transportation channel block contains the following items for each transportation channel: name, label, type, origin node label, destination node label, lower bound and upper bound on throughput capacity, fixed cost, variable cost, length, transit time, and shipment size. Finally, the last data block contains various scalar data items, some of which might be aggregates of stie or channel data. Examples are

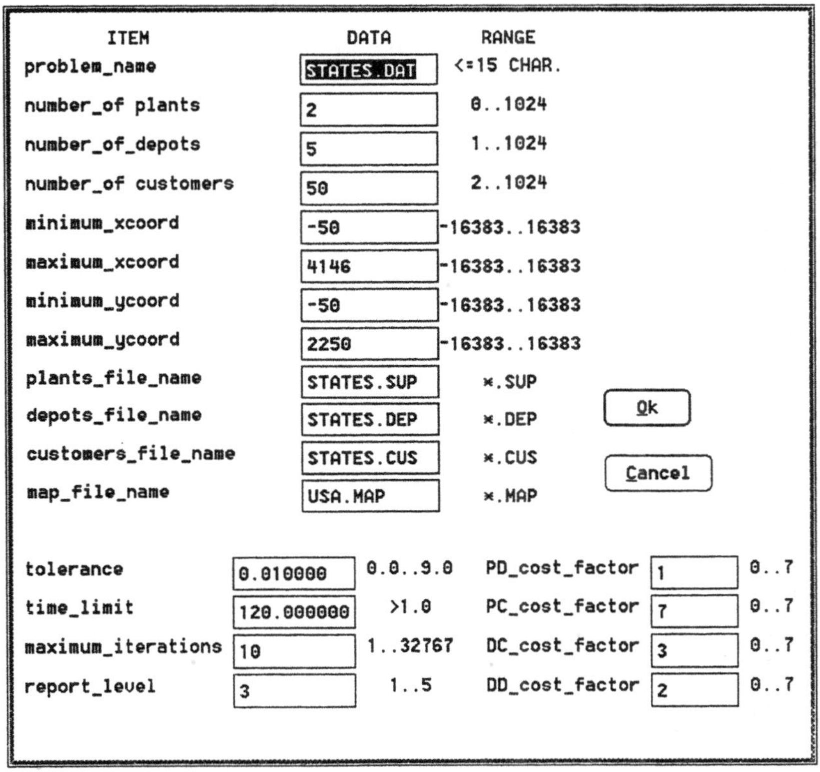

ITEM	DATA	RANGE
problem_name	STATES.DAT	<=15 CHAR.
number_of plants	2	0..1024
number_of_depots	5	1..1024
number_of customers	50	2..1024
minimum_xcoord	-50	-16383..16383
maximum_xcoord	4146	-16383..16383
minimum_ycoord	-50	-16383..16383
maximum_ycoord	2250	-16383..16383
plants_file_name	STATES.SUP	*.SUP
depots_file_name	STATES.DEP	*.DEP
customers_file_name	STATES.CUS	*.CUS
map_file_name	USA.MAP	*.MAP

Ok Cancel

tolerance	0.010000	0.0..9.0	PD_cost_factor	1	0..7
time_limit	120.000000	>1.0	PC_cost_factor	7	0..7
maximum_iterations	10	1..32767	DC_cost_factor	3	0..7
report_level	3	1..5	DD_cost_factor	2	0..7

Fig. 6 Project summary data entry screen

straight line to real (over-the-road) distance adjustment factors, average depot to customer cost factors, average depot fixed and variable cost.

For example, the rapid analysis assumes a complete network from plants to customers, from depots to depots and from depots to customers. The costs are derived based on the straight line Euclidean distance multiplied by the real distance adjustment factor and the appropriate cost factor. Total system cost would be based on the transportation cost and average depot costs. Solutions are obtained by iterative location-allocation heuristics. The location problem is solved with the iterative Weiszfeld algorithm, see Francis and White [5]. The allocation problem is solved with a minimum cost network flow algorithm. The more detailed mixed integer programming model uses the individual channel costs and depot costs. At the current time general purpose mixed integer programming packages are used such as LINDO and ZOOM. An illustration of a typical output screen is given in Fig. 7.

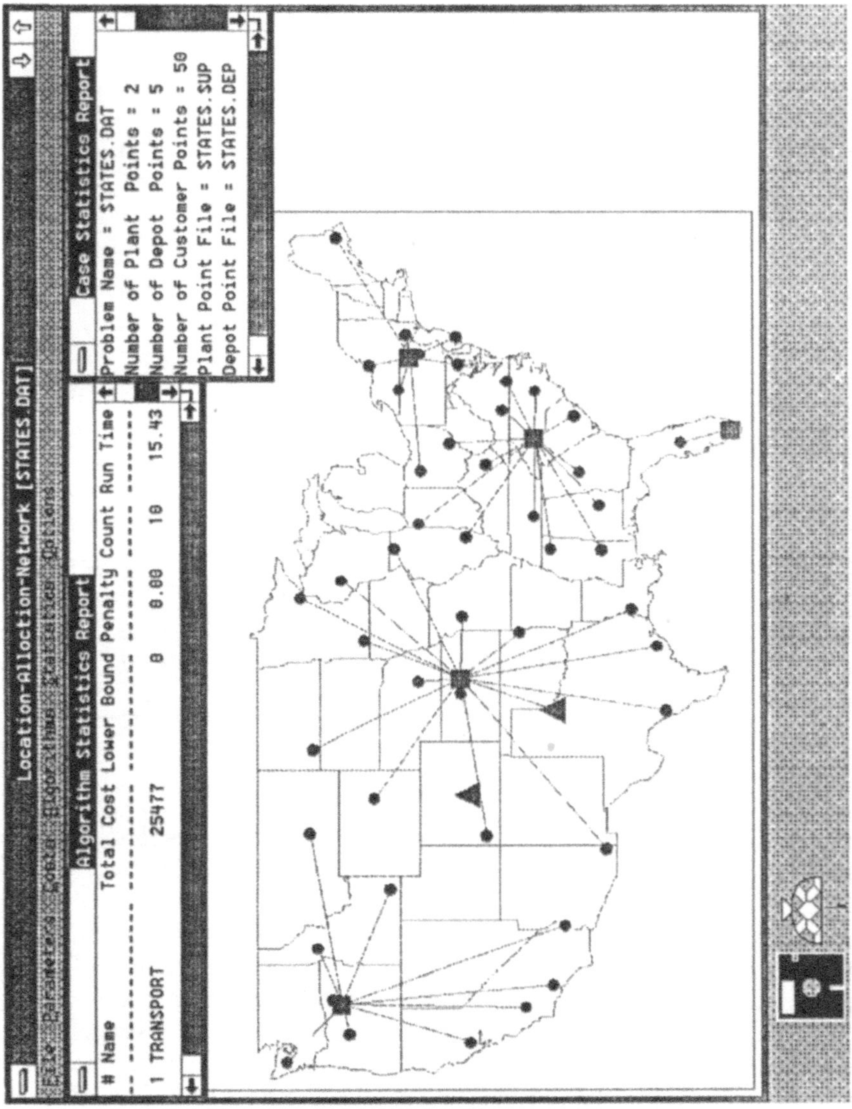

Fig. 7 CIMPEL output screen example

A further description of the functional requirements and the overall program structure of CIMPEL is given in Goetschalckx [8].

5 CONCLUSIONS

The primary research contribution is the development of the structure for a CAE design tool, which maximizes the flexibility and power of the tool, while at the same time minimizing the programming and training time. The main principle of this EWS is the integration of existing special purpose programs through the concept of a neutral data base. Standard commercially available programs were used as much as possible for each of the EWS modules. This assures easy portability, training, and availability. In addition, a number of customized rapid prototyping analysis tools have been developed, which allow the engineer to reserve computationally expensive optimization and simulation program to be exercised only on a limited number of alternative designs. The structure of the CAE design tool has been proven effective and portable based on two prototype implementations. One is an engineering workstation for designing AGVS systems inside a plant, the second is a modeling environment for designing strategic distribution systems.

6 ACKNOWLEDGEMENTS

This research has been supported in part by the member companies of the Material Handling Research Center of the Georgia Institute of Technology. Additional support was provided by the Intel Corporation through an equipment grant.

7 REFERENCES

1. Aikens, C.H., "Facility Location Models for Distribution Planning," *European Journal of Operational Research,* Vol. 22, pp. 263-279 (1985).
2. Bohlander, R.A., W.D.Holcombe, and J.W. Larsen, "An Advance AGVS Control System: An Example of Integrated Design and Control," *Proceedings of the 1990 Material Handling Research Colloquium*, Hebron, KY (June 19-21, 1990).
3. Bookbinder, J.H. and K.E. Reece, "Vehicle Routing Considerations in Distribution System Design," *European Journal of Operational Research,* Vol. 37, pp. 204-213 (1988).
4. Fox, M.L., "Integrating Logistics Systems Across Functional Lines Gives Companies a Competitive Edge," *Industrial Engineering,* Vol. 19, No. 6, pp. 34-43 (1987).
5. Francis, R.L. and J.A. White, *Facility Layout and Location,* Prentice-Hall, Englewood Cliffs, NJ (1974).

6. Geoffrion, A.M. and G.W. Graves, "Multicommodity Distribution System Design by Benders Decomposition," *Management Science*, Vol. 20, No. 5, pp. 822-844 (1974).
7. Geoffrion, A.M., G.W. Graves, and S. Lee, "Strategic Distribution Systems Planning: A Status Report," Chapter 7 in *Studies in Operations Management*, A. Hax (ed.), North Holland Elsevier, Amsterdam & New York (1974).
8. Goetschalckx, M., "A Computer Integrated Logistics Workstation: Modeling Framework and Program Structure," *Proceedings of the Material Handling Focus '88*, Research Forum, Atlanta, GA (Sept. 13-14, 1988).
9. Goetschalckx, M. and L.F. McGinnis, "An Engineering Workstation for Computer Aided Engineering of Material Flow Systems," *Industrial Engineering*, Vol. 21, No. 6, pp. 34-38 (1989).
10. Golden, B.L. and A.A. Assad (editors), *Vehicle Routing: Methods and Studies*, Elsevier Science Publishers, New York, NY (1988).
11. Kearney, A.T., *Measuring and Improving Productivity in Physical Distribution*, National Council of Physical Distribution Management (NCPDM), Oak Brook, IL (1984).
12. McGinnis, L.F. and M. Goetschalckx, "An Engineering Workstation for Computer Aided Engineering of Material Handling Systems," *Proceedings of Manufacturing International '88*, Atlanta GA, Vol. III, pp. 137-142 (1988).
13. McGinnis, L.F., "Computer Aided Facilities Design Revisited: A Prototype Design Workstation for AGV Systems," *Proceedings of the 1990 Material Handling Research Colloquium*, Hebron, KY (June 19-21, 1990).
14. Perl, J. and S. Sirisponsilp, "Distribution Networks: Facility Location, Transportation and Inventory," *International Journal of Physical Distribution and Materials Management*, Vol. 18, No. 6, pp. 18-26 (1988).
15. Robeson, J.F. and R.G. House (editors), *The Distribution Handbook*, The Free Press, New York, NY (1985).
16. Silver, E.A. and R. Peterson, *Decision Systems for Inventory Management and Production Planning*, 2nd Ed., John Wiley and Sons, New York, NY (1985).
17. Tompkins, J.A. and J.A. White J.A., *Facilities Planning*, John Wiley and Sons, New York, NY (1984).
18. Warner, E., "Smart Maps: New Route to Profits," *High Technology*, Vol. 8, No. 12, pp. 20-23 (1988).

TRANSPORTATION WITH COMMON CARRIER AND PRIVATE FLEETS: SYSTEM ASSIGNMENT AND SHIPMENT FREQUENCY OPTIMIZATION

Randolph W. Hall
University of California
Michael Racer
Memphis State University

ABSTRACT

In this chapter, a shipment can either be served by a private carrier or a common carrier. The cost of using the private carrier declines as the space between stops decreases. Therefore, if more stops are assigned to private carriage, the cost of adding another stop will decline. The methodology presented in this chapter for assigning stops to private and common carriage accounts for this cost interaction through use of a square-root route length approximation. Techniques are presented for systems with (1) fixed shipping cycle for all stops with a variable fleet size, (2) fixed shipping cycle for all stops with a fixed fleet size, and (3) variable shipping cycles with a variable fleet size. In the last case, the optimal time between shipments accounts for both transportation and inventory costs.

1 INTRODUCTION

Many manufacturers and distributors utilize private fleets, or contracted transportation service, for the purpose of collecting and delivering shipments for their facilities. In addition to offering greater control over goods movement, private fleets may reduce costs over common carrier prices. Where common carriers typically require that shipments be processed at consolidation terminals, private fleets can transport shipments directly from origin to destination via multiple stop routes. Hence, sorting and circuity costs may be eliminated or reduced.

Even if a company establishes a private fleet, it may not be economical to forego common carriage completely. Due to its scale economies, a common carrier may be able to offer a lower price, for small shipments in particular. To take an extreme example, few companies would use a private fleet to retrieve small packages or letters from their vendors. These can be more economically handled by the postal service, package services, or less-than-truckload (LTL) carriers. At another extreme, remote shipments, whether large or small, may also be good candidates for common carriage, especially when the private carriage fleet size is small.

In the design of a private carrier system, one needs to assess which suppliers or customers (which will be called stops) should be served by the private carrier and which should be served by common carrier. For many stops, the decision is obvious, but for some, the decision will depend on how other stops in its vicinity are served. The cost per stop of a private carrier system serving a fixed region size declines as the number of stops increases. Common carrier prices, on the other hand, usually do not depend on stop spacing. Hence, private carriage is most attractive in regions where stop density (stops/area) is large. Furthermore, as more stops are included in a private carriage system, the cost for making additional stops will decline. A critical stop density may be needed before any stops are served by private carriage.

This chapter develops methodologies for assigning stops to private and common carriage systems. Three related techniques are developed for the situations: (1) fixed service frequency, unconstrained fleet capacity, (2) fixed service frequency, constrained fleet capacity, and (3) variable service frequency, unconstrained fleet capacity. For (1) and (2), the objective will be to minimize total transportation cost. For (3), the objective will be to minimize total inventory and transportation cost. All techniques will exploit the square-root approximation for vehicle route length developed by Daganzo [3]. Routes will either serve pickups or serve deliveries, not both.

Prior research on the subject seems to be limited. Heuristics have been developed by Ballou and Chowdhury [1] and Jordan [8]. Both begin with an initial assignment of stops to systems, and initial vehicle routes for private carriage. They then attempt to reduce cost through small changes in the assignment and, subsequently, the routes. Both utilize exact stop locations (rather than stop densities) and exact shipment sizes. Racer [9] has developed a technique based on the Generalized Assignment Problem (GAP). In solving a standard vehicle routing problem, GAP assigns stops to seed points, which represent vehicle routes. Racer includes an additional seed point representing common carriage.

2 VARIABLE FLEET SIZE MODEL

In this section, the fleet capacity will be allowed to vary in accordance with the quantity of freight assigned to private carriage. Costs will be approximated from Daganzo's route length model, which represents the average cost to serve a set of stops per transportation cycle. It is assumed that all stops are served in each cycle. The following general parameters are used:

α = cost per vehicle mile traveled
β = fixed cost per stop
γ = fixed cost for initiating a tour
A = size of region served
N = number of stops served by private carrier
M = total number of stops in region
ρ = stop density for private carriage = N/A
v = vehicle capacity
ω = overhead cost per shipment

The stop specific parameters are:

b_i = cost for common carriage shipment, stop i
d_i = distance to stop i
q_i = shipment size.

The common carriage cost is defined by a rate schedule, which is usually a function of q_i and d_i, as well as freight classification.

The decision variables are integer, representing whether or not each stop is included on a private carrier tour:

$$X_i = \begin{cases} 0 & \text{stop served by common carrier} \\ 1 & \text{stop served by private carrier} \end{cases}$$

The total transportation cost per cycle can now be expressed as:

$$C = \left[\alpha k\sqrt{\Sigma X_i}\,\sqrt{A} + \beta\Sigma X_i\right] + \left[\frac{\Sigma X_i(\gamma+2\alpha d_i)q_i}{v}\right] + [\Sigma(1-X_i)b_i] + \omega M \qquad (1)$$

The four terms in Equation (1) represent local delivery cost, line-haul cost, common carriage cost, and overhead cost, respectively.

A necessary condition for optimality is that cost cannot be reduced if a stop is reassigned from one transportation system to the other. The change in cost due to a reassignment can be approximated from the derivative of Equation (1) with respect to X_i, evaluated at X_i =1. If the derivative is negative, then a stop should be served by private carriage, and if the derivative is positive, then a stop should be served by common carriage. For a given value of $\rho=\Sigma X_i/A$, this condition amounts to the following:

$$\frac{\partial C}{\partial X_i}(1)=\frac{\alpha k}{2\sqrt{\rho}}+\beta+\frac{(\gamma+2\alpha d_i)q_i}{v}-b_i \quad \begin{matrix}\le 0\text{: private}\\ > 0\text{: common}\end{matrix} \tag{2}$$

Equation (2) indicates that private carriage becomes more attractive as ρ increases, due to the reduction in local transportation cost. Therefore, a break-even point may exist for ρ, above which a stop should be served by private carrier, below which a stop should be served by common carrier.

2.1 Private Carriage Preferred

$$N/A=\rho\ge\tilde{\rho}_i=\left[\frac{\alpha k/2}{b_i-\beta-(\gamma+2\alpha d_i)q_i/v}\right]^2,\quad b_i\ge\beta+\frac{(\gamma+2\alpha d_i)q_i}{v} \tag{3}$$

where $\tilde{\rho}_i$ is the break-even point on ρ for stop i. For some stops, the optimal transportation system is independent of ρ. Specifically:

$$b_i<\beta+\frac{(\gamma+2\alpha d_i)q_i}{v}: \qquad \tilde{\rho}_i=\infty \tag{4}$$

$$b_i>2\alpha d_i\left[\frac{q_i}{v}\right]^{+}+(\gamma+\beta): \qquad \tilde{\rho}_i=0 \tag{5}$$

The second condition follows from an upper bound on private carrier cost, assuming that shipment i is the sole stop on a tour.

Equations (3)-(5) establish a ranking among stops, such that if $\tilde{\rho}_i\le\tilde{\rho}_j$, stop j will only be served by private carriage if stop i is also served by private carriage, and stop i will only be served by common carriage if stop j is served by common carriage. Therefore, to find the optimal solution, one need only compare different cut-off points within the ranking.

Without loss of generality, suppose that stops are sequenced in ascending order according to $\tilde{\rho}_i$. Also let $P(\rho)$ represent the set of stops assigned to private carriage by Equations (3)-(5) when stop density is ρ. Then a characteristic of an optimal cut-off point is the following:

$$\tilde{\rho}_i\le\rho^*=|P(\tilde{\rho}_i)|/A<\tilde{\rho}_{i+1}, \qquad i\,\varepsilon\,P \tag{6}$$

In words, if ρ^* is optimal, all of the stops assigned to private carriage must have break-even points at or below ρ^*, and all of the stops assigned to common carriage must have break-even points above ρ^*. Equation (6) is a necessary, but not sufficient, condition for optimality. If more than one solution satisfies Equation (6), the costs of these solutions must be compared to identify the optimum, as illustrated in the following example.

2.2 Example

Stops were randomly located according to a uniform distribution over a circle of radius 100 miles, the center of which was the shipment destination and fleet base. Shipment sizes were randomly generated according to a uniform distribution with range [0, 10,000] pounds. Parameters used follow:

$$\alpha = \$1.40/\text{mile} \quad \beta = \$35 \quad \gamma = \$35 \quad \omega = \$20 \quad v = 20{,}000 \text{ lbs.}$$

$$b_i = \begin{cases} 19 + .0012q_i + .00014q_i d_i; & q_i < 1000 \text{ lbs} \\ 20 + .0012q_i + .00012q_i d_i; & 1000 \le q_i < 5000 \text{ lbs.} \\ 30 + .0012q_i + .00011q_i d_i; & 5000 \le q_i \end{cases} \tag{7}$$

The common carrier rate structure includes a large weight based component, which reflects the cost of sorting freight at terminals.

The solution to a 25 stop problem was based on the break-even points provided in Table 1. In addition to ρ=0, three densities satisfy Equation (6): ρ=.00025, .00029 and .00032. The last value has the lowest cost and is the optimal solution. The optimal assignment of stops to systems is shown in Fig. 1a. For the example, the figure shows that shipment weight has a greater impact on assignment than shipment distance. Later, when we consider a limited capacity scenario, shipment distance will have a significant impact.

Seventy-five more stops were simulated and added to the original 25 stops to create a 100 stop problem, the solution for which is provided in Fig. 1b. Where 10 stops were assigned to private carriage in the original problem, 5 additional stops from the original problem are assigned to private carriage in the expanded problem. Expanding the total number of stops causes the spacing between stops to decline, making private carriage attractive for more stops.

3 FIXED FLEET SIZE

Private carriage might be implemented either through contracting for transportation service or through the actual purchase of a private fleet. In the former case, fleet capacity could be varied from day to day, in response to fluctuations in shipment sizes. In the latter case, fleet size, and transport capacity, might be restricted. In any given transport cycle, it may then be impossible to transport all of the shipments generated from suppliers in P, or there may be surplus capacity remaining after the shipments from P are accommodated, in which case the capacity could be allocated to some of the

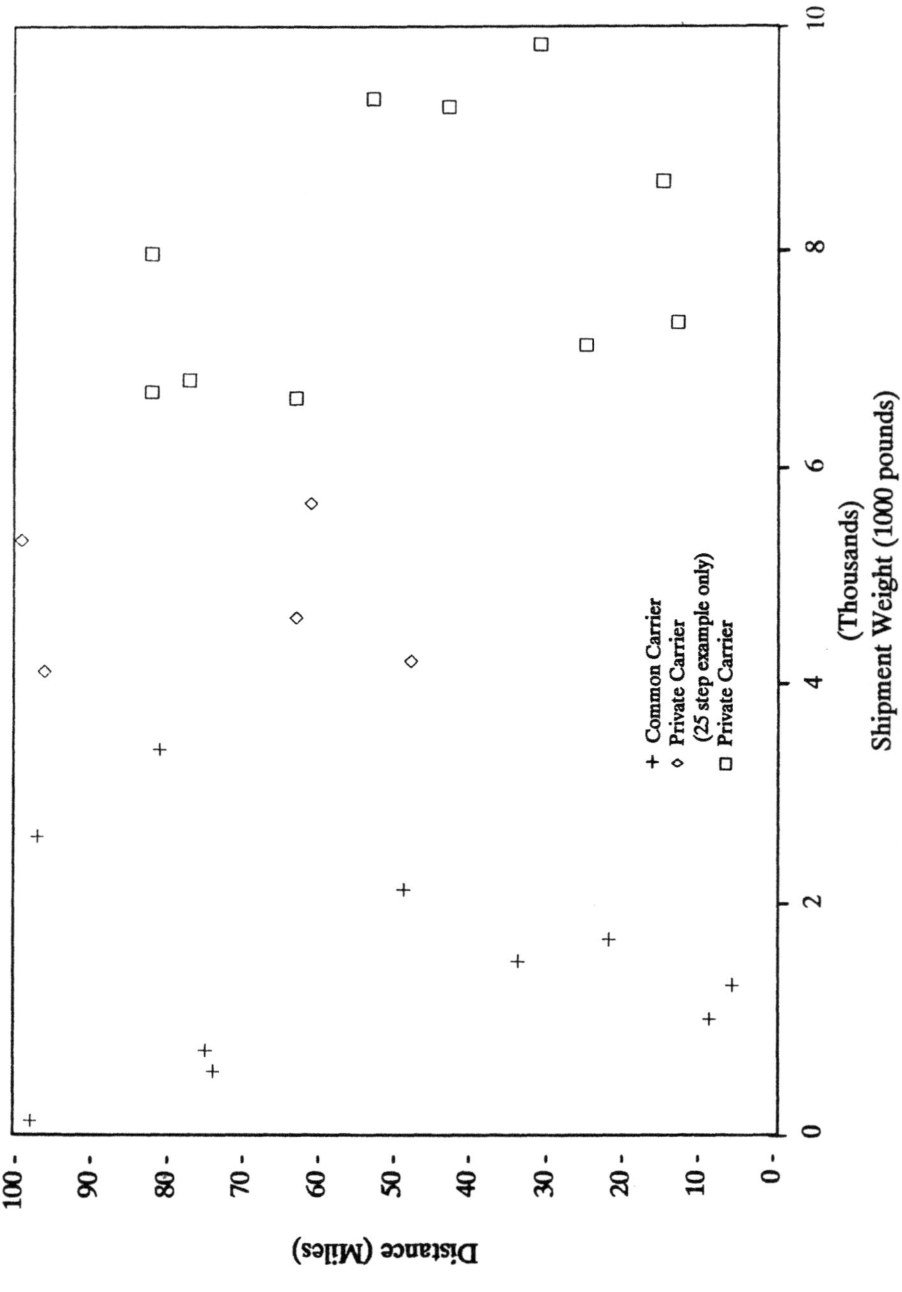

Fig. 1a System assignment, 25 stops, variable capacity

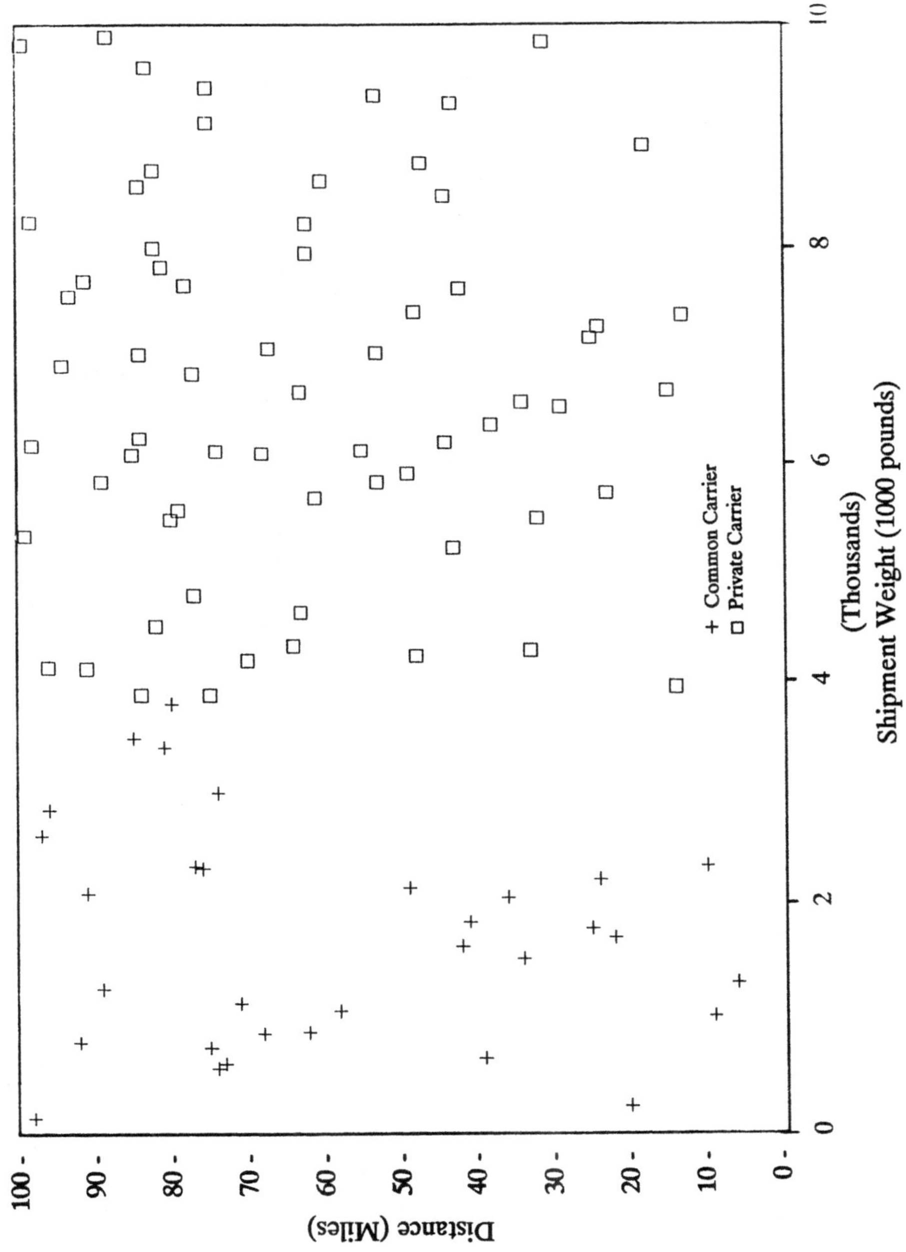

Fig. 1b System assignment, 100 stops, variable capacity

Table 1 Break-even points for 25 stop example

Customer	Distance (miles)	Ship. Size (pounds)	Break-even Density($\tilde{\rho}_i$)	Density
1	31	9901	.000092	.00003
2	43	9327	.000114	.00006
3	53	9391	.000121	.00010
4	13	7411	.000152	.00013
5	25	7191	.000176	.00016
6	82	8689	.000179	.00019
7	15	6712	.000191	.00022
8	82	7979	.000215	.00025*
9	63	6669	.000274	.00029*
10	77	6825	.000291	.00032*
11	61	5698	.000383	.00035
12	99	5319	.000620	.00038
13	63	4643	.000985	.00041
14	48	4260	.001170	.00045
15	96	4128	.001880	.00048
16	81	3422	.003330	.00051
17	97	2625	.015200	.00054
18	49	2194	.024700	.00057
19	22	1770	.127000	.00060
20	75	737	∞	.00064
21	9	1072	∞	.00067
22	34	1569	∞	.00070
23	74	553	∞	.00073
24	6	1373	∞	.00076
25	98	90	∞	.00080

*Solution satisfying Equation (6)

Costs: ρ = 0: $2786
ρ = .00025: $2373
ρ = .00029: $2313
ρ = .00032: $2253

common carrier stops. Suppose that the number of tours is constrained not to exceed the value F, representing the fleet size. Also suppose that all of the cost parameters considered in the previous section are reduced, to account for a variable cost component. That is, costs that depend solely on the fleet size, and not on the number of miles driven, would be excluded. Then total cost can be expressed as:

$$\min C_2 = \left[\alpha' k \sqrt{\Sigma X_i}\, \sqrt{A} + \beta' \Sigma X_i \right] + \left[\frac{\Sigma X_i(\gamma' + 2\alpha' d_i) q_i}{v} \right] + \left[\Sigma(1 - X_i) b_i \right] + \omega M \tag{8}$$

$$\text{s.t.} \quad vF \geq \Sigma X_i q_i. \tag{9}$$

When fixed costs are excluded from the objective function, private carriage would likely be preferred for most stops, and the capacity constraint would be tight. The constraint can be incorporated in the objective function, through use of a Lagrange multiplier, λ:

$$\min C'_2 = C_2 - \lambda(vF - \Sigma X_i q_i) \tag{10}$$

As in the previous section, a break-even point can be established by evaluating the derivative of C'_2 with respect to X_i at $X_i=1$:

$$\frac{\partial C'_2}{\partial X_i}(1) = \left[\frac{\alpha' k}{2\sqrt{\rho}} + \beta' + \frac{(\gamma' + 2\alpha' d_i) q_i}{v} \right] - b_i + \lambda q_i \tag{11}$$

The break-even point now depends on two variables: ρ and λ. Increasing ρ has the effect of decreasing local cost, making private carriage more attractive. Increasing λ has the effect of increasing capacity cost, making common carriage more attractive.

For any given value of ρ, Equation (11) establishes a ranking among stops, such that a stop should only be included in *P* if:

$$\lambda \leq \tilde{\lambda}_i = \frac{\left[\frac{\alpha' k}{2\sqrt{\rho}} + \beta' + \frac{(\gamma' + 2\alpha' d_i) q_i}{v} - b_i \right]}{q_i} \tag{12}$$

where $\tilde{\lambda}_i$ is the break-even point on λ for stop i. For any value of λ, $X_i = 1$ if $\tilde{\lambda}_i \geq \lambda$ and $X_i = 0$ if $\tilde{\lambda}_i < \lambda$. λ should be set at a value that guarantees all capacity is consumed or, if this is not possible, λ should be set at zero. Because $\Sigma X_i q_i$ must be non-increasing as λ increases, this one-dimensional search is quite simple.

To complete the optimization, ρ^* must also be identified. This can be accomplished through a second one-dimensional search for ρ. Or ρ^* can be approximated by the following:

$$\rho^* \approx \frac{vF}{(\Sigma q_i)/M} / A \tag{13}$$

That is, fleet capacity might be assumed to be consumed by N stops generating average shipment sizes.

The methodology was applied to the 100 stop data used in the previous section. All of the cost parameters were adjusted to omit fixed costs, resulting in the following:

$$\alpha' = \$.60/\text{mile} \quad \beta' = \$17.5 \quad \gamma' = \$17.5 \quad \omega' = \$20$$

Common carrier prices were identical to the previous problem. Fleet capacity was set to 440,000 pounds, which approximates the combined weight of shipments assigned to private carriage in the previous section. The solution, shown in Fig. 2a, indicates that a sharp weight-based division no longer exists. The Lagrangean term in Equation (10) has increased the weight cost for private carriage shipments, making private carriage less of a "bargain" for heavy shipments. This is further revealed in Fig. 2b, in which all shipment sizes are doubled over the previous example. The figure illustrates that random fluctuations in shipment sizes can have a drastic impact on the assignment. Now, the division is made almost entirely on the basis of distance.

The change in the dividing line is largely an artifact of how the capacity constraint is formulated. With a weight based constraint, private carriage is naturally more attractive for distant stops. A more realistic constraint might be based on the total driving time to serve the stops, which would depend on the product of shipment weight and distance. Such a constraint could easily be accommodated in Equation (10). The solution methodology would be the same.

4 VARIABLE SERVICE FREQUENCY

The previous section assumed that the service frequency for the stops cannot be altered when, in reality, it should be adjusted to reflect inventory and transportation costs. This section will examine how to simultaneously optimize service frequency and select stops for private carriage. First, service frequency will be assumed to be the same for all private carriage stops, and then this constraint will be relaxed to allow differing service frequencies. Both parts allow a variable fleet capacity.

4.1 Identical Service Frequencies

To account for inventory costs, additional variables need to be defined:

f_i = demand rate for stop i
h_i = inventory holding cost for stop i, per unit time
$b_i(q)$ = common carrier transportation cost for stop i, for shipment size q
T = time between dispatches for private fleet
T_i = time between shipments for stop i.

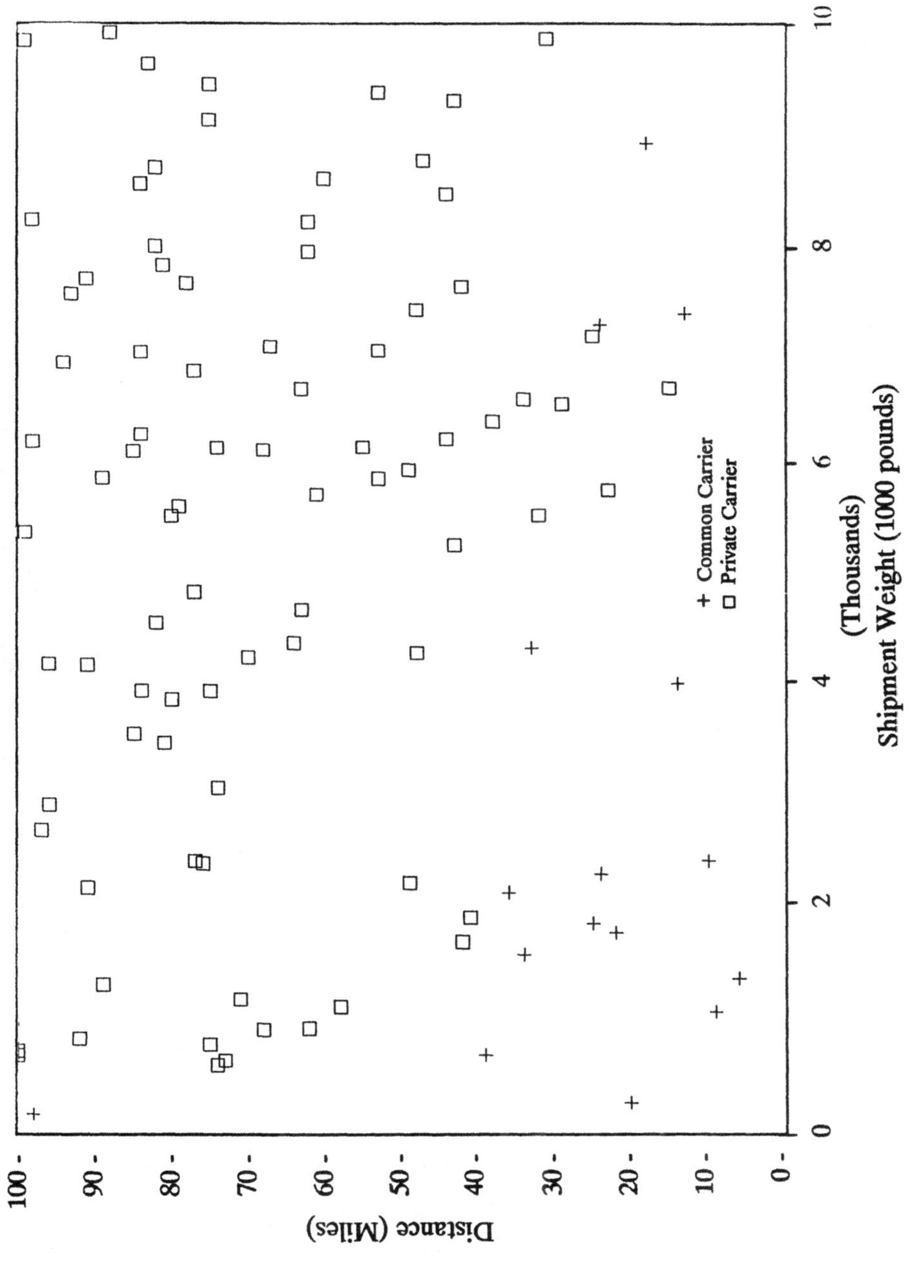

Fig. 2a System assignment, fixed capacity (large)

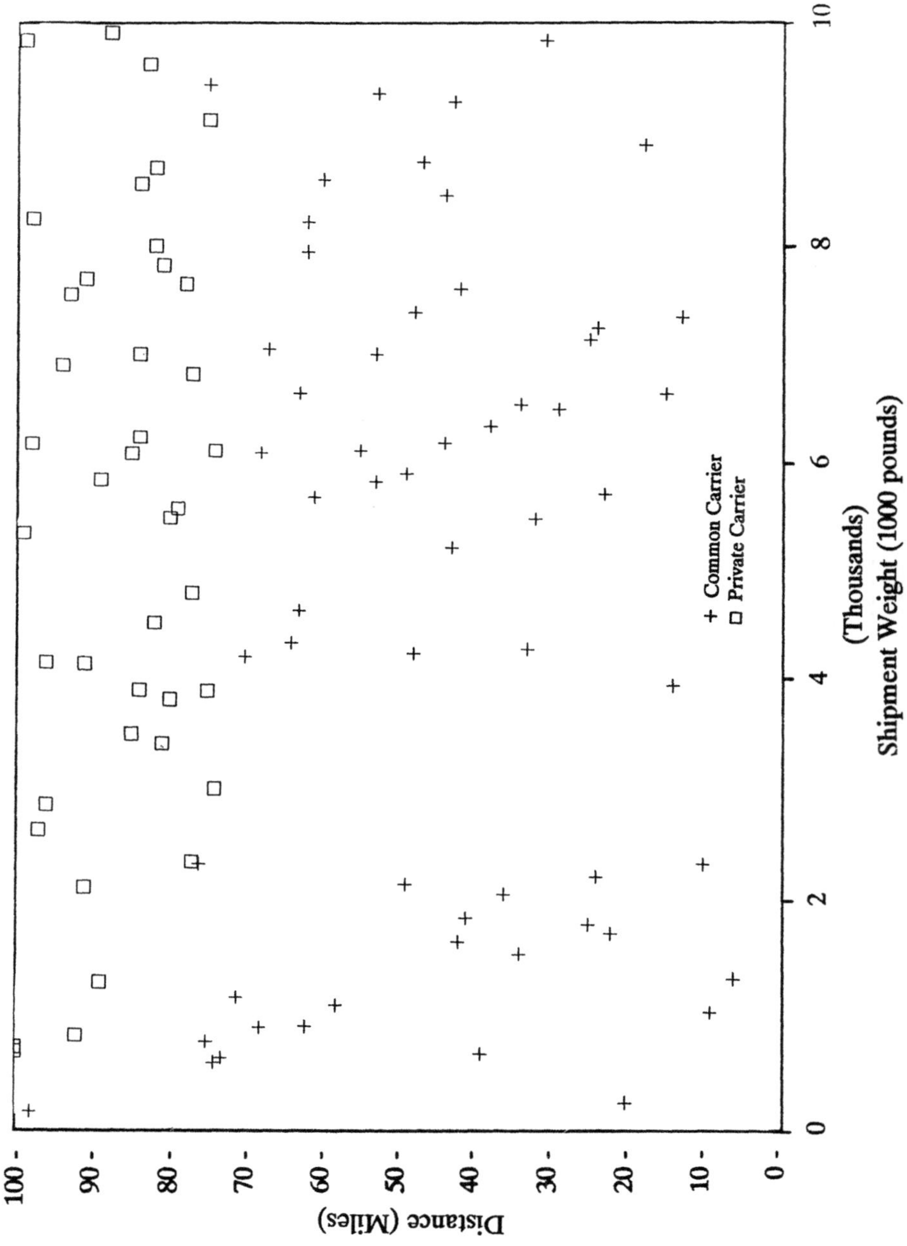

Fig. 2b System assignment, fixed capacity (small)

The objective function will now be expressed as a cost per unit time, accounting for both transportation and inventory. In the following formulation, inventory costs are assumed to be incurred at both the origin and the destination (see [2]), and the interval between shipments is assumed to be constant:

$$C_3 = \Sigma X_i \left[T h_i f_i + \frac{\alpha k \sqrt{A}}{\sqrt{\Sigma X_i}\; T} + \frac{\beta}{T} + \frac{(\gamma + 2\alpha d_i) f_i}{v} + \frac{\omega}{T} \right]$$

$$+ \Sigma(1-X_i) \left[\frac{b_i(f_i T_i)}{T_i} + h_i f_i T_i + \frac{\omega}{T_i} \right] \tag{14}$$

T_i can be optimized independently of T and X_i. The solution will depend on the common carrier price structure, which is highly case specific. To simplify subsequent discussion, c_i will be defined to equal the optimal cost per unit time for stop i, given that it is served by common carrier:

$$c_i = \frac{b_i(f_i T_i^*)}{T_i^*} + h_i f_i T_i^* + \frac{\omega}{T_i^*} \tag{15}$$

Equation (14) can also be simplified by expressing T^* as a function of the X_i. T^* is obtained by differentiating C_3 with respect to T, and substituting $A\rho = \Sigma X_i$:

$$T^* = \sqrt{\frac{(A\rho)\,[\alpha k / \sqrt{\rho} + \beta + \omega\;]}{\Sigma\, h_i f_i X_i}} \tag{16}$$

Substituting T^* in Equation (14) provides the following cost expression:

$$C_3 = 2\sqrt{\Sigma X_i \left[\frac{\alpha k \sqrt{A}}{\sqrt{\Sigma X_i}} + (\beta + \omega) \right] (\Sigma X_i h_i f_i)} + \frac{\Sigma X_i (\gamma + 2\alpha d_i) f_i}{v}$$

$$+ \Sigma(1-X_i) c_i \tag{17}$$

In the case where $h_i f_i$ is identical among commodities, the derivative of C with respect to X_i can be written as:

$$\frac{\partial C_3}{\partial X_i}(1) = 2\sqrt{\left[\frac{\alpha k}{\sqrt{\rho}}+(\beta+\omega)\right][h_i f_i]} - \left(\frac{\alpha k}{2\sqrt{\rho}}\right)\left(\frac{1}{T^*}\right) + \frac{(\gamma+2\alpha d_i)f_i}{v} - c_i \qquad (18)$$

As in the previous sections, Equation (18) establishes a ranking among stops, dependent on ρ, which can be used to assign stops to transportation systems. In the interest of brevity, no example is provided.

If $h_i f_i$ differs appreciably among stops, then the optimization of Equation (17) is more difficult. However, if $h_i f_i$ differs appreciably, it would not be optimal to use the same shipping frequency among all stops. Fortunately, as will be shown in the following section, if shipping frequency is allowed to differ among stops, the optimization problem becomes easier to solve.

4.2 Differing Shipping Frequencies

As indicated in [6] and [4], stops that generate large quantities of goods should be served more frequently than stops that generate small quantities. This can be accomplished by establishing two shipping intervals representing the time between fleet dispatches to a region (T) and the time between shipments for each stop (T_i). T_i must be constrained to be an integer multiple of T. That is, a stop can only ship by private carrier when there are vehicles in its region.

The optimization problem can now be expressed as:

$$\min C_4 = \Sigma X_i \left[h_i f_i T_i + \frac{\frac{\alpha k}{\sqrt{\rho}}+(\beta+\omega)}{T_i} + \frac{(\gamma+2\alpha d_i)f_i}{v} \right] + \Sigma(1-X_i)c_i \qquad (19)$$

$$\text{s.t.} \quad \rho - \frac{\Sigma X_i(T/T_i)}{A} = 0 \qquad (20)$$

$$T_i \geq T \qquad (21)$$

$$T_i / T \quad \varepsilon \quad \{\text{positive integers}\} \qquad (22)$$

$$T, T_i \geq 0 \qquad (23)$$

ρ accounts for the proportion of stops visited in each delivery cycle. The integer constraint has a relatively minor effect on the optimization of T and the optimization of ρ. Neglecting this constraint, the problem can be formulated as the following Lagrangean:

$$\min C'_4 = C_4 + \lambda\left[\rho - \frac{\Sigma X_i(T/T_i)}{A}\right] + \mu_i[T_i - T] \tag{24}$$

where λ and μ_i are Lagrange multipliers.

For given values of μ_i, ρ and λ, the optimal value of T_i can be expressed as:

$$T_i^* = \sqrt{\frac{\frac{\alpha k}{\sqrt{\rho}} + \beta - \lambda\left(\frac{T}{A}\right)}{f_i h_i + \mu_i}} \approx \sqrt{\frac{\frac{\alpha k}{\sqrt{\rho}} + \beta}{f_i h_i}} \tag{25}$$

To account for the integer constraint, T_i^* can be rounded to one of the adjacent values of mT, where m is a positive integer. The effects of the two Lagrange multipliers approximately cancel each other. Therefore, T_i^* can be expressed, approximately, as a function of ρ. For fixed values of X_i, [6] approximately solves for T_i, ρ and T, through a two-dimensional search over ρ and T, and insertion of the resulting values in Equation (25) to obtain T_i^*.

By a similar process, X_i^* can be determined at the same time as T_i^*. The change in cost due to reassigning a stop can be approximated by evaluating derivative of C'_4 with respect to X_i at $X_i=1$. Also substituting λ^* for λ.

$$\frac{\partial C'_4}{\partial X_i}(1) \approx 2\sqrt{\left[\frac{\alpha k}{\sqrt{\rho}} + \beta + \omega\right] h_i f_i} - \frac{1}{2}\left(\frac{T}{T_i}\right)\left(\frac{\alpha k}{\sqrt{\rho}}\right)\sum_i \frac{X_i}{NT_i} + \frac{(\gamma + 2\alpha d_i)f_i}{v} - c_i \tag{26}$$

If Equation (26) is less than zero, then the stop should be included on a private carrier route. This calculation depends on three variables -- ρ, T and $\Sigma(X_i/NT_i)$ -- the latter being the average shipping frequency among the stops using private carriage. As a practical matter, this average is likely to be insensitive to ρ and to T, and might simply be approximated. Further, the search for T would likely be restricted to a few practical values, such as one day, 1/2 week, one week, etc. Nevertheless, solving Equation (26) amounts to a three dimensional search over ρ, T, and $\Sigma X_i/NT_i$. For each solution considered, T_i and X_i are evaluated as follows:

1. If Equation (26) is less than zero, assign to private carriage; otherwise assign to common carriage.
2. If stop is assigned to private, use Equation (25) to calculate T_i^*. If stop is assigned to common, calculate T_i^* by optimizing Equation (15).

A solution would not be optimal unless Equation (20) is satisfied. A search procedure would be to make an initial guess for ρ, T and $\Sigma X_i/NT_i$, then complete the two steps above for each stop. Based on the solution, ρ and $\Sigma X_i/NT_i$ can be updated, and the process iterated, until the solution converges. The search process must be repeated for each value of T considered.

4.3 Example

The same data set was used a third time to demonstrate the variable frequency model. Costs parameters were set at their full values. Common carriage costs were based on an optimal solution to Equation (15), using the pricing structure provided earlier. The shipment size data was interpreted as demand per week (f_i), and an inventory holding cost of h=$.01/pound-week was used for all stops. With h set at this value, Equation (25) is approximately one week for an average stop.

The optimal solution was to dispatch vehicles every week. The assignment of stops to systems, and the shipping intervals for private carriage stops, are shown in Fig. 3. Compared to Fig. 1b, note that more stops are served by private carriage. By allowing smaller stops to be served less frequently, they are now able to create sufficient shipment sizes to justify private carriage. Every stop that is served once per two weeks in Fig. 3 was previously assigned to common carriage. However, none of the private carriage stops is served less frequently than once per two weeks. When T_i^* is 3 weeks or greater in the example, the stop is assigned to common carriage.

5 SUMMARY AND CONCLUSIONS

This chapter has developed methodologies for assigning stops to common carriage and private carriage systems, based on Daganzo's square-root route length approximation. In all cases, a stop is assigned to private carriage only when a break-even stop density is achieved, meaning that the local transportation cost is sufficiently small to justify private carriage. The break-even point is sensitive to shipment weight, because of large weight based charges imposed by common carriers. The assignment also depends on whether it is possible to achieve the break-even stop density, which depends on the number of stops located in the vicinity of the stop. As the number of stops increases, a higher proportion of the stops is served by private carriage.

The cost approximation does not account for specific customer locations and the fact that some stops will be more closely spaced than others. In related research, the second author of this chapter has found that the error introduced causes the objective function to be 2-3% larger than optimal. On the other hand, the Ballou and Chowdhury, and Jordan heuristics may be too sensitive to this information because they do not account for the fact that vehicle routes will change from dispatch to dispatch (depending on actual shipment sizes). Because actual shipment sizes will differ from assumed shipment sizes, the actual cost will likely be much more than predicted. Because the density approximation is

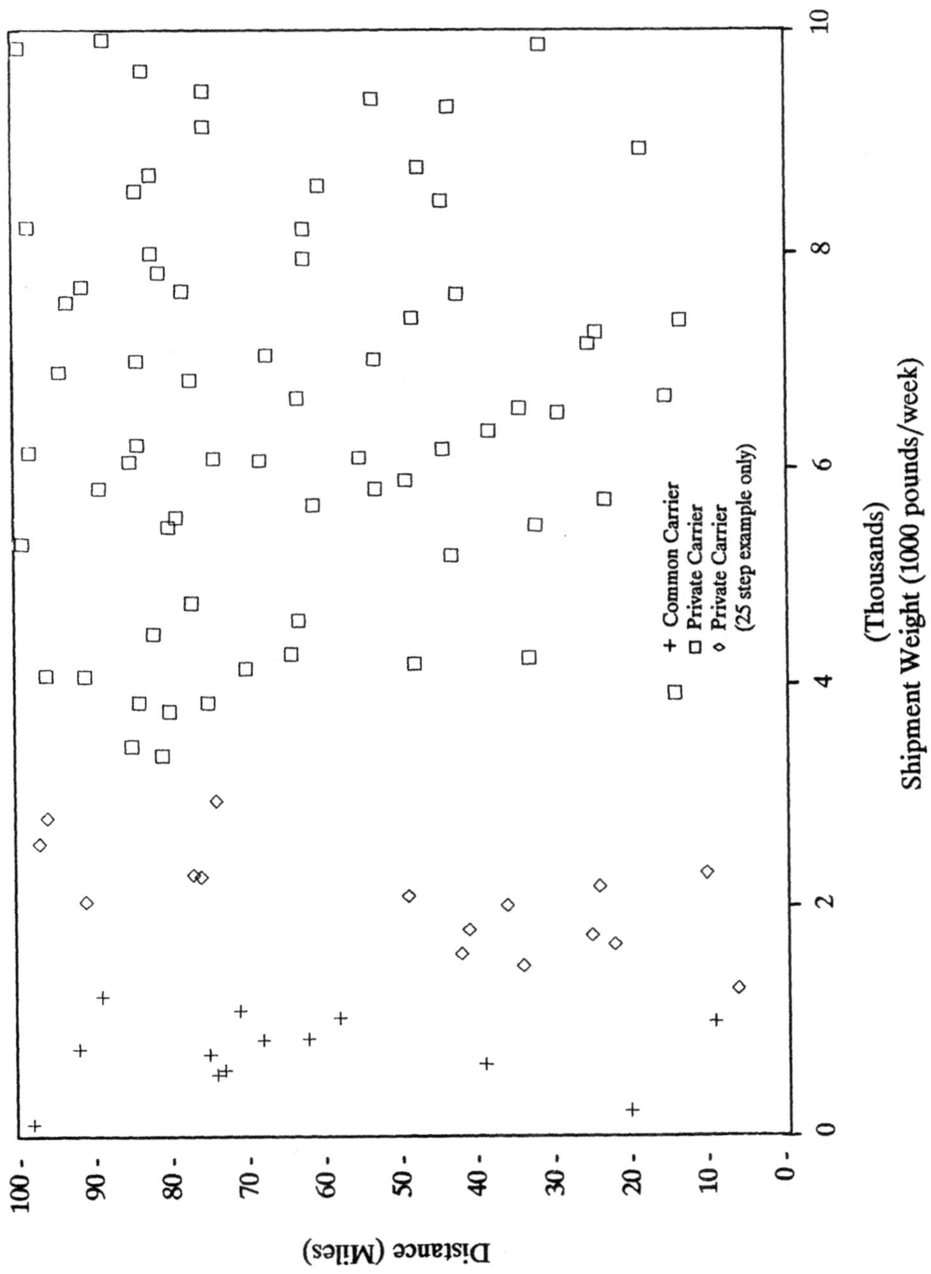

Fig. 3 System assignment, variable service frequency

an average result, it may provide the more cost-effective solution in the long run. (In practice, it may still be worthwhile to develop a localized density measure, specific to a subregion or perhaps an individual stop.)

The cost approximation also does not account for the fact that filling vehicles to capacity might require that shipments be split into multiple vehicle loads. In practice, there is usually nothing to prevent splitting shipments, and splitting shipments can reduce cost [5]. Nevertheless, the square-route approximation may underestimate cost when shipments cannot be split. For a discussion of this issue, see [7].

In addition to assigning stops, the techniques presented in this chapter can be used to select fleet sizes. For instance, the fleet size could be set to accommodate all of the stops assigned to private carriage on an average day, using one of the two variable fleet size models. As an alternative, the fixed fleet size model could be applied to simulated shipment sizes. The fleet size yielding the minimum expected cost would then be chosen.

6 REFERENCES

1. Ballou, R.H. and M. Chowdhury "MSVS: An Extended Computer Model for Transport Mode Selection," *Logistics and Transportation Review*, 16, pp. 325-338 (1980).
2. Blumenfeld, D.E., L.D. Burns, J.D. Diltz and C.F. Daganzo "Analyzing Trade-Offs Between Transportation, Inventory and Production Costs on Freight Networks," *Transportation Research*, 19B, pp. 361-380 (1985).
3. Daganzo, C.F. "The Length of Tours in Zones of Different Shapes," *Transportaiion Research*, 18B, pp. 135-146 (1984).
4. Daganzo, C.F. "Supplying a Single Location from Heterogeneous Sources," *Transportation Research*, 19B, pp. 409-420 (1985).
5. Dror, M. and P. Trudeau "Savings by Split Delivery Routes," *Transportation Science*, 23, pp. 141-145 (1989).
6. Hall, R.W. "Determining Vehicle Dispatch Frequency When Shipping Frequency Differs Among Suppliers," *Transportation Research*, 19B, pp. 421-431 (1985).
7. Hall, R.W. "Vehicle Packing," *Transportation Research*, 23B, pp. 103-121 (1989)
8. Jordan, W.C. "Analyzing Shipping Strategies for Suppliers Shipping to a Single Plant," General Motors Research Laboratories, GMR-6652 (1989).
9. Racer, M.J. "Coordinating Inbound and Outbound Vehicle Routes Within a Decentralized Decision Environment," Ph.D. Thesis, University of California at Berkeley (1990).

OPTIMAL CONTAINER LOCATION IN MINILOAD AS/R SYSTEMS

D.J. Medeiros
Bahram Emamizadeh
The Pennsylvania State University

ABSTRACT

A procedure for determining optimal storage locations for containers in a miniload automated storage and retrieval system is presented. Dominance conditions are developed to restrict the search space. An example demonstrates that turnover based assignment is not optimal for miniloads implementing dual cycles.

1 INTRODUCTION

A miniload is an AS/R system designed for small parts handling and storage. The system consists of a storage/retrieval machine, a rack system, and a control system. The S/R machine travels in an aisle to store and retrieve containers of parts from the rack. Containers are brought to and removed from one or more pickup and deposit (P&D) stations which typically are located at the end of the aisle. These P&D stations may be used for order picking, or may contain an interface to another material handling system, such as a conveyor or AGV system, which moves the containers to their destination. Miniload AS/R systems often operate in dual cycle mode: on one trip into the aisle, a container is stored and the next one is retrieved. Dual cycles are preferred because they allow greater system throughput.

Miniload AS/RS are frequently used in order-picking applications because such systems provide efficient space utilization as well as improved material tracking and control. Principal distinguishing characteristics of miniloads in these applications are storage and retrieval of less than unit load quantities, and often, the use of dedicated storage locations for containers. Each container may store one or several related part numbers.

The design of a miniload AS/R system for a particular application presents several problems. These include:

1. Determining the size of the system, including number of aisles, length and height of the rack structure, and size and location of openings.
2. Assigning part numbers to containers.
3. Location of containers in the rack system.

Once a system is designed, changes are infrequent, and would be made as a result of observed changes in throughput requirements, part mix, customer demand, etc. Optimal or near-optimal solutions might results in significantly increased throughput over the life of the system.

Several authors have addressed these problems. Medeiros, Smith and Enscore [1] developed a methodology for initial system sizing, for a miniload. Wenzel [2,3] considered the problems of stock colocation and container location, and suggested that, due to conflicting objectives, no storage policy could be optimal. He recommends locating the highest activity container closest to the order picking station, and also discusses a storage policy which assigns items in the same bill of material to one or several adjacent containers.

Goetschalchx [4] studied storage policies for various order picking operations. For a single command order picking system, he showed that turnover based assignment is the optimal policy. Turnover rate is defined as:

$$f_i = d_i/(q_i v_i)$$

where:

d_i is the demand for product i

q_i is the inventory level for product i

v_i is the size of one unit of product i.

Products are assigned, by decreasing turnover rate, to the closest location, and each product is assigned a zone of size $q_i\ v_i$. For a one-dimensional miniload system (one container high or one container long) using dual cycles, Goetschalchx also proves that turnover-based assignment is optimal given that each container is dedicated to a single product type.

Rosenblatt and Roll [5] developed a simulation model for the design of a warehouse which included a storage cost associated with the type of storage policy chosen. Storage policies ranged from total random storage to complete dedicated storage.

Hausman, Schwarz and Graves [6] compared randomized storage and dedicated storage for an AS/R system using single cycles. The storage area was assumed to be square in time, i.e. travel time to the farthest column and the farthest row was the same. Using a continuous approximation to represent the storage area, they computed the expected travel time of the S/R machine under the two storage policies, and showed that turnover based assignment minimized expected one-way travel time. This work was then extended to a dual cycle system with zoned storage [7], where the performance of different zone geometry was analyzed. Further work [8] analyzed system performance under stochastic conditions.

Bozer and White [9] developed several travel-time models for automated AS/R systems, assuming the Chebyshev norm for travel time and a random storage policy.

Several authors have developed models for stock location in a warehouse. Heskett [10,11] introduced the Cube Per Order (CPO) index, a ratio of item space requirement to its demand. Francis [12] presented a cost model for assigning facilities to locations in the context of a warehouse layout problem. The cost function includes the cost of moving an item in and out of the warehouse plus the perimeter cost. Both authors' work leads to the conclusion that turnover based assignment should be used. Mallete and Francis [13] introduced a factoring assumption and demonstrated that the CPO rule is optimal if the assumption holds and products have equal size. The factoring assumption is equivalent to stating that product demands are independent, i.e., items have an equal probability of being retrieved from any loading dock in the warehouse.

Of the three problem described above (system layout, assignment of parts to containers, location of containers) much research effort has been devoted to the third problem, probably because of its applicability in general warehousing problems as well as automated storage and retrieval systems. It has been conjectured that turnover based assignment will result in good (though not optimal) solutions for miniloads using dual cycles.

We present a methodology for location of containers in a miniload AS/R system, under the following assumptions:

- system layout has been defined; travel time to each location in the rack is known.
- the S/R machine uses dual cycles.
- items have been assigned to containers; container demand is known.
- all containers are the same size, and thus can be located in any rack opening.
- container demands are independent, i.e.

$$\delta_{ij} = \lambda_i P_j$$

where:

δ_{ij} is the expected number of times the S/R machine retrieves container j immediately after container i.

λ_i is the expected number of retrievals of container i

P_j is the probability that container j will be retrieved, i.e.

$$P_j = \lambda_j \,/\, \sum_{k=1}^{M} \lambda_k$$

2 CONTAINER LOCATION MODEL

The objective of container location is to minimize the expected travel time of the S/R machine, and thus maximize system throughput. Under dual cycles, the total expected traveling time is the summation of the time that the S/R machine travels between the P&D location and the containers and the time that the S/R machine travels between the containers. Specifically, when the S/R machine makes a trip, it leaves the P&D location to store a container. From there it moves to another location to retrieve the next container and then returns to the P&D location. This problem may be formulated as follows:

$$\text{Min} \sum_{i=1}^{M} \sum_{j=1}^{M} \sum_{k=1}^{M} \sum_{l=1}^{M} (\delta_{ij} + \delta_{ji}) x_{ik} x_{jl} d_{kl} + \sum_{i=1}^{M} \sum_{k=1}^{M} 2\lambda_i \, x_{ik} d_{0k}$$

$$\text{s.t.} \quad \sum_{k=1}^{M} x_{ik} = 1 \qquad i = 1,\ldots, M$$

$$\sum_{i=1}^{M} x_{ik} = 1 \qquad k = 1,\ldots, M$$

$$k_{ik} = 0 \text{ or } 1$$

where:

x_{ik} is 1 if container i is in location k, and 0 otherwise.

d_{kl} is the travel time between locations k and l.

d_{0k} is the travel time between the P&D and location k.

There are two parts to the objective function. One is the total expected traveling time between the containers (due to interleaving), and the other is the total expected traveling time between the P&D location and the containers. The first constraint requires that each container be assigned to one location and the second constraint ensures that each location is occupied by exactly one container. The problem is a quadratic set-covering problem. Bazzara [13] presented a similar formulation for plant layout. He proposed a branch and bound solution method. His lower bound is the sum of three costs: the interaction cost among departments which have been assigned a location, a lower bound on the interaction cost between the assigned departments and the unassigned departments, and a lower bound on the interaction cost between the unassigned departments. The latter two costs are found by solving a linear set covering problem.

We propose the use of a depth first branch and bound method to solve this problem. Efficiency of the solution procedure is dependent on the establishment of good lower bounds, as well as development of a good initial assignment. Because turnover based assignment has been shown to be optimal for a single cycle system, the travel time for this assignment is used as an initial lower bound. Containers are ranked by decreasing demand, and storage locations by increasing travel time. The containers are then assigned, in ranked order, to the locations. A lower bound on a partial assignment must then be established.

Given that some containers have been assigned to some locations, a lower bound on the overall time of retrieval is given by $t_1 + t_2$, where:

t_1 is the time to retrieve the already assigned containers.

t_2 is the lower bound on the interleaving time from unassigned containers to assigned ones, plus the retrieval time of the unassigned containers.

Locations of the assigned containers are known; total retrieval time is calculated as:

$$t_1 = \sum_{i \in S} \sum_{j \in S} \sum_{v=1}^{M} \sum_{w=1}^{M} (\delta_{ij} + \delta_{ji}) X_{iv} X_{jw} d_{vw} + \sum_{i \in S} \sum_{v=1}^{M} \lambda_i X_{iv} d_{0v}$$

where S is the set of containers which have been assigned to a location.

A lower bound for unassigned containers is then calculated by solving the following problem:

$$t_2 = \min \sum_{i \notin S} \sum_{v \in A} b_{iv} X_{iv}$$

$$\text{s.t.} \quad \sum_{v \in A} X_{iv} = 1 \qquad \forall\ i \notin S$$

$$X_{iv} = 0 \text{ or } 1$$

where

b_{iv} is the lower bound on the retrieval time when container i is assigned to location v.

A is the set of available locations.

The lower bound on retrieval time when container i is assigned to location v is calculated as:

$$b_{iv} = \lambda_i d_{0v} + \sum_{j \in S} (\delta_{ji} + \delta_{ij}) d_{vw} + \sum_{\substack{k \neq i \\ k \notin S}} l_k$$

where:

d_{vw} is the time between location v and the already determined location of assigned container j, which is w.

$l_k = \min_{y \in A} (\delta_{ik} d_{vy})$

The lower bound on interleaving time between the container to be located and the unassigned container k is l_k. The variable b_{iv} is a lower bound on the retrieval time when container i is assigned to location v, and consists of travel time from the P&D station to location v, interleaving time between container i and the assigned containers, and a bound on interleave time between container i and the unassigned containers.

A single set of constraints is used in the formulation for t_2. These constraints require that each container be assigned to an available location, but do not prevent multiple containers from occupying the same location. A lower bound, rather than a feasible solution, is sought, and removal of the second set of constraints enhances speed of computing the lower bound.

In addition to the lower bound, a dominance rule was developed to further reduce computational requirements. Given two locations, v and w, and two containers i and j to be located, container i may be said to dominate container j with respect to locations v and w if by placing container in i location v and container j in location w the objective function has a smaller value than if the opposite assignment were made. Thus, for any rack with two empty locations, v and w, if container i dominates container j with respect to these two locations, then container i should be placed in location v.

Given an existing layout, two containers i and j, and 2 empty locations, v and w, the procedure is as follows:

1. Calculate the overall retrieval time when container i is in location v and container j is in location w; that is:

$$T_1 = \sum_{\substack{k=1 \\ k \neq ij}}^{M} \sum_{\substack{u=1 \\ u \neq v,w}}^{M} 2\lambda_k X_{ku} d_{0u} + 2\lambda_i d_{0v} + 2\lambda_j d_{0w}$$

$$+ \sum_{\substack{k=1 \\ k,t \neq ij}}^{M} \sum_{t=1}^{M} \sum_{\substack{u=1 \\ u,n \neq v,w}}^{M} \sum_{n=1}^{M} (\delta_{kt} + \delta_{tk}) X_{ku} X_{tn} d_{un}$$

$$+ \sum_{\substack{k=1 \\ k\neq ij}}^{M} \sum_{\substack{u=1 \\ u\neq v,w}}^{M} [(\delta_{ki}+\delta_{ik})X_{ku}d_{uv} + (\delta_{kj}+\delta_{jk})X_{ku}d_{uw}]$$

$$+ (\delta_{ij}+\delta_{ji})d_{vw}$$

where k and t are the containers and u and n are the locations.

2. Calculate the overall retrieval time when container i is in location w and container j is in location v.

$$T_2 = \sum_{\substack{k=1 \\ k\neq ij}}^{M} \sum_{\substack{u=1 \\ u\neq v,w}}^{M} 2\lambda_k X_{ku} d_{0u} + 2\lambda_i d_{0w} + 2\lambda_j d_{0v}$$

$$+ \sum_{\substack{k=1 \\ k,t\neq i}}^{M} \sum_{t=1}^{M} \sum_{u=1}^{M} \sum_{\substack{n=1 \\ u,n\neq v,w}}^{M} (\delta_{kt}+\delta_{tk})X_{ku}X_{tn}\, d_{un}$$

$$+ \sum_{\substack{k=1 \\ k\neq ij}}^{M} \sum_{\substack{u=1 \\ u\neq v,w}}^{M} [(\delta_{ki}+\delta_{ik})X_{ku}d_{uw} + (\delta_{kj}+\delta_{jk})X_{ku}d_{uv}]$$

$$+ (\delta_{ij}+\delta_{ji})d_{wv}$$

3. To show that container i dominates container j, $T_1 < T_2$ must hold for all container assignments.

Define U_{ij} as follows:

$$U_{ij} = \delta_{ij} + \delta_{ji}$$

Therefore,

$$T_i - T_2 = 2(\lambda_i - \lambda_j)(d_{0v} - d_{0w}) + \sum_{\substack{k=1 \\ k\neq ij}}^{M} \sum_{\substack{u=1 \\ u\neq v,w}}^{M} (U_{ki} - U_{kj})(d_{uv} - d_{uw})$$

The first part of this equation is a constant, while the second part depends on the locations of the other containers. If the second part is maximized and the result of the equation is still negative, then container i dominates container j with respect to locations v and w. Define:

1. $\delta(k, i, j) = U_{ki} - U_{kj}$ $k = 1,...,M$ and $k \neq i, j$
2. $d(u, v, w) = d_{uv} - d_{uw}$ $u = 1,...,M$ and $u \neq v, w$

By multiplying the largest values of d (u, v, w) by the largest values of δ (k, i, j), the maximum value of the summation is obtained.

Before discussing implementation issues, we make the following observations:

1. Given two locations, v and w, if container i dominates all other containers with respect to locations v and w, then the optimal assignment can not have container i in location w.

2. If container i dominates all other containers with respect to location v and all other locations, then in the optimal layout product i is placed in location v.

These can be easily proved; starting with container i in a location other than v, a simple interchange of 2 containers is sufficient to cause a decrease in the objective function.

The dominance rule was implemented as follows:

Step 1. Number the locations based on their closeness to the P&D station. If there is a tie between locations, then the location with the smaller total traveling time to other locations is preferred. Order the containers in decreasing sequence of their demands.

Step 2. Using the dominance conditions, find the locations which container 1 can occupy. Let A_v represent the set of these locations. If the Mth location is in the set A_v, then go to step 4, else place container 1 in the set I, set k=k+1 and go to step 3.

Step 3. For container k in the sequence test the dominance conditions given each of the set of locations that the containers in the set I can occupy. Place the additional locations in the set A_v, and place container k in the set I. If k=M or the Mth location has been placed in A_v then go to step 4; otherwise set k=k+1 and repeat step 3.

Step 4. Stop. The dominance rule has been implemented and the domain for each product is found.

Fig. 1 illustrates an example dominance tree. In this example, the dominance conditions are first evaluated for container 1. Container 1 can occupy locations 1 and 2. Then given that container 1 is in location 1, it is found that container 2 can occupy locations 2 and 3. Similarly when container 1 is in location 2, it is found that container 2 can occupy locations 1 and 3. Therefore, container 2 can occupy locations 1, 2, and 3.

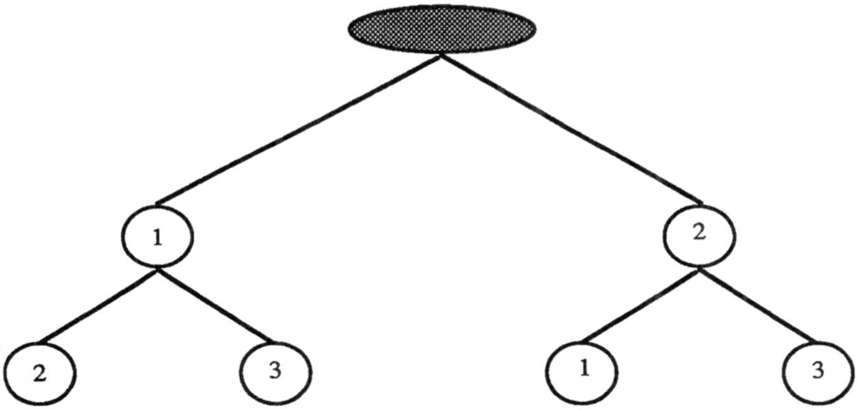

Fig. 1 Example of dominance tree

There is a minor modification to the dominance conditions at this point. Since the products in set I are fixed in the specific locations, their contribution to the retrieval time must be evaluated. Therefore, the formula for the modified dominance condition will be:

$$t_1 - t_2 = 2(\lambda_i - \lambda_j)(d_{ov} - d_{0w}) + \sum_{\substack{k \in I \\ k \neq ij}} \sum_{\substack{u \in A_v \\ u \neq v,w}} (U_{ki} - U_{kj})(d_{uv} - d_{uw})$$

$$+ \sum_{\substack{k \notin I \\ k \neq ij}} \sum_{\substack{u \notin A_v \\ u \neq v,w}} (U_{ki} - U_{kj})(d_{uv} - d_{uw})$$

The procedure terminates when all the locations in the miniload system are included in the set A_v. A further search for dominance conditions will determine locations in the lower part of the dominance tree. However, when a container with higher demand can not dominate the other containers, it is unlikely that any further dominance can be found. Therefore, the procedure terminates as soon as the last location is included in the domain of a container.

The dominance rule has two applications in the branch and bound procedure. The major contribution of this rule is that it reduces the search region for the branch and bound method since many locations are not considered for calculation of the lower bound. The second contribution is in terms of the lower bound itself. It was shown that part of the lower bound was the lower bound on the interleaving time between the unassigned containers. This lower bound was defined as l_k. The dominance rule restricts the locations in which the unassigned containers can be placed. This will increase the value of l_k, and will usually result in faster fathoming.

3 EXAMPLE

Consider a rack system with 9 locations and 9 containers (Fig. 2). For illustration, a matrix of travel times was generated, as shown in Table 1.

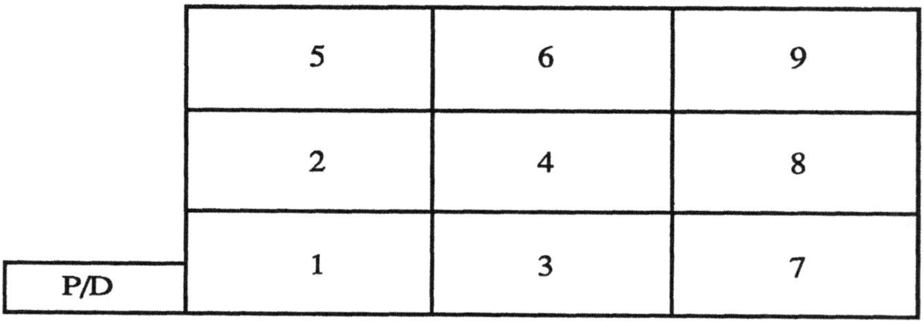

Fig. 2 Layout for the example miniload system

The demand for containers was generated randomly from a uniform distribution between 0 and 1100. The container interactions, U_{ij}, were calculated. Table 2 shows these interactions. Index zero in both tables refers to the P&D station.

The optimal solution for this example is shown in Fig. 3. Without the dominance conditions, 15,548 nodes were searched before obtaining the solution. After implementation of the dominance rule the optimal solution was obtained after evaluating 97 nodes. As can be seen, turnover based assignment is not the optimal solution for this example.

4 CONCLUSION

A formulation of the container location problem in a miniload system was presented. The dominance rule developed appears to be effective in reducing the number of nodes in the branch and bound tree. It was shown, through an example, that turnover based assignment is not optimal for a miniload running dual cycles.

Set covering problems are known to be np-complete. Therefore, it is likely that heuristics will be required for large problems. The dominance conditions developed in this paper could be used, in combination with a heuristic search procedure, to generate an improved, though not necessarily optimal, container assignment.

Table 1 Traveling time between locations in the example

	0	1	2	3	4	5	6	7	8	9
0	0	8	10	12	12	15	15	15	15	15
1	8	0	5	3	5	10	10	7	7	10
2	10	5	0	5	3	5	5	7	7	10
3	12	3	5	0	5	10	3	10	7	7
4	12	5	3	5	0	5	10	3	5	10
5	15	10	5	10	5	0	5	5	3	5
6	15	10	5	3	10	5	0	10	5	3
7	15	7	7	10	3	5	10	0	5	10
8	15	7	7	7	5	3	5	5	0	5
9	15	10	10	7	10	5	3	10	5	0

Table 2 Table of interaction among different products for the example

	0	1	2	3	4	5	6	7	8	9
0	0	1062	1025	914	795	774	669	503	67	44
1	1062	0	227	202	176	171	148	111	14	9
2	1025	225	0	194	168	164	142	106	14	9
3	914	196	189	0	147	143	123	93	12	8
4	795	166	161	143	0	121	105	79	10	6
5	774	161	156	139	121	0	101	76	10	6
6	669	137	132	177	102	99	0	64	8	5
7	503	99	96	85	74	72	62	0	6	6
8	67	12	11	10	9	8	7	5	0	0
9	44	8	7	6	6	5	5	3	0	0

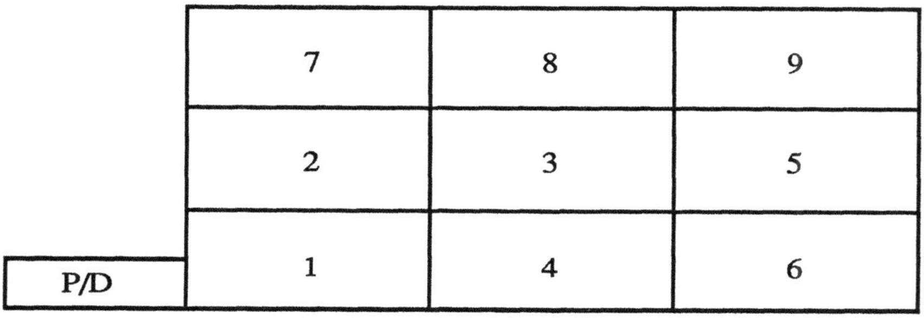

Fig. 3 Optimal container locations for the example

5 REFERENCES

1. Medeiros, D.J., A.D. Smith, and E.E. Enscore, "Economic Design of Miniload AS/R Systems," Working Paper 87-152, Department of Industrial and Management Systems Engineering, The Pennsylvania State University (1987).
2. Wenzel, C.D., "Maximize Your Mini: The Case for Softloading," *Proceedings of the Third International Conference on Automation in Warehousing*, Vol. 1, pp. 213-223 (1979).
3. Wenzel, C.D., "Optimizing Miniload Systems," *Material Handling Engineering*, pp. 80-87 (June 1980).
4. Goetschalchx, M., *Storage and Retrieval Policies for Effective Order Picking Operations*, Ph.D. thesis, Georgia Institute of Technology. University Microfilms International, Ann Arbor (1983).
5. Rosenblatt, M.J. and Roll, Y. Warehouse Design with Storage Policy Considerations. *International Journal of Production Research*, Vol. 22, No. 5, pp. 809-821 (1984).
6. Hausman, W.H., Schwarz, L.B. and Graves, S.C. Optimal Storage Assignment in Automatic Warehousing Systems. *Management Science*, Vol. 22, No. 6, pp. 629-638 (February 1976).
7. Graves, S.C., Hausman, W.H. and Schwarz, L.B. Storage-Retrieval Interleaving in Automatic Warehousing Systems. *Management Science*, Vol. 23, No. 9, pp. 935-945 (May 1977).
8. Schwarz, L.B., Graves, S.C. and Hausman, W.H. Scheduling Policies for Automatic Warehousing Systems: Simulation Results. *AIIE Transactions*, Vol. 10, No. 3, pp. 260-270 (September 1978).
9. Bozer, Y.A. and White, D. Travel-Time Models for Automated Storage/Retrieval Systems. *AIIE Transactions*, Vol. 16, No. 4, pp. 329-337 (December 1984).
10. Heskett, J.L. Cube Per Order Index: A Key to Warehouse Stock Location. *Transportation and Distribution Management*, Vol. 3, pp. 27-31 (April 1963).
11. Heskett, J.L. Putting the Cube-Per-Order Index to Work in Warehouse Layout. *Transportation and Distribution Management*, Vol. 4, pp. 23-30 (August 1964).
12. Francis, R.L. On Some Problems of Rectangular Warehouse Design and Layout. *The Journal of Industrial Engineering*, Vol. 18, No. 10, pp. 595-604 (October 1967).
13. Bazzara, M.S. Computerized Layout Design: A Branch and Bound Approach. *AIIE Transactions*, Vol. 7, No. 4 pp. 432-438 (December 1975).

14. Malette, A.J. and Francis, R.L. A Generalized Assignment Approach to Optimal Facility Layout. *AIIE Transactions*, Vol. 4, No. 2, pp. 144-147 (June 1972).

SMALL PARTS ORDER PICKING: ANALYSIS FRAMEWORK AND SELECTED RESULTS

Gunter P. Sharp
Kyung Il-Choe
Chang S. Yoon
Georgia Institute of Technology

ABSTRACT

This chapter presents a general analysis framework and the major issues for designing and operating order pick systems for less-than-case retrieval. The interplay among stock keeping unit data analysis, order data analysis, overall system structure, equipment types for subsystems, and operating strategies, is emphasized. Selected results are presented for parametric analysis, batching and zoning using ride-and-pick, proximity batching of small orders with no order splitting, and correlated storage assignment. Future research opportunities are outlined.

1 INTRODUCTION

Order picking is one of the most costly and critical material handling activities. In distribution systems there is increasing emphasis on better delivery time and accuracy standards. In manufacturing the move to smaller lot sizes, point-of-use delivery, and cycle time reductions, make efficient order picking crucial to being competitive. A recent study identified order picking as the number one area for improvement in warehousing [38].

This chapter is concerned with the specification of a general framework for analyzing order pick systems (OPS) and the identification of the major issues involved. The chapter focuses on small parts, or less-than-case retrieval, where the quantity requested per stock keeping unit (SKU) on an order is less than 10.0 liters (0.35 cu. ft.). Some of the concepts presented apply also to case picking and whole pallet retrieval.

Seven types of factors affect the design of order pick systems, as shown in Fig. 1.

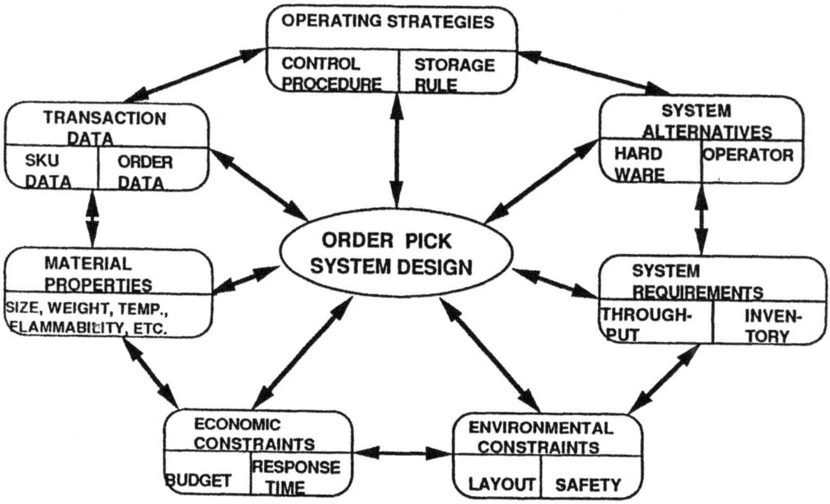

Fig. 1 Factors affecting order pick system design

1. Material properties of items stored,
2. Transaction data for SKUs and orders,
3. Operating strategies for storage assignment and retrieval,
4. Specification of hardware for subsystems and design of operator work areas,
5. System requirements with respect to throughput, inventory capacity, accuracy, and system response time,
6. Constraints based on building layout and operator safety, and
7. Budget constraints related to investment and total annual cost.

Each of the seven types incorporates several factors, as presented in detail later. The design of an OPS thus represents a multifaceted problem of great dimensional complexity.

Some combinations of factors are more likely to exist than others. By focusing primarily on transaction data, Choe [10] classifies OPS into four categories for analysis purposes, as shown below. The sub-factors of order size (SKU line items per order), number of orders, and extent of advance information, result in $2^3 = 8$ combinations when 2 levels of each subfactor are considered.

MSS - Many small orders with statistical information about the composition of orders. A stock room serving a manufacturing facility might fit into this category. The primary criterion is often response time, with cost a secondary criterion.

MSA - Many small orders with advance information about the composition of orders. A typical example of this type of system is a mail order retailer. The primary criterion is often cost.

MLA - Many large orders with advance information about the composition of orders. A typical example here would be a factory warehouse supplying dealers. Criteria are usually cost and accuracy.

FLA - Few large orders with advance information. A typical application would be a warehouse supplying retail stores. Criteria are usually cost and accuracy.

Precise definitions of the descriptors are not possible. It is suggested that the number of orders be compared to the number of packing/shipping lanes. If during a process cycle a packing/shipping lane can be dedicated to an order, we say there are <u>few</u> (F) orders. If a packing/shipping lane is to be shared by several orders during a process cycle, we say there are <u>many</u> (M) orders.

Order size can be determined by cubic volume and/or number of line items. It is suggested that a <u>small</u> (S) order be defined as one that contains 10 or fewer SKU line items. If the quantity per SKU line item is less than 10.0 liters (0.35 cu. ft.), then a small order is also limited roughly to 0.1 cu.m. (3.5 cu. ft.). A <u>large</u> (L) order, containing more than 10 SKUs, would often have a volume greater than 0.1 cu.m. In some situations a large order might fill a truck. These definitions allow for some awkward, in-between situations. We believe that the demarcation based on number of SKUs is more important since it relates more to operating strategies than a split based on cubic volume.

The difference between <u>advance</u> (A) information and <u>statistical</u> (S) information relates to the ability to preprocess the order data for more efficient operation. In the MSS example of a stock room serving a manufacturing facility, the requirement for fast response (e.g. 20 minutes) probably would preclude the types of batching strategies used by a catalog retailer, who usually has several hours at night for data processing and a late-afternoon shipping deadline. If there is adequate time for preprocessing, we say there is <u>advance</u> information. Otherwise, the information is <u>statistical</u>.

The scenarios of a few small orders, with statistical or advance information (FSS and FSA), do not provide much incentive for analysis and are thus excluded. The situation of a few large orders with statistical information (FLS) is relatively unusual. The situation of many large orders with statistical information (MLS), it can be argued, can be approached with many of the same strategies as MLA. Thus, the four categories listed should suffice for analysis purposes. The reason for classifying is that some operating strategies and equipment types are more suitable in some circumstances than in others.

Previous research in the analysis, design, and operation of pallet storage system is applicable to OPS. Some of the strategies developed for pallet systems can also be applied to an OPS, and an OPS may have a pallet system for reserve storage. Francis [15] investigates the circumstances when activity-based storage yields reduced cycle times. Rosenblatt [32] presents a procedure for solving the combined problem of warehouse size, internal layout, and storage policy. Ashayeri [2] reviews the field of warehouse design optimization. Goetschalckx [19] presents algorithms for order picking in a wide aisle using a powered vehicle.

Research related to mechanized and automated equipment for order picking includes that by Kanet [23] on pallet systems, White [39] on part-to-picker versus picker-to-part, Stecke [36] on machine interference models, Houmas [21] on various equipment types, Bredenbeck [7] on vertical storage systems, and Mardix [28], Kim [24], and Koulamas [25] on carousel systems.

Research related to operating strategies for walk-and-pick and ride-and-pick systems includes that by Kunder [26] on tour lengths, Bartholdi [3] and Ratliff [30] on routing, Elsayed [12,13,14] on batching, Francis [15] on storage assignment, and Dichtl [11], Frazelle [16,17], and Mutel [29] on dependent-item storage assignment.

Relatively little has been published on design of the operator work area and on performance standards. Busser [8] describes a work station for retrieval from carousels. Riaz-Khan [31] presents a model for establishing labor efficiency standards.

Work related to conveyor sortation systems includes that by Armstrong [1], Santana [33], and Bozer [5,6]. Gupta [20] describes a kitting matrix, an alternative to a conveyor system.

Algorithms for palletizing different size boxes have been developed by Steudel [37], Smith [35], Carpenter [9], and Malstrom [27].

The overall structure of the OPS within a warehouse has been addressed by Jones [22], Shimizu [34], and Goetschalckx [19].

The next section of this chapter presents the general analysis framework for OPS. The third section contains some recent results for selected operating strategies and some equipment comparisons. The fourth section describes future research opportunities that seem to offer promise for significant efficiency gains.

2 GENERAL ANALYSIS FRAMEWORK

The seven types of factors listed previously form the basis of the analysis framework.

2.1 Material Properties

The size, weight, stackability, suitability for nesting, and environmental requirements of the SKUs must be considered in selecting equipment for storage/retrieval and handling. For example, high-security items should be stored in a closed system, such as storage drawers, vertical carousel, or miniload, and not in an open system, such as shelving. Fragile items might be damaged by a high-speed sortation system. Diversity in size and weight of items can also lead to problems in using high-speed sortation. Material properties thus restrict the compatible equipment types and influence the total cost of the OPS.

2.2 Transaction Data

Transaction data for SKUs include:

- total number of different SKUs stored
- number of different SKUs retrieved during a shift
- total quantity retrieved per SKU per shift, and distribution
- quantity retrieved per SKU line item (when the SKU appears on an order)
- structural (seasonal) and random variability of above items.

Transaction data for orders includes:

- total number of orders per shift
- size distribution by SKUs
- size distributions by quantity and volume
- distributions by priority classes and shipping methods
- similarities among orders.

The availability of transaction data for preprocessing has already been mentioned. Generally, the more time that is available before an order must be shipped, the greater the opportunity to improve productivity by batching, sequencing, and work load balancing. Such preprocessing requires a sophisticated management information system (MIS).

Figs. 2 and 3 show an example order profile. Such information could be either advance for the next shift, or statistical based on past orders. The large number of single-item orders suggests that these might be retrieved using a strategy of batching by SKU. The type of information shown by such an order profile is invaluable in formulating operating strategies.

2.3 Operating Strategies

The key to efficient operation of an OPS is in selecting proper operating strategies for subsets of SKUs and orders. Strategies fall into three general categories:

SKU Line Items per Order	Order Volume, liters								Orders		Line Items	
	<= 3	3-15	15-30	30-60	60-150	150-300	300-600	600-1500	Total	%	Total	%
1	1400	950	300	230	190	100	30	20	3220	49	3220	12
2 - 5	280	670	430	370	330	200	120	65	2465	37	8628	34
6 - 10	1	30	45	80	170	90	80	35	531	8	4248	17
11 - 15	0	5	10	20	70	85	80	65	335	5	6030	24
26 - 50	0	0	0	1	10	11	20	25	67	1	2546	10
51 - 75	0	0	0	0	0	0	2	5	7	0	441	2
75 - 100	0	0	0	0	0	0	0	2	2	0	175	1
Total	1681	1655	785	701	770	486	332	217	6627		25288	
%	25	25	12	11	12	7	5	3		100		100

Fig. 2 Example order profile

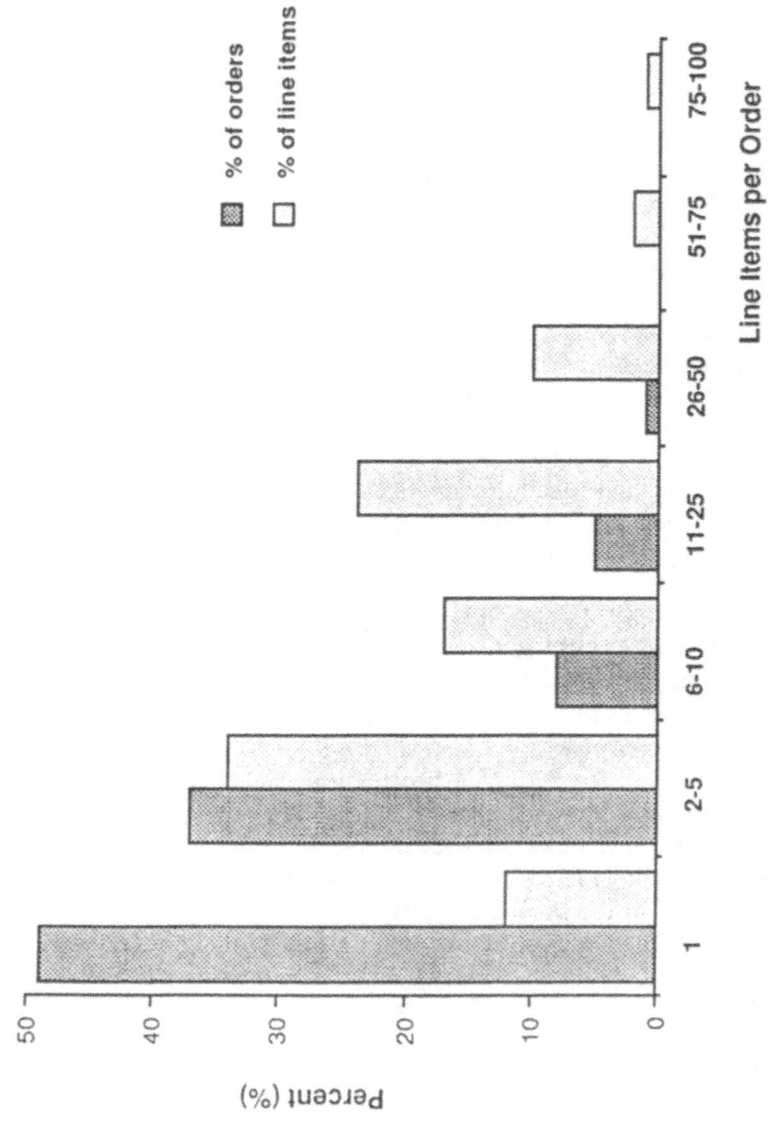

Fig. 3 Example plot of order profile

Storage rules include the following:

- random location assignment
- assignment based on adjusted turnover [15]
- assignment based on location visits between replenishments
- assignment based on time in system for each unit
- single versus multiple stocking locations
- location based on dependent item demand [17],

Some of the storage rules developed initially for pallet systems, such as adjusted turnover, lose much of their effectiveness in a multi-command operation.

Control procedures for retrieval include:

- single-order-pick (SOP)
- batching and sort-while-pick (SWP)
- zoning with progressive picking of an order
- zoning with simultaneous picking of an order
- batching and zoning
- batching and zoning with time windows
- batching without order splitting.

The selection of a particular control procedure might limit the compatible equipment types. Multiple units of mechanized or automated equipment, such as carousel or miniload, make it difficult to implement a single-order-pick strategy and thus usually are operated with a zoning strategy. Extensive batching of orders might require subsequent sortation using a conveyor loop system. Also, more complex control procedures place a greater burden on the MIS.

There is a third, broader issue in operating strategies, and that is in specifying how each SKU should flow through the OPS. Fig. 4 shows the general structure of an OPS, including back-up storage areas, sortation, receiving, and so forth. The thick arrows represent likely flows, the thin arrows other possible flows.

Let us examine the flow of a hypothetical, high-activity product. The product is received in pallets and stored in a pallet reserve area (e.g. the upper levels of pallet rack). It is then moved to a case pick area (e.g. the lower levels of pallet rack). Individual cases are removed from the case pick area and placed in the small parts pick area (e.g. gravity flow rack holding cases). The order pickers selectively retrieve items from the flow racks and place them on conveyors which carry the items to sortation and accumulation area A. There the items of each order are sorted using a bar code scanner and packed into totes. If all the items of an order are contained in such totes, they might then go to a unitizing area to be palletized or combined into mixed unit loads (mul), and then to shipping, or directly to shipping.

Some customers might order this high-activity product in case lots. These might be retrieved directly from the case pick area and sent to sortation and accumulation area B, where cases and totes from area A are sorted by order. The equipment for area B, for sorting cases and totes, might be different

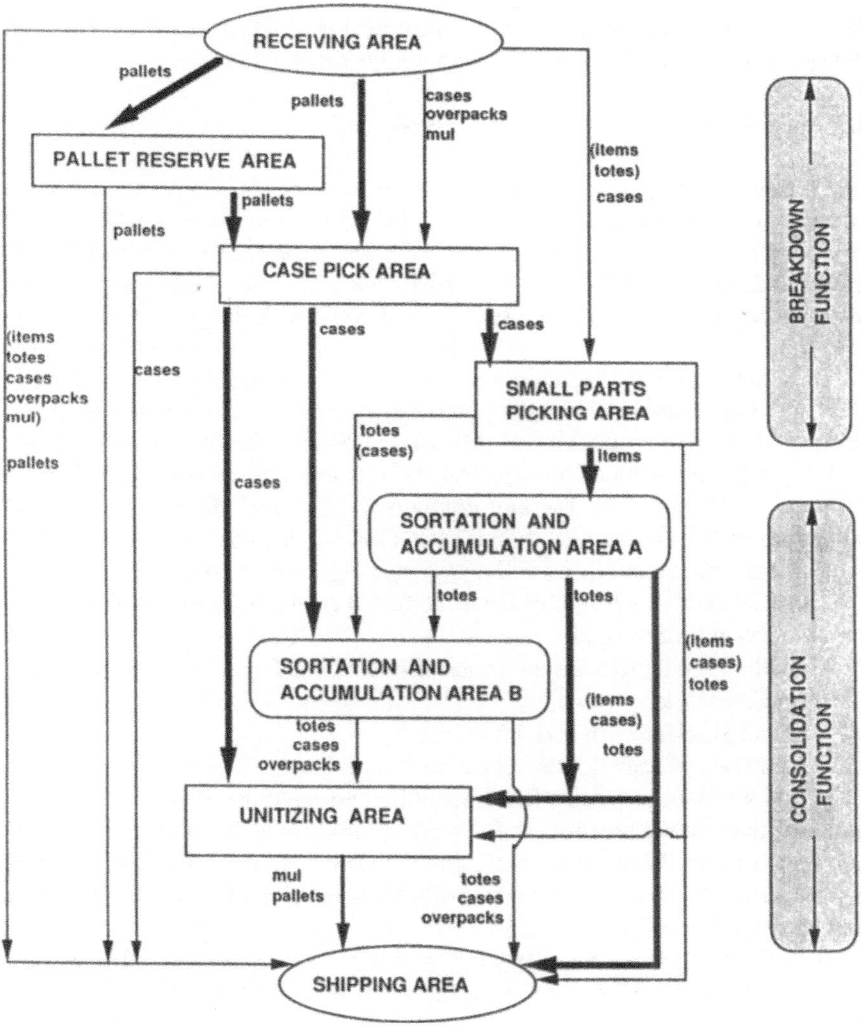

Fig. 4 General structure of an order pick system

from that in area A, which is for sorting smaller items. It is conceivable that the same equipment could be used for both purposes, either simultaneously or at different times.

Fig. 4 should not suggest that every OPS should have this structure and all the diverse flows. Instead, it shows the many possible ways that products can flow through an OPS. The question of which products for which orders should flow in which ways is in the realm of specifying the operating strategy. We designate this as the specification of order pick system structure. Closely related are the questions of multiple locations for an SKU within an area and the space allocation for each SKU within an area.

2.3 Specification of Hardware for Subsystems

Each area shown in Fig. 4 that is selected for inclusion in an OPS can be considered a subsystem or subsystems. The latter can occur when different equipment types, such as gravity flow rack and aisle-captive person aboard, are used for different SKUs in the order pick area. For each subsystem we must specify an equipment type that is compatible with the operating strategy selected.

In some instances the equipment type depends little on the operating strategy. For example, the pallet reserve area could be block stacking, pallet rack, or deep-lane rack with fork lift trucks. Since inflows and outflows are pallets, only storage rules are applicable. But the storage rules are likely to produce similar benefits for any equipment type, including an automated storage/retrieval system (AS/RS) if such a system makes sense.

Typical equipment types for the case pick area include the following:

- pallet rack to a height of 2.0 m. (6 ft.-7 in.) with a walk-and-pick or ride-and-pick vehicle
- pallet rack up to a height of about 10.0 m. (33 ft.), with a mobile order picker truck
- block stacking with fork lift trucks
- gravity flow rack holding cases, with a take-away conveyor along the rack face (a physically feasible but less popular alternative).

Some of the control procedures for retrieval that were listed earlier can also be applied here. When case picking is for replenishment of the small parts picking area, the control procedures will likely be limited to single-SKU-pick and batch picking of several SKUs.

It is in the small parts picking area that the selection of hardware becomes difficult. First, the equipment for the small parts picking area depends on the specification of the order pick system structure. Second, it depends on the transaction data analysis. Third, it is inextricably linked with the operating strategies selected. Fourth, it influences the type of equipment used for subsequent sortation. Fifth, effective use of many equipment-strategy combinations requires a sophisticated MIS.

These interrelationships are shown in Fig. 5, which portrays a design procedure for order pick systems. The procedure consists of three main stages. The input stage consists of:

- SKU transaction data analysis
- order transaction data analysis
- specification of order pick system structure.

For a given set of results in the input stage, we then proceed to the selection stage, which consists of:

- specification of hardware for subsystems
- specification of operating strategies for storage rules and control procedures for retrieval
- calculating the physical transformation of items
- calculating the information transformation of work lists.

The middle box in Fig. 5 indicates that these four design tasks must be performed together for each option. For example, a five-aisle person-aboard system with batch picking in five zones, one per aisle, will have a different information transformation of the order data into pick lists than a 20-aisle shelving system with single-order-pick strategy. These two systems will most likely result in different numbers and sizes of loads coming out of the small parts picking area. Such physical transformation of items depends on the equipment, the operating strategy, and the results from the input stage. Fig. 6 shows the possible and likely ways that items can be transformed in an OPS.

After performing the tasks in the selection stage for different options, we proceed to the evaluation stage. An important task here is subsystem reconciliation. An example is the use of a pallet rack for both pallet reserve area and case pick area: there must be a proper balance between spaces required and time needed for each function. Another example is when single-order-pick and batch picking are performed using the same equipment. A third example is the capacity matching of a sortation system with the output of small parts picking area.

Some type of multi-criteria evaluation procedure is usually employed in the evaluation stage. It is not unusual for the design procedure to be partially repeated, as shown by the feedback arrows to the input and selection stages in Fig. 5.

The schematic version of the design procedure in Fig. 5 highlights the interplay among SKU data analysis, order data analysis, specification of OPS structure, equipment types for subsystems, and operating strategies. An understanding of these interrelationships is crucial to the design of an efficient OPS. It should be noted that specification of the OPS structure should precede the specification of equipment and operating strategies. Conceptually, it is a higher-level decision, and there are many equipment types and operating strategies to satisfy each function portrayed in Fig. 4.

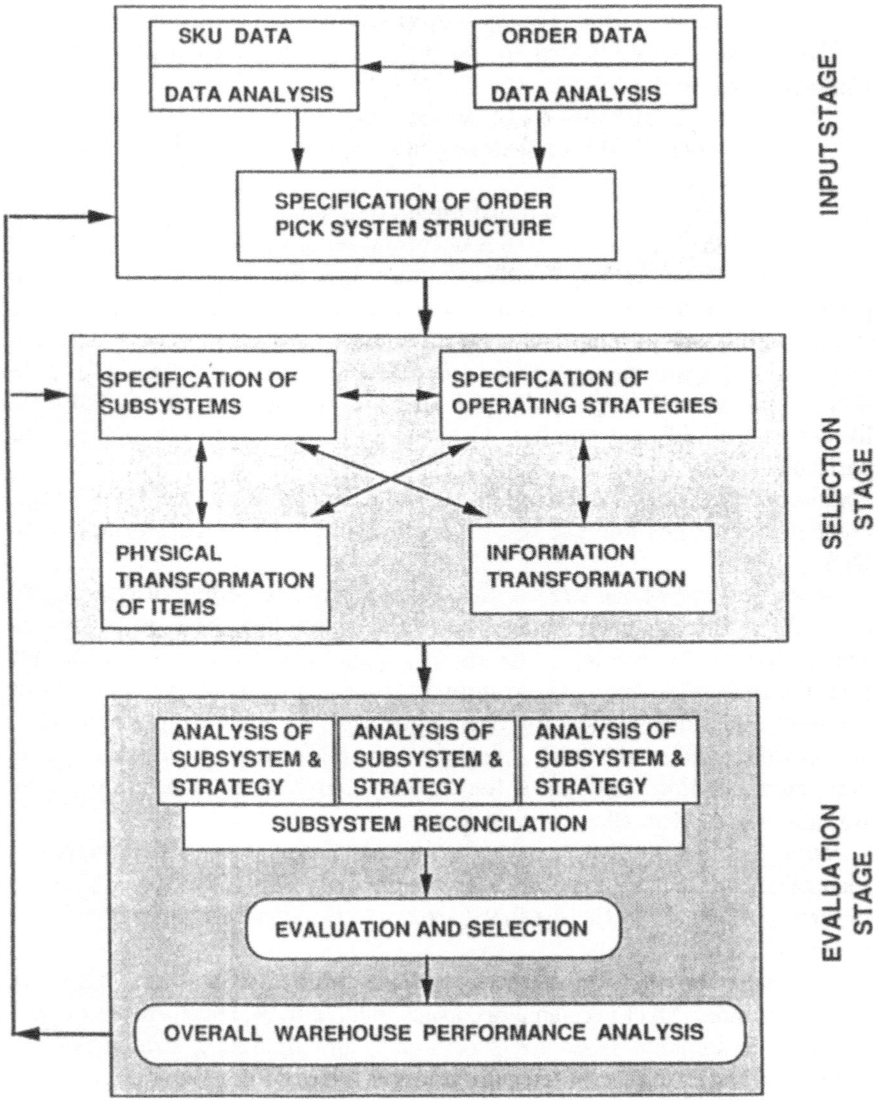

Fig. 5 Design procedure for order pick systems

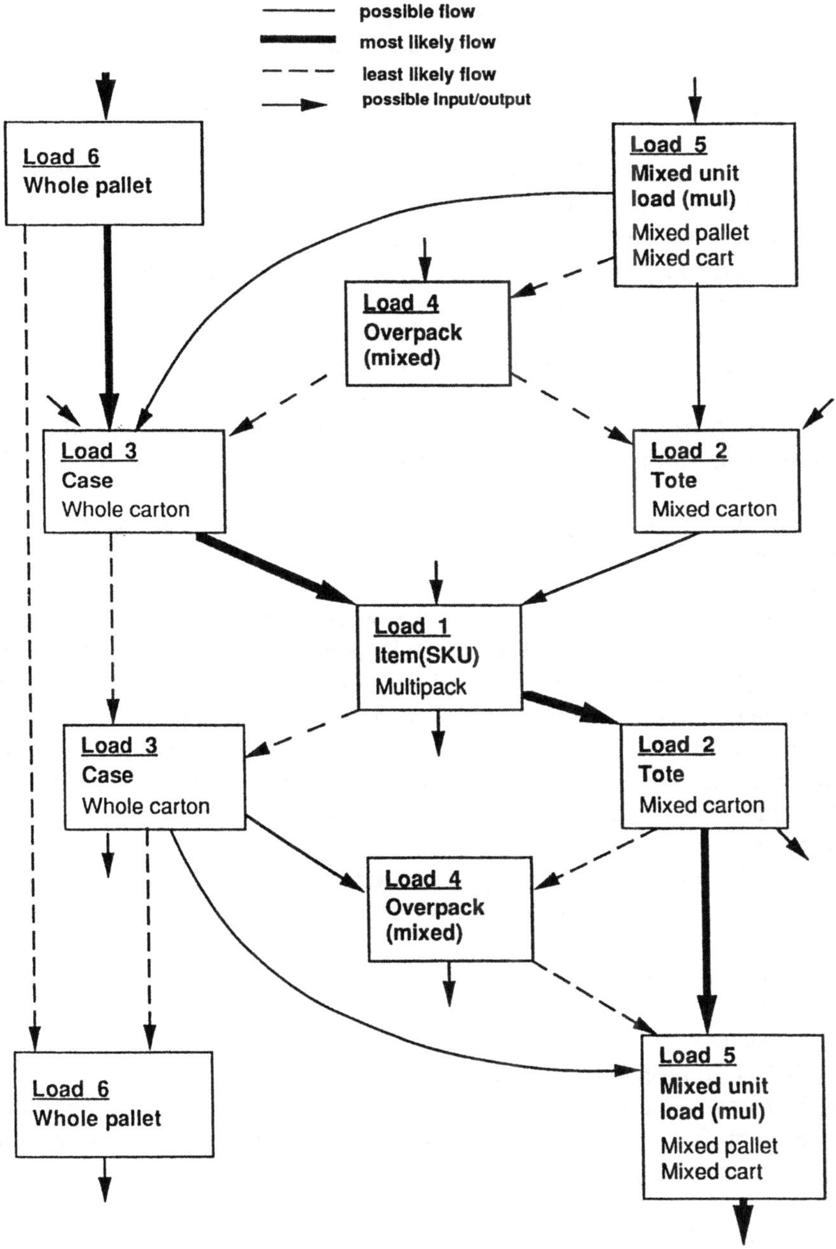

Fig. 6 Product transformation in order pick system

For the small parts picking area the equipment types can be classified into three categories:

- part-to-picker: horizontal carousel, horizontal carousel with robotic extraction of bin, rotary rack, vertical carousel, miniload, vertical storage system
- horizontal picker-to-part: shelving, gravity flow rack, storage drawers; travel can be by walking or by a powered vehicle; product transport can be by hand, by a push-cart, by a powered vehicle, by unpowered conveyor, or by powered conveyor
- horizontal and vertical picker-to-part: shelving, gravity flow rack, storage drawers; travel can be by aisle-captive vehicle that moves horizontally and vertically simultaneously, or by a vehicle designed primarily to move sequentially in horizontal and vertical directions.

For the part-to-picker systems a primary strategy issue is how to minimize picker idle time. Typical solutions include assigning an operator to two machines (carousels), increasing the number of queue positions (miniload), and making storage assignments and sequencing the retrievals to minimize machine time.

For picker-to-part systems a primary strategy issue is to reduce travel. Solutions usually involve storage assignments by activity and some type of order batching.

For any type of system the time the operator spends at the pick location can be reduced by proper design of the work area. Good lighting, devices for counting and weighing, graphic displays, and indicator lights at the pick face, can reduce the time spent extracting an item by 50% while at the same time improving accuracy. Such work area design is easier to implement for part-to-picker systems and more difficult in the horizontal picker-to-part systems. Computer-aided order picking does require a more complex and expensive MIS, however.

For the <u>sortation area</u> the equipment types include the following:

- powered or gravity conveyor with mechanical or manual divert
- tilt-tray conveyor
- carousel system
- gravity flow rack
- shelving.

Partial or complete sortation can also be done sometimes during the order pick process. This type of sortation does not really belong to either sortation area A or B, but falls under specification of equipment and strategies for small parts order picking.

2.4 System Requirements

The requirements of an OPS should be specified in preliminary form at the beginning of the design process. Requirements should be specified for orders per shift, volume of items shipped per shift, storage capacity, system response time, auxiliary processing such as retail pricing or labeling, accuracy, returns and restocking, cycle counting, stock reassignment, etc. In some situations it is appropriate to specify requirements with respect to detailed SKU data and order data. In nearly all situations it is vital to specify peak/average ratios for weekly, daily, and hourly operation. As the design process proceeds, it is not unusual to refine the system requirements.

2.5 Specify Environmental Constraints

At the beginning of the design process it is also necessary to specify building layout constraints, noise restrictions for the operators' environment, and desirable manual lift and reach limits. Building restrictions might be expressed in total square feet or in specific spaces available. It should be mentioned, however, that many equipment types for the case pick area, the small parts picking area, and the sortation area, are quite flexible with respect to room dimensions and layout. The performance of a miniload system might suffer if ceiling height is inadequate. Non-contiguous areas can be connected by conveyor or fork lift trucks. If limited floor space is a factor, many equipment types for small parts order picking and sortation can be installed on mezzanines. Manual lift and reach limits should be established considering the labor force available for the system.

2.6 Specify Economic Constraints

Budgetary constraints are usually imposed on a system design. Such constraints may be direct in terms of limitations on investment, total annual cost, cost per item retrieved, labor force, etc. Or they may be indirect in the sense that the lowest cost solution, on an equivalent annual after-tax basis, should be selected. Other constraints might relate to time to install the system. A more elaborate system evaluation scheme would involve trade-offs among desired levels of investment, annual cost, labor force, time to install, and other factors such as accuracy, flexibility, maintainability, etc.

The general analysis framework for the design of an OPS is seen to be not a hierarchical process, but rather a hybrid process with feedback and correction, as suggested by Fig. 5:

1. Specify material properties, system requirements, environmental constraints, and economic constraints.
2. Perform analysis of SKU and order data, specify OPS structure.
3. Specify system equipment alternatives and operating strategies, perform calculations for physical transformation of items and information transformation.
4. Perform analysis of subsystems, reconcile subsystems, perform overall evaluation and selection.

In the next section we present some selected results on specific options with respect to equipment and strategies.

3 SELECTED RESULTS

In this section we present selected results from recent research efforts that focus on the small parts picking area. These efforts deal with four concepts: parametric analysis for selecting equipment, batching and zoning using ride-and-pick, proximity batching of small orders with no order splitting, and correlated storage assignment.

3.1 Parametric Analysis for Selecting Equipment

In an effort to determine under which circumstances various equipment types are preferred, Houmas extended his earlier research [21] and performed a limited parametric analysis. The equipment types considered for small parts picking area include: bin shelving, storage drawers, gravity flow rack, carousel, and miniload. The selection criterion is annual equivalent after-tax cost. List prices are used for equipment. Other important cost factors are $15/hour labor cost, $270/sq.m. ($25/sq. ft.) building construction costs, 5-year depreciation for equipment, 46% marginal tax rate, and 15% discount rate.

For picker-to-part systems it is assumed that orders are batched in such a way so that an order picker passes each SKU location 5 times per hour. Picker-to-part systems are configured up to a height of 1.83 m. (6 ft.), with a conveyor along the pick face for taking away the items selected. Downstream sortation is not considered directly in the analysis, since it would be similar for all systems.

For part-to-picker systems using one picker per machine (carousel, miniload), it is assumed that pick lists of 10 items each are sequenced to minimize machine travel. A second operating strategy for carousels, called double-carousel, is to have one picker select alternately from two carousels; on each carousel the items to be retrieved are sequenced first-come-first-served. In each situation the configuration of machines is determined so as to minimize the number of machines.

The following factors are varied parametrically:

- cubic volume of inventory per SKU location, 1.8 - 142 liters (107 cu.in.-5 cu. ft.), 5 values
- location visits per SKU per 8-hour shift, 0.008 - 51, 9 values
- processing time at the pick face per SKU line item, 10-30 seconds, 3 values.

For any scenario all SKUs are considered to be similar and the processing time is the same for all equipment types. Fig. 7 shows results for 142 liters inventory per SKU and 20 second processing time. This graph shows that equipment choice depends very much on activity.

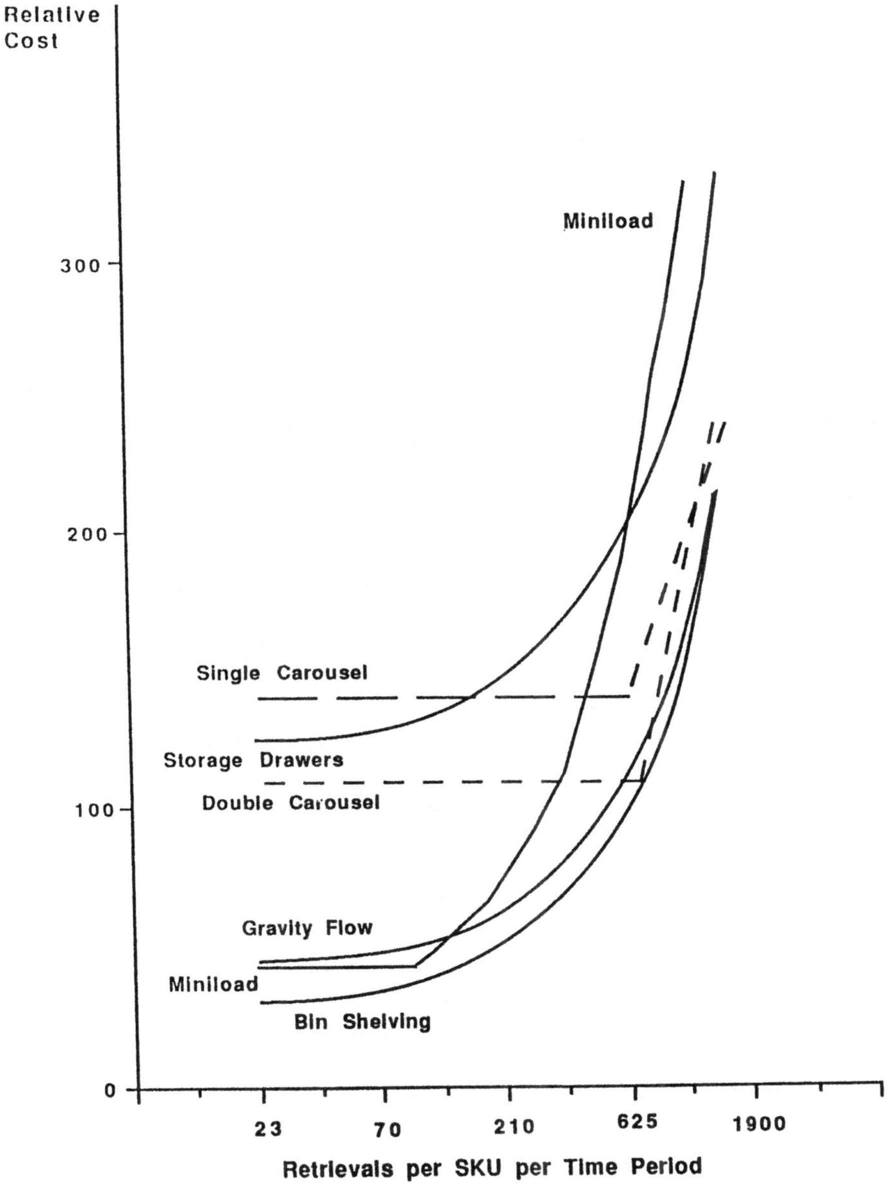

Fig. 7 Example results from parametric analysis

The overall results suggest the following preferred equipment types:

	Small volume	Medium volume	Large volume
Low activity	shelving, storage drawers,	shelving	miniload
Med. activity	storage drawers, shelving	shelving, carousel	shelving
High activity	storage drawers, shelving	carousel	flow rack

These results hold only for the given values of the fixed parameters, and cannot be generalized without further evaluations. They also depend on the assumption of the same processing time per SKU for each system. Nevertheless, the results show that it is possible to perform this type of analysis, given enough cost data, and the patience to develop reasonably consistent assumptions about operating strategies across different equipment types.

3.2 Batching and Zoning with Time Windows

Choe [10] has conducted a preliminary comparison of three operating strategies for a ride-and-pick system:

- System A, single-order-pick
- System B, sort-while-pick, orders are assigned in batches to pickers, no zoning
- System C, batching and zoning with a downstream accumulation/sortation system, the number of zones is equal to the number of pickers.

The example is a rack system with the following characteristics:

- 36 aisles
- 40 SKUs per aisle
- 1.22 m. (4 ft.) modular bin spacing
- 100 m./min. (328 ft./min.) vehicle velocity
- 2 SKU line items/order, on average
- 1.5 units/SKU line item, on average
- 6 sec. processing time per unit.

Routing through the aisle system is one way: a picker may enter an aisle from either direction, but may not turn around in an aisle. The length of the time window determines how many orders a picker will retrieve on a tour.

The performance criteria of interest are the service time/order and the capacity of the system. For an arrival rate of 200 orders/hour and a staff of 9 pickers, the service times (minutes) per order as a function of the time window are as follows:

	Time Window, Mins.					
	10	20	30	40	50	60
System A	2.59	2.59	2.59	2.59	2.59	2.59
System B	1.42	1.10	0.97	0.84	0.81	0.76
System C	0.94	0.64	0.58	0.52	0.47	0.44

These service times consist of three components: travel time, picking time, and synchronization time or wave delay (System C only). The potential advantages of batching and zoning are clearly evident for the example data, even recognizing the need for additional sortation with System C. The capacity of the system is inversely related to the service time per order.

3.3 Proximity Batching with No Order Splitting

Gibson [18] has examined proximity batching techniques that are suitable for the situation of no order splitting in a walk/ride-and-pick system with selective one-way travel in the aisle. Orders are grouped into batches of 50 SKU line items. The average order size is 5 SKUs, so each batch contains about 10 orders. The storage region is a square of area 1.0, containing 10 aisles; each aisle has 40 locations per side, resulting in 800 total locations. The start/end point of each tour is a corner of the region.

The baseline strategy simply groups orders into batches using a first-come-first-served (FCFS) discipline. Under this strategy, a tour that visits up to 50 locations will nearly always visit every aisle, resulting in a tour length of about 12 (10 aisles * 1.0 plus 2 cross aisles * 1.0).

Activity based (skewed) storage assignment is investigated for the following distribution

Class	Storage Locations	Retrieval Activity
A	10%, 1 aisle	52%
B	40%, 4 aisles	41%
C	50%, 5 aisles	7%

Activity zoning with the baseline FCFS batching gives an average tour length of about 10, a 16% reduction, as shown in Fig. 8. (The number of orders available for preprocessing has no effect when the FCFS batching is used.)

The intelligent batching technique tries to combine orders that have SKU locations in the same aisles (proximity batching). When intelligent batching is combined with skewed storage assignment, tour length reductions up to 45% are possible. Obviously, the more orders that are available for preprocessing, the greater are the reductions, as shown in Fig. 8. The extreme right value for 1,200 orders to be preprocessed corresponds to 6,000 SKU line items that are combined into 120 batches of 50 SKU line items each.

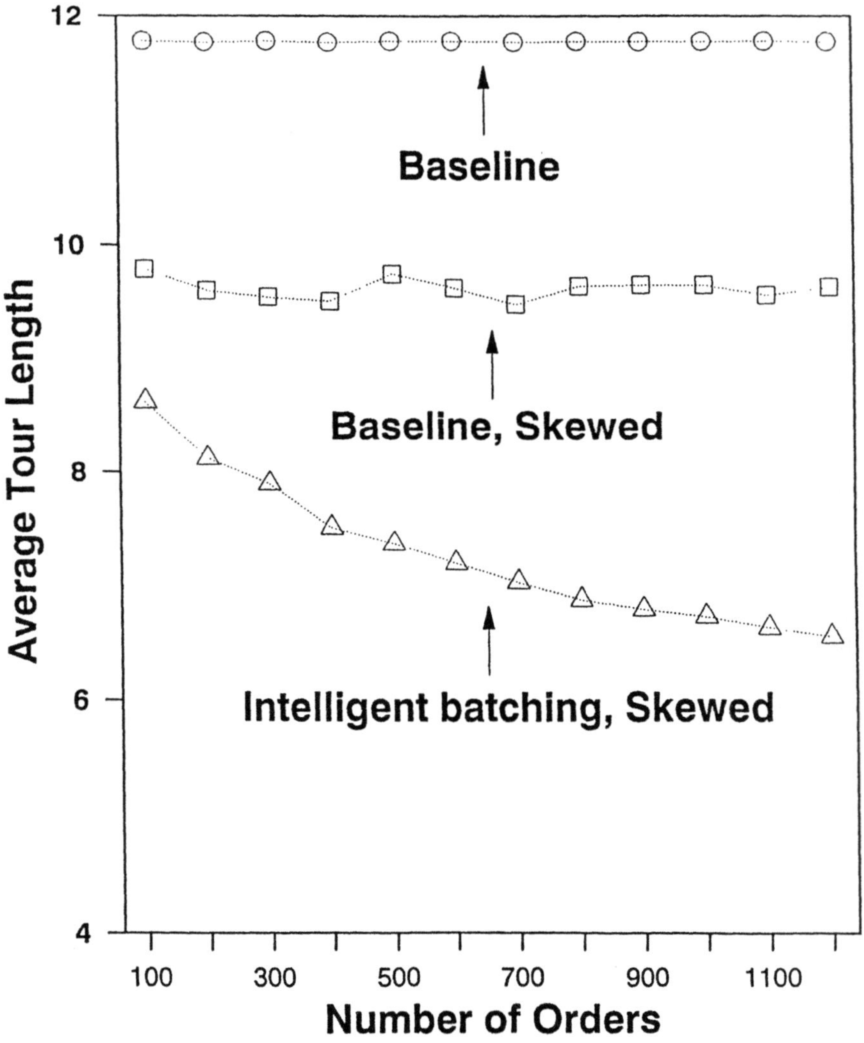

Fig. 8 Results for proximity batching, no order splitting

3.4 Correlated Storage Assignment

A storage assignment technique that takes advantage of similarities among orders should make it possible to reduce travel time and SKU location visits in a scenario where order batching is used. Dependent item storage assignment applies to both walk-and-pick and mechanized systems. Frazelle [17] has developed a cluster-first-zone-second (CFZS) procedure that groups SKU into clusters and then assigns clusters to areas in the storage region. The SKUs in a cluster have a relatively high probability of being requested on the same order.

A typical result is for a company that distributes sporting goods, men's and women's apparel, and some food items. The distribution of SKU line items per order is

$$1, 66\%; \quad 2, 16\%; \quad 3, 7\%; \quad 4, 4\%; \quad 5 \text{ or more}, 7\%$$

The activity distribution is

Class	Storage Locations	Retrieval Activity
A	5%	47%
B	5%	23%
C	20%	25%
D	70%	5%

A sample of 5,000 orders was used with the CFZS procedure to generate about 300 clusters of SKUs and to make storage assignments. A second sample of 5,000 orders was then used with Gibson's proximity batching technique to generate tours for a walk-and-pick system in a square region of area 1.0.

The results indicate that potential savings of about 55% can be achieved compared to random storage assignment and FCFS batching. Shown below are average tour lengths as a function of batch size.

Orders per batch	5	10	25	50
Random storage, FCFS batching	11.7	11.7	11.7	11.7
Random storage Intelligent batching	4.65	5.21	6.25	7.41
CFZS, Intelligent batching	3.72	3.99	4.52	5.16

These four selected results illustrate the potential benefits that are possible from careful specification of equipment and operating strategies. They also provide evidence that it is possible to draw some general conclusions about an OPS, despite the numerous factors involved in design and operation. Finally, they indicate potential areas for future research.

4 FUTURE RESEARCH OPPORTUNITIES

The performance characteristics of equipment types such as carousel, miniload, and vertical storage device, have been determined, and thus there is little need for additional research in this area. However, some newer, high throughput devices, such as rotary rack and multi-shuttle miniload, have not been fully examined. In general, though, there are few research opportunities in the area of equipment performance and configuration. Nevertheless, the question of which equipment type to use under what conditions remains an open one. Parametric analysis, either numerical or analytical, is a relatively unexplored research area.

Another equipment-related area that offers potential is the design of work stations for order pickers and for packers. Work station improvements in conjunction with computer-aided order picking can improve productivity by a factor of two and more. Packaging often requires as much labor as picking, which points to a potential area of research.

The specification of operating strategies is a fertile area for investigation. Questions related to storage assignment, batching procedures, number of zones, length of time window, dispatching of order pickers, etc., need to be addressed. A closely related problem is the workload balancing between small parts picking area and sortation area. The information transformation of orders into pick lists, using any type of batching method, represents a statistical process. A characterization of this process would help an analyst specify a batching technique.

The pick-versus-reserve question is only one facet of the specification of the OPS structure. How much of each SKU should be stored in the pick area? Which SKUs should be stored also in the case pick area, in the reserve area?

Dynamic operation of an OPS involves the questions of stock reallocation and dynamic stocking. Stock reallocation occurs because SKUs change activity, newer SKUs push out older ones, or external factors related to space, costs, etc. Dynamic stocking is the method of storing in the small parts picking area only those SKUs that will be selected during the next shift. The method offers the possibility of drastically reduced storage requirement and more efficient picking.

5 SUMMARY

The major thrust of this chapter is the presentation of a general analysis framework for the design of an order pick system for small parts. The key point is the interplay among transaction data analysis, specification of OPS structure, equipment specification, and selection of operating strategy.

A second contribution is the presentation of selected research results for parametric analysis, batching procedures, and correlated storage assignment. Potential future research opportunities are also discussed.

6 ACKNOWLEDGEMENTS

This research was supported by the National Science Foundation under grant CDR-8300965 and by United States Industry through the Material Handling Research Center at the Georgia Institute of Technology.

7 REFERENCES

1. Armstrong, R.D., W.D. Cook, and A.L. Saipe, "Optimal Batching a Semi-Automated Order Picking System," *J. Opl. Res. Soc.*, 30, pp. 711-720 (1979).
2. Ashayeri, J. and L.F. Gelders, "Warehouse Design Optimization," *Euro. J. O. R.*, 21, pp. 285-294 (1985).
3. Bartholdi, III, J.J. and L.K. Platzman, "Heuristics Based on Spacefilling Curves for Combinatorial Problems in Euclidean Space," *Management Sc.*, 34-3, pp. 291-305 (1988).
4. Bozer, Y.A., *Optimizing Throughput Performance in Designing Order Picking Systems*, Ph.D. thesis, Georgia Institute of Technology, Atlanta, Georgia (1985).
5. Bozer, Y.A., M.A. Quiroz, and G.P. Sharp, "An Evaluation of Alternative Control Strategies and Design Issues for Automated Order Accumulation and Sortation Systems," *Material Flow*, 4, pp. 265-262 (1988).
6. Bozer, Y.A. and G.P. Sharp, "An Empirical Evaluation of a General Purpose Automated Order Accumulation and Sortation System Used in Batch Picking," *Material Flow*, 2, pp. 111-131 (1985).
7. Bredenbeck, J.E., W.T. Shirk, and J.C. Majure, "Small Parts Handling Needs Are Met By Installing a Vertical Storage System," *Ind. Eng.*, 21-7, pp. 34-38 (July 1989).
8. Busser, B. and C. Keller, "Carousel Automation for Increased Productivity," *Material Handling Focus '88*, Atlanta, Georgia (Sept. 1988).
9. Carpenter, H. and W.B. Dowsland, "Practical Considerations of the Pallet-Loading Problem," *J. Opl. Res. Soc.*, 36, pp. 489-497 (1985).
10. Choe, K.-I., *Analysis of Small Parts Order Picking Systems*, Ph.D. thesis, Georgia Institute of Technology (1990).
11. Dichtl, E. and W. Beeskow, "Optimal Allocation of Commodities in Storehouses by Means of Multi-Dimensional Scaling," *Zeitschrift f. Optns. Res.*, 24, B51-864 (1980).
12. Elsayed, E.A., "Algorithms for Optimal Material Handling in Automatic Warehousing Systems," *Int. J. Prod. Res.*, 19, pp. 525-535 (1981).

13. Elsayed, E.A. and R.G. Stern, "Computerized Algorithms for Order Processing in Automated Warehousing Systems," *Int. J. Prod. Res.*, 21, pp. 579-586 (1983).
14. Elsayed, E.A. and O.I. Unal, "Order Batching Algorithms and Travel-Time Estimation for Automated Storage/Retrieval Systems," *Int. J. Prod. Res.*, 27-7, pp. 1097-1114 (1989).
15. Francis, R.L. and J.A. White, *Facility Layout and Location*, Prentice-Hall, Englewood Cliffs, NJ (1974).
16. Frazelle, E.H., *Stock Location Assignment and Order Picking Productivity*, Ph.D. Thesis, Georgia Institute of Technology, Atlanta, GA (1989).
17. Frazelle, E.H. and G.P. Sharp, "Correlated Assignment Strategy Can Improve Any Order-Picking Operation," *Ind. Eng.*, 21-4, pp. 33-37 (April 1989).
18. Gibson, David, *Order Batching Without Splitting Orders*, M.S. Thesis, Georgia Institute of Technology, Atlanta, GA (1990).
19. Goetschalckx, M. and J. Ashayeri, "Characterization and Design of Order Picking Systems," *Logistics World*, 2-2, pp. 99-106 (June 1989).
20. Gupta, N.K., "Kitting Matrix Adds Accuracy to Small Part Picking System," *Ind. Eng.*, 35-1, pp. 35-38 (Jan. 1982).
21. Houmas, C.G. and G.P. Sharp, *Comparing Small Parts Storage Systems*, Report MHRC-TD-86-04, Georgia Institute of Technology, Material Handling Research Center, Atlanta, GA (1986).
22. Jones, P.S., *A Least-Cost Equipment Selection Technique for Distribution Warehouses*, Stanford Research Institute, Menlo Park, CA (1971).
23. Kanet, J.J. and R.G. Ramirez, "Optimal Stock Picking Decisions in Automatic Storage and Retrieval Systems," *OMEGA Int. J. Mgmt. Sci.*, 14-3, 239-244 (1986).
24. Kim, W.B. and E. Koenigsberg, "The Efficiency of Two Groups of N Machines Served by a Single Robot," *J. Opl. Res. Soc.*, 38, pp. 523-538 (1987).
25. Koulamas, C.P. and M.L. Smith, "Look-ahead Scheduling for Minimizing Machine Interference," *Int. J. Prod. Res.*, 26-9, 1523-1533 (1988).
26. Kunder, R. and T. Gudehus, "Mittlere Wegzeiten beim eindimensionalen Kommissionieren," (Average Travel Times in One-Dimensional Order Picking), *Zeitschrift f. Optns. Res.*, 19, B53-B72 (1975).

27. Malstrom, E.M., H.D. Meeks, and J.R. Fleming, "A Robotic Pallet-Loading Algorithm for Dynamic Carton Distribution," *Proc. 7th Int. Conf. on Automation in Warehousing*, San Francisco, CA, pp. 161-168 (1986).
28. Mardix, I. and G.P. Sharp, *Cost and Efficiency Analysis of the Carousel Storage System*, Report MHRC-TR-85-08, Georgia Institute of Technology, Material Handling Research Center, Atlanta, GA (1985).
29. Mutel, B. and D. Anciaux, "Study of Storage and Picking Policies of Items in Large Delivery-Type Warehouse," *10th Int. Conf. on Automation in Warehousing*, Dallas, TX (1989).
30. Ratliff, H.D. and A.S. Rosenthal, "Order Picking in a Rectangular Warehouse: A Solvable Case of the Traveling Salesman Problem," *Optns. Res.*, 31-3, pp. 507-521 (May-June 1983).
31. Riaz-Khan, M., "An Efficiency Measurement Model for Computerized Warehousing Systems," *Int. J. Prod. Res.*, 22, pp. 443-452 (1984).
32. Rosenblatt, M.J. and Y. Roll, "Warehouse Design With Storage Policy Considerations," *Int. J. Optns. Res.*, 22, pp. 809-821 (1984).
33. Santana, J.L. and L.K. Platzman, "Real-Time Scheduling of an "Open Warehouse" Sorter-Palletizer System," *Proc. 18th IEEE Conf. on Decision and Control*, Fort Lauderdale, FL, pp. 537-539 (1979).
34. Shimizu, A., "Engineering Analysis of Automated Warehouse," *Proc. 4th Int. Conf. on Automation in Warehousing*, Tokyo, S1-5: 1-11 (1981).
35. Smith, A. and P. De Cani, "An Algorithm to Optimize the Layout of Boxes in Pallets," *J. Opl. Res. Soc.*, 31, pp. 573-578 (1980).
36. Stecke, K.E. and J.E. Aronson, "Review of Operator/Machine Interference Models," *Int. J. Prod. Res.*, 23-1, pp. 129-151 (1985).
37. Steudel, H.J., "Generating Pallet Loading Patterns: A Special Case of the Two-Dimensional Cutting Stock Problem," *Management Sci.*, 25, pp. 997-1004 (1979).
38. Warehouse Education and Research Council, Survey, Oak Brook, IL (1984).

SECTION FOUR
ECONOMICS AND AUTOMATION

This section of the book examines some of the less traditional areas of material handling research. These concerns, economic justification, automation, and reliability, are hardly peripheral issues. As material handling follows material processing into the low-labor/high-automation environment, there is an increased concern for the magnitude of the capital expenditure and the justification of those expenditures. This concern is compounded by the fact that material handling is often viewed as not adding value.

Material handling is, like data handling and electric power, something that links together entire facilities, therefore electro-mechanical reliability becomes more important because of its impact on the entire production process.

Material handling research addresses these issues. The chapters by Choobineh, Cullinane, and Sullivan and Brimson, tackle the question of economic justification of automation head-on. The Dickerson and the Slutzky, et. al. chapters address automation from a general and specific perspective, respectively. Since one of the few tools available at the present time for analyzing automation is queuing theory, the dual chapters by Toro-Ramos provide an analysis of the capacity of automated functions where a single material handling device serves an automated process.

The Lander's chapter is one of those rare reviews that present real material handling reliability information, and sets an environment in which reliability data can contribute to operational strategy.

Stephen L. Dickerson

JUSTIFICATION OF MANUFACTURING SYSTEMS

F. Choobineh
University of Nebraska-Lincoln

ABSTRACT

Possibility distributions are recommended for explicit representation of uncertainty in models used for justification of manufacturing systems. A procedure is presented that is capable of obtaining and combining the possibility distributions of strategic and economic aspects of an investment situation. A numerical example is presented, and a method of ranking alternatives is discussed.

1 INTRODUCTION

A prevailing view in recent years has been that the implementation of a programmable manufacturing system improves the firm's competitive advantage in its market place. Although this view is not universally accepted, nevertheless, its subscribers have been disadvantaged by the absence of an effective justification procedure that aids their investment decision making. This disadvantage is prevalent in spite of numerous publications on the justification of manufacturing systems. These publications have been helpful in developing a keen awareness among the analysts and decision makers that the strategic as well as economic aspects of acquiring programmable manufacturing systems need to be explicitly considered.

Incorporation of the strategic aspects of manufacturing systems into the justification process has further exposed the weaknesses of the existing procedures. These weaknesses are identified with three interrelated stages of a justification process, namely, identification, estimation, and modeling.

The weakness in identification is the inability to properly identify all relevant costs and benefits of a proposed system prior to its implementation.

Specifically, those costs and benefits that go beyond the boundaries of the project (i.e. those resulting from interaction of the proposed system with the rest of the enterprise) are the most difficult to identify.

The weakness in estimation is the inability to properly estimate the values of costs, benefits, economic lives, and discount rate. This weakness is manifested by the reality that most often it is unknown how the values of these parameters will change over time and/or how their values should be measured. Specifically evident is the lack of know-how in measuring strategic and intangible costs and benefits. Unfortunately, the weaknesses in identification and estimation have been mutually reinforcing such that the philosophy of "if you can't measure it, ignore it" has become a common practice.

The weakness in modeling is the inability to develop and solve a precise economic model of the situation under study. The primary contributor to the modeling weakness is the presence of uncertainty and the inability to effectively account for it. Uncertainty is a phenomenon that abundantly exists in the planning and justification of manufacturing projects. Uncertainty in decision making results from the lack of information about the systems parameters in order to make a go/no-go decision.

The majority of practitioners treat uncertainty as follows. First, they use single estimates for the parameters of an economic model that are used in a chosen measure of performance; furthermore, they compensate for the aggregate uncertainty of the parameters and the strategic aspects of the project by adjusting the acceptable value (hurdle value) of the measure of performance. Second, in order to set the hurdle value they employ sensitivity analyses to provide a better understanding of the combined effects of the parameters' uncertainties on the selected measure of performance. A go/no-go decision on the project is made based on the relative value of the project performance measure with respect to its hurdle value and subject to availability of funds.

These practices are documented by numerous industrial surveys (e.g. [1,10,12,14]) showing the popularity of the pay-back period and the internal rate of return as measures of economic performance of a project. Also, surveys have shown that the acceptable pay-back period is decreased and the acceptable rate of return (hurdle rate) is increased in the face of increasing project uncertainty. Although several articles [3,5,9,15] have pointed out the theoretical shortcomings of these two procedures, the practitioners have not abandoned their use.

Alternately, scientists' suggestion for explicitly modeling uncertainty, in general, has been to assume that uncertainty is caused by aleatory processes and consequently have prescribed probabilistic approaches for modeling uncertainty. Although probability theory is very powerful, the precise solution of the economic model when parameters are assumed to be random variables requires calculating the probability distribution of products and ratios of random variables that are generally cumbersome and computationally intractable. Thus, simplifying assumptions are made to ease the computational intractability which results in a less precise model that can be solved precisely.

Insistence of the practitioners on the adjustment of the pay-back period and/or the hurdle rate of return as a means of compensating for the uncertainty of the project is an indication that either they do not understand the probabilistic methods and/or they do not believe that they are practical to use. In either case, alternative methods of modeling uncertainty are needed that are theoretically sound and practical to use.

A secondary contributor to the modeling weakness is the lack of effective know-how for explicitly incorporating strategic variables in justification models. Although recently promising approaches have been proposed [5], still the knowledge of effectively integrating economic and strategic criteria into a single model is in its infancy.

Because of the nature of manufacturing environment - i.e. volatility of the market and technology - identification and estimation weaknesses cannot be totally eliminated. However, they may be reduced by using knowledgeable experts for identifying and estimating input variables to the justification model. Experts should understand the technology of the proposed project, intricacies of the existing plant, and the behavior of the market. Furthermore, these experts can be more effective if they are supported by nontraditional cost accounting systems, such as an activity based cost accounting system [4].

Unlike the first two weaknesses, more opportunity exists for eliminating the modeling weakness. First, to improve modeling, a model that is consistent with the principle of engineering economy, e.g. incorporating the time-value of money, must be considered. Second, all available expert generated information about the values and behavior of every parameter must be fully incorporated into the model.

In this chapter a justification process is presented that is capable of considering and combining monetary and nonmonetary criteria. Uncertainty will explicitly be represented via possibility distributions in the justification process. The proposed process is demonstrated through a numerical example.

2 USE OF POSSIBILITY DISTRIBUTIONS IN THE JUSTIFICATION PROCESS

Practitioners would feel most comfortable in providing a range of possible values for every parameter used in a justification model. The range or an interval is the simplest form of representing uncertainty associated with a parameter. Furthermore, if an expert feels that there is enough information to assign weightings, bounded by zero and one, that represent the likelihood of occurrence of the values of the specified range, then a possibility distribution for possible values of the parameter can be created. Consequently, an interval can be viewed as a special case of a possibility distribution where every value in the interval is assigned a weight of one. Under the possibility paradigm, specification of a single value for a parameter is permissible since a single value is a degenerate interval where its upper and lower limits are the same. Behrens and Choobineh [2] give a discussion on the advantages of using possibility distribution vs. probability distribution in economic justification of projects.

The theory of possibility [19] is an extension of the theory of fuzzy sets introduced by Zadeh [18]. A possibility distribution for parameter X, $\pi(X)$, can be represented as the union of a number of ordinary (crisp) intervals. These intervals are referred to as α-cuts of $\pi(X)$. An α-cut is defined as $X_\alpha = \{x \in R^+ \mid \pi(x) \geq \alpha, 0 \leq \alpha \leq 1\}$. A possibility distribution is called normal if it is possible to obtain an α-cut with $\alpha=1$ from the distribution. Furthermore, a possibility distribution is convex if and only if each of its α-cuts is continuous. The support of a possibility distribution is defined as SUPP ($\pi(X)$) = $\{x \mid \pi(x) > 0\}$.

Since a possibility distribution can be viewed as a union of a set of crisp intervals, the interval arithmetic can be used to obtain the α-cuts of the possibility distribution of the chosen measure of the performance. Moore [13] presents the necessary background for interval arithmetic. Unfortunately, the distributivity property does not hold under the interval arithmetic procedures. However, Dong and Shah [7] have presented a method for interval analysis called the vertex method that overcomes the subdistributivity of the interval arithmetic.

In justification process, two types of parameters are encountered--monetary and nonmonetary. Parameters with monetary value are used in a function that generates a single economic measure of performance. For example, this measure could be the net present value (NPV), and if all parameters are identified by possibility distributions, then there exists a possibility distribution for the NPV that needs to be determined.

Nonmonetary parameters include strategic as well as some tactical parameters and they are often treated informally. However, the fact that they cannot be easily represented in monetary units does not diminish their importance in final decision. The decision maker(s) may have an intuitive appreciation for the behavior of these parameters and are often influenced by them. Thus, because of the influence of nonmonetary parameters, they should be treated formally in the justification process.

For each alternative, the recommended justification process combines an aggregate possibility distribution of all nonmonetary parameters with the possibility distribution of the economic measure of performance. The steps of the justification process are as follows:

1. Treat each nonmonetary parameter as a criterion and obtain the possibility distribution of each criterion by interfacing with the appropriate expert(s). The domain of these distributions should be the same, e.g. the interval [0,1].

2. If desired, establish a measure of relative importance among the nonmonetary criteria by assigning a weight to each criterion.
3. Obtain the weighted average of all nonmonetary possibility distributions. This will be referred to as the project nonmonetary possibility distribution.
4. Obtain the possibility distributions of all monetary parameters and determine the possibility distribution of the monetary measure of performance. Normalize this distribution such that the largest point in its support is the same as the largest possible point in the support of the nonmonetary possibility distribution, e.g. one.
5. In order to obtain a possibility distribution for the project, combine the project nonmonetary and the normalized monetary possibility distributions by an appropriate connective that considers the relative importance of these criteria.

Finally, rank the possibility distributions of the alternatives by the method of Tseng and Klein [16] and select the appropriate alternative(s). This ranking method is explained in the example section.

Step 4 calls for normalizing the monetary possibility distribution such that the supremum of its support is equal to the largest possible supremum of the nonmonetary possibility distribution. For example, if in Step 1 the interval [0,1] is selected to be the domain for the nonmonetary values, then the largest possible supremum of the nonmonetary possibility distribution would be one.

A plausible reasoning for specifying the appropriate connective mentioned in Step 5 is as follows. A project (P) needs to satisfy monetary (M) "as well as" nonmonetary (N) criteria before it becomes a candidate for funding. The type of connective used to represent the term "as well as" depends on the importance of one criterion relative to the other. Generally, this relative importance depends on the nature of the project. For example, if the project is in a segment of the company's business that is targeted for substantial growth (no growth), then the nonmonetary criterion is more (less) important than the monetary criterion. Thus, a connective should allow the possibility distribution of the more important criterion to play a more dominant role in shaping the possibility distribution of the project than that of the less important criterion. One way of accomplishing this is by using a maximum operator as the connective and accounting for the relative importance by degragating the possibility distribution of the less important criterion according to the value of the relative importance. More precisely, let the scalar $\beta \geq 1$ be the importance of N relative to M, the possibility distribution of the project is obtained by

$$\pi_P(x) = \max_{\forall x} (\pi_N(x), (\pi_M(x))^{\beta})$$

Increasing the value of ß places more importance on the nonmonetary criterion, and consequently the monetary criterion is deemphasized (since $0 \leq \pi_M\ (x) \leq 1$) accordingly.

For this application using the maximum operator as a connective for the possibility distributions of the two criteria has a number of advantages over using a minimum operator as suggested by Yager [17]. First, the possibility distribution of the more important criteria is not altered. Second, since the support of the monetary criterion may contain negative values whereas the support of the nonmonetary criterion contains non-negative values, the maximum operator allows for the negative values to be reflected in the project possibility distribution. Both of these advantage over the minimum operator allows for a more effective comparisons among the alternative projects.

Finally, two important points need to be delineated. First, it should be noted that the term "risk" has been avoided, and it will not be used as a criterion in the justification model. Risk is the manifestation of taking an action in an uncertain environment [6] and thus far the discussion has focused on the modeling aspect of the justification process not the selection and implementation of a project. The issue of risk should be dealt with after the possibility distribution of all projects have been determined, and when one is ready to select a project by comparing projects' possibility distributions. Second, as pointed out by Kaplan [11] the true cost of capital should be used in the analysis rather than a "risk adjusted" discount rate. In the next section the proposed process is demonstrated through an example.

3 A NUMERICAL EXAMPLE

The proposed justification procedure is demonstrated by using the example presented in the reference [8]. However, the single estimates provided in the example are replaced by their appropriate possibility distributions. For ease of computation the form of possibility distributions used is restricted to trapezoidal shapes. An example of a trapezoidal possibility distribution is shown in Fig. 1, and it will be represented by the quadruplet (a,b,c,d). When b = c, the possibility distribution is a triangle; when a = b and c = d, the possibility distribution represents an interval; when a = b = c = d, no uncertainty exists about the value possibility distribution of Figure 1 is convex, normal, and its support is the interval [a,d].

Table 1 shows the possibility distributions for the parameters of themodel.For example, the initial investment is a negotiated fixed price of $725,000 which represents a cash outflow (-) with no uncertainty associated with it. The uncertainty associated with the models' parameters are reflected in their

Table 1 Possibility distributions for the cash flows (monetary factors) of the example

	For	In (+)	Possibility Distributions			
	Year	Out (+)	a	b	c	d
INVESTMENT						
CAPITAL FACILITIES:						
Vendor-supplied System	0	-	$ 700,000	$ 700,000	$ 700,000	$ 700,000
Accessories	0	-	$ 25,000	$ 25,000	$ 25,000	$ 25,000
PROCESS DEVELOPMENT:						
Engineering	0	-	$ 91,200	$ 102,600	$ 125,400	$ 136,800
Installation	0	-	$ 13,500	$ 15,000	$ 15,000	$ 16,500
Tooling/Fixtures, Other	0	-	$ 13,500	$ 15,000	$ 15,000	$ 16,500
TRADITIONAL SAVINGS						
Direct labor	1-10	+	$ 94,052	$ 112,862	$ 137,942	$ 156,753
Scrap (material)	1-10	+	$ 7,128	$ 7,967	$ 8,805	$ 10,063
Energy	1-10	+	$ 313	$ 313	$ 313	$ 1,013
NONTRADITIONAL SAVINGS						
OVERHEAD REDUCTION:						
Management	1-10	+	$ 6,198	$ 6,972	$ 8,522	$ 9,296
Scheduling	1-10	+	$ 6,870	$ 7,728	$ 9,446	$ 11,163
Production Floorspace	1-10	+	$ 2,851	$ 3,564	$ 3,564	$ 4,277
Maintenance	1-10	+	$ 2,416	$ 1,416	$ 1,416	$ 416
Sustaining Engineering	1-10	+	$ 3,652	$ 2,952	$ 2,952	$ 2,252
INVENTORY REDUCTION:						
Raw Materials	1-10	+	$ 9,200	$ 11,500	$ 11,500	$ 13,800
Work-In-Process	1-10	+	$ 16,400	$ 20,500	$ 20,500	$ 24,600

Table 1 (continued.)

	For	In (+)	Possibility Distributions			
	Year	Out (+)	a	b	c	d
Finished Goods	1-10	+	$ 2,450	$ 3,500	$ 3,500	$ 4,550
W.O. Material Preparation	1-10	+	$ 29,783	$ 35,039	$ 35,039	$ 40,295
Inventory Floorspace	1-10	+	$ 31,882	$ 31,882	$ 38,966	$ 38,966
QUALITY IMPROVEMENTS:						
Rework Time	1-10	+	$ 8,748	$ 9,777	$ 10,807	$ 11,836
Inspection Time	1-10	+	$ 31,993	$ 35,004	$ 40,274	$ 43,285
Material Handling (replacement parts)	1-10	+	$ 35,211	$ 39,123	$ 39,123	$ 43,035
Inspection/Rework Floorspace	1-10	+	$ 3,420	$ 3,420	$ 3,780	$ 3,780
Rework-produced Scrap	1-10	+	$ 1,421	$ 1,921	$ 1,921	$ 2,421
Management of Inspection/Rework	1-10	+	$ 5,834	$ 6,315	$ 7,413	$ 7,894
BEFORE-TAX CASH FLOW PER YEAR:						
	0	-	$ 843,200	$ 857,600	$ 880,400	$894,800
	1-10	+	$ 286,359	$ 332,393	$ 376,421	$ 423,033
Depreciation			$ 84,320	$ 85,760	$ 88,040	$ 89,480
Taxable Income			$ 196,879	$ 244,353	$ 290,661	$ 338,713
Income Taxes			$ 98,440	$ 122,177	$ 145,330	$ 169,356
After-Tax Cash Flow (ATCF)			$ 187,920	$ 210,217	$ 231,090	$ 253,676
Net-Present-Value (NPV)			$ 48,324	$ 307,370	$ 590,129	$ 625,460

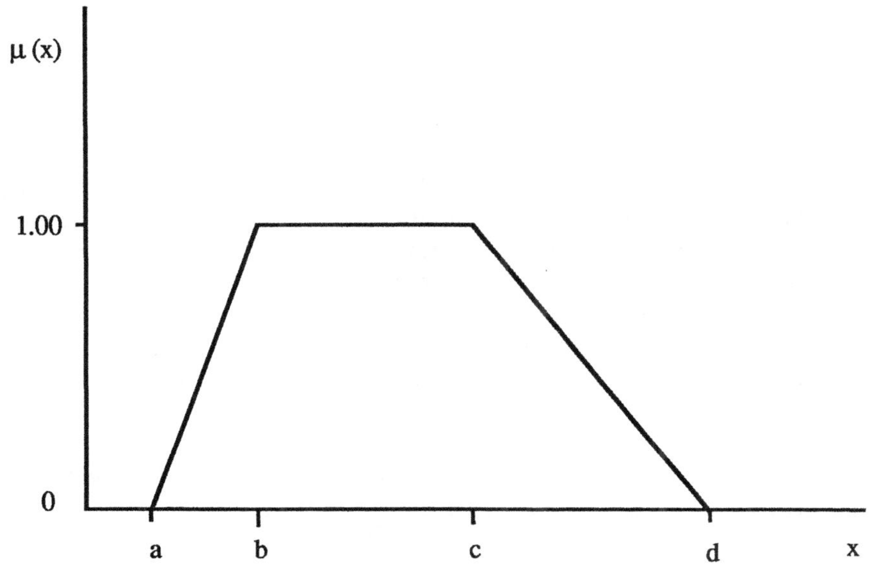

Fig. 1 A general trapezoidal possibility distribution used in the example

possibility distributions. The possibility distribution of the before-tax cash flow for each of the years one through ten is obtained by the methods of interval arithmetic [13]. Also, assuming a 50% tax rate and using the straight line depreciation method the possibility distribution of the annual after tax cash flow (ATCF) is obtained by using the vertex method [7].

Since the cost of capital varies over time, discount rate (I) is represented by the possibility distribution (0.09, 0.12, 0.12, 0.15). The possibility distribution of the NPV is obtained by employing the vertex method [7] on the following NPV relation,

$$\text{NPV} = \sum_{n=0}^{10} \text{AT}\underline{\text{CF}}_n (1 + \underline{\text{I}})^{-n},$$

where an underline indicates a fuzzy number. The possibility distribution of the NPV is calculated to be (48324, 307370, 590129, 625460).

Table 2 shows the nonmonetary factors considered in this example. All possibility distributions are defined on the unit interval [0,1], where 1 represents the highest score and zero represents the lowest score. A weight for each factor is also used. The nonmonetary possibility distribution for the project is (0.69, 0.81, 0.83, 0.91). This possibility distribution is obtained from

$$\sum_{i=1}^{8} w_i \underline{C}_i \;/\; \sum_{i=1}^{8} w_i ,$$

where $\underline{C}_i$ is the possibility distribution of the ith criterion and w_i is its weight.

In order to combine the monetary and nonmonetary possibility distributions, the monetary possibility distribution needs to be normalized. For this purpose, a possibility distribution (a,b,c,d) is normalized to be (a/d, b/d, c/d, 1). Using this approach the normalized monetary possibility distribution for the example is (0.077, 0.491, 0.943, 1.00).

Combining the trapezoidal possibility distributions of N and M by the max operator and assuming that they are of equal importance, i.e. $\beta = 1.0$, results in

Table 2 Possibility distributions for the nonmonetary factors of the example

Nonmonetary Factors	Weight	Possibility Distributions a	b	c	d
Quality Contribution to Market Share/Profitability	1.0	0.6	0.7	0.7	0.8
New Product Introduction On Time/High Quality	1.0	0.7	0.8	0.8	0.9
Development of Engr./Mgmt. Expertise	0.9	0.8	1.0	1.0	1.0
Ability To Meet Customer Due Dates	0.7	0.9	1.0	1.0	1.0
Quality of Work Life	0.6	0.6	0.8	0.8	0.9
Expandability to Other Products	0.5	0.8	0.8	0.8	1.0
Ease Transition of New Product to Mature State	0.4	0.4	0.5	0.6	0.7
Process Flexibility	0.3	0.6	0.7	0.8	0.9
Sum of Weighted Factors		3.75	4.39	4.46	4.89
Project Nonmonetary Possibility Distribution		0.69	0.81	0.83	0.91

the project possibility distribution of (0.69, 0.81, 0.83, 0.91). In this case, the project possibility distribution is the same as that of monetary criterion and therefore it is trapezoidal. However, using a max operator on two trapezoidal possibility distribution does not always result in a trapezoidal possibility distribution unless one distribution is contained in the other. A relative ranking of projects may be required when more than one project is being considered for funding. A suitable pair-comparison index and a ranking procedure based on that index is described by Tseng and Klein [16]. The ranking index is based on comparisons of the areas under the possibility distributions of two projects. The ranking index between project A and B is defined as R(A,B), where

$$R(A,B) = \frac{\text{Sum of areas where A overlaps, is to the right, and is to the left of B}}{\text{Area of A and Area of B}}$$

Project A dominates B if R(A,B) > 1/2, project A is indifferent to B if R(A,B) = 1/2, and project B dominates A if R(A,B) < 1/2. If there are n projects, n(n-1)/2 pair comparisons is needed and projects are ranked based on the number of times each project dominates other projects. Finally, R(A,B) can also be viewed as a relative risk index. For example, if the possibility distribution of the status queue is available, then the ranking index of an alternative relative to the status quo can be viewed as risk of implementing that alternative.

4 CONCLUSION

Explicit representation of uncertainty in models used for justification of manufacturing systems improves the understanding of decision makers about the range of possible outcomes of a project, and thus may lead to more effective decision making process. This would be true only if simplifying assumptions about the situation under study are not made for the sake of explicitly representing uncertainty in the model. This implies that decision making would improve if an approximate solution to an exact problem rather than an exact solution to an approximate problem is obtained.

By using possibility distribution to represent uncertainty one has the ability to capture and model experts opinions about the values and behavior of parameters of an investment situation. Furthermore, such a model can be solved without simplifying assumptions detrimental to the validity of the model. The proposed justification process represents one way of utilizing the flexibility of possibility distribution paradigm for alleviating some of the shortcomings of the existing techniques, and we believe that the proposed process can improve the ability of decision makers to make wiser investment decisions.

5 REFERENCES

1. Bavishi, V.B., "Capital Budgeting Practices at Multinationals," *Management Accounting*, August, pp. 32-35 (1981).
2. Behrens, A.M. and F. Choobineh, "Can Economic Uncertainty Always be Described by Randomness?" Proceedings of IIE Conference, Toronto, pp. 116-120 (1989).
3. Benzion, V. and J. Yagil, "On the Discounting Formula for a Stream of Independent Risky Cash Flows," *The Engineering Economist*, 32, pp. 337-345 (1987).
4. CAM-I, Inc., "Management Accounting in Advanced Manufacturing Environments," Arlington, TX, 32 Pages (1988).
5. Canada, J. and W. Sullivan, *Economic and Multiattribute Evaluation of Advanced Manufacturing Systems*, Prentice Hall, (1989).
6. Choobineh, F., "Modeling Uncertainty in Economic Justifications of Manufacturing Systems," Proceeding of IIE Integrated System Conference, Atlanta, pp. 605-611 (1989).
7. Dong, W. and H.C. Shah, "Vertex Method for Computing Functions of Fuzzy Variables," *Fuzzy Sets and Systems*, 14, pp. 65-78 (1987).
8. Falter, B. and F. Choobineh, "CAM Economic Justification - A Case Study in Electronic Industry," in *Productivity and Quality Improvement in Electronics Assembly*, McGraw-Hill, pp. 523-535 (1988).
9. Fama, E.F., "Risk-Adjusted Discount Rates and Capital Budgeting Under Uncertainty," *Journal of Financial Economics*, 5, pp. 3-24 (1977).
10. Grablowsky, B.J. and W.L. Burns, "The Application of Capital Allocation Techniques by Small Business," *Journal of Small Business Management*, 18, 3, pp. 50-67 (1980).
11. Kaplan, S.K., "Must CIM be Justified by Faith Alone?" *Harvard Business Review*, March-April, pp. 87-95 (1986).
12. Kim, S.H. and T. Crick, "Foreign Capital Budgeting Practices Used by the U.S. and Non-U.S. Multinational Companies," *The Engineering Economist*, 29, 3, pp. 207-215 (1984).
13. Moore, R.E., *Methods and Applications of Interval Analysis*, SIAM, Philadelphia (1979).
14. Pike, R., "Do Sophisticated Capital Budgeting Approaches Improve Investment Decision-Making Effectiveness?" *The Engineering Economist*, 34, 2, pp. 149-161 (1988).
15. Robicheck, A.A. and S.G. Myers, "Use of Risk-Adjusted Discount Rates," Journal of Finance, 21, pp. 727-730 (1966).
16. Tseng, X.Y. and C.M. Klein, "New Algorithm for the Ranking Procedure in Fuzzy Decision Making," *IEEE Trans. on Systems, Man, and Cybernetics*, 19, 5, pp. 1289-1296 (1989).

17. Yager, R.R., "Fuzzy Design Making Including Unequal Objectives," *Fuzzy Sets and Systems*, 1, pp. 87-95 (1978).
18. Zadeh, L.A., "Fuzzy Sets," *Info. Control,* 8, pp. 338-353 (1965).
19. Zadeh, L.A., "Fuzzy Sets as a Basis for a Theory of Possibility," *Fuzzy Sets and Systems*, 1, pp. 3-28 (1978).

ECONOMIC JUSTIFICATION: RESEARCH CONCERNS

Thomas P. Cullinane
Northeastern University

ABSTRACT

Economic justification of automated material handling and manufacturing systems is a frequently encountered barrier to the implementation of these systems. The purpose of this chapter is to describe the economic justification problem and to outline where current practices are inadequate. A discussion of some of the efforts being made to develop improved justification practices is presented. Individual efforts are described in terms of the issues they address and the progress they are making.

1 INTRODUCTION

Systems designers frequently face problems in economically justifying automated manufacturing and material handling systems. Much to the surprise of some engineers and designers, developing the design of an automated system can be an easy task when compared with building a case for economically justifying the system. Although in some instances it may be logically obvious that an automated material handling system is an absolute prerequisite for realizing the potential gains of a manufacturing system, the inability to quantify the benefits and to transform these benefits into an economic justification for the system will often prevent management from approving the funds for the system. In many companies only those projects that can stand up under economic evaluation receive approval.

There is far reaching agreement among most analysts that the traditional economic evaluation techniques and the cost accounting procedures of today are vastly inadequate when it comes to economic justification of automated manufacturing and material handling systems. Although some professional accounting and manufacturing groups such as CAM-I International have formal consortiums with groups of companies to address these problems, a great deal of work still needs to be done to change accounting systems such that they address those areas in which substantial benefits are realized through automation. Adopting new accounting practices involves a very long term effort. As Robert Kaplan of the Harvard Business School recently stated, "Developing a new cost accounting system is much more extensive than acquiring a piece of equipment. It's a major investment, one you make for on-going information, not for the special nature of acquisition, and one can cost hundreds of thousands of dollars."

Experience gained throughout the 1980's has shown that the main reasons the traditional discounted cash flow methods of economic analysis do not work is management's inability to use these techniques realistically and the inadequacy of these tools for analyzing systems that have the major benefits imbedded in areas that are not directly quantifiable through accounting data.

In his 1986 landmark paper titled "Must CIM Be Justified on Faith Alone" [1] Professor Robert Kaplan outlines that the discounted cash flow methods often go wrong for a number of reasons but primarily because management sets "hurdle rates" for justifying the investment much too high. Two of the other areas that are also recognized as shortcomings in management's application of discounted cash flow methods are management's unwillingness to consider the real cost of maintaining the "Status quo" and a bias towards considering small incremental investments as opposed to projects that require major changes. The literature abounds with reports supporting the fact that many of the gains to be realized through implementation of automated material handling systems and automated manufacturing systems simply cannot be accurately evaluated by looking only at the impact the system may have on the cash flows and costs associated with direct materials and direct labor. Additional analysis work must be performed to take into account those factors which can be very important to the economic justification of an automated system. Some of the factors include:

- Improved quality of working life
- Reduced training requirements
- Reduced supervision requirements
- Improved product quality
- Increased system's flexibility

None of these factors appear as line items in the traditional accounting statements. In order to factor these nontraditional economic variables into the traditional discounted cash flow methods, some judgement will be required. This introduces a problem when evaluating new technologies, since most managers do not feel comfortable assigning costs and estimating the benefits of new and untested technologies. Another shortcoming in the analysis methods that typically precede and support the gathering of data for use in the discounted cash flow methods is that they deal only with the primary system being evaluated. These approaches fail to recognize that a change in one area of a facility or plant can have a significant impact in other areas as well. These secondary or ripple effects of the costs and benefits associated with a project must be considered as part of the justification process. Many of the new technologies, such as flexible manufacturing, robotics, guided vehicles, and others, save large amounts of time and money through the elimination of redundant, overlapping, or opposing activities. These benefits are the result of taking an integrated approach to planning and designing the enterprise. Many of these benefits never needed to be considered when the discounted cash flow methods were applied to "focused projects". If the new technologies which touch many areas of the enterprise are to be justified, then these new technologies must be evaluated with techniques that will fairly judge their economic viability.

The challenge in justifying projects economically does not require that we eliminate all of the traditional methods of economic analysis. The challenge is to modify and embellish the existing economic modeling tools to the extent that they reflect the critical costs and benefits associated with automated systems and to develop realistic guidelines for management to follow when applying these tools. This will require an understanding of the functions performed within an integrated handling and manufacturing system and an assessment of the impact a given system will have on existing operations and systems. Such an approach requires a thorough analysis of all the economic consequences of a change in the method of handling materials or producing product.

2 PROGRESS TOWARDS IMPROVED ANALYSIS METHODS

During the past four years many firms that produce automated systems and manufacturing consultants have been working on developing an approach to economic evaluation that will alleviate or reduce the problems discussed above. Most of these firms recognize that economic evaluation can no longer be focused on one machine or one work center; it must take into account the whole enterprise. For this reason the approaches that are being developed do not just address the topic of economic analysis but also involve a broad range of other enterprise based concerns such as total quality management, material flow analysis, functional integration, and others.

The Digital Equipment Corporation has been working on economic justification tools to be used internally and by their consultants. The approach Digital has taken is to examine economic justification problems from the viewpoint of the total enterprise. Digital has viewed economic analysis as requiring a broad set of tools or a single broad based package of tools that will enhance the analysts' ability to make sound economic decisions at all levels of enterprise planning.

A software package titled Information Systems Investment Strategies (ISIS) is used by IBM consultants as part of their programs for economically evaluating systems. The IBM Information Investment Strategies process and the Associated Software is actually a decision support tool which assists the decision maker in determining the impact of various investments while taking into account the firm's positions in the market and its financial resources. The software requires extensive input data. Publicly available data is combined with company economic data to form a proforma model of the firm. The ISIS processes data help an analyst to identify applications which will yield high returns on investment. The metrics utilized to assess the acceptability of a program are return on equity, return on assets, and earnings per share.

The consulting firm of Ernst and Whinney has developed many economics based tools for examining technology based investments. They have specifically developed a methodology for computer Integrated Manufacturing (CIM) justification that begins with strategic planning and moves systematically through to implementation. The strategic audit, which is the first step in the process, quantifies the profitability impact of certain market factors and competitive position factors. Utilizing both the PIMS and Wharton economics databases operational improvements and marketing considerations are identified and made part of the model. The projects are characterized by their relationship with other programs, their costs, the time required, and the changes they cause in process performance. The programs are quantified and evaluated using the tools of operation research such as simulation, learning curves and mathematical modeling. The proposed project budgets and costs are estimated both in terms of direct costs and the potential support costs associated with interprogram interfacing. The software "works" the accounting data to gain a full economic understanding of a project.

The consulting firm of Booz-Allen and Hamilton performed a study to determine the feasibility of attempting to measure the return on information systems investments. They concluded "that measuring the return on information systems investments is both doable and essential". They also conclude "the benefits of developing a systems evaluation process are worth the efforts". Booz-Allen is in the process of converting what they have learned from this study into a package for justifying many types of computer based systems. Their current approach is a much expanded version of the traditional discounted cash flow methods.

Coopers and Lybrand works closely with companies following a well documented procedure for developing critical success factors, identifying business strategies and matching business strategies with CIM strategies. The methodology they utilize in their work is called CIM Plan. This methodology is aimed at providing a decision framework for discovering opportunities for improvement, identifying technologies that will satisfy the opportunities for improvement, and cost justifying the proposed technologies. The Coopers and Lybrand approach looks at both the discounted cash flow and the intangible economic factors. The CIM justification process takes into account risk factors, operational considerations, executive commitment, education requirements, and other Design Team considerations. The financial justification process is on revenue enhancement from a cost reduction and increased cash flow viewpoint.

Arthur Andersen Company utilizes a package of methods in their consulting work titled Method/1. In general, economic evaluation is found throughout the Method/1 packages. The evaluation methodologies fall into the categories of discounted cash flow, intangible analysis, and cost benefit analysis. The Method/1 packages are very comprehensive with regard to enterprise analysis but utilize very traditional economic models.

In general the work that is going on in industry is leading the way towards better economic evaluation tools. There are some efforts to expand approaches to take intangibles into account but a lot of work remains to be done.

3 SUMMARY

Economic evaluation of automated material handling and manufacturing systems is a major barrier to the implementation of these systems. During the past five years many companies have been working on improving their economic evaluation tools. Although progress has been made, a great deal of work remains to be done. Research issues still remain around the problems of quantifying intangibles, assessing hurdle rates, and evaluating how a change in a system will impact the enterprise.

4 REFERENCES

1. Kaplan, Robert S., "Must CIM Be Justified by Faith Alone?", *Harvard Business Review* (March-April 1986).

COMMENTS ON THE AUTOMATION OF MATERIAL HANDLING

Stephen L. Dickerson
Georgia Institute of Technology

1 INTRODUCTION

This chapter is a collection of fundamental ideas from several years involvement with the Material Handling Research Center. It results from the interactions of a controls oriented mechanical engineer with many materials handling professionals both in academia and industry. Most, but not all, of these professionals have industrial engineering backgrounds. As a consequence, the material handling community has a strong orientation toward operations issues and rather less concern with the technology of machinery used in material handling. This chapter discusses some very basic issues relative to automation of material handling with emphasis on some perceived misconceptions.

The particular points that the chapter attempts to make are:

1. Manufacturing is labor intensive. In fact, labor is almost the sole cause of product costs.
2. Material handling is the largest component of labor if material handling is taken to include all planning and execution of material motions.
3. The intrinsic nature of material handling motions make them more difficult than processes to automate. Processes are themselves nearly totally automated in modern manufacturing.
4. Automation, the use of machines to perform formerly manual functions, needs to be addressed with the unique strengths and weaknesses of the technology included in the design of the products and manufacturing environment. This is called "structure for automation" here.
5. Flexible automation is not inherently more expensive than fixed automation even in cases of mass production.

2 MANUFACTURING AND DISTRIBUTION IS LABOR INTENSIVE

The role of labor, much of it in material handling, is not well understood. It is a cliche that labor is a small part of total manufacturing costs. If material handling functions (planning and making motions), and maintenance functions are labelled "indirect" labor and not counted in labor then this cliche is true in the case of most modern manufacturing. However, if all forms of labor that contribute to the manufacturing process are included then labor is almost all of the cost.

The argument that labor is essentially all of the cost follows and is trivially true. To produce a manufactured product one needs

1. PEOPLE,
2. EQUIPMENT,
3. SPACE (buildings),
4. ENERGY, and
5. MATERIALS (any and all purchased components except 1 thru 4)

However, the production of items 2 thru 5 require the same five items, plus perhaps raw materials as they exist in nature. Let us suppose for the sake of illustration that for each of these items labor was 20% of the total for each production unit, e.g., factory, construction company, energy company, etc.

Now recalling that equipment, space, energy, and materials must also be produced, they also have labor contents. To produce equipment, for example, we must again purchase equipment, space, energy, materials and labor. If we attempt to account for all labor as a fraction of the total, the result is

$$\begin{aligned}\text{labor fraction} &= .2 + .8(.2+.8(.2+.8(.2+.8(...)))...)) \\ &= .2+.2(.8)+.2(.8)\text{^}2+.2(.8)\text{^}3... \\ &= .2/(1-.8) = 1\end{aligned}$$

As logic would tell us, all costs are "labor", if natural resources in their natural state are considered free. We must also recognize that (1) if the fruits of labor are deferred (that is saved or invested) the laborer will demand a higher price for his labor and (2) the market puts labor costs measured in dollars/hour at greatly different values depending on the talents and activities of the laborer. Of course, this differential often results from costs, including time, involved in training to have that skill. None the less, if production of goods and services required no labor they would all be "free" except for the design costs that would need to be spread over all units.

The bottom line is that in order to reduce the costs of production, human effort must be removed from the production process. That effort might be reduced in some upstream stage of production; materials, energy, etc.; or in labor content at the production unit of interest. In fact, we need to reduce labor content in all goods and service producing activities to raise the average standard of living.

3 THE ROLE OF MATERIAL HANDLING IN MANUFACTURING

If labor is essentially all of manufacturing costs, in what activities are people engaged? Lets consider the case of factories, although construction, mining, retailing, etc. could be considered. Assume that manufacturing is the conversion of raw materials to discrete products. There are many intermediate products or parts in the process. Hence, for an individual factory, purchased items, that is other manufactured items, are typically a large part of costs as discussed previously.

An ongoing manufacturing operation can be thought of as consisting of two main activities, TRANSFORMATION of materials, and MOTION of materials. When we ask what people actually do in factories, consider whether or not the people are involved in the actual TRANSFORMATION of material, that is the shaping, mixing, coating, hardening, or other operations that actually change the material form, or are they involved with MOVING of material. Because people are poor, relative to machines, at supplying energy, and at the precision motions necessary to make interchangeable parts, we have succeeded in almost total automation of the transformation processes themselves. But the moving of material between processes requires a scale of motion and a planning of motion not readily duplicated by machines. Assembly is a special case where hand-eye coordination is especially important. Here two or more objects must be precisely brought together. Because it is difficult to provide the precise position control between processes, the assembly of parts together requires hand-eye coordination.

The point of the preceding paragraph: most labor in factories involves moving parts, not transforming materials.The reason is not some conspiracy to automate the transformations and leave the movement manual. Rather the automation of material handling motions is more difficult. The reason for this involves the scale of motions required and the need for hand-eye coordination at those scales.

By way of example in construction, and aircraft and ship manufacture, where the scale of motion is large many of the transformation operations, e.g. sawing a piece of lumber, have remained manual.

For completeness, there is one important activity in production that is neither TRANSFORMATION nor MOTION. That activity is MAINTENANCE. It can be, and here is, argued that MAINTENANCE will become the dominant cost in manufacturing after we succeed in automating material handling. The rising relative cost of maintenance, to keep the machines that do the transformation and motion working, is one of the driving forces behind the emphasis on QUALITY. A product of high quality can be defined as one that requires a minimum of maintenance, routine or otherwise. The demand for quality is an attempt to cut an increasingly large fraction of total cost. The cost of unscheduled maintenance is especially high because of the interruption of an otherwise well scheduled production process.

4 THE INHERENT DIFFICULTY OF MATERIAL HANDLING AUTOMATION

This chapter has argued that most manufacturing costs are associated with material handling, when assembly, machine loading, and logistics are included as a part of material handling. This situation is largely the result of the relative success in automation of those manufacturing processes that shape and modify parts. Material handling is less automated because of the inherent difficulty of such automation.

The difficulties are two fold. It is difficult to generate by computer a PLAN of when and what motions to make, and then be able to EXECUTE the planned motions reliably. Generating by computer the requirements for motion needs something akin to an expert system, while the ability to make the motions requires an ability to achieve hand-eye coordination. It has been difficult to build machines that perform the human like functions of planning and hand-eye coordination. In fact, these functions were prohibitively expensive by machine until the advent of low cost digital computation.

The next few paragraphs look ahead, to predict the form and future of computer controlled automation in material handling. Both of the difficulties, planning and motion execution, are at least addressable using advances in technology that utilize micro-computers, machine vision, electronic information transfer, and electric servo systems. Except for servo systems, the inherent cost of these items has trended toward zero (relative to other costs) [1,2,3]. The bottlenecks are software, very generally defined, and a shortage of personal knowledge of the hardware. There is the potential that very general software can be created that would be readily tailored to a wide variety of situations. This reduces the unit software cost because the cost of duplicating software is trending toward zero.

4.1 Planning

Planning is essentially the specification of actions that need to be made to achieve efficient production. In the case at hand, those actions are motions. The plan depends on two things, the demand for product, and the "state" of the "factory." Factory is in quotes because a factory in this case can consist of a machine, a work cell, a literal factory, or an enterprise. Each of these defines a level of planning. Each should be thought of as including the functions of maintenance, transformation, and motion accomplished by a combination of people and machines. The state of the factory, refers to the then current physical arrangement including the work in progress and the capabilities of the system (is it in any sense broken?). The demand for product is not just the current, unfilled demand, but also the history and projection for demand.

A PLANNER is that entity that SCHEDULES motions. To be at all good at planning, the planner must "know" the state of the system and the demand for product. It must also be able to envision the consequences of various alternative plans and pick ONE that will result in efficient production. The production can be efficient only in the sense of expected value, because of changes in demand, malfunctions of various types, and "noise" in the available information. Because such scheduling problems, treated as combinatorial mathematical problems combined with simulations, are impossibly difficult to solve even with modern computers, one must rely on the judgement of an expert.

To excel at the planning function by computer remains one of the research and development needs of material handling automation.

4.2 Hand-Eye Coordination

The ability of humans to move relative to their surroundings is the the second capability which is very difficult to achieve by machine. A person is able to grasp an object that is described only generally and move it relative to other objects with a high degree of precision, e.g., "sharpen the yellow pencil on the desk using the sharpener on the bookcase". This does not require the judgement or even the intelligence that is the hallmark of people. Rather it requires the motor and perception skills that is characteristic of higher animals. However, it is a largely unsolved problem in engineering.

4.3 Role of Sensing in Material Handling

Computer driven automatic control of material handling requires two types of information about material to be handled...identification and location.

Identification information is the knowledge of WHAT item is being handled. Knowledge of the geometric position, generally relative to other objects, is the location information. That is, we must "have the RIGHT THING, in the RIGHT PLACE, at the right TIME." Fortunately, knowledge of time is easily acquired by automatic equipment. Identification and location information are much more difficult to acquire, hence the traditional reliance on the abilities of the human eye and mind.

There are two fundamental ways of a computer acquiring identification and location information. One way is entirely open loop. An object is placed somewhere, the computer remembers that fact. We then depend on that memory not failing and require that no other actions, not known to computer, change the location or identity. If the machines that do the placing, and subsequent motions, are all sufficiently precise then automation can proceed without any feedback information on location or identification.

The second method is to have a way of "closing the loop" by providing a sensing of identification and location. Because people are so able to sense with their eyes both location and identification, people are the dominant choice to close the loop in material handling. However, because of the availability of more cost-effective systems, electronic sensing of both location and identification is becoming increasingly popular. The use of this feedback information greatly reduces the need for geometric precision in material handling equipment and reduces the dependence on total recall and unerring knowledge of past actions.

It is worth noting that both electronic sensing and human sensing of identity and location are greatly facilitated by what is herein called "structure." A person can only tell a can of Coke from a can of Sprite by the words and design on the can. In fact, this is an extreme case of structure for assistance to human identification. The manufacturer goes to great lengths to make the identification quick and unmistakable by the customer. On the other hand, all such cans have a relatively small portion of the surface devoted the the universal product code (UPC) to ASSIST AUTOMATIC DEVICES in identifying the product. It would be "nice" if no such UPC was needed and more general industrial vision systems were used for distinguishing among all grocery items. That is, the manufacturer would only need to use the structure intended for the human and not have a separate structure for the machine identification. At some future data, that technology may in fact be cost effective. (Of course, it would also be "nice" if people were good at instantly reading the UPC.) Clearly, in manufacturing and material handling, a common way of providing structure for automatic identification is bar codes in various forms.

Structure in assisting humans in acquiring location (e.g. aisle markings, retro-reflective road markings, and indoor lighting) and identification information is universal and not just limited to soft drink cans. It is realistic to expert that structure for machine location and identification information is desirable for cost-effective solutions to material handling problems.

5 THE ROLE OF STRUCTURE FOR AUTOMATION

It is well recognized that automating a formerly manual task is fraught with dangers and often leads to failures. We often take a task that was "structured" for human performance and ask that a machine be a direct substitute for the human. Although often possible from a technological standpoint, the cost, including particularly the costs of unreliability, are likely to be high. The alternative approach, to structure the design of the product, and the factory/warehouse environment to assist machines is often fruitful. Some successful examples:

1. the bar code and UPC. The RF tag is a recent addition
2. design for automation, mostly automatic assembly
3. wire guided vehicles
4. fiduciary marks in electronics manufacture

In each of these cases we add something to the product (items 1 and 4), something to the environment (item 3), or change the design of the product (item 2) in such a way as to facilitate automation. It is true that there is some cost associated with these additional constraints on the product or environment. However, they can be very small costs. One goal of technology is to find ways of reducing these costs even further. For example, smaller areas for ID information, and software guided vehicles with minimal structural change in a facility.

There are failures also in attempting to provide structure. An obvious way of resolving the location problem in material handling is to never loose control of the position of a part. This leads to specialized pallets and trays to maintain the position of parts when moving through the production process. Such an approach is feasible, but often results in more cost than the value of the labor saved.

The conclusion being that structural changes in the product or environment can greatly facilitate material handling automation, with the bar code/UPC being the most universally accepted example. The author believes that machine vision systems, initial disappointments notwithstanding, will turn out to be elements in many automation systems utilizing structure.

Too often structural changes are overlooked as a way out of otherwise costly automation.

6 FLEXIBLE AUTOMATION IS NOT MORE EXPENSIVE THAN FIXED AUTOMATION

The popular wisdom is that flexible automation is used for those production situations where one expects to see a high variability of products, generally with modest production runs of any one product. U.S. manufacturers have been criticized for building too few flexible production facilities [4], and dedicating those that have been built to a small set of products. However, there is no justification now, and will be less in the future, for undertaking the expense of custom automation machines. The standardized robot, AS/RS, and AGV, will be less expensive than simpler, yet customized, special purpose machines. Furthermore, the maintenance of a few machines, used repetitively in a factory, is less burdensome than trying to maintain an array of special automation devices for which there is no ready supplier of documentation and spare parts. Consideration must also be given to the effect of special equipment on the overall software environment, including communications between machines and computer. A standard product is more likely to have a reduced level of bugs.

I have heard several complaints from manufacturers of AGVs, and AS/RS equipment that too often manufacturers want something special.This is really a result of lack of universal use of automatic material handling. The same tailoring of equipment to tasks was commonplace in the early days of various tasks in process automation. However, the use of standard automation equipment is prevailing in the process part of automation, which is now highly automated. Standard machine tools for metal cutting, injection molding, stamping, forging, etc. is the norm.

A good example of flexibility being cheaper than fixed automation, even if the flexibility is never needed is the personal computer/workstation. The personal computer/personal workstation is an example of flexibility far in excess of normal use. It pays to buy such a computer, even if it is only used for word processing using a single program. Why? Because it is the cheapest and most reliable processing device. It is cheap because it is sold into a mass market. The mass market exists because there is a great diversity of applications. The point being that flexibility creates a mass market, but that applications do not need to take advantage of that flexibility to justify the flexible equipment.

Some examples in material handling that must be considered:

1. An AGV instead of a conveyor
2. A free ranging AGV instead of a wire or stripe guided machine
3. A robot instead of a fixed transfer mechanism
4. A vision system instead of precision locating trays
5. Combined voice and data communications instead of separate systems
6. Wireless communications instead of wired

In each of these cases, the probability is that the more flexible mode will become cheaper than the more fixed mode for an increasing fractions of all applications, regardless of whether the flexibility is in fact needed in the application.

7 SOME COMMENTS ON TECHNOLOGICAL APPLICATIONS TO MATERIAL HANDLING

Two words of caution. First, an engineering solution must be economical. What can be done from a technological standpoint, and what should be done from the standpoint of reduced cost of production are two different things. The most efficient production results in the highest average qualify of life.

Second, while it is wise to look at how people and animals do something, an engineering solution may be much different. For example, no fish or bird has propellers; no bird flies with fixed wings, no animal has wheels, there are no crankshafts in nature, microelectronics processes signal much differently than nervous systems. The end result is that engineered machines ARE OFTEN ABLE TO PERFORM BETTER THAN ANIMALS. The appropriate technology often does not emulate nature. Furthermore, there is a need for an environment structured for the technology. We build roads for cars. Roads are the structure we provide so that the machine, an automobile, can function well. We could try to build walking machines instead of cars, and use trails but that wouldn't be very efficient.

8 CONCLUDING REMARKS

Widespread use of automation in material handling is only appropriate if it can be economically justified. The author if of the opinion that automation is really no different than other capital investments, it must meet the criteria that implementation means more efficient production. Of course, efficiency is more than low unit cost, it deals with quality, and timeliness as well. A product of low quality or produced "late" is of less value.

The essential conclusions of this chapter are that:

1. Material Handling, including assembly and logistics, has a greater potential contribution to reducing manufacturing costs than any other activity.
2. Automation of material handling is greatly dependent on the ability to build machines that can PLAN and that can achieve HAND-EYE coordination.
3. Micro-computers, industrial vision systems, advances in communications, and advances in servo-systems offer the promise of such machines.

9 ACKNOWLEDGEMENTS

The author appreciates the comments of Dr. Edward Frazelle on the issue of labor content in manufacturing costs, the work of Mr. Yi-Tzer Lin on the trends in manufacturing automation for the MHRC, the program managers in the MHRC for enlightening me over the years on material handling issues, and the support of the MHRC on various material handling automation projects.

10 REFERENCES

1. Ramchandran, Jaikumar, "Postindustrial Manufacturing,"*Harvard Business Review*, pp. 69-76 (Nov. 1986).
2. Peled, Abraham, "The Next Computer Revolution," *Scientific American*, pp. 57-64 (Oct. 1987).
3. Stuck, B.W., "Imaging Technologies: The Next Decade, "*EEE Communications Magazine*, pp. 80-86 (July 1987).
4. Ranta, Juffa, "The Impact of Electronics and Information Technology on the Future Trends and Applications of CIM Technologies," *Technological Forecasting and Social Change*, pp. 231-260, Vol. 35 (1989).

INDUSTRIAL LIFT TRUCK RELIABILITY

Thomas L. Landers
Waseem Mohammed Qureshi
University of Arkansas

ABSTRACT

Material handling systems are designed to be repairable. There may be improvement in reliability early in equipment life and/or deterioration as the system is operated. Reliability is also a function of explanatory variables, such as operating environment or maintenance procedures. This chapter presents modeling approaches for reliability of repairable equipment including improvement or deterioration trends and/or explanatory variables. A case study is discussed for industrial lift trucks including modeling of reliability as stochastic processes. Time-to-first-failure for the lift trucks is shown to be distributed exponential with operational environment being an explanatory variable forming proportional hazards.

1 INTRODUCTION

Reliability is an important engineering consideration throughout the lives of material handling equipment and systems. Although most systems are designed to be repairable, most reliability theory and practice has concentrated on methods of analysis for time-to-first-failure and is therefore more appropriate for non-repairable systems [1]. Recent developments have substantially increased the power of models for time-to-first-failure, by permitting consideration of explanatory variables such as environment [2,5,7]. The limited work on modeling of reliability for repairable systems has treated the failure patterns as stochastic processes, primarily homogeneous Poisson processes (HPP). However, engineers rarely apply these models and sometimes tend to misapply them [1].

In recent years, theorists have devoted more attention to the problem of repairable systems. Techniques have been developed to estimate the parameters of a nonhomogeneous Poisson process (NHPP) [3,10]. More recently, the NHPP models have been extended to account for the effects of explanatory variables, on the failure process [4,8]. Researchers have also proposed non-parametric methods of estimating reliability as a function of failure history (i.e., counting processes) [9,11].

This chapter presents some powerful models for repairable systems and reports the results of reliability analyses for a fleet of 4000-pound capacity lift trucks powered by gasoline internal combustion engines. A better understanding of lift truck reliability can lead to improved management of the lift truck fleet, from acquisition through applications, repair and spares inventory planning, and finally in replacement decisions.

We begin by describing some useful reliability models. Then we report the preliminary results of on-going research in the behavior of lift truck reliability.

2 TERMINOLOGY AND NOTATION

CMTBF	- Cumulative mean time between failures
F/L	- forklift truck
HPP	- Homogeneous Poisson Process
IID	- Statistically independent and identically distributed
IMTBF	- Instantaneous mean time between failures
MTBF	- Mean time between failures
MTTF	- Mean time to failure
m.l.e.	- Maximum likelihood estimate
NHPP	- Nonhomogeneous Poisson Process
PWP	- Prentice, Williams and Peterson
ROCOF	- Rate of occurrence of failures
T	- time-to-failure random variable
t	- observation on the random variable T
h(t)	- hazard function for the random variable T
R(t)	- complementary cumulative distribution function for the random variable T
N	- discrete random variable for number of failures
n	- observation on the random variable N
X_i	- interarrival time between the i-1st and ith events
T_i	- ith ordered observation on the random variable T
$\rho(t)$	- intensity function of a Nonhomogeneous Poisson Process
δ	- shape parameter of the NHPP power-law intensity; Weibull shape factor
θ	- Weibull scale parameter
υ	- parameter of a HPP and scale parameter of the power law intensity
μ	- extreme-value location parameter ($\mu = \ln \theta$)
$\underset{\sim}{z}$	- vector of explanatory variables
$\underset{\sim}{\beta}$	- vector of regression coefficients estimated by the method of maximum likelihood
ρ_{os}	- baseline intensity in a stratum defined by failure count

r	- number of individuals observed ($j = 1,\dots,r$)
T_m	- time to the last failure on a unit
U_L	- Laplace statistic
W_m^2	- Modified Cramer-von Mises statistic
Λ	- Likelihood ratio statistic
$E[y]$	- Expected value of a random variable y
$STDERR(\hat{y})$	- Standard error of an estimator y
γ	- Significance level ($0 < \gamma < 1$)
D	- Chi-square test statistic for $H_o: \delta_1 = ... = \delta_r$
$X^2_{r,\gamma}$	- Chi-square critical value for r degrees of freedom and γ significance level

3 PROPORTIONAL HAZARDS MODELS

The proportional hazard model applies to the case of time-to-first-failure. It is a generalization of traditional methods, such as the parametric Weibull and the nonparametric product-limit (Kaplan-Meier) [6] models. The proportional hazards approach permits explicit incorporation of explanatory (concomitant) variables, such as environment, in the reliability model. These models require that the explanatory variables have a multiplicative effect on the hazard function. The different possible values of the explanatory variable(s) stratify the population into sub-populations having hazard functions that are proportional over time. For a random variable T, the proportional hazard function is expressed as

$$h(t;\underset{\sim}{z}) = h_0(t)\exp(\underset{\sim}{z}\,\underset{\sim}{\beta}) \tag{1}$$

Cox developed a distribution-free proportional hazards model [2]. The well-known and widely useful exponential and Weibull distributions are also in the class of proportional hazards models. The Weibull distribution is known to apply to the time-to-first-failure for mechanical equipment, and is a good potential candidate for modeling reliability of material handling equipment such as forklift trucks. The Weibull hazard function is

$$h(t;\theta;\delta) = (\delta / \theta)(t / \theta)^{\delta-1} \tag{2}$$

The Weibull proportional hazard function is formed by making the scale parameter θ a function of explanatory variables:

$$h(t;\theta(\underset{\sim}{z},\underset{\sim}{\beta});\delta) = \delta\exp(-\underset{\sim}{z}\,\underset{\sim}{\beta})[t\exp(-\underset{\sim}{z}\,\underset{\sim}{\beta})]^{\delta-1} \tag{3}$$

The Weibull reliability function is

$$R(t;\theta(\underset{\sim}{z},\underset{\sim}{\beta});\delta) = \exp\left\{-[t\exp(-\underset{\sim}{z}\,\underset{\sim}{\beta})]^{\delta}\right\} \qquad (4)$$

The exponential distribution is a special case of the Weibull distribution where $\delta=1$. Graphical plots of ln(t) versus ln{-ln[R(t)]} are linear (a property of the Weibull distribution) and parallel (a property of proportional hazard functions). Fig. 1 illustrates Weibull proportional hazards plots for two values of one explanatory variable.

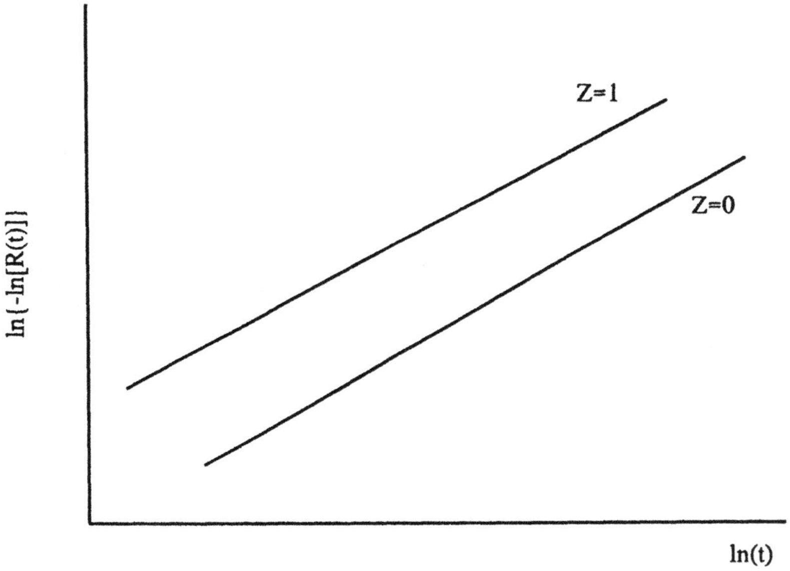

Fig. 1 Weibull proportional hazards

4 STOCHASTIC PROCESSES

For a repairable system, it is not sufficient to model time-to-first-failure. We are interested in the subsequent failure patterns over time. The sequence of failure and repair events in the continuous time domain forms a stochastic point process.

In general, a stochastic process can be represented along the time line, as in Fig. 2, where

- X_i is the interarrival time between the i-1st event and the ith event (X_m is the last recorded interarrival in the process)
- T_i is the time interval measured from time zero to the occurrence of the ith event (T_i is the time to the last recorded event in the process).

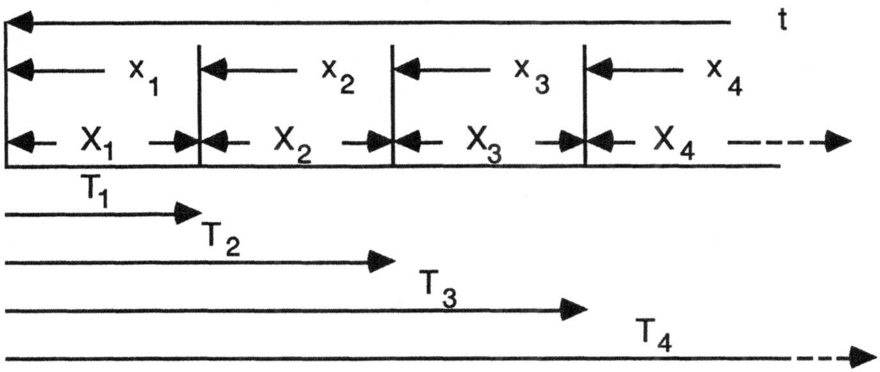

Fig. 2 Sample path of a stochastic point process [2]

There are several classes of stochastic point processes, two of which are widely used in reliability

(a) Renewal Processes

(b) Poisson Processes

A renewal process is a sequence of IID, non-negative random variables X_1,X_2, X_3,..., which with probability one are not all zero. The renewal process arises when, upon failure, a system is either replaced or repaired to "same-as-new" condition. The distribution of X could be the exponential, Weibull, normal, or lognormal, to name a few.

In a non-terminating Poisson process, the number of events N in a time interval t is a random variable having the Poisson distribution

$$\{N = n \mid \rho(t)\} = \frac{e^{-t\rho(t)}(t\rho(t))^n}{n!} \tag{5}$$

where $\rho(t)$ is the intensity or rate of occurrence of events. When the events are failures, the intensity is called a rate of occurrence of failures (ROCOF). The state of the system at any time is given by n = 0,1,2,... A homogeneous Poisson process has the following characteristics:

(a) The number of failures N in a time interval t is a discrete random variable distributed Poisson, with parameter υ.

(b) The time between failures X is a continuous random variable distributed exponential, with parameter θ.

Thus, the HPP is a special case of both the Poisson and renewal processes. Additionally, there is a numerical relationship between υ and θ:

$$\upsilon = \frac{1}{\theta} \tag{6}$$

It is important to note that this relationship is numerical equality, not definitional equivalence.

The nonhomogeneous Poisson process is a generalization from the HPP to the case of a Poisson process with time-dependent intensity function $\rho(t)$. For the NHPP, the number of failures in an interval is distributed Poisson. However, the successive interarrival times (X_i) are not identically distributed. The intensity could take on any form. The power law intensity function is

$$\rho(t) = \upsilon \delta t^{\delta - 1} \tag{7}$$

where δ is the shape parameter. Note that the HPP arises as a special case where $\delta = 1$.

The expected number of failures in the interval (t_1, t_2) is

$$E[N(t_2) - N(t_1)] = \int_{t_1}^{t_2} \upsilon \delta t^{\delta - 1}\, dt$$

$$= \upsilon (t_2^{\delta} - t_1^{\delta}) \tag{8}$$

For the special case of $t_1 = 0, t_2 = t$

$$E[N(t)] = \upsilon\, t^{\delta} \tag{9}$$

The expected time of the nth failure is

$$E[t_n] = \{ \frac{n}{\upsilon} \}^{1/\delta} \tag{10}$$

The IMTBF is

$$\mathrm{IMTBF}(t) = \frac{1}{\rho(t)} = [\upsilon \delta t^{\delta - 1}]^{-1} \tag{11}$$

The CMTBF is

$$\mathrm{CMTBF}(t) = \frac{t}{E[N[t]]} = \frac{t}{\upsilon t^{\delta}}$$

$$= \frac{t^{1-\delta}}{\upsilon} \tag{12}$$

5 PROPORTIONAL INTENSITY MODELS

Proportional intensity models are generalizations of the stochastic process that explicitly treat the relationships of explanatory variables to the stochastic process. Proportional intensity models also generalize the proportional hazard model to the case of repeated (counting) events in the time domain.

Lawless [8] defines the proportional intensity Poisson process model for individuals with covariate vectors of explanatory variables $\underset{\sim}{z}$:

$$\rho_{\underset{\sim}{z}}(t) = \rho_o(t)g(\underset{\sim}{z};\underset{\sim}{\beta}) \tag{13}$$

where $\underset{\sim}{\beta}$ is a vector of regression coefficients estimated by the method of maximum likelihood. The convention is to use an exponential form for $g(\underset{\sim}{z};\underset{\sim}{\beta})$ to provide flexibility and assure a non-negative intensity function. Then the proportional intensity function is

$$\rho_{\underset{\sim}{z}}(t) = \rho_o(t)\, e^{\underset{\sim}{z}\underset{\sim}{\beta}} \tag{14}$$

In the proportional intensity model, the explanatory variables are incorporated through the scale parameter υ, as follows. Let

$$\upsilon = e^{\beta_o} \tag{15}$$

Then

$$g(\underset{\sim}{z},\underset{\sim}{\beta}) = e^{\beta_o + z_1\beta_1 + z_2\beta_2 + \dots}$$

and when all explanatory variables $z_1, z_2, \dots$ are equal to zero the baseline scale parameter υ results. When some explanatory variables are not equal to zero $g(\underset{\sim}{z};\underset{\sim}{\beta})$ takes on a value proportional to υ, hence the name proportional intensities. Plots of $g(\underset{\sim}{z};\underset{\sim}{\beta})$ for different values of the covariates $\underset{\sim}{z}$ should have roughly the same shape. The power law proportional intensity model is

$$\rho(t) = \delta\, t^{\delta-1}\, e^{\underset{\sim}{z}\underset{\sim}{\beta}} \tag{16}$$

The time-to-first-failure for a power-law NHPP is a random variable distributed Weibull. However, subsequent failures are not IID Weibull.

Williams [11] has proposed an approach for proportional intensity modeling based on the semi-parametric Cox model (Equation (1)) [2]. Cox proposed a heuristic method for estimating $\underset{\sim}{\beta}$ which is independent of the baseline hazard. Williams generalized the Cox model to the case of repeated events such as the failures of a repairable system. Williams considered time and history-dependent covariates and strata. A model reported by Prentice, Williams and Peterson (PWP) [9] addressed a special case quite relevant to repairable systems. The PWP model considers time-dependent covariates and a history-dependent stratification variable (the failure count, or number of previous failures). The model is

$$\rho\{t \mid N(t), \underset{\sim}{z}(t)\} = \rho_{os}(t - t_{n(t)})\, e^{\underset{\sim}{z}\underset{\sim}{\beta}_s} \tag{17}$$

where

s	denotes the stratum determined by failure count
$(t - t_{n(t)})$	is the failure time measured from the previous failure time (backward recurrence time).

For the special case of constant covariates,

$$\rho\{t \mid N(t), \underset{\sim}{z}(t)\} = \rho_{os}(t - t_{n(t)})\, e^{\underset{\sim}{z}\underset{\sim}{\beta}_s} \tag{18}$$

The PWP modeling approach involves:

(a) Stratification based on failure count.

(b) Approximations of the baseline intensity function in stratum s using the baseline hazard function for time-to-next-failure.

(c) Estimation of $\underset{\sim}{\beta}_s$ in each stratum by Cox's method of partial likelihood.

If $\beta_1 = \beta_2 = \beta_k$ for k strata, then one β is estimated across strata. The PWP model essentially provides a piece-wise analysis of a stochastic process, by modeling time-to-failure-n for all units in the risk set for failure n (those which have already experienced failure n-1). Note that the PWP is not a renewal process, because different baseline hazard functions are allowed in the different strata.

6 LIFT TRUCK RELIABILITY STUDY

We have performed reliability studies on a fleet of industrial lift trucks at a military logistics depot. This study examined a sample of 19 lift trucks (4000 pound capacity, internal combustion engine). Nine (9) were operated in a shop environment, and the remaining ten (10) were operated in a warehouse environment. The lift trucks were equipped with elapsed time (operation hour) meters and maintenance actions were thoroughly documented. The data was failure truncated and the observed lives ranged from 2500 to 8000 hours.

Model development was based on the results of formal statistical tests. Refer to existing sources in the literature (such as Lawless [7,8], Asher and Feingold [2], Crow [3], and Rigdon and Basu [10]) for the test statistics and decision rules.

7 STOCHASTIC PROCESSES

Table 1 summarizes the failure history for one of the units (equipment number 4727) operated in the shop environment.

The second column of Table 1 gives the (global) cumulative operating times to the successive failures while the third column gives the (local) interarrival time of each failure measured from the previous failure. For a stochastic point

Table 1 Failure history for F/L 4727

Failure Number	Failure Times Global	Local
1	227	227
2	1632	1405
3	1952	320
4	2651	699
5	3288	637
6	3307	19
7	3388	81
8	3756	368
9	3825	69
10	4093	268
11	4122	29

process such as in Table 1, we are interested in determining whether there is a renewal process, such as the HPP. The local times seem to indicate a trend over time toward more frequent failures (shorter interarrival time), which contradicts an assumption of the HPP.

The Laplace test statistic is an asymptotic normal variate which converges rapidly for as few as four failures, and gives a formal test of trend. The null hypothesis is H_0: HPP. For F/L 4727 the Laplace statistic was calculated to be $U_L = 1.99$. At the two-sided 5% confidence level, we would reject the null hypothesis and conclude that there is a trend. Additionally, a positive value of the Laplace statistic arises for an increasing ROCOF.

Table 2, column 5 contains the Laplace statistics for all 19 trucks. For the shop sample, the statistics are predominantly positive, with four of the nine significant at the two-sided 20% level. No statistic was computed for F/L 4710 because the sample size was too small. For the warehouse environment, the statistics were scattered about and clustered near the zero value, indicating an HPP renewal process.

We had speculated that there might be differences in the environment and/or operating patterns in the warehouse versus the shop. Such a difference would indicate that organization (or environment) was an explanatory variable, potentially leading to proportional hazards for the distribution of time-to-first-failure (Equation (1)) or a proportional intensity function for the stochastic process (Equations (14) or (18)). The Laplace test statistics indicated an increasing ROCOF for the shop sample and a constant ROCOF for the warehouse sample, ruling out the possibility of a proportional intensity model. We treated the two samples as coming from two different populations, and thus two separate stochastic processes. The distribution of time-to-first-failure was examined for proportional hazards, and that analysis is discussed in a separate section of this chapter.

Table 2 Laplace test statistics

Organization	Equipment Number	Operating Hours	Number of Failures	U_L	$\hat{\delta}$
Shop	4706	5002	10	1.48	1.77
	4708	6433	9	0.15	1.47
	4710	2824	2	-	-
	4721	2496	8	1.10	1.93
	4722	7997	15	2.88	2.93
	4724	4534	7	1.67	3.11
	4725	6651	12	-0.34	1.41
	4727	4122	11	1.99	1.89
	4743	3185	6	0.75	1.98
Warehouse	4728	6497	14	-2.03	
	4733	7963	13	0.04	
	4734	5286	6	1.21	
	4756	6892	11	-1.52	
	4757	5336	9	1.22	
	4763	6993	9	0.52	
	4765	6353	14	0.19	
	4768	5484	15	0.60	
	4772	7971	10	-0.12	
	4774	5778	12	-0.44	

Assuming an HPP for the lift trucks operated in the warehouse, the estimate of MTBF is

$$\text{MTBF} = \frac{\text{Total Operating Time}}{\text{Total Number of Failures}} \tag{19}$$

$$= \frac{64553}{113} = 571$$

For each shop truck we estimated the shape parameter δ in Equation (7), using Crow's method [3]. Table 2, column 6 gives the estimators of δ for all shop trucks except 4710. All estimates were greater than one, indicating increasing ROCOFs. Crow [3] proposed a Chi-square test of equality for shape parameters, with r-1 degrees of freedom:

H_o : $\delta_1 = \dots = \delta_r$

H_a : at least one δ not equal.

The test statistic was calculated to be D = 5.11. The Chi-square critical values for a two-sided 10% significance level are:

$$\chi^2_{7,.95} = 2.167 \qquad \chi^2_{7,.05} = 14.067$$

Since 2.167 < 5.11 < 14.067, we failed to reject the null hypothesis and proceeded to treat the shop units as samples from a single population. The pooled estimates of the parameters for a NHPP with power-law intensity, Equation (7), were

$$\hat{\delta} = 1.7 \qquad \hat{\upsilon} = 4.13 \times 10^{-6}$$

The intensity function is

$$\hat{\rho}(t) = 7.02 \times 10^{-6}\, t^{0.7}$$

From the intensity function, we can estimate the IMTBF. For example, at 6000 operating hours,

$$\hat{\rho}(6000) = 7.02 \times 10^{-6}(6000)^{0.7} = 0.0031$$

$$\text{IMTBF}\ (6000) = (0.0031)^{-1} = 323 \text{ hours}$$

The Cramer-von Mises test provides a formal basis for evaluating the model goodness-of-fit. The test statistic was calculated to be $W^2_m = 0.1157$. Crow [3] performed simulations to compile tables of the critical values for the Cramer-von Mises test. The test value of 0.1157 was well within the critical values for γ=.01 through γ=.20 indicating failure to reject the null hypothesis of a NHPP with power law intensity function.

8 PROPORTIONAL HAZARDS

We examined organization (warehouse versus shop) as an explanatory variable potentially forming proportional hazards in the distribution of time-to-first-failure. The product-limit method provided a nonparametric method of estimating the reliability function in each stratum. Fig. 3 contains the plots of ln{-ln[R(t)]} versus ln(t) for the two strata (shop (0), warehouse (1)). Accounting for the small sample sizes, the plots appear to be roughly linear and parallel, indicating a possible Weibull proportional hazards model. To further examine the model, the Weibull parameters were estimated within each stratum. Software routines provided estimates of the extreme-value parameters (Table 3) from which the Weibull parameters may be determined by simple transformations.

Since the m.l.e.s of $\hat{\sigma}$ were dispersed about the value $\hat{\sigma} = 1$, we tested the hypothesis H_0: $\hat{\sigma} = 1$ in each stratum. The likelihood ratio statistic provides a

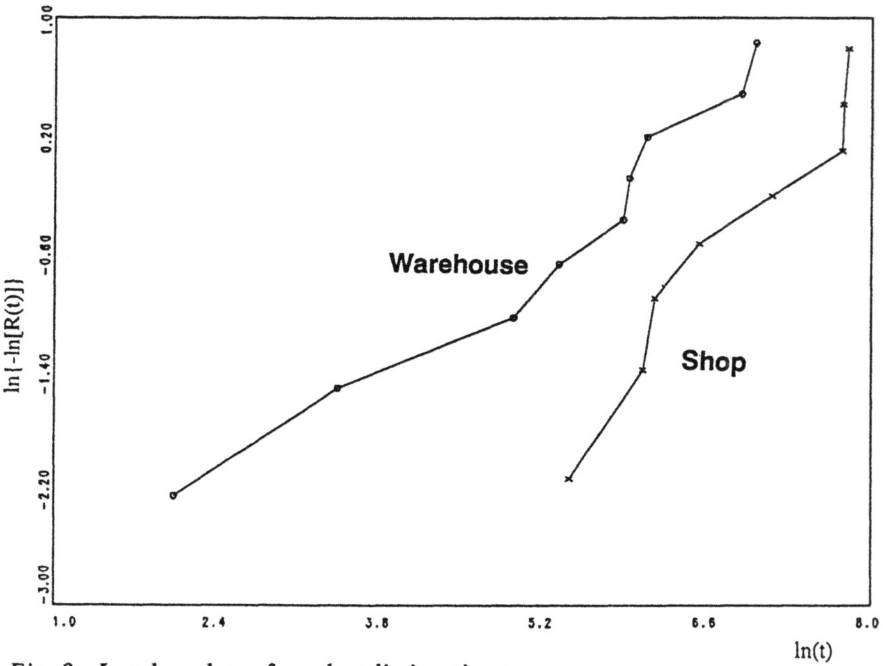

Fig. 3 Log-log plots of product-limit estimates

means for testing the null hypothesis, by measuring the relative change in the maximized log likelihoods under restricted ($\hat{\sigma} = 1$) and unrestricted ($\hat{\sigma}$) models. Table 4 contains the values of the maximized log likelihood functions.

Table 3 Extreme-value estimates

Organization	$\hat{\mu}$	Stderr($\hat{\mu}$)	$\hat{\sigma}$	Stderr($\hat{\sigma}$)
Shop(0)	7.35	0.237	0.678	0.189
Warehouse(1)	6.252	0.396	1.190	0.303

Table 4 Maximized log-likelihood functions

Organization	Unrestricted	Restricted
Shop(0)	-10.98	-11.82
Warehouse(1)	-17.71	-17.96

The resulting likelihood ratio statistics were

$$\Lambda_0 = 2\,[\,-10.98 - (-11.82)] = 1.68$$

$$\Lambda_1 = 2\,[\,-17.71 - (-17.96)] = 0.50$$

These statistics are asymptotically distributed Chi-square with one degree of freedom, and are generally accepted as powerful for small sample sizes. The critical value is $\chi^2_{1,0.05} = 3.84$, so we failed to reject the null hypothesis in either stratum.

The extreme value parameters σ and μ are related to the Weibull parameters δ and θ by

$$\delta = \sigma^{-1} \qquad \theta = \exp(\mu) \tag{20}$$

A Weibull distribution with δ=1 reduces to the special case of an exponential distribution. When a Weibull distribution consists of two strata with equal shape parameters then proportional hazards exist. Consequently, we proceeded to model the time-to-first-failure using an exponential proportional-hazards model with one explanatory variable (organization). The resulting estimates and standard errors were:

$$\hat{\beta}_0 = 7.25 \qquad \hat{\beta}_1 = 0.91$$

$$\text{STDERR}(\hat{\beta}_0) = 0.33 \quad \text{STDERR}(\hat{\beta}_1) = 0.46$$

From Equation (20) the exponential scale parameter is

$$\theta(\underset{\sim}{z}, \underset{\sim}{\beta}) = \exp(\beta_0 + z_1\beta_1)$$

$$= \exp(7.25 - 0.91z_1) \tag{21}$$

The MTTFs for the two strata are

$$\text{MTTF}_0 = \hat{\theta}_0 = \exp[7.25 - 0.91\,(0)] = 1408 \text{ hours}$$

$$\text{MTTF}_1 = \hat{\theta}_1 = \exp[7.25 - 0.91\,(1)] = 567 \text{ hours}$$

The estimated MTTF_1 (567 hours) for lift trucks in the warehouse corresponds closely to the estimated MTBF_1 (571 hours) in Equation (19), as expected for a HPP. Conversely, while for shop lift trucks the MTTF_0 (1408 hours) was initially high compared to the reliability of warehouse trucks, a deterioration process appears to exist with decreasing IMTBF over time. For example, the IMTBF_0 (6000) = 323 hours.

9 FAILURE MODES

Since there appeared to be a deterioration process in the sample of shop lift trucks, we studied the maintenance records to identify major failure modes. Most of the failures were disbursed among a wide variety of components and faults. However, for the shop lift trucks, four dominant components accounted for 45% of all failures. Fig. 4 isolates these four components and compares the shop and warehouse failure distributions. The shop trucks experienced a substantially higher incidence of problems with the starters and the steering links. The shop and warehouse trucks were subject to the same preventive maintenance policies, and maintained by the same organization. The differences in the estimated reliabilities and observed failure modes could be due to differences in operating environment or operator practices. These questions will be addressed in future research.

10 CONCLUSIONS

This chapter has demonstrated the potential of reliability tools to evaluate and model the reliability of material handling equipment such as industrial lift trucks. We have emphasized that material handling equipment are repairable items and that the HPP, while an accurate model in some instances, is not adequate to describe equipment deterioration (or improvement) over time. In some cases, the reliability model should explicitly include the effects of important explanatory variables. The resulting reliability models are useful for identifying reliability problem areas and trends. These models also provide a basis for decision-making in operations, maintenance and equipment replacement.

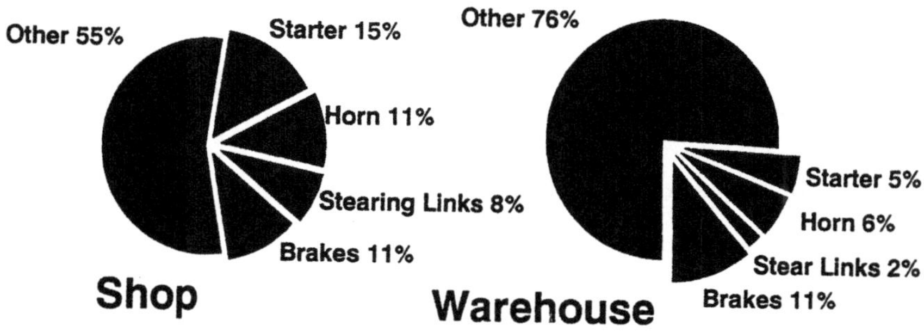

Fig. 4 Failure distributions, by component

11 REFERENCES

1. Ascher, H. and H. Feingold, *Repairable System Reliability*, Marcel Dekker, Inc., New York (1984).
2. Cox, D.R., "Regression Models and Life Tables (with discussion)," *Journal of the Royal Statistical Society* , B34, pp. 187-220 (1972).
3. Crow, L.R., "Reliability Analysis for Complex Repairable Systems," in F. Proschan and R. J. Serfling (Eds.), *Reliability and Biometry*, SIAM, Philadelphia, pp. 379-410 (1974).
4. Engelhardt, M. and L.J. Bain, "Statistical Analysis of a Compound Power-Law Model for Repairable Systems," *IEEE Transactions on Reliability,* R-36, pp. 391-396 (1987).
5. Kalbfleisch, J.D. and R.L. Prentice, , *The Statistical Analysis of Failure Time Data*, John Wiley and Sons, New York (1980).
6. Kaplan, E. L. and P. Meier, "Nonparametric Estimation from Incomplete Observations," *Journal of the American Statistical Association*, 53, pp. 457-481 (1958).
7. Lawless, J.F., *Statistical Models and Methods for Lifetime Data*, John Wiley and Sons, New York (1982).
8. Lawless, J.F., "Regression Methods for Poisson Process Data," *Journal of the American Statistical Association*, 82, pp. 808-815 (1987).
9. Prentice, R.L., B.J. Williams and A.V. Peterson, "On the Regression Analysis of Multivariate Failure Time Data," *Biometrika*, 68, pp. 373-379 (1981).
10. Rigdon, S. E., and A.P. Basu, "The Power Law Process: A Model for the Reliability of Repairable Systems," *Journal of Quality of Technology*, 21, 4, pp. 251-260 (1989).
11. Williams, B.J., Proportional Intensity Models for Multiple Event Times, Ph.D. Dissertation, University of Washington (1980).

INTELLIGENT ROBOTS FOR FLEXIBLE PACKAGING

G. D. Slutzky
E. L. Hall
R. L. Shell
University of Cincinnati

ABSTRACT

Problems, such as automated palletizing, bin-picking, automated storage and retrieval, automated kitting of parts for assembly, and automated warehousing are currently being considered for expert solutions. The use of expert robots or automated systems often consist of specialized computer programs, manipulators and integrated sensors has been demonstrated with a variety of applications. Games such as robot checkers, peg games, etc. have been developed as test demonstrations. Actual solutions for automated palletizing, pit-carb basket loading, etc. have also been developed for industrial applications at our Center. The generic concepts for flexible packaging will be described in this chapter, some unsolved problems discussed, and some applications briefly described. The significance of this work lies in the broad application of flexible packaging to a variety of generic industrial problems. Solutions can improve product quality, reduce waste, and eliminate the potential for human injuries.

1 INTRODUCTION

Most robots have been used in applications where major intelligence was not a necessity and little modification was needed with existing automation. From this unintelligent beginning the use of robots in the industrial arena, the problem of integrating their usage with existing processes has proven to be a difficult task. These tasks are more difficult to achieve because of a lack of intelligent systems and training of end-users. As new tasks become available for robot and non-robotic utilization the development of "intelligent" devices has emerged and awakened interest in expert system algorithms and the employment of multiple sensory systems such as vision, tactile, bar code, and ultrasonic. One example area is flexible packaging which includes such problem areas as automated palletizing, bin picking, automated storage and retrieval, automated kitting of parts for assembly, and automated warehousing. According to Matson and White, there is a worldwide importance for automated palletizing, [1].

A major advantage for a flexible manufacturing systems is their ability to accommodate a range of production rates between single products and high rate products. Groover points out two categories of mass production: (1) quantity production and (2) flow production, [2]. In both categories, material packaging and processing steps are important. In quantity production of a single item, bin-picking and bin-packing are often the least automated process in the entire plant. In flow production, the "flow" is only maintained if raw materials are available in sufficient amounts and if sufficient transfer devices are available. Again many of the operations required are labor intensive, especially in the packaging operations.

In order to investigate these problems in more detail, a description of an "intelligent" robot is essential. Generally, it is easy to see intelligence or the lack of it in humans, but trying to define this in terms of machines conjures up science fiction themes. At present, robots cannot display original thought leading to complex solutions. However, with the use of expert system algorithms, a robot can appear to "think" in a logical manner and actually solve difficult problems which appear to require intelligence. The components of an intelligent robot are shown in Fig. 1.

The use of expert robots which consist of specialized computer programs, robots, and sensors has now been demonstrated in a variety of situations. A game-playing robot can serve as a good indicator of an intelligent robot system, [3]. Game playing represents a form of intelligence acceptable to everyone. A game usually has a set of rules defining a competition, including the permissible set of actions. The elements of a game are usually given and easily understood, and the set of actions required are also defined. The game rules define the computer and sensor problems. The set of actions defines the robot manipulator requirements, [4]. If a well known game is chosen for a demonstration, it can rightly be called a "trivial pursuit" since the rules of the game and actions are common knowledge. However, the accomplishment of a game by an automated machine, especially for the first time, is neither commonplace nor lacking in importance, since it demonstrates the capabilities of robots and other automation for real world applications.

2 SOME PREVIOUS INTELLIGENT SYSTEMS

The early development of intelligent robots can be traced to research at the Massachusetts Institute of Technology in the early 1950s, [5]. Shannon in 1955 implemented a maze-following electronic mouse. Approximately 10 years later, a robot equipped with vision was utilized by Minsky to stack blocks. Further refinements of this idea have led to the development of the expert system for palletizing random size and weight parcels at the University of Cincinnati, [6].

In 1970, mobile robots were developed at Stanford University with the design of Shakey by Nitzan.

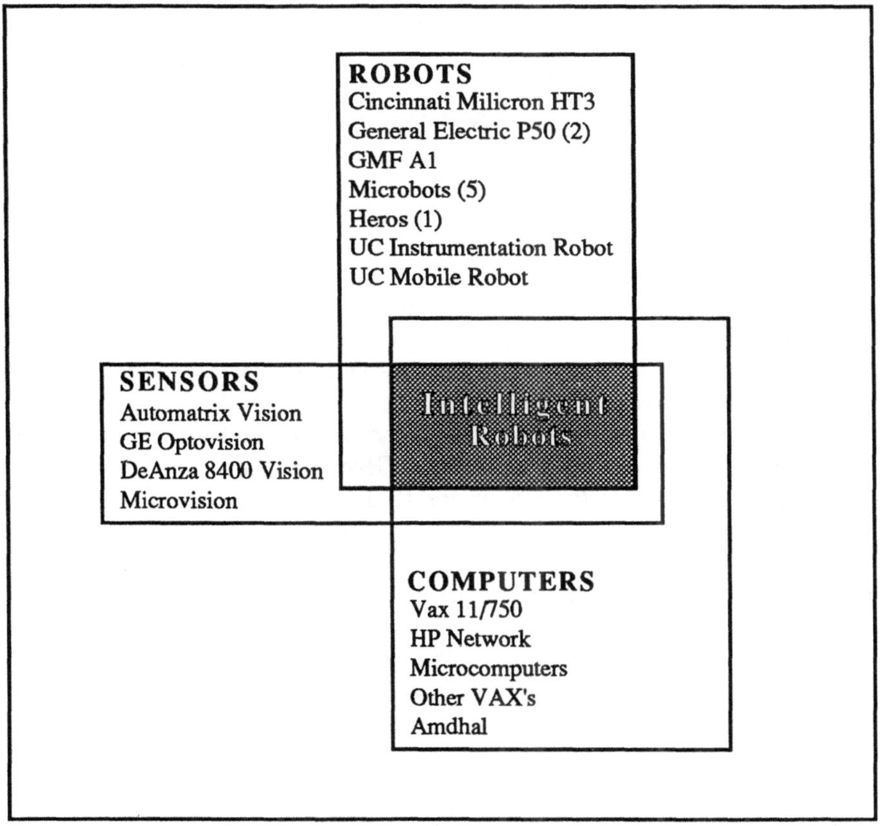

Fig. 1 Intelligent robot system components

Recent examples of game playing robots are found in Hall's book on robotics including a robot solution to Rubik's cube and a checker playing robot, [7]. Other systems which have not been as well documented include a blackjack playing vision guided robot developed and demonstrated by Cincinnati Milacron, several mobile robot competitions called the "seeker contest" sponsored by the Southeastern regional IEEE, the micromouse competition sponsored by the Western regional IEEE, and a recent ping pong robot competition which originated in Great Britain but also conducted in Europe and the United States. A vision guided, dexterous manipulator which reads sheet music and plays an organ was developed at Waseda University and demonstrated at the World's Fair in Japan in 1984 might also be called a game playing robot since the act of performing music has a set of rules and actions.

The University of Cincinnati has been conducting research in intelligent robots since 1983. Several of the systems from the Center for Robotics Research will be examined to illustrate the requirements for intelligent robot systems.

2.1 A Demonstration Manufacturing Cell

The game is made up of a triangular piece of wood with 15 drilled holes arranged in a bowling pin configuration. Wooden pegs are placed into all but one of the holes and the game begins. A move consists of the robot gripper picking up a peg, jumping an adjacent peg, placing the peg into the empty hole on the opposite side of the jumped peg, and removing the jumped peg from the board. This operation continues until only one peg is left or the player is unable to make more jumps. A robot or human player wins the game if only one peg is left standing on the board.

A multi-robot workcell consisting of a burn stamp stand, a General Electric P50 (5 axis) robot, an A40 (4 axis) robot, a General Motors Fanuac A1 (4 axis) robot, and a General Electric Optovision II imaging system was designed and constructed to manufacture the peg solitaire game shown in Fig. 2.

A typical workcell consists of the P50 robot manufacturing the game board, the A40 robot playing the game, and the A1 robot using the burn stamp to place the rules on the board. A conveyor then brings the packaged game to a predetermined destination. Implementing the vision system and artificial intelligence algorithm for peg detection, the robot plays the game to demonstrate the marriage of artificial intelligence to a robotic system to produce an "intelligent guided" robot systems, [4]. The technology and techniques developed to build and then find a solution to this game establish the capabilities of an intelligent robotics system to solve a "real world" problem.

To communicate to the other components of the workcell, an input/output box contains a set of switches so that any interface condition can be simulated. These switches are linked to the interface capabilities of the robot controller and the vision system computer. Mounted on the robot wrist are three independent end-effectors or grippers. The first is simply an electric drill for boring the holes; the second can grasp the wood preforms; and the last has specialized fingers for gripping the wooden pegs during the game playing process.

The final processing step consists of placing a preform on the burn stand stamp to have the rules and University of Cincinnati logo imprinted on the game board. After loading the board, one peg is removed and the game is placed under the vision system to determine what solution of the game to play. Since the board is symmetrical, a computer simulation is executed to determine the number of possible game solutions, there are five fixed solutions for this game.

The procedure of how to play them are stored as separate robot programs and the algorithm to determine the missing peg triggers the input/output binary code which is sent to the robot controller in order to solve the game. Then quickly, precisely, and with no trial and error learning the robot solves the game. It is capable of achieving this goal faster than its human counterparts.

Fig. 2 Workcell for manufacturing peg solitaire game

2.2 Space Filling Applications

Space filling techniques represent a new approach which can be applied to industrial applications in the area of flexible packaging . Many individuals have addressed this issue in order to do pallet loading, [8,9,10]. For example, the simple task of stacking random parcels onto a pallet seems very trivial to a human. The human implements vision to determine available locations and a high-level knowledge system to decide the most appropriate space for a given box. To accomplish this procedure using an automated system requires many steps to be executed. A key decision of this scheme is how to represent the pallet volume in computer memory.

The illustration of a three dimensional volume may be done in several ways such as surfaces or solid representation, [11]. In the solid representation the pallet and parcel volumes are depicted by discrete volume elements or voxels. This technique allows for a discrete computer representation which can be thought of as a three dimensional array V(x,y,z) which is equal to 0 for empty regions and 1 for full regions. In order to apply region filling to the pallet, a relational database about the boxes must be established. The pallet volume could be filled with one box or with a multiple of other box volumes. These boxes are multiples of the unit box referred to as a volume element or voxel.

From the previous discussion about intelligent robots, the next generation of robots must contain a computer utilizing some high-level language for the controller. These expert system robots are beginning to be developed for a variety of tasks. One such system was devised at the University of Cincinnati for the random palletizing of mixed size and weight parcels. Most robots are able to stack cartons of the same unit load or several different unit loads without additional software. The Center for Robotics Research has developed an algorithm in OPS5 code running under LISP on a VAX 11/750 computer that has approximately 400 rules used to stack a good pallet of mixed size and weight parcels in a random delivery system. Because a heuristic approach is incorporated, a "good" solution may not be the optimal solution. But, based on existing practice in the distribution industry it will pass inspection as a good pallet. Some samples of these rules are interlocking and intermeshing of cartons, toxicity of contents, fragility, and crushability. The voxel elements establish the knowledge database scheme for determining stacking relationships. The expert system will also incorporate multiple sensors into its design, [11].

At the Institute for Advanced Manufacturing Sciences (IAMS), the Center for Robotics Research has established a new laboratory for flexible packaging. Fig. 3, shows the layout for this laboratory. The palletizing cell consists of a Cincinnati Milacron electric articulated robot, Buschman carousel conveyor with vision and bar code sensors for box identification, an automated storage and retrieval system (ASRS) and automated guided vehicle (AGV) produced by Litton, and a workstation to house a computer system for networking with the Center for Robotics Research VAX 11/750 host computer.

A scenario of how this system might work consists of the following:

1) A computer interface must be established between host computer and robot controller.
2) A bill-of-lading must be input to generate the set formations based on voxel criteria.
3) Boxes are picked from the automated storage retrieval system and delivered by the automated guided vehicle to the conveyor.
4) A bar code or vision system is employed to identify boxes and trigger stacking operation.

5) As determined by the stacking algorithm boxes are picked up and placed on the pallet by the robot.
6) Full pallets are removed by the automated guided vehicle.

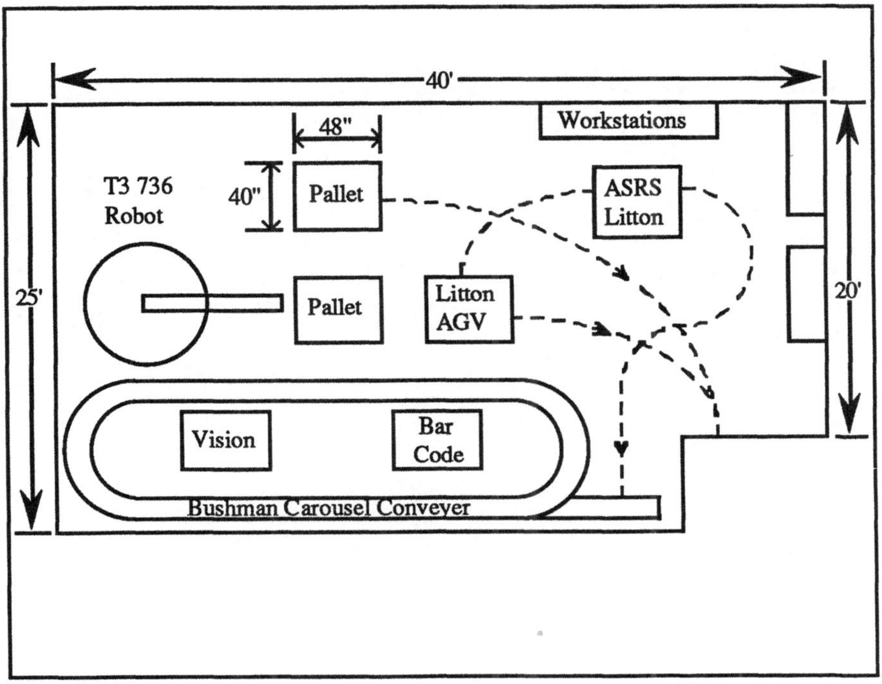

Fig. 3 Layout for flexible palletizing laboratory

To generalize palletizing, a robot or automated palletizer must be able to locate an object, determine how to pick up that object, and decide where to stack the object based on an expert system space filling algorithm. The long term goal of this research is to develop a useful algorithm for random palletizing for mixed size, content, and shape objects, both as a basic science problem, and to perform experimental applications to illustrate its usefulness for industrial applications. Techniques devised by other researchers will augment this effort in applications such as bin-picking. Improving on the earlier work is an increase in complexity as a specific solution for random palletizing is generalized for all shapes, sizes, contents on a single pallet. A specific solution to palletizing cartons is shown in Fig. 4.

The short term goal is to improve the logic and design of the existing region filling expert system for palletizing. Additional shapes will be added to existing parcels to begin the generalization process of the algorithm. Another development will be the start of rewriting the existing code into a more transportable language such as C. This will permit an increase in operational efficiency of the overall system.

Fig. 4 Automated palletizing system for randomly arriving parcels

2.3 Space Emptying Applications

Robots and robotics technology having been successfully implemented for many tasks. We now face the challenge of using them in more complicated applications. It is apparent that a robot has to function in a much more adaptable manner with the environment. Complex part geometries, novel data structures, and advanced artificial intelligence tasks, such as learning, are needed to accomplish a difficult task. In industrial applications, a robot should be capable of picking up a workpiece from a bin, orienting it, placing it into a container or machine, and performing some operation such as welding, painting, or etching on it. One of the remaining obstacles to the widespread introduction of industrial robots is their inability to identify and grasp parts that are not precisely positioned. The dilemma of picking randomly oriented parts from a bin could be solved if vision capability were available to analyze the mixture of parts. Intensive research in vision around the world guarantees that improved visual processing capabilities will shortly allow robots to work with overlapping objects on cluttered backgrounds with unconstrained conditions, [12].

The area of inquiry being studied at the University of Cincinnati involves "bin-picking" using moment and instantaneous invariants. This bin-picking problem requires both object recognition and manipulation. Three dimensional object recognition using perspective moment invariants has been recently investigated by K. T. Park and E. L. Hall, [13,14,15]. To acquire three dimensional data from images, we will use surface measurements, which have been intensively studied at the University of Cincinnati by J. Nurre and E. Hall with National Science Foundation support. The moment invariants approach is reasonable for three dimensional object recognition because the moments can be global or local features of an object. Instantaneous invariants are powerful tools for the study of the inverse motion in a robot mechanism. The solutions to the inverse kinematic problems are important in the control of kinematically simple robot manipulation and we believe it can be extended to more complex designs.

To manipulate an object, a robot must know where it is (position and orientation), what it is (recognition), and where to move it or what to do with it (path location). The picking problems may be considered in terms of increasing complexity or difficulty. The easiest question is picking up an isolated part from a flat two dimensional surface, and the most difficult may be picking up an arbitrary object from an unstructured environment. Picking up an object can involve several steps. As an example, consider the task of pulling a nail from a board; first, tentative objects, nail heads, must be pinpointed in terms of position and orientation; then, a force control must be used until the nail is released. This problem is easily accomplished by most humans, but it presents a number of difficult technical challenges for robots, because it requires coordinated vision, sensing, force control, and manipulation, none of which have been developed to a sufficient degree to make this type of task subject to economically viable automation.

A goal of this research is improved understanding of the theoretical development of three dimensional object recognition and the theoretical background of manipulation under the control of a robot vision system for bin-picking problems. Experimentation of recently developed surface measurement techniques will be used to explore computer perspective moment invariants for three dimensional object recognition and manipulation. This requires both object recognition and manipulation. To grasp an object, it is necessary to determine the orientation and position of objects relative to a robot gripper. The movement of a robot gripper relative to objects and the relationships between objects is also important. First, the research will consider an image and ground plane because most objects are supported by the ground, a table or other objects. They provide the constraints for determining the position and orientation of an object. Next the research will develop the relationship between the image and the ground plane. Finally, perspective invariants will be determined and used for bin-picking with a robot.

A principal problem of the moment invariant approach is in acquiring three or more dimensional data for object recognition. In order to obtain surface measurements previously developed imaging techniques from the Center for Robotics Research will be utilized. Also, three dimensional object recognition often involves object manipulation because parts of the object are hidden from a simple image view. Several tentative recognition decisions may be required. In three dimensional recognition using three dimensional data, models of the object are compared to some unknown image, which requires an implementation with parallel processing. The advantage of invariant recognition is that it is possible to generate the same orientation, and ambiguity may be removed by additional region analysis derived from edge detection and line segmentation.

To put theory into practice, experiments with the performance of bin-picking under realistic conditions will be tested in laboratory applications for industry. Under common lighting conditions, a CCD TV camera connected to the moment generator system will be mounted on a General Motors Fanuac A1 robot. The detection of edges, points and vertices from the image will be performed and displayed on a workstation, and stored in the central processor. Using the VAX 11/750, Gould DeAnza image processor, and CCD TV camera, three dimensional data will be acquired. Then, orientation, position, and perspective moment invariants will be determined for recognition and manipulation. Finally, the robot will be commanded to pick up an object from a bin on the table or ground and place it in a predetermined site.

3 INTELLIGENT SYSTEMS

As shown by the previous discussion, robot "intelligence" has been demonstrated by a variety of games and applications. In a game, perhaps a necessary ingredient is "uselessness." However, in the applications, a necessary justification of the machine is its "usefulness." What is it about these machines that makes their design such a challenge in both situations?

Perhaps the underlying mystery is that we are not cognizant of human intelligence, although we may feel confident in being able to "know it when we see it." A characteristic of all the games is that they can be described by astoundingly simple arithmetic or logical relations. The most challenging games have such an enormous state space that exhaustive enumeration of all possible solutions is not practical, say on the order of 2^{64}. Since we have computers with 2^{32} memory locations, it seems likely that in just a few years, the games with 2^{64} states may become deterministic and, thus, lose their mystery. Examples of such games are chess, checkers, and a 64 ring tower of Hanoi. Simulated randomness can always be inserted into game decisions to retain their amusement value, but the problems they represent will be theoretically solved.

The same cannot be said about human intelligence except perhaps by psychologists or sociologists. In a demonstration of the peg solitaire robot to a group of psychology professors, one of the authors first passed the game around to the humans with the challenge that in Psychology I, he was taught that "intelligence was what an intelligence test measured" and here was an opportunity to test their IQ's. None of the audience was able to solve the game. They were then shown the robot playing the perfect game. So, if intelligence is what an intelligence test measures, machines can come out ahead. The important point is that, in attempting to build intelligent robots, we may come closer to understanding human intelligence. We can, at least, present some interesting challenges to currently accepted practices.

Adaptability is another feature of intelligence in man and machine. The most intelligent robot today can be simply unplugged. One minor malfunction can incapacitate the entire machine. Redundancy of function is at least a partial solution to this. The human arm has more than seven degrees of freedom which may be used to move the hand to any point in three dimensional space with any of three possible orientations which requires only six. Redundant degree of freedom machines are now being built. Self diagnosis and maintenance are also being studied. To accomplish these diagnostic decisions, sensors of the state of the machine are needed for full adaptability, sensors of the state of nature are required. The work in robot vision, tactile sensors, and other sensors is progressing; however, the problem called "multisensor fusion" has also been recognized. If more than one sensor is used to determine a state, then it is possible to have incongruence of the sensor outputs. A human in space may encounter very little or no input from the balance sensors in the ear, be touching the wall as indicated by tactile and kinesthetic sensors in an arm, and see the interior of the spacecraft upside down. Vision, since humans receive over 70% of their information through their eyes, would probably be believed; however, unless at least one other sensor becomes congruent with the visual, the human might still have a difficult time moving about. For example, even an earth-bound human cannot be mobile using sight alone (consider an intoxicated person).

Even though we know sensors are essential to adaptability, they alone are not enough, and confusion can arise when the sensor inputs are not congruent. The expedient solution for incongruent sensors is to not move.

Learning is another feature of intelligence and a great deal of recent interest in "neural networks" can be observed. Learning has classically been divided into two types: Supervised learning with a teacher and unsupervised learning from experience. A recent joke has one person ask another: How do you develop good judgment? Answer - experience. And, how does one develop experience? Answer - bad judgment.

The mathematical theory of statistical and structural pattern recognition is well developed though recent advances are quite impressive, [16]. In the past, unsupervised learning or clustering has been used but the time required has often been considered impractical. With the emphasis on parallel computing, these solutions are not of much more interest.

To end this discussion, we are reminded of the ancient question: How does one find the lions in the desert? Obviously, build a sieve that lets sand through but not lions. Then simply filter out the sand and what remains will be lions. We are currently attempting to build a sieve for intelligent machines.

4 CONCLUSIONS

As the industrial base of the United States improves, new methods for continuing manufacturing excellence must also progress. At the University of Cincinnati, research in intelligent robots for flexible packaging is being conducted. Areas of investigation under this theme include automated palletizing, bin-picking, automated storage and retrieval, automated kitting of parts for assembly, and automated warehousing. One example of an expert system, is the region filling application for random palletizing of mixed size and weight parcels. This system successfully utilized artificial intelligence and a robot to stack cartons for the distribution industry.

Another experimental area is a new method for solving the bin-picking problem. A new technique for obtaining surface measurements has been developed so that moments and invariants can be examined to determine position and orientation of parts.

Previous examples of intelligent game playing robots and mobile robots have served as bases for the development of expert systems. The newly developed expert system manufacturing techniques can increase production and flow. Cost of manufacturing goods should be less and quality improvement could be significant.

5 ACKNOWLEDGEMENTS

This work was partially supported by NSF Grants ECS8609524 and DMC8806313.

6 REFERENCES

1. Matson, J.O. and J.A. White, "Operation Research and Material Handling." *European Journal of Operational Research*, Vol. 11, no. 4, pp. 309-318, (Dec. 1982).
2. Groover, M., *Automation, Production Systems, and Computer-Aided Manufacturing*, pp. 22-23, Prentice-Hall, Inc., New Jersey (1980).

3. Slutzky, G. and E. Hall, "Vision Guided Intelligent Robot Design and Experiments," SPIE Vol. 848 *Intelligent Robots and Computer Vision:Sixth in a Series*, p. 66, Cambridge, MA (1987).
4. Slutzky, G., J. Roning, F. Wu, and M. Kozak, "An Intelligent Robot Work Cell," Society of Manufacturing Engineers, Technical Paper MS86-773, Conferences Robotics Research, Scottsdale, AZ (August 1986).
5. Hall, E. and B. Hall, *Robotics: A User-Friendly Introduction*, Holt, Rhinhart, and Winston, New York (1985).
6. Mazouz, K., E. Hall, and R. Shell, "Expert System for Flexible Palletizing of Mixed Size and Weight Parcels," *Proceedings, Society of Photo-Optical Instrumentation Engineers, Intelligent Robots and Computer Vision Conference*, Cambridge, MA (November 6-10, 1987).
7. Reich, E. et al., "Robot to Solve Rubik's Cube," SAE Technical Paper 830341, *SAE International Congress and Exposition*, Detroit, MI (February 1983).
8. Tsai, R. and E. Malstrom "Robotics Unitization of Warehouse Pallet Loads," *International Industrial Engineering Conference Proceedings*, Chicago, Il., pp. 222-227 (1985).
9. Puls, F. M. and J.M.A. Tanchoco, "Robotic Implementation of Pallet Loading Patterns," *International Journal of Production Research*, Vol. 24, No. 3, pp. 635-645 (1986).
10. Anon, "Robots Picks, Counts, and Palletizes Parts," *Modern Material Handling*, Vol. 39, No. 10, pp. 65-75 (July 6, 1984).
11. Slutzky, G., Mazouz, A. K., and Hall, E., "An Expert System for Palletizing Mixed Size and Weight Parcels," Final Report to Institute for Advanced Manufacturing Sciences, Cincinnati, Ohio (1986).
12. Klaus, B. and B. Horn, *Robot Vision*, The MIT Press, Massachusetts (1980).
13. Hall, E. and K. Park, "Form Recognition Using Moment Invariants," *Proceeding of First International Symposium on Science of Form*, Tsukuba, Japan (November 1986).
14. Park, K. and E. Hall, "Form Recognition Using Moment Invariants for Three-Dimensional Perspective Transformation," *Proceeding of SPIE*, Vol. 726, Intelligent Robot and Computer Vision: Fifth in a Series (1986).
15. Park, K. and E. Hall, "Moment invariants for perspective transformation," *Proceeding of SPIE, Fourth International Symposium*, The Hague, The Netherlands (March 1987).
16. Kasner, E. and J.R. Newman, "Pastimes of Past and Present Times," *The World of Mathematics*, Vol. 4, pp. 2393-2414, Tempus Books of Microsoft Press, Redmond, WA (1988).

STRATEGIC INVESTMENT EVALUATION OF INTEGRATED MANUFACTURING SYSTEMS

William G. Sullivan
Virginia Polytechnic Institute and State University
James A. Brimson
Coopers and Lybrand

1 INTRODUCTION

The decision to invest in capital equipment is one of the most critical decisions an executive can make. A prudent capital investment in machinery and advanced technologies will increase production, reduce costs, and provide a competitive advantage. The future health of the company is closely tied to the investment decision.

Traditional investment justification techniques were developed for a labor-intensive environment in which the impact of an investment was localized in the department implementing the technology. For a company to be an industry leader today it must be able to integrate advanced technologies judiciously into its plant. This requires that companies upgrade their investment management practices so that they can (1) identify those areas within the manufacturing entity where the introduction of enhanced technology will have the greatest impact on improving performance; (2) evaluate the impact of the investment on cost and performance; (3) prepare an analysis and plan to identify and manage investment risk; and (4) integrate the planning assumptions from the analysis into the cost-management system to help manage the realization of the benefits.

The traditional role of manufacturing was to build a product in the most cost-effective manner possible. The "optimal" investment was often the one that most reduced cost. Factors such as quality, flexibility, and time reductions were looked upon as qualitative factors to be considered only after the quantitative analysis of cost reduction was undertaken. It is not surprising that in this environment, productivity was typically increased through the substitution of technology for labor.

Today many companies are struggling with traditional justification technologies to justify their adoption of advanced manufacturing technologies. It is a common complaint among manufacturing executives that traditional ROI and payback analysis often show a return insufficient to warrant an investment; yet they believe the company should make the investment to remain competitive. As a consequence, many advocate that the survival of their firms requires investment in advanced manufacturing technologies. There are increasing calls to disregard quantitative analyses of capital investments in favor of a more intuitive approach to evaluating investment opportunities or the development of new approaches to capital investment analysis. Other schools of thought believe that the traditional investment methodologies are adequate but claim that they are being used incorrectly.

2 UNDERSTANDING COST DRIVERS

Getting to the rationale for investment analysis requires an understanding of cost drivers. A cost driver is a factor whose occurrence creates cost. The chain of activities necessary to manufacture a product requires significant interaction among distinct functional departments. Each activity is constrained by the output from each previous activity. The event that activates a business process is known as its driver. For example, the process of developing a product too often requires numerous repetitions of design and redesign to profitably deliver the product to the customer. Many of the activities and costs are incurred to accommodate or rectify limitations inherent in the interaction of specialized departments.

An effective investment management system identifies the few cost drivers with the greatest impact on the thousands of activities that occur in even the smallest operational unit. Of all these activities only a handful are critical and have a significant impact on the success of the organization.

Attacking cost drivers is important because an investment might be misdirected toward solving a symptom rather than a problem. For example, a company is able to cost justify an automated storage and retrieval system based on current levels of inventory. A more rational strategy would be to eliminate the cause of inventory rather than to search for more efficient methods of moving inventory.

Examples of cost drivers include:

- Engineering change orders
- Floor space layout
- Forecast errors/master schedule changes
- Inventory levels
- Product design and lack of interchangeable components
- Multiple bills of materials

Investments should be driven by explicitly stated strategic objectives. A firm's strategic plan should provide guidance in the investment management process to help select among investment opportunities. By doing this, it encourages the firm to work toward a uniform goal.

3 SELECTION OF THE INVESTMENT

3.1 Present Value and Payback Only

Manufacturing excellence extends beyond lowest cost to include quality, flexibility, throughput time, and customer responsiveness. No longer does the least expensive product necessarily command the dominant market share. Financial analysis that focuses on cost reduction to the exclusion of other performance measures can result in an unsatisfactory portfolio of investments.

Traditional techniques are based on the philosophy of maximizing profitability through improved returns on fixed assets as seen on the left side of Fig. 1. Based on this philosophy, various principles for performing financial evaluations can be articulated (e.g., investment decisions should be made in light of cash flow differences among alternatives). All capital investment techniques have the objective of ensuring the firm's financial success.

To compete effectively, companies must strive to manufacture innovative products at a low cost while simultaneously maintaining high quality and providing outstanding customer service (short lead times). At the same time, they must structure their manufacturing facilities to provide the needed flexibility to meet short product life cycles and increasing international competition. This philosophy leads to the need for techniques that increase competitiveness through continuously improving quality and reducing waste (right-hand side of Fig. 1).

3.2 Ranking Projects by Present Value and Payback Criteria (Assumes a Company Strategy of Highest Financial Return)

The criteria used to select projects for a capital budget represent an implicit strategy. This strategy is based on the project-ranking criteria established in the investment analysis. The assumption of the traditional model is that the projects offering the highest financial return will be those selected. However, this strategy is often not consistent with the company's strategic plan. The primary reason is that in more cases than not the implicit strategy is to continue doing what the firm has done in the past. The reason for this bias is that the highest short-term financial returns are normally generated by improving existing operations rather than by making significant investments in new areas where the return is to be realized in the future.

The fallacy of this approach is that investment analysis should be forward-looking rather than focused on the past. A company ought to make investments that enhance its probability of remaining competitive. If future performance exceeds or is as good as past performance, so much the better. However, businesses are subject to business cycles and an investment should not be rejected simply because it does not exceed past financial performance measures.

3.3 Present Value and Payback Often Ignore Capital Decay

Changes in product and manufacturing technologies are occurring at an increasing rate. Traditional investment justification techniques typically assume a constant future market share, with the alternative of not making an investment being zero impact on future operations. This assumption ignores capital decay--lost sales owing to technologically obsolete products and processes. Companies that do not keep pace with technological advancements often face loss of market share. Competitors can invest in a technology and potentially gain a competitive advantage. In other words, a company cannot afford to stand still technologically.

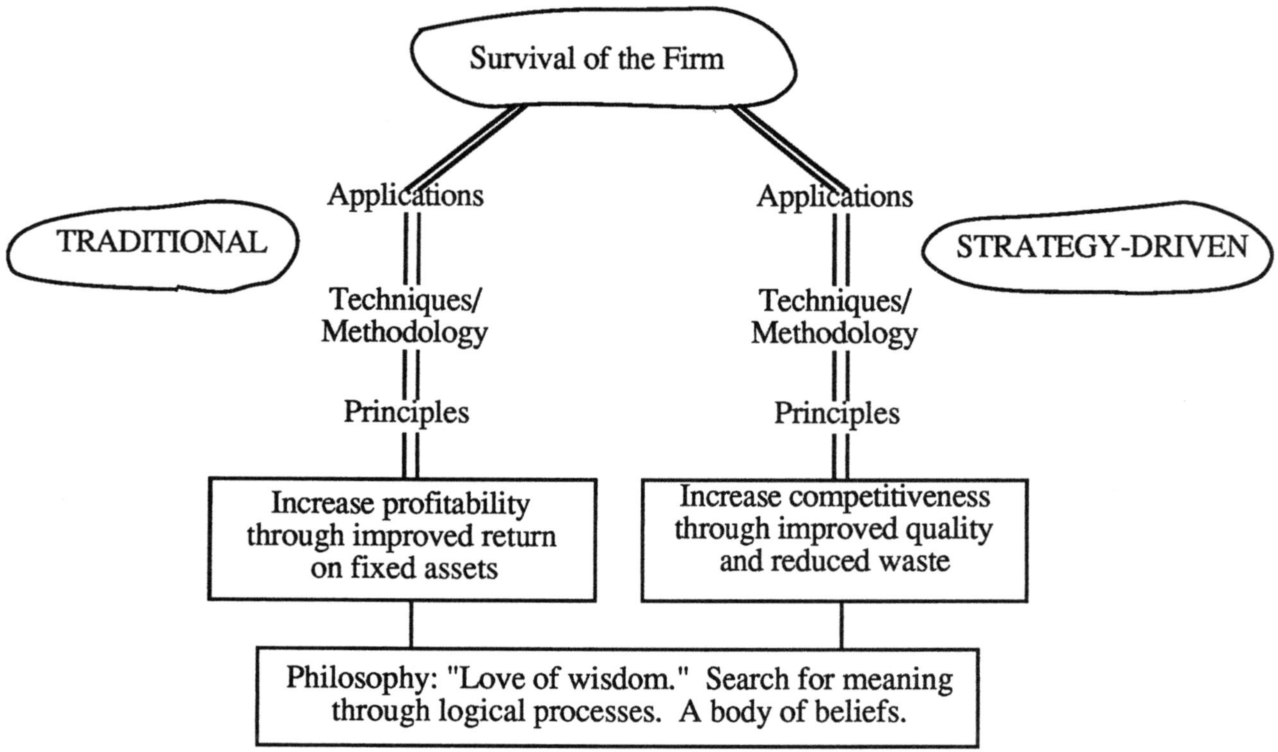

Fig. 1 Basic building blocks of traditional and strategy-drvien investment analysis

3.4 Traditional Investment Justification Assumes an Investment Time Continuum in Which There Is No Timing Impact

Investment in advanced manufacturing technologies often provides a "window of opportunity" dictated by a company's competitive strategy. The "window of opportunity" is manifest in two ways. First, a company must decide whether to be an industry leader, a follower, or compete in niche markets. To be an industry leader it must drive technology. An industry leader strategy has a high probability of gaining market share by using technology as a competitive weapon.

Another approach is to be technology responsive. Under this philosophy, a company chooses not to change its manufacturing processes until the technologies have proved their value in the marketplace. The key to success with this strategy is that a company cannot be far behind once the new process has proved successful. A company must be able to implement the new processes rapidly.

Finally, a company can choose to be technology reactive. This strategy implies that a company waits to enter the market until a product is available "off the shelf."

Each of these strategies implies that the timing of the investment is an integral part of strategy. Most investment analysis techniques do not incorporate the timing considerations that affect the worth of an investment.

Second, the timing of investments is often dictated by new products, changed manufacturing strategies, and other requirements. The timing of the investment may be dependent on the new product introduction cycle. A company that introduces a product requiring new or upgraded manufacturing processes must have these technologies in place prior to competing for market share for that product. Because most advanced manufacturing technologies require a significant amount of time to implement, the investment process must recognize these time-dependent relationships.

3.5 The Risk Factor Is Not Explicitly Incorporated In Most Investment Analysis

Risk is an important distinguishing feature of an investment decision. Despite a sophisticated investment appraisal there is no guarantee that the investment will achieve its anticipated benefits. The decision to make an investment pivots on the expectation that a profit will be made. Today's investment analysis techniques do not realistically deal with risk.

4 ACTIVITY-BASED INVESTMENT MANAGEMENT

Activity-based analysis is the key to implementing a continuous improvement philosophy (right-hand side of Fig. 1). One of its chief purposes is better understanding and control of "factory overhead" which traditionally comprises 30-40 percent of a manufactured product's cost. The key features of an activity-based analysis include:

- Continuously challenging how activities are performed, with the goal of improving performance.

- Linking investment opportunities to the company's strategic objectives.
- Highlighting nonvalue-added and inefficient activities as candidates for improvement.
- Identifying cost drivers as the critical control point for reducing costs.
- Facilitating an understanding of the total cost of all activities affected by an investment.
- Evaluating the total performance of an activity (time, quality, flexibility) rather than myopically focusing on cost.
- Evaluating the relationship among activities to determine their synergistic effects.
- Evaluating capital investments relative to "efficient operations" rather than existing cost structures, thus decreasing the probability of choosing and implementing an inappropriate investment.
- Coupling the accounting system with the investment analysis to facilitate corrective action.

Accordingly, the principles of investment management are listed in Fig. 2.

Activity analysis forms the basis of new investment management concepts. Before investment opportunities can be identified, it is essential to consider just what activities and functions comprise a company, how they are performing, and whether they are profitable or not. The approach analyzes in detail the impact of a new technology on the cost, performance, and interdependencies of key activities. Activity-based investment analysis is iterative, starting with the initial estimates of market conditions, process improvements, etc. and progressing through several phases of refinement as the investment is implemented. The investment analysis process systematically decomposes company objectives into a base set of activities that are impacted by the investment. This conversion facilitates an in-depth analysis of the investment by defining a baseline set of activities against which to measure change.

The principles that underlie classical engineering economy and activity- based investment management are compared in Fig. 3. For instance, principles 2a and 2b of engineering economy roughly correspond to principle 2 of investment management. The main differences between the two sets of principles in Fig. 3

- **Continually improve to eliminate non-value added activities**
- **Link corporate strategy to manufacturing**
- **Identify and control cost drivers**
- **Trace costs to activities**
- **Identify best practices and external targets**
- **Monitor progress in meeting performance targets**
- **Evaluate total performance**

Fig. 2 Investment management principles

What Are the Principles of Engineering Economy?

Principle 1: Prospective outcomes of feasible alternatives should be consistently developed from a defined viewpoint (perspective).

Principle 2a: The choice (decision) is among alternatives, which need to be identified and then defined for subsequent analysis.

Principle 2b: Only differences in expected future outcomes among alternatives are relevant to their comparison and should be considered in the decision.

Principle 3: A common unit of measurement to enumerate prospective outcomes will make easier the analysis and comparison of the feasible alternatives.

Principle 4: Selection of a preferred alternative (decision making) requires the use of a criterion (or criteria). The decision process should consider outcomes enumerated in the monetary unit, and those expressed otherwise should be carefully and completely described.

Principle 5: Uncertainty is inherent in estimating future outcomes of alternatives and should be recognized in their analysis and comparison.

Principle 6: Improved decision making results from an adaptive process; to the extent practicable, initial projected outcomes of the selected alternative and actual results achieved should be subsequently compared.

What Are the Principles of Investment Management?

Principle 1: Sound investment management decisions must be tied to long-term strategic plans and short term goals.

Principle 2: Investments should be considered to be interrelated elements of an integrated strategy, rather than individual projects.

Principle 3: A consistent methodology should be used to evaluate investment alternatives.

Principle 4a: Investments should be evaluated on qualitative and quantitative, non-financial criteria, in addition to traditional financial criteria.

Principle 4b: The evaluation and selection of an investment opportunity should be through a consensus-driven group process.

Principle 5: The investment management process should assess the risk associated with both the total investment strategy and the individual projects.

Principle 6: The cost management system must provide the cost performance data required for ongoing monitoring of the investment.

Fig. 3 The principles of engineering economy juxtaposed against those of investment management

are enumerated in Fig. 4, and they form the basis for new thinking by engineers on the subject of justifying integrated manufacturing systems.

4.1 Defining Company Activities

An alternative to the traditional method of managing costs by organizational unit is to control the costs of the activities irrespective of functional boundaries.

Comparing the principles of engineering economy with those of investment management provides interesting insight into these areas where engineers can contribute:

(1) emphasis on linking corporate strategy with investment analysis (i.e. what new activities are anticipated that will give rise to new investment opportunities?)

(2) emphasis on developing performance measures and aspiration levels ("targets") that reinforce the attainment or corporate goals (i.e. what financial and/or non-financial measures are appropriate in appraising investment opportunities?)

(3) emphasis of examining portfolios of discrete investment projects (i.e. what synergies among mutually exclusive bundles of funded and existing projects might exist and how do we estimate their consequences?)

(4) emphasis on monitoring the "quality" of investment decisions so that true control, through feedback, is integral to the investment management process (i.e. continual improvement is a reality rather that a cliche).

(5) emphasis on consensus-driven decision processes involving many organizational units and managerial levels.

(6) a world-view is imperative as contrasted with the more traditional domestic view.

Fig. 4 Differences in principles between engineering economy and investment management

Activities are defined as the repetitive tasks performed by each specialized group with a company as it executes its business objectives. An activity describes what an enterprise does; that is, the way employees assigned to a specialized group spend their time. Under activity-based investment analysis, resources are consumed in executing activities.

Activity-based analysis permits improved investment management by viewing each activity that can be continually improved. To illustrate the concept of activities, first consider the major activities of a typical overhead department such as material control. These activities would include:

Incoming material handling;
In-process material handling;
Outgoing material handling;
Raw material storage;
WIP storage;
Finished goods storage;
Management and administration; and
Training.

The concept of continual improvement would involve evaluating each activity to determine, first, whether the activity is necessary and, second, whether its performance can be improved. For example, in a material control department, all activities associated with moving and storage material are non-value added. Capital investment should not be made for the purpose of automating nonvalue-added activities. Instead, such activities should be eliminated or minimized.

The goal of activity-based investment management is to compute the cost of performing existing (as is) activities. An investment in a new technology such as a computer system or a new type of manufacturing would change the factors of production needed to perform an activity. The cost behavior pattern of the improved activity (to be) would then be estimated. Where the material control department's activities are concerned, products that require extensive material handling would consume more material handling resources than products using cellular manufacturing. By understanding the cost of material handling, these costs can be traced to products based on distance moved.

4.2 Defining Business Processes and Functions

Once the activities pertaining to each organization are defined, "business processes" characterizing a larger enterprise can be identified. A business executes a set of related and dependent activities. The activities are related because a specified event initiates the first activity in the process and this in turn generates the need for subsequent activities. Business processes often transcend organizational units. Consider, for example, the business process called "manufacturing the product." In this case activities are structured to succeed one another in a relatively fixed way and lead toward a particular result. Moreover, processes describe how the different activities relate to each other. Clearly, each process groups the related activities and describes the flow of information among them.

There are certain characteristics of business processes that are important to investment analysis. First, by understanding activity relationships, a company can focus on the events that trigger a business process. By controlling the relatively few initiating events, the cost of all subsequent activities can be reduced or eliminated. Second, understanding activity relationships facilitates the streamlining of business processes by identifying redundant and unnecessary activities. The business process, in total, can be restructured so as to reduce costs and improve efficiency. Third, business processes provide a basis for comparing the performance of one company with that of other similar organizations or with that of a theoretical standard.

4.3 Linking the Strategic Plan to Activities

An important input to the investment justification process that is provided by the company's strategic plan is the product forecast. The product forecast is a projection of products and product lines for which a company will compete over a five-to ten-year planning horizon. It provides visibility into future manufacturing requirements. A graphic illustration of a product strategy appears in Fig. 5.

From Fig. 5, we see that a company is currently selling products A, B, and C. Slowly, over time, demand for these products decreases and is replaced by a demand for later products, in other words, D and E. The product forecast for both

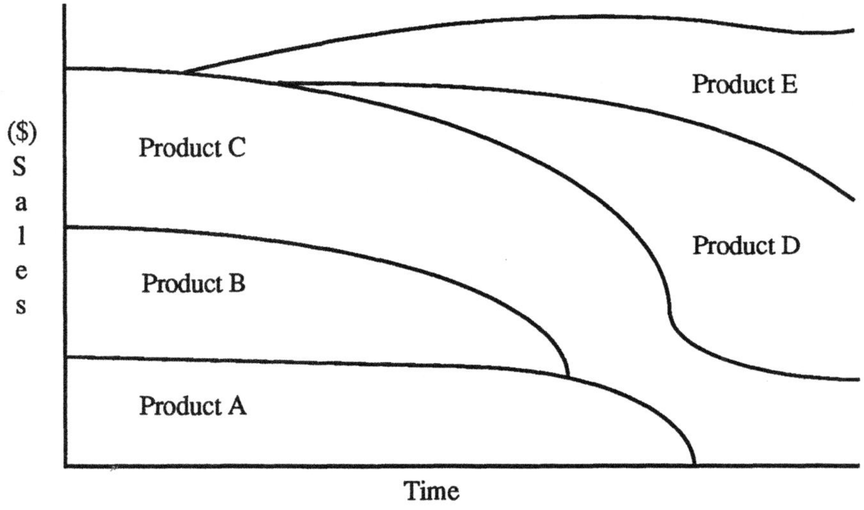

Fig. 5 Sales strategy for existing and new products

existing and new products provides a basis for identifying the required manufacturing processes. This can be accomplished by building a matrix (seen in Fig. 6) that maps products together into their necessary manufacturing activities.

To illustrate this approach, assume that a new product, Product E, is to be introduced that requires surface mount technology. Clearly this technology has significantly different manufacturing processes from those used in automatic and manual insert technology. Furthermore, this technology must be installed before the company can begin full-scale production of Product E. Hence, the effect of this new product on company operations will be assessed in terms of its activity structure.

Manufacturing Activities	Products				
	A	B	C	D	E
Component preparation	X	X	X	X	
Manual insert	X	X			
Automatic insert			X	X	
Hand solder	X				
Flow solder		X	X	X	
Surface mounted					X
Cleaning	X	X	X		
Inspection	X	X	X	X	
Testing	X	X	X	X	X
Burn-in	X	X	X	X	

Fig. 6 Products versus manufacturing activities matrix

Another important input from the strategic plan is the company's competitive strategy. Does the company want to drive technology, does it want to be responsive, or does it want merely to be reactive? These strategies are translated into specific measurable goals as illustrated in Fig. 7.

Activity-based investment analysis links the strategic plan to the activities through quantitative performance measures. An evaluation starts with a general strategy (such as whether a company wants to dominate a particular industry), and then this objective is decomposed into specific, measurable goals that are affected by specific activities. Assume a company's goal is to increase its market share from 20 percent to 25 percent. The current level of company performance in terms of sales, cost, quality, throughput, flexibility, and other criteria is evaluated. Management would assess these performance factors to determine how they influence market share and then establish specific improvement goals.

Activity-based analysis decomposes each of the performance goals (such as throughput) into the specific activities pertaining to it. Each level of decomposition relates the measure to its impact on market share. The impact of changing activities in terms of their influence on throughput, quality, and flexibility is analyzed and tangible measures are then developed. Target improvements are established, and a performance measurement system monitors progress in achieving these performance targets. Such a proactive approach to investment management (Fig. 2) helps ensure that capital expenditure objectives are realized.

5 SUMMARY

The challenge facing engineers is how to use activity-based financial and non-financial information for decision making purposes. Based on differences noted

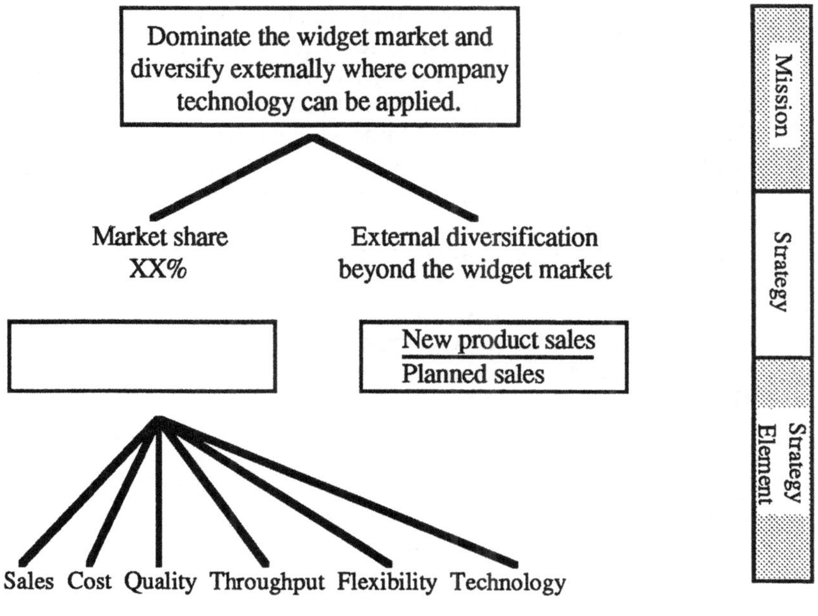

Fig. 7 Example of a company's competitive strategy

in Fig. 4, research is currently underway in four application areas (abandonment, make-buy, design to cost and resource acquisition). Appropriate methodologies are currently being developed which utilize activity-based accounting data and provide means by which numerous "intangible" factors can be reduced to quantitative measures. Better decision making involving integrated manufacturing systems will be the ultimate result.

6 GLOSSARY OF TERMS

1. **Activities**- any set of one or more repetitive tasks performed as various functions are accomplished.
2. **Activity basis**- this is a common denominator for a natural expense category that is used to distribute cost to activities (e.g. machine hours, distance traveled, student-weeks). (It is volume-based in the sense that it is numerical and easy to count.)
3. **Business process**- a set of related and dependent activities. They often transcend organizational units.
4. **Capital decay**- lost sales and market position caused by technologically obsolete products/processes.
5. **Competitive strategy**- how a company plans to meet the enemy (through leadership in the market, by reacting to change, or by following lock-step with the industry leader).
6. **Cost driver**- a factor, such as a bill of material's size, inventory or poor quality, whose occurrence creates cost. Drivers give rise to critical control points for reducing waste.
7. **Cost elements**- cash flows due to, for example, hardware and software purchases, business travel and utilities.
8. **Externally driven targets**- a benchmark against which to measure activity improvement efforts.
9. **Factors of production**- resources such as capital, land, labor and managerial skills required to create real wealth in the form of manufactured items, agricultural goods and extracted natural resources. (Brimson also incorrectly calls "energy" a factor of production.)
10. **Natural expense categories**- these are aggregations of general ledger accounts to be used for purposes of accumulating activity costs by cost center. For example, "energy" and "information systems" are natural expense categories.
11. **Non-value added activity**- an activity that does not add value to a product but consumes a resource (usually capital).
12. **Performance measures**- metrics by which changes in an activity's status can be evaluated (cost, quality, time, flexibility, etc.).
13. **Resources**- inputs to an activity that enable its accomplishment.
14. **Risk**- the possibility of not achieving forecasted benefits (objectives) of an investment project. Benefits pertain to schedule, technical performance, etc.
15. **Synergistic benefits**- complimentary (usually additive) effects when two or more investment projects are implemented that would be missing if only one of them were implemented.

7 REFERENCES

1. Brimson, James A., *Activity-Based Investment Management*, American Management Association, New York (1989).
2. Canada, John R. and William G. Sullivan, *Economic and Multi-attribute Evaluation of Advanced Manufacturing Systems*, Prentice Hall, Englewood Cliffs, N.J. (1989).

PERFORMANCE MODELING FOR A SINGLE MATERIAL HANDLING DEVICE WITH RANDOM SERVICE REQUESTS - PURE BLOCKING

Zulma R. Toro Ramos
University of Puerto Rico
Leon F. McGinnis
Georgia Institute of Technology

ABSTRACT

In this chapter, a manufacturing cell in which a single device provides movement between a number of distinct locations is studied. We propose an approach for analyzing the performance of the material handling device based on finite networks of queues. In this chapter, the case in which a full input buffer at a delivery location blocks the sending location is considered. Numerical results obtained from the proposed approach are compared with simulation results for the specific case of a robot tending several machines. A companion paper, [10], addresses the situation in which an alternative temporary storage area is provided for input buffer overflows.

1 INTRODUCTION

There are a number of material handling applications where a single device provides movement between a number of distinct locations. Examples include: (1) the storage/retrieval (S/R) machine in an automated storage/retrieval (AS/R) system; (2) a single automated guided vehicle (AGV) in a Flexible Manufacturing System (FMS); (3) a single lift truck on a shop floor; (4) a robot tending several machines; (5) a single crane serving a shop floor. The common feature among such material handling applications is that the sequence of movements is dictated by the durations of processing operations, and may appear to the material handling device "randomly", i.e. they are not deterministically preprogrammed.

In system design, it is important to determine the maximum cell throughput rate, considering the capacities of both the work centers and material handling device (MHD). The most common methodology currently in use for analyzing this kind of system is simulation. Despite the progress in developing simulation languages and tools, simulating manufacturing cells remains an expensive, tedious exercise with potential for modeling errors. In this chapter, we propose an alternative approach, based on networks of queues models.

The proposed approach is illustrated using the specific case of a robot tending several machines. A formal statement of the problem is presented in Section 2. Section 3 describes a general strategy based on a network of queues model for solving the problem. Section 4 presents detail methodologies for estimating the parameters required in the solution of the network of queues model. In Section 5, an experimental study is described and numerical results obtained from the strategy in Section 3 are compared with simulation results. The last section presents the conclusions.

2 GENERAL PROBLEM

Fig. 1 shows a schematic of the material flow within a manufacturing cell with a single MHD. In this case there is a system input point through which jobs enter the system and there is an output point through which jobs leave the system. Each work center within the cell has an input buffer where the MHD places the jobs to be processed by the work center and an output buffer where the work center places the completed jobs to be picked up by the MHD.

We are concerned with cells in which the sequence of transfers between locations is not known a priori. This situation may arise because of a random mix of part types, and variations on processing times between different parts. It may also arise even with a single part type when the processing time for an operation is variable.

Suppose that we know the cell physical configuration including locations of work center buffers and system input and output points, production ratios, process times, part routes, and buffer capacities. In addition, we know the move times for the MHD. Now the question is: what is the maximum cell throughput rate considering both work center and MHD capacities?

There are two cases to consider in addressing the above question. In the first case a full input buffer at a work center blocks the sending work center. This case will be referred to as "pure blocking". In the second case a full input buffer at a work center causes a job to be diverted to a temporary storage area. The second case will be referred to as "blocking with recourse". In this chapter, the "pure blocking" case is considered. The case of blocking with recourse is considered in [10].

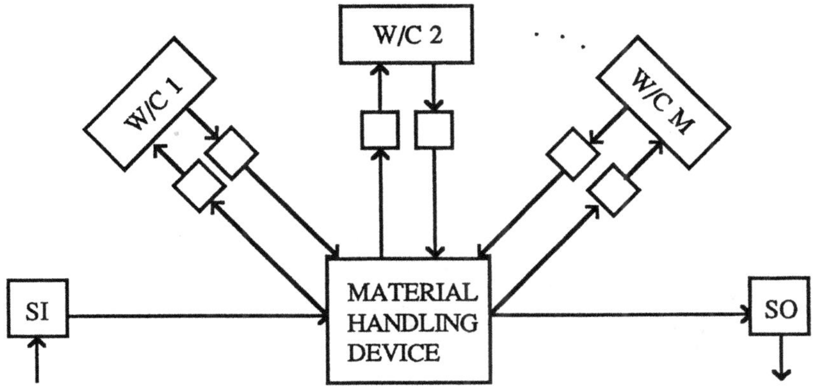

Fig 1 Material flow schematic

3 MODELING STRATEGY

As a job progresses through its process route, it receives a number of "services" which can be classified as: (1) movement by the MHD; (2) wait in a work center input; (3) process at a work center; and (4) wait in work center output. When viewed in the aggregate, each of these services has a duration which may be described as a random variable. If the routes of the job types are known, and an estimate of the production rate for each job type is given, then we can estimate the rates of flow between the work centers.

The solution strategy presented in this chapter is based on modeling the job arrivals at the system input by a Poisson process and all service times by exponential distributions. A number of researchers have argued for the robustness of these assumptions ([6], [7], [8], and [9]).

A queueing network (QN) model that may be used to represent the manufacturing cell under study is the central server model [6], [9], [11], [12]. This model requires all jobs completed by a work center to be routed to the central server (in this case, the MHD) before proceeding to another work center. Fig. 2 presents a schematic of the network of queues model for a manufacturing cell with M work centers. In Fig. 2 there are five different types of node: work center input buffer nodes, work center output buffer node, MHD node (central server), and system input and output point nodes.

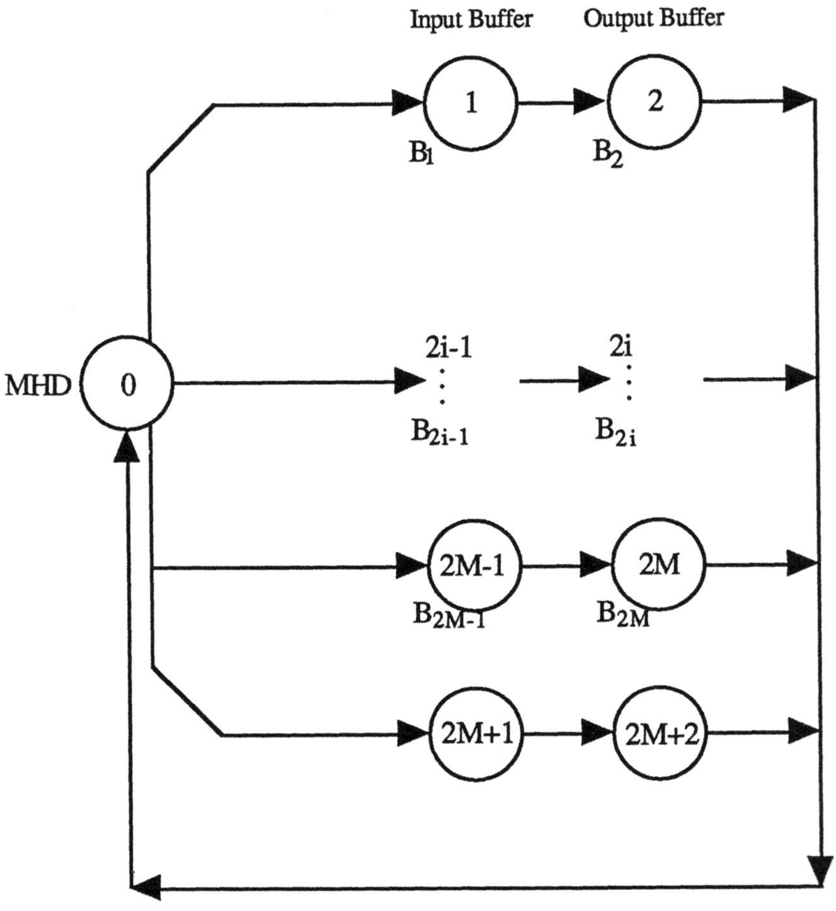

Fig 2 Network of queues model for the manufacturing cell

Each work center in the queueing network model has an associated buffer capacity, B_k, number of servers, s_k, and service rate μ_k. If the node corresponds to a work center input buffer (k=2i-1, i=1,...,M), then B_k is a design parameter, s_k is the number of servers at the work center, and μ_k is the service rate at the work center. If the node corresponds to a work center output buffer (k=2i, i=1,...,M), then B_k is a design parameter, $s_k=1$, and μ_k depends upon the operation of the MHD.

For the system input point, node k=2M+2 in the queueing network, $B_k=\infty$, $s_k=1$, and $\mu_k=\infty$. For the system output point, node k=2M+1 in the queueing network, $B_k=0$, $s_k=1$, and $\mu_k=\infty$. Finally, for the MHD, node k=0 in the queueing network, $B_k=\infty$, $s_k=1$, and μ_k depends on the operations of the MHD.

We assume that all queue disciplines are first-come-first-served, and that, at a node, all servers have identical, independently distributed exponential service times. Note that there is not an explicit node in the model for the work centers themselves. Their contribution to system behavior is captured in the work center input buffer nodes.

We assume that the work center mean service rates are known, so the mean service rates for the work center input buffer nodes in the queueing network are known. Because we assume pure blocking, the routing probabilities are determined from process routes and production ratios.

In order to use this queueing network approach, we must be able to estimate the mean service rates at the MHD and the work center output buffer nodes, and we must be able to compute the steady state solution. In Section 4, we show how to use revised queueing network models to estimate the mean service rate at the output buffer. The methodology for computing steady state solutions is addressed in Section 5.

4 SERVICE RATE ESTIMATION

Consider first the service rate at the MHD. A typical service request can be described as a pair, (origin location, destination location). Service time consists of four elements:

(1) empty travel time: from the location where the previous job was delivered to the origin location (also called "deadhead travel time");
(2) pick up time: time to load the job on the MHD;
(3) loaded travel time: from the origin location to the destination location; and
(4) deposit time: time to unload the job from the MHD.

We propose to estimate μ_0 by:

$$\mu_0 = \frac{1}{\text{DHTT} + \text{LTT} + 2\text{P}} \tag{1}$$

where:

DHTT = expected deadhead travel time
LTT = expected loaded travel time
P = average pickup or deposit time.

Note that P is a parameter, based on configuration and MHD specifications.

Define:

t_{ij}: travel time between locations i and j per move
f_{ij}: average number of moves between locations i and j per job

The t_{ij} values are computed from distance data and MHD performance specifications. The f_{ij} values are computed from the process routes and production ratios. Then:

$$LTT = (\sum_i \sum_j f_{ij} t_{ij})/F \quad [\text{hrs/move}] \tag{2}$$

where:

$$F = \sum_i \sum_j f_{ij}$$

LTT is easy to estimate because the required loaded travel is specified in the problem data. Estimating DHTT is not so easy, because the amount of unloaded travel depends upon the operation of the MHD. Rather than attempt to a priori optimize MHD operations, we make the following simplifying assumption. Service requests arrive to the MHD in an essentially random manner, and are served in a first-come-first-served order.

The loaded travel of the MHD creates both "supplies" of and "requirements" for unloaded MHD travel. In order to estimate the DHTT, we must estimate how the supplies will be distributed among the requirements. We will use a method often referred to as factoring (see [3], [4], and [5] for additional discussions). Let

g_{ij}: average number of deadhead moves between locations i and j per job.

Factoring estimates the g_{ij} values by:

$$g_{ij} = \sum_k f_{ki} \left[\frac{\sum_k f_{jk}}{F} \right] \quad [\text{moves/job}] \tag{3}$$

Now, DHTT is estimated in the same fashion as LTT:

$$DHTT = (\sum_i \sum_j g_{ij} t_{ij})/F \quad [\text{hrs/move}] \tag{4}$$

Now consider the service rate at the work center output buffer. When a job reaches the front of the output buffer queue, it has an associated service request that is in a service request queue (SRQ) for the MHD. Once this service request reaches the front of the SRQ, the MHD is dispatched to the output buffer location, incurring some deadhead travel time. Thus, the total "service time" seen by a job in the output buffer is the sum of service request delay, deadhead travel time and pickup time.

In modeling the manufacturing cell, we face a dilemma: "should the deadhead travel time be included in the service time of the output buffer, or in the service time of the MHD?" If we include it in the former, we will understate, perhaps significantly, the MHD service time because all output buffers can be busy at the same time. If we include it in the latter, we will overstate the output buffer service rate, hence understate the buffer utilization and the probability of work center blocking.

Because here we are most concerned with predicting the behavior of the MHD, we have elected the latter option. A partial justification is that, in a well designed system, we would want incidences of work center blocking to be rare.

The problem now is to estimate the delay experienced, not by a physical job being serviced, but by a request for service in a queue of requests at the MHD dispatcher. Not surprisingly, we propose to use a queueing network model.

4.1 Service Request Delay Estimation

A network of queues model that can be used to represent the service request processing at the MHD is illustrated in Fig. 3. Note that this model is the model in Fig. 2 with the output buffer nodes for each work center removed.

The justification for this model is that in a real system as soon as a job is placed in the output buffer of a work center it generates a service request that will join the MHD service request queue. The job is physically in the output buffer, however, the service request generated by that job will join the MHD service request queue. The service request delay is equal to w_0, the service request waiting time at the MHD service request queue. Thus, the model describes what happens to the service request, rather than what happens to the job.

Assuming that a method is available to compute steady state solutions for finite buffer queueing networks, this model will provide the necessary estimate for the delay experienced by a job in the output buffer queue while its associated service request waits in the service request queue.

5 ANALYZING THE QUEUEING NETWORK MODELS

The queueing network models proposed for estimating the service request delay and evaluating MHD utilization incorporate limited buffer capacities. Most well-known methods of analysis do not allow finite buffer capacities. However, Chin [2] has shown that, under a relatively mild simplifying assumption, tree-like queueing networks with finite buffers can be analyzed. Consider a "branch" in Fig. 2, i.e., the nodes 2i-1 and 2i, and arcs from node 0 to 2i-1, 2i-1 to 2i, 2i to 0. Suppose that saturation (i.e., the node is full) at either 2i-1 or 2i reduces the probability of job transfers into the branch.

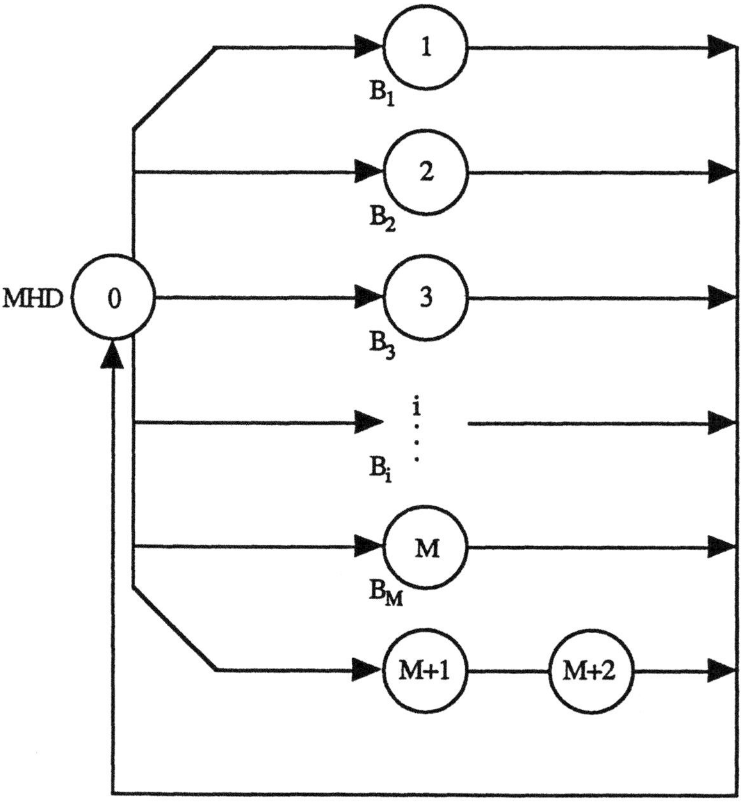

Fig. 3 Network of queues model for service requests

Chin [2] shows that this assumption allows the derivation of a product-form solution.

In describing the product-form solution for tree-like queueing networks with finite buffers the following notation will be used.

B :	normalization constant
C :	network state which is described by the 2M+2 tuple $(c_1,c_2,\ldots,c_{2M+2})$
c_i :	number of jobs at node i
D_i :	set of nodes that are downstream of node i, i.e., $D_i = \phi$ for $i \in F$
F :	set of "leaf" nodes
I_i :	collection of nodes that are <u>immediately</u> downstream of node i, i.e., $I_i = \{k{:}i \rightarrow k\}$ for $i \notin F$ and $I_i = \{0\}$ for $i \in F$
p_k :	routing probability into node k
$i \rightarrow j$:	implies that it is possible for a job to move from node i to node j in one step

For tree-like queueing networks with finite buffers the product-form equilibrium probability of having c_0 containers at node 0, c_1 containers at node 1, and so on, is given by Equation (5) which presents Theorem 4.1 from [2].

$$\pi(C) = B\,\Phi^{-1}(C) \prod_{i=0}^{2M+2} h_i(C) \qquad C \in \mathcal{L} \tag{5}$$

In Equation (5) $\Phi^{-1}(C) = \prod_{i=0}^{2M+2} \prod_{x=1}^{c_i} \phi_i^{-1}(C)$ where $\phi_i^{-1}(C) = \mu_i \min(c_i, s_i)$ is the departure rate from node i when the network is in state C and there are c_i customers at node i. The term $h_i(C)$ in Equation (5) is a nonnegative function that depends on c_k, $k \in D_i$ for $i=0,1,3,5,\ldots,2M-1,2M+1$, $h_i=1$ for all $i \in F$, and gives the impact of the status of the nodes that are downstream of node i on the departure rate from node i. In general $h_i(C)$ is given by Equation (6). In Equation (6) $c_{(k)} = c_k + \sum_{i \in D_k} c_i$.

$$h_i(C) = \prod_{k \in I_i} \prod_{x=1}^{c_{(k)}} \psi_{(k)}(x\text{-}1) \prod_{x=1}^{c_{(i)}-c_i} \psi^i(x\text{-}1) \tag{6}$$

In Equation (6) $\psi_{(k)}(x\text{-}1)$ is a function that defines the transition rate into node k and the last term

$$\prod_{x=1}^{c_{(i)}-c_i} \psi^i(x\text{-}1)$$

is a normalization factor term for load balancing routings. Under the assumption of fixed routes $\psi_{(k)}(x\text{-}1) = p_k$ and $\psi^i(x\text{-}1) = 1$. Therefore, for the network of queues model in Fig. 2 $h_i(C)$ is given by Equation (7).

$$h_i(C) = \prod_{k \in I_i} p_k^{\,c_{(k)}} \tag{7}$$

For the network model in Fig. 2, p_k is equal to one for $k=0,4,6,8,\ldots,2M$ and p_k, for $k=1,3,5,\ldots,2M-1$ are computed from production ratios and part routes. Once the p_k for $k=0,1,\ldots,2M+2$ are computed, the solution to the network model in Fig. 2 may be computed easily by a fairly straightforward application of the convolution algorithm developed by Buzen [1]. See Chin

[2] and Toro Ramos [9] for details of the supporting theory and computational techniques.

For analyzing the manufacturing cell under consideration we propose the following procedure.

Step 1. Estimate μ_0 using the approach presented in Section 4.
Step 2. Solve the QN model in Fig. 3 assuming $\mu_{M+2}=\infty$. To determine the population size (N) of the network to be used, N has to be increased up to the point where the system throughput does not increase any more even if N is increased. Obtain $\mu_{2i}=1/w_0$ for $i=1,...,M$.
Step 3. Solve the QN model in Fig. 2. Obtain the maximum cell throughput rate.

Note that for the system input node, 2M+2 in Fig. 2, assuming $\mu_k = \infty$ allows the determination of the maximum cell throughput rate. Thus, in Step 3 the maximum throughput rate is determined by increasing the population size (N) of the network up to the point where the system throughput does not increase any more even if N is increased.

The proposed approach is illustrated in the next section for the specific case of a robot tending several machines. Manufacturing cells with two and three work centers are considered.

6 EXPERIMENTAL DESIGN

In this section, the performance of the modeling strategy presented in Section 3 along with the methodologies presented in Section 4 and 5 are evaluated. This performance evaluation involves a computational study to answer the following questions:

1) How well does the methodology in Section 4 estimate the expected MHD service time, $E[\tau_0]=1/\mu_0$?
2) How well does the model for service request processing at the MHD presented in Section 4.1 estimate the service request delay, SRD?

To answer these questions the $E[\tau_0]$ estimated by the methodology in Section 4 and the service request delay estimated by the methodology in Section 4.1 are compared to results from a discrete event simulation model for a robot tending a group of work centers.

The factors investigated in the study are: (1) number of work centers served by the robot, (2) input and output buffer size, and (3) interarrival time and service time distribution. For the first factor two levels are considered, three and six work centers. Input and output buffer sizes of one, two, three, five and ten are considered. Three interarrival and service time distributions including deterministic, exponential, and hyperexponential are considered. Thus, the total number of cases considered in the computational study is thirty.

Tables 1 and 2 show the data used in the computational study for the two and three work center cases, respectively. The expected MHD service time results for the two and three work center cases are included in Tables 3 and 4, respectively.

On the average the percentage of error for the estimate of the MHD service time obtained from the proposed methodology is around -22%. This is an acceptable performance for a preliminary cell design phase. Further, from an analysis of variance of the percentage of error it can be concluded that

Table 1 Two work centers case data

Work Center	Mean Service Time	Number of Servers
1	1.00	1
2	1.00	1

Job Routing

Job	Work Center	Probability of Arrival
1	1	0.3333
2	2	0.3333
3	1,2	0.3333

Table 2 Three work centers case data

Work Center	Mean Service Time	Number of Servers
1	0.75	1
2	1.00	1
3	1.00	1

Job Routing

Job	Work Center	Probability of Arrival
1	1,3	0.25
2	2,3	0.25
3	1,2	0.25
4	1,2,3	0.25

Table 3 $E[\tau_0]$ *for two work centers case*

Deterministic

B	Q.N.	Sim.	C.I.	% Error
1	0.2348	0.2986	(0.2975, 0.2997)	-21.36
2	0.2348	0.3004	(0.2997, 0.3011)	-21.84
3	0.2348	0.3004	(0.2997, 0.3011)	-21.84
5	0.2348	0.3004	(0.2997, 0.3011)	-21.84
10	0.2348	0.3004	(0.2997, 0.3011)	-21.84

Exponential Distribution

B	Q.N.	Sim.	C.I.	% Error
1	0.2348	0.3038	(0.3024, 0.3052)	-22.71
2	0.2348	0.3006	(0.2999, 0.3013)	-21.89
3	0.2348	0.2998	(0.2988, 0.3008)	-21.68
5	0.2348	0.3004	(0.2993, 0.3015)	-21.84
10	0.2348	0.2994	(0.2987, 0.3001)	-21.58

Hyperexponential Distribution

B	Q.N.	Sim.	C.I.	% Error
1	0.2348	0.3030	(0.3018, 0.3042)	-22.51
2	0.2348	0.3010	(0.2995, 0.3025)	-21.99
3	0.2348	0.3010	(0.3001, 0.3019)	-21.99
5	0.2348	0.2994	(0.2983, 0.3005)	-21.58
10	0.2348	0.2986	(0.2979, 0.2993)	-21.37

Table 4 $E[\tau_0]$ *for three work centers case*

Deterministic

B	Q.N.	Sim.	C.I.	% Error
1	0.2250	0.2972	(0.2966, 0.2978)	-24.29
2	0.2250	0.2964	(0.2953, 0.2975)	-24.09
3	0.2250	0.2964	(0.2953, 0.2975)	-24.09
5	0.2250	0.2964	(0.2953, 0.2975)	-24.09
10	0.2250	0.2964	(0.2953, 0.2975)	-24.09

Exponential Distribution

B	Q.N.	Sim.	C.I.	% Error
1	0.2250	0.2956	(0.2949, 0.2963)	-23.88
2	0.2250	0.2932	(0.2916, 0.2948)	-23.26
3	0.2250	0.2914	(0.2907, 0.2921)	-22.79
5	0.2250	0.2900	(0.2900, 0.2900)	-22.41
10	0.2250	0.2896	(0.2889, 0.2903)	-22.31

Hyperexponential Distribution

B	Q.N.	Sim.	C.I.	% Error
1	0.2250	0.2938	(0.2922, 0.2954)	-23.42
2	0.2250	0.2932	(0.2922, 0.2942)	-23.26
3	0.2250	0.2914	(0.2891, 0.2937)	-22.79
5	0.2250	0.2896	(0.2882, 0.2910)	-22.31
10	0.2250	0.2884	(0.2877, 0.2891)	-21.98

this performance is affected by the number of work centers in the system but it is not affected neither by the input and output buffer size nor by the interarrival and service time distribution.

The underestimation of the MHD service time by the proposed methodology is due to the assumption that a full input buffer at a work center blocks the sending work center. This is, the factoring approach employed for estimating the MHD service time does not consider the possibility for a flow between two points to be reduced due to the delivery work center being full; however, the simulation model is based on such an assumption. Note that when the delivery point is a system input point the blocking of a sending work center will never occur. Thus, the factoring approach overestimates the frequency of short moves, e.g. moves between work centers, and underestimates the frequency of long moves, e.g., moves between work centers and the system output point(s).

The results for the service request delay for two and three work centers are included in Tables 5 and 6, repectively. Based on an analysis of variance of the percentage of error for the SRD estimate it can be concluded that the performance of the proposed methodology is affected by all of the factors studied including the number of work centers in the system, the input and output buffer size, and the interarrival and service time distribution.

Not surprisingly, the proposed methodology overestimates the SRD for deterministic interarrival and service times, and it underestimates the SRD for hyperexponential interarrival and service times. In general, it also underestimates the SRD for exponential interarrival and service times.

Table 5 SRD for two work centers case

Deterministic

B	Q.N.	Sim.	C.I.	% Error
1	0.3554	0.2216	(0.2120, 0.2312)	60.38
2	0.3237	0.1950	(0.1924, 0.1976)	66.00
3	0.3078	0.1950	(0.1924, 0.1976)	57.85
5	0.2938	0.1950	(0.1924, 0.1976)	50.67
10	0.2855	0.1950	(0.1924, 0.1976)	46.41

Exponential Distribution

B	Q.N.	Sim.	C.I.	% Error
1	0.3554	0.3178	(0.3028, 0.3328)	11.83
2	0.3237	0.3142	(0.2528, 0.3756)	3.02
3	0.3078	0.3022	(0.2433, 0.3611)	1.85
5	0.2938	0.3054	(0.2394, 0.3714)	- 3.80
10	0.2855	0.2920	(0.2252, 0.3588)	- 2.23

Hyperexponential Distribution

B	Q.N.	Sim.	C.I.	% Error
1	0.3554	0.3400	(0.3091, 0.3709)	4.53
2	0.3237	0.4574	(0.3645, 0.5503)	-29.23
3	0.3078	0.4998	(0.3444, 0.6552)	-38.42
5	0.2938	0.5070	(0.2655, 0.7485)	-42.05
10	0.2855	0.4782	(0.3570, 0.5994)	-40.30

Table 6 SRD for three work centers case

Deterministic

B	Q.N.	Sim.	C.I.	% Error
1	0.3866	0.2350	(0.2318, 0.2382)	64.51
2	0.3508	0.2248	(0.2212, 0.2284)	56.05
3	0.3349	0.2248	(0.2212, 0.2284)	48.98
5	0.3227	0.2248	(0.2212, 0.2284)	43.55
10	0.3177	0.2248	(0.2212, 0.2284)	41.33

Exponential Distribution

B	Q.N.	Sim.	C.I.	% Error
1	0.3866	0.4502	(0.4307, 0.4697)	-14.13
2	0.3508	0.6884	(0.5520, 0.8248)	-49.04
3	0.3349	0.6524	(0.5934, 0.7114)	-48.67
5	0.3227	0.6932	(0.4179, 0.9684)	-53.45
10	0.3177	0.7684	(0.5714, 0.9654)	-58.65

Hyperexponential Distribution

B	Q.N.	Sim.	C.I.	% Error
1	0.3866	0.4686	(0.4214, 0.5158)	-17.50
2	0.3508	0.8890	(0.7134, 1.0646)	-60.54
3	0.3349	1.1392	(0.7780, 1.5004)	-70.60
5	0.3227	1.1100	(0.7031, 1.5169)	-70.93
10	0.3177	1.0742	(0.7901, 1.3583)	-70.42

The underestimation of the SRD for exponential and hyperexponential interarrival and service times is due to two reasons: (1) the underestimation of the MHD service time and (2) the simplifying assumption made to obtain a product-form solution to the finite buffer network of queues model. With the simplifying assumption the probability of job transfers into a work center input buffer is reduced when either such an input buffer or its corresponding output buffer is full. As the work center input and output buffer size decreases this reduction increases and the service request will remain longer at the MHD request queue. Thus, this explains the behavior of the service request delay observed in Tables 5 and 6.

7 CONCLUSIONS

A methodology based on finite networks of queues has been presented for analyzing the performance of a material handling device (MHD) that provides movement between a number of distinct locations in a manufacturing cell where a full input buffer at a delivery work center blocks the sending work center.

Results from the proposed methodology have been compared to results obtained from a discrete event simulation for a robot tending several machines. Based on this comparison it can be concluded that the proposed methodology can be used for estimating both the MHD service time and the service request delay (SRD) at the MHD during a preliminary design phase. However, two things should be remembered: if the interarrival and service times are deterministic the true value of SRD will be lower than the estimate obtained from the proposed methodology and if the interarrival and service times are hyperexponential then the estimate of SRD will be a lower bound on the true value of SRD.

In obtaining a product-form solution to the finite buffer network of queues model employed by the proposed methodology we assume that when any node in a branch is saturated (i.e., node is full) the probability of job transfers into the branch is reduced. The results show that this simplifying assumption contributes to the underestimation by the proposed methodology of the service request delay (SRD) for exponential and hyperexponential interarrival and service times.

8 REFERENCES

1. Buzen, J.P, "Computational Algorithms for Closed Queueing Networks with Exponential Servers," *Communications of the Association for Computing Machinery*, 16, 9, pp. 527-531 (1973).
2. Chin, K.E., *Networks of Queues with State Dependent Flows*, Ph.D. Dissertation, School of Industrial and Systems Engineering, Georgia Institute, of Technology, Atlanta, GA (1986).
3. Egbelu, P.J., "The Use of Non-Simulation Approaches in Estimating Vehicle Requirements in an Automated Guided Vehicle Based Transport System," *Material Flow*, 4, pp. 17-32 (1987).
4. Kuhn, A., "Efficient Planning of AGVS by Analytical Methods," *Proceedings of the 2nd International Conference on Automated Guided Vehicle Systems and 16th IPA Conference* (1983).
5. Potts, R.B. and R.M. Oliver, *Flows in Transportation Networks*, Academic Press, New York and London (1976).
6. Solberg, J.J., "A Mathematical Model of a Computerized Manufacturing System," *Proceedings of the 4th International Conference on Production Research*, Tokyo (1977).

7. Suri, R., "Robutness of Queueing Networks Formulas," *Journal of the Association for Computing Machinery*, 30, 3, pp. 564-594 (1983).
8. Suri, R. and Y.C. Tay, "Error Bounds for Performance Prediction in Queueing Networks," *Association for Computing Machinery Transactions on Computer Systems*, 3, 3, pp. 227-254 (1985).
9. Toro Ramos, Zulma R., *Network of Queues and Decomposition Methods for Modeling Manufacturing Systems*, Ph.D. Dissertation, School of Industrial and Systems Engineering, Georgia Institute of Technology, Atlanta, GA (1988).
10. Toro Ramos, Zulma R. and L.F. McGinnis, "Performance Modeling for a Single Material Handling Device with Random Service Requests, Part II: Blocking with Recourse". Submitted to *IIE Transactions* (1989).
11. Yao, D.D. and J.A. Buzacott, "Modeling the Performance of Flexible Manufacturing Systems," *International Journal of Production Research*, 23, pp. 945-959 (1985).
12. Yao, D.D. and J.A. Buzacott, Models of Flexible Manufacturing Systems with Limited Local Buffers," *International Journal of Production Research*, 24, 1, pp. 107-118 (1986).

PERFORMANCE MODELING FOR A SINGLE MATERIAL HANDLING DEVICE WITH RANDOM SERVICE REQUESTS - BLOCKING WITH RECOURSE

Zulma R. Toros Ramos
University of Puerto Rico
Leon F. McGinnis
Georgia Institute of Technology

ABSTRACT

In this chapter, a manufacturing cell is studied in which a single device provides movement between a number of distinct locations and temporary storage is provided between work centers. For the cell under consideration, a bootstrapping strategy is proposed for analyzing the performance of the material handling device. The bootstrapping strategy is based on a finite network of queues model. The proposed strategy is evaluated by means of a computational study. Results from the proposed strategy are compared to results obtained from a high-fidelity discrete event simulation for a microload automated storage/retrieval (AS/R) system.

1 INTRODUCTION

A methodology to estimate the service time and service request delay for a single material handling device (MHD) serving a manufacturing cell with no temporary storage area was considered in [12]. While there are many examples of such cells, there also are many examples in which the MHD also provides temporary storage between work centers. This chapter extends the approach presented in [12] to include temporary storage.

In the case being considered a work center may incur idle time due to blocking or starving. A work center is blocked when it finishes processing a job and there is no available space in the work center output buffer to place the job. A work center is starved when it finishes processing a job and there are no new jobs in the work center input buffer waiting to be processed by the work center.

When a work center input buffer is saturated (i.e., the input buffer is full) the jobs that were supposed to go to this work center may be diverted to a temporary storage area. This introduces an additional complication in analyzing the performance of the MHD since both the routing probabilities and the service times at the MHD and the work center output buffers are affected by the frequency of saturation and the resulting MHD transfers to and from the temporary storage area. We cannot use the finite buffer queueing network (FBQN) model in [12] to obtain the routing probabilities because our analytic procedures require routing probabilities as input. We propose to "bootstrap" estimates of the routing probabilities using two additional queueing network models.

The bootstrapping strategy is illustrated using the specific case of the storage/retrieval (S/R) machine in an automated storage/retrieval (AS/R) system. A formal statement of the problem is presented in Section 2. Section 3 describes the modifications to the FBQN model necessary to accommodate the temporary storage area. An estimation of the temporary storage area service rate is presented in Section 4. Section 5 presents the bootstrapping strategy for estimating the routing probabilities. In Section 6, the experimental design used in the computational study is described and the numerical results obtained using the proposed approach are compared with the results obtained from a high-fidelity discrete event simulation model. Conclusions about the work presented are included in Section 7.

2 GENERAL PROBLEM

The material flow within the manufacturing cell under consideration is illustrated Fig. 1. In this figure there is a system input point and a system output point. Each work center has an input buffer and output buffer, both with limited capacity. Additionally, there is a temporary storage area which holds jobs that were supposed to be delivered to a work center whose input buffer was saturated. Once a space become available at such a work center input buffer a request is generated for the MHD to remove the job from temporary storage and deliver it to the work center input buffer.

Our objective is to answer the question: what is the maximum cell throughput rate considering both work center and MHD capacities? To answer this question, it is not enough to know: (1) cell physical configuration including locations of work center buffers, system input and output points, and temporary storage area, (2) production ratios, (3) process times, (4) part routes, and (5) buffer capacities. In addition, we must know the frequency of saturation at the work center input buffers and the resulting MHD operations to the temporary storage area. The bootstrapping strategy in Section 5 addresses this issue.

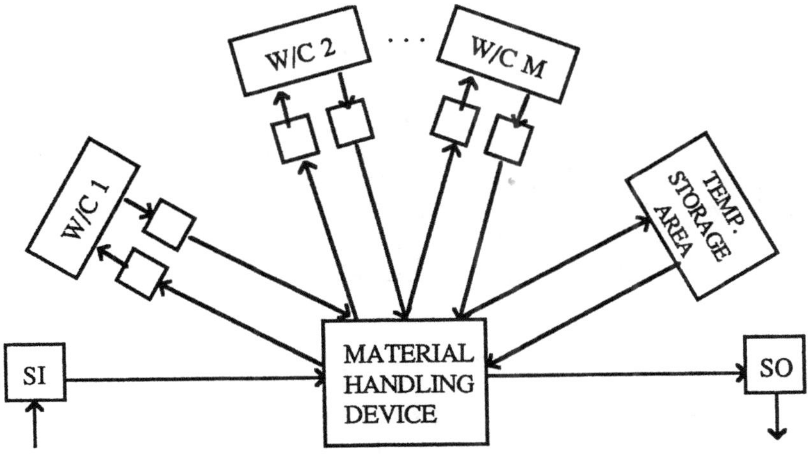

Fig. 1 Material flow schematic

3 MODIFICATIONS TO FBQN MODEL

In the manufacturing cell under consideration, as a job progresses through its process route, it receives five types of "services". The four types described in [12] and the fifth type which is wait in temporary storage.

The physical configuration of the temporary storage may take one of several forms. In an FMS using automated guided vehicles (AGVs), for example, it may be an automated carousel-type device for storing fixtured parts on pallets [4]. Alternatively, it may take the form of sets of load stands strategically located throughout the system. In a cell based on an automated storage/retrieval system (AS/RS) the rack locations nearest to a work center may be dedicated (or "zoned") to that work center. Fig. 2 illustrates such a configuration.

A key modeling question, then, revolves around the form of the temporary storage. The subsequent presentation addresses the case of distributed storage areas, where there are storage locations assigned to each work center. The case of centralized temporary storage could be analyzed in a similar way.

It is important to note that the case of distributed temporary storage cannot be accurately analyzed simply by adding a work center's temporary storage locations to its input buffer. The reason is that every storage to the temporary storage area generates a material handling cycle in addition to those implied by the production rates and process routes. The cummulative effect of these additional cycles may be to create a bottleneck at the MHD. This is precisely the effect that we wish to predict.

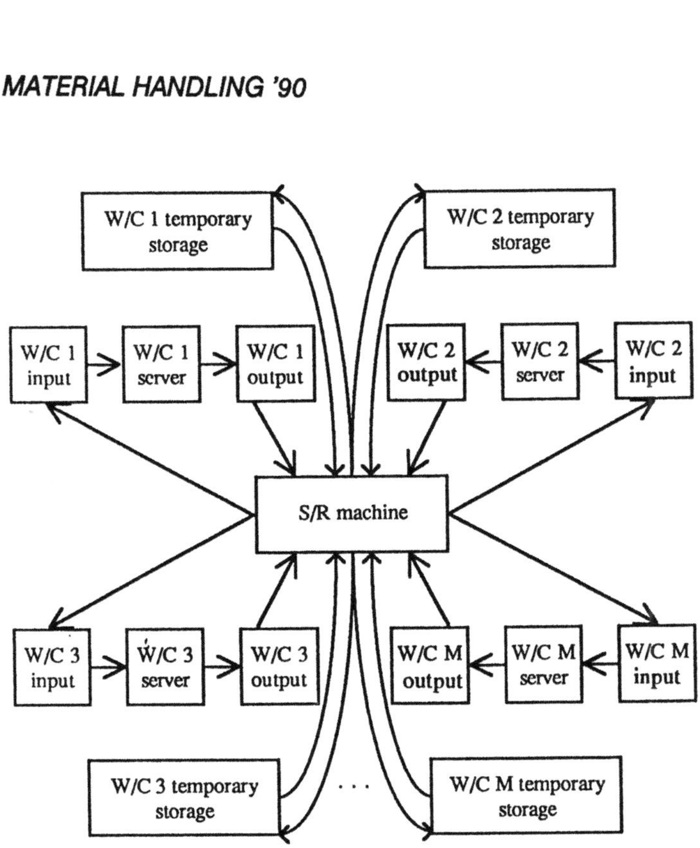

Fig. 2 AS/RS based manufacturing cell

Fig. 3 shows a schematic of the finite buffer queueing network (FBQN) model for a manufacturing cell with M work centers. In Fig. 3 there are six different types of nodes: work center input buffer node, work center output buffer node, MHD node (central server), system output point node, system input point node, and temporary storage area node.

Each work center in the queueing network model has an associated buffer capacity, B_k, number of servers, s_k, and service rate, μ_k. If the node corresponds to a work center input buffer ($k=3i-2$, $i=1,...,M$), then B_k is a design parameter, s_k is the number of servers at the work center, and μ_k is the service rate at the work center. If the node corresponds to a work center output buffer ($k=3i-1$, $i=1,...,M$), then B_k is specified as a design parameter, $s_k=1$, and μ_k depends upon the operation of the MHD. If the node corresponds to a work center temporary storage area ($k=3i$, $i=1,...,M$), B_k is the number of locations in the storage area, $s_k=1$, and μ_k depends on both the service rate at the associated work center and the operation of the MHD.

For the system input point, node $k=3M+2$ in the queueing network, $B_k=\infty$, $s_k=1$, and $\mu_k=\infty$. For the system output point, node $k=3M+1$ in the queueing network, $B_k=0$, $s_k=1$, and $\mu_k=\infty$. Lastly, for the MHD, node $k=0$ in the queueing network, $B_k=\infty$, $s_k=1$, and μ_k depends on the operations of the MHD.

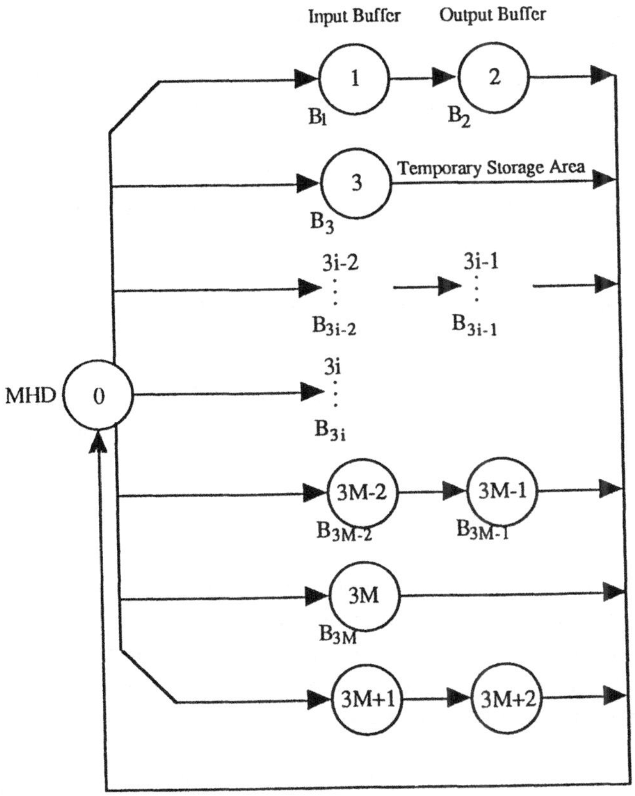

Fig. 3 Network of queues model for the manufacturing cell

We assume that all queue disciplines are first-come-first-served, and that, at a node, all servers have identical, independently distributed exponential service times. The first-come-first-served queue discipline assumption for a temporary storage zone results in newly arriving jobs going into temporary storage if there are already jobs at such a temporary zone. We also assume that the work center service rates are known, so the service rates for the work center input buffer nodes in the queueing network are known. Because we assume blocking with recourse, the routing probabilities need to be estimated based on the frequency of saturation at the work center input buffers and the resulting MHD operations to the temporary storage area. In addition to the parameters estimated in [12], for this case we need to estimate the mean service rate at the temporary storage area nodes.

4 TEMPORARY STORAGE AREA SERVICE RATE ESTIMATION

When a job reaches the front of the temporary storage area queue, it has to wait until a job is processed at the associated work center so a space become available at the work center input buffer and a service request is generated for the MHD to remove the job in the first position of the temporary storage area queue. Then the job has to wait until the service request generated by the work center input buffer reaches the front of the service request queue (SRQ) for the MHD. When this happens the MHD is dispatched to the temporary storage area location, incurring some deadhead travel time. Hence, the total "service time" seen by a job in the temporary storage area is the sum of service time at the associated work center, service request delay, deadhead travel time and pickup time.

Here we face the same dilemma as in [12] with the output buffer service time. "Should the deadhead travel time be included in the service time of the temporary storage area, or in the service time of the MHD?" In this case, we have elected to include it in the service time of the MHD, for two reasons. Firstly, here we are most concerned with predicting the behavior of the MHD and secondly for this type of cell the utilization of the temporary storage area is not a critical issue.

5 BOOTSTRAPPING STRATEGY

The key difficulty in modeling a manufacturing cell where blocking with recourse occurs is to determine how often jobs are rerouted to the temporary storage area. Two events cause rerouting of jobs to the temporary storage area. First, if a destination work center input buffer is saturated then the incoming job is rerouted to the temporary storage area. Secondly, if there is space available in the input buffer but an older job is waiting at the temporary storage area, due to our assumption of first-come-first-served, the incoming job needs to go into the storage area and the older job delivered to the work center input buffer. This last event will be referred to as temporary storage area being busy.

Now, we know that in modeling blocking with recourse we need to estimate the saturation probabilities at the input buffer nodes and the probability for a temporary storage area node being busy. With this information the resulting MHD operations to the temporary storage area can be estimated.

The bootstrapping strategy employs an iterative refinement of the required parameter estimates. In describing the strategy, the following notation will be used.

α^t_k : routing probability adjustment factor for work center k at iteration t $\quad k=1,\dots,M$

β^t_k : probability for node k being busy at iteration t $\quad k=3i, i=1,\dots,M$

p^t_{ij}: routing probability from node i to node j at iteration t $\quad i=0,1,2,\dots,3M+2$ $\quad j=0,1,2,\dots,3M+2$

sp^t_k: saturation probability for node k at iteration t
k=3i-2, i=1,...,M

w^t_k: waiting time at the queue of node k at iteration t
k=0,1,2,...,3M+2

Fig. 4 illustrates an outline of the bootstrapping strategy. In this strategy we use two network of queues models. The first one, models the job processing within the manufacturing cell. The output from this model includes sp^t_k and β^t_k. The second model represents the service request processing at the MHD and its output is w^t_k. The service request delay at the MHD queue required in the estimation of the service rate at the output buffer nodes and the temporary storage area nodes is equal to w^t_0. The bootstrapping strategy iterates between the job processing model and the service request processing model until the difference between two successive values of μ_0 is very small.

The first order approximation of the routing probabilities used in the bootstrapping strategy is obtained from process routes and production ratios assuming that the probability of going into a temporary storage area node is zero. The first order approximation for the service rate at the MHD, node 0, is obtained using the procedure presented in [12]. The first order approximation for the service rate of an output buffer node is μ_k=G (k=3i-1, i=1,...,M), where G is a big number greater than the system arrival rate, and of a temporary storage area node $\mu_k=\mu_i$ (k=3i, i=1,...,M). The service rate of an input buffer node is $\mu_k=\mu_i$ (k=3i-2, i=1,...,M) and of the system output and input point node is μ_k=G (k=3M+1,3M+2, i=1,...,M). With this input then the FBQN model for job processing in Fig. 3 is solved obtaining new estimates for sp_k (k=3i-2, i=1,...,M) and β_k (k=3i, i=1,...,M). Using these new estimates the routing probabilities are updated as follows:

$$p^t_{0j} = \begin{cases} p^0_{0j} \left[\dfrac{1}{1+\alpha^t_k} \right] & \text{for } j=3k-2 \\[2ex] p^0_{0j} \left[\dfrac{\alpha^t_k}{1+\alpha^t_k} \right] & \text{for } j=3k \end{cases} \qquad (1)$$

where

$$\alpha^t_k = sp^t_j + \beta^t_{j+2} - sp^t_j \beta^t_{j+2} \qquad j=3k-2,\ k=1,...,M \qquad (2)$$

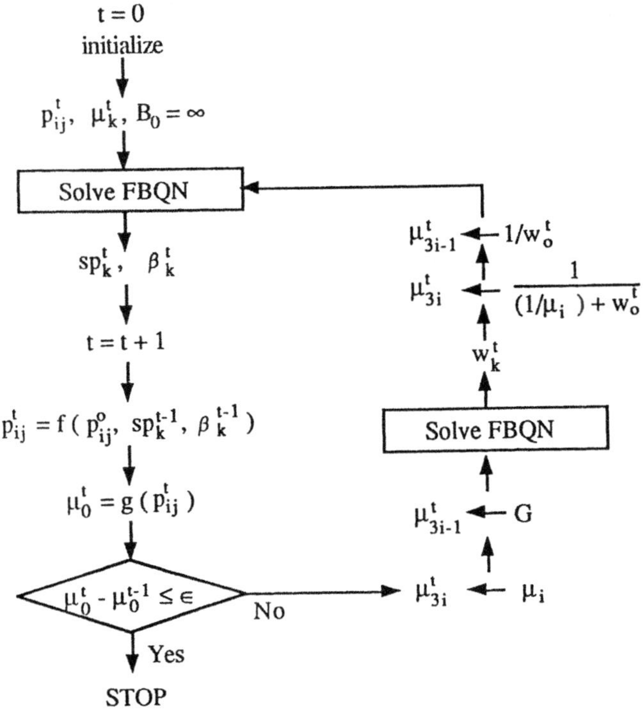

Fig. 4 Outline of bootstrapping method

Note that α^t_k in Equation (2) represents the probability for a job that has center k in its route being rerouted to the temporary storage area associated with work center k. In Equation (3) we are assuming that the event of saturation at input buffer node j is independent of the event of temporary storage node j+2 being busy, thus, α^t_k clearly is an approximation.

With the new estimate of the routing probabilities the service rate at the MHD, μ_0, can be updated. This estimation involves the same procedure as for pure blocking. The only transformation required for the estimation of the loaded travel time is given by Equation (3).

$$f^t_{ij} = Fp^t_{ij} \tag{3}$$

In [12], we described a method for estimating the deadhead travel. Since we now have estimates for f^t_{ij}, the same method can be used in the present case. If the stopping criteria is not met at this point then the model for request processing at the MHD is solved to obtain new estimates of w^t_k.

Although it has not been proved analytically, in our empirical tests this procedure has always converged and taken a small number of iterations to do so. In fact, in all cases studied the procedure has converged in at most four iterations.

6 EXPERIMENTAL DESIGN

This section presents the performance evaluation of the proposed bootstrapping strategy. In this evaluation the following questions will be addressed.

1) How well does the bootstrapping strategy estimate the expected MHD service time, $E[\tau_0]=1/\mu_0$, when the interarrival time and the service time at the work centers are exponentially distributed and other than exponentially distributed?
2) How well does the bootstrapping strategy estimate the service request delay (SRD)?

To answer these questions the results obtained from the bootstrapping strategy are compared to results from a high-fidelity discrete event simulation model for the microload AS/R system developed in [7].

In the study, three factors are investigated: (1) number of work centers in the system, (2) input and output buffer size, and (3) interarrival time and service time distribution. Systems with three, six, and nine work centers are investigated. Input and output buffer sizes of one, two, three, five, and ten are considered. The interarrival and service time distributions investigated are: deterministic, exponential, and hyperexponential. Thus, forty-five different cases are considered in the computational study.

The data used in the computational study for the three, six, and nine work center cases is illustrated in Tables 1, 2, and 3, respectively. Tables 4, 5, and 6 show the expected MHD service time results for the three, six, and nine work center cases, respectively.

Table 1 Three work centers case data

Work Center	Mean Service Time	Number of Servers
1	2.00	1
2	1.00	1
3	2.00	1

Job Routing

Job	Work Center	Probability of Arrival
1	1,3	0.25
2	2,3	0.25
3	1,2	0.25
4	1,2,3	0.25

Table 2 Six work centers case data

Work Center	Mean Service Time	Number of Servers
1	4.00	1
2	2.00	1
3	4.00	1
4	6.00	1
5	4.00	1
6	6.00	1

Job Routing

Job	Work Center	Probability of Arrival
1	1,4	0.25
2	2,3,5,6	0.25
3	2,3	0.25
4	1,2,5	0.25

Table 3 Nine work centers case data

Work Center	Mean Service Time	Number of Servers
1	4.00	1
2	2.00	1
3	4.00	1
4	6.00	1
5	4.00	1
6	4.00	1
7	6.00	1
8	6.00	1
9	4.00	1

Job Routing

Job	Work Center	Probability of Arrival
1	1,4,7,9	0.25
2	2,3,5,6	0.25
3	2,3,9	0.25
4	1,2,5,6,8	0.25

Table 4 $E[\tau_0]$ for three work centers case

Deterministic

B	B.S.	Sim.	C.I.	% Error
1	0.4476	0.4477	(0.4472, 0.4482)	-0.02
2	0.4466	0.4478	(0.4473, 0.4483)	-0.27
3	0.4458	0.4478	(0.4473, 0.4483)	-0.45
5	0.4447	0.4478	(0.4473, 0.4482)	-0.69
10	0.4445	0.4477	(0.4472, 0.4483)	-0.71

Table 4 (continued)

Exponential Distribution

B	B.S.	Sim.	C.I.	% Error
1	0.4476	0.4695	(0.4644, 0.4746)	-4.66
2	0.4466	0.4496	(0.4475, 0.4557)	-0.67
3	0.4458	0.4473	(0.4464, 0.4483)	-0.34
5	0.4447	0.4452	(0.4441, 0.4463)	-0.11
10	0.4445	0.4444	(0.4427, 0.4460)	0.02

Hyperexponential Distribution

B	B.S.	Sim.	C.I.	% Error
1	0.4476	0.5019	(0.4766, 0.5271)	-10.82
2	0.4466	0.4645	(0.4572, 0.4717)	- 3.85
3	0.4458	0.4581	(0.4477, 0.4685)	- 2.69
5	0.4447	0.4468	(0.4440, 0.4496)	- 0.47
10	0.4445	0.4428	(0.4416, 0.4440)	0.38

Table 5 $E[\tau_0]$ for six work centers case

Deterministic

B	B.S.	Sim.	C.I.	% Error
1	0.5327	0.5291	(0.5276, 0.5305)	0.68
2	0.5306	0.5325	(0.5310, 0.5340)	-0.36
3	0.5286	0.5326	(0.5307, 0.5345)	-0.75
5	0.5258	0.5324	(0.5310, 0.5339)	-1.24
10	0.5236	0.5325	(0.5309, 0.5341)	-1.67

Exponential Distribution

B	B.S.	Sim.	C.I.	% Error
1	0.5327	0.5344	(0.5249, 0.5439)	-0.32
2	0.5306	0.5228	(0.5186, 0.5271)	1.49
3	0.5286	0.5210	(0.5182, 0.5237)	1.46
5	0.5258	0.5217	(0.5197, 0.5237)	0.79
10	0.5236	0.5202	(0.5184, 0.5219)	0.65

Hyperexponential Distribution

B	B.S.	Sim.	C.I.	% Error
1	0.5327	0.5578	(0.5389, 0.5767)	-4.50
2	0.5306	0.5501	(0.5268, 0.5734)	-3.54
3	0.5286	0.5278	(0.5070, 0.5486)	0.15
5	0.5258	0.5214	(0.5129, 0.5300)	0.84
10	0.5236	0.5182	(0.5168, 0.5196)	1.04

Table 6 $E[\tau_0]$ for nine work centers case

B	B.S.	Sim.	C.I.	% Error
Deterministic				
1	0.5396	0.5328	(0.5317, 0.5339)	1.28
2	0.5385	0.5329	(0.5321, 0.5336)	1.05
3	0.5374	0.5335	(0.5321, 0.5350)	0.73
5	0.5288	0.5322	(0.5314, 0.5330)	-0.64
10	0.5288	0.5333	(0.5318, 0.5347)	-0.84
Exponential Distribution				
B	B.S.	Sim.	C.I.	% Error
1	0.5396	0.5374	(0.5347, 0.5401)	0.41
2	0.5385	0.5334	(0.5303, 0.5364)	0.96
3	0.5374	0.5325	(0.5305, 0.5345)	0.92
5	0.5288	0.5335	(0.5323, 0.5346)	-0.88
10	0.5288	0.5321	(0.5307, 0.5334)	-0.62
Hyperexponential Distribution				
B	B.S.	Sim.	C.I.	% Error
1	0.5396	0.5432	(0.5308, 0.5556)	-0.66
2	0.5385	0.5307	(0.5283, 0.5332)	1.47
3	0.5374	0.5332	(0.5165, 0.5398)	0.79
5	0.5288	0.5290	(0.5267, 0.5313)	-0.04
10	0.5288	0.5290	(0.5274, 0.5306)	-0.04

The percentage of error for the bootstrapping strategy estimate of the MHD service time is less than 11% for all the cases studied. Thus, we can conclude that the bootstrapping strategy performs very good in estimating the MHD service time regardless of the number of work centers in the system, the interarrival and service time distribution, and the input and output buffer size.

Although, in general the bootstrapping strategy performs very good, it underestimates the MHD service time in some cases, specifically when the input and output buffer size is small. Therefore, we can conclude that when the bootstrapping strategy is used for estimating the MHD service time its performance is affected by the input and output buffer size. This is, the performance of the bootstrapping strategy improves as the input and output buffer size increases. This result suggests that the bootstrapping strategy underestimates the frequency with which the jobs are rerouted to the temporary storage area, α_k.

Tables 7, 8, and 9 show the service request delay results for the three, six, and nine work center cases, respectively. From an analysis of variance of the percentage of error for the SRD we can conclude that the performance of the bootstrapping strategy is affected by the interarrival and service time distribution. However, this performance is not affected neither by the number of work centers in the system nor by the input and output buffer size.

As it was expected, the bootstrapping strategy overestimates the SRD when the interarrival and service time distributions are deterministic and it underestimates the SRD when these distributions are hyperexponential. The SRD is also underestimated by the bootstrapping strategy when the interarrival and service times are exponential.

When the bootstrapping strategy is used for estimating the SRD its performance improves as the input and output buffer size increases. This is in part due to the underestimation of the expected MHD service time for small input and output buffer sizes and it could be partially due to an underestimation of the frequency with which service requests are generated at both the output buffers and the temporary storage zones.

Table 7 SRD for three work centers case

Deterministic

B	B.S.	Sim.	C.I.	% Error
1	0.5602	0.2244	(0.2216, 0.2272)	149.64
2	0.5274	0.2241	(0.2211, 0.2271)	135.34
3	0.5117	0.2240	(0.2210, 0.2269)	128.44
5	0.4976	0.2239	(0.2208, 0.2270)	122.24
10	0.4925	0.2240	(0.2208, 0.2272)	119.87

Exponential Distribution

B	B.S.	Sim.	C.I.	% Error
1	0.5602	1.2477	(1.1411, 1.3542)	-55.10
2	0.5274	1.0060	(0.8933, 1.1187)	-47.57
3	0.5117	1.9165	(0.7941, 1.0388)	-44.17
5	0.4976	0.7270	(0.6613, 0.7927)	-31.55
10	0.4925	0.6847	(0.5788, 0.7906)	-28.07

Hyperexponential Distribution

B	B.S.	Sim.	C.I.	% Error
1	0.5602	2.1798	(1.6439, 2.7157)	-74.30
2	0.5274	2.0450	(1.5717, 2.5183)	-74.21
3	0.5117	2.2175	(1.4048, 3.0301)	-83.41
5	0.4976	1.7936	(1.2775, 2.3096)	-72.26
10	0.4925	1.2967	(1.1262, 1.4672)	-62.02

Table 8 SRD for six work centers case

Deterministic

B	B.S.	Sim.	C.I.	% Error
1	0.9567	0.3575	(0.3374, 0.3777)	167.61
2	0.8925	0.3374	(0.3322, 0.3426)	164.52
3	0.8566	0.3387	(0.3280, 0.3493)	152.91
5	0.8211	0.3355	(0.3227, 0.3482)	144.74
10	0.7994	0.3383	(0.3282, 0.3485)	136.30

Table 8 (continued)

Exponential Distribution

B	B.S.	Sim.	C.I.	% Error
1	0.9567	2.2316	(1.8877, 2.5754)	-57.13
2	0.8925	1.7046	(1.2506, 2.1585)	-47.64
3	0.8566	1.4160	(1.1310, 1.7010)	-39.51
5	0.8211	1.2154	(0.9954, 1.4354)	-32.44
10	0.7994	1.0046	(0.8122, 1.1969)	-20.43

Hyperexponential Distribution

B	B.S.	Sim.	C.I.	% Error
1	0.9567	3.7067	(2.7069, 4.7066)	-74.19
2	0.8925	4.9102	(2.7118, 7.1085)	-81.82
3	0.8566	3.2970	(1.0648. 5.5292)	-74.02
5	0.8211	3.2711	(1.3205, 5.2217)	-74.90
10	0.7994	1.7137	(1.2162, 2.2111)	-53.35

Table 9 SRD for nine work centers case

Deterministic

B	B.S.	Sim.	C.I.	% Error
1	0.8988	0.3581	(0.3434, 0.3727)	150.99
2	0.8386	0.3616	(0.3415, 0.3816)	131.91
3	0.8117	0.3575	(0.3423, 0.3728)	127.05
5	0.7940	0.3590	(0.3532, 0.3649)	121.17
10	0.7909	0.3557	(0.3399, 0.3715)	122.35

Exponential Distribution

B	B.S.	Sim.	C.I.	% Error
1	0.8988	1.8463	(1.4879, 2.2046)	-51.32
2	0.8386	1.5792	(1.3552, 1.8031)	-46.90
3	0.8117	1.3515	(1.1918, 1.5111)	-39.94
5	0.7940	1.1328	(0.9598, 1.3057)	-29.91
10	0.7909	1.0846	(0.8852, 1.2839)	-27.08

Hyperexponential Distribution

B	B.S.	Sim.	C.I.	% Error
1	0.8988	2.7729	(1.5463, 3.9994)	-67.59
2	0.8386	2.4396	(1.9497, 2.9294)	-65.63
3	0.8117	2.7565	(1.4570. 4.0559)	-70.55
5	0.7940	2.2215	(1.2821, 3.1608)	-64.26
10	0.7909	2.0714	(1.2368, 2.9059)	-61.82

7 CONCLUSIONS

A bootstrapping strategy has been presented for analyzing the performance of a material handling device that provides both movement and temporary storage between a number of distinct locations in a manufacturing cell. The bootstrapping strategy is based on a finite network of queues model.

Results from the bootstrapping strategy are compared to results obtained from a high-fidelity discrete event simulation for the microload automated storage/retrieval (AS/R) system. From the comparison we conclude that the bootstrapping strategy can be used for estimating the MHD service time regardless of the number of work centers in the system, the interarrival and service time distribution, and the input and output buffer size. On the other hand, in using the bootstrapping strategy for estimating the service request delay it should be kept in mind that this delay is overestimated by the strategy when the interarrival and service times are deterministic while it is underestimated when the interarrival and service times are hyperexponential. The degree of overestimation for deterministic interarrival and service times varies between 115% and 170% whereas the degree of underestimation for hyperexponential interarrival and service times varies between 50% and 85%. The bootstrapping strategy service request delay estimate when the interarrival and service times are exponential is acceptable for a preliminary cell design phase.

In the bootstrapping strategy for estimating the service request delay we assume that the MHD service time is independent of request class membership. This is, the MHD service time is the same for all of the requests regardless which point in the system is generating the request. From this assumption, a common service request delay results and therefore, the estimate for the service rate at all of the work center output buffers results in the same value which is not necessarily realistic. Therefore, to improve the performance of the bootstrapping strategy we suggest the definition of request classes based on the points that generates the requests (i.e., system input point(s), work center output buffers, work center temporary storage areas). The definition of request classes allows the estimation of the service request delay per class and this estimate per class could be used to obtain more precise estimates for the service rate at the work center output buffers.

In this chapter we have discussed the analysis of a system, that is, we assume as given the cell physical configuration including locations of work center buffers, system input and output points, and temporary storage area. However, in the design of the type of manufacturing cell under consideration all of these parameters may be independent and need to be determined. Therefore, the development of procedures and methodologies to design such cells is open to further research.

8 REFERENCES

1. Buzen, J.P.,"Computational Algorithms for Closed Queueing Networks with Exponential Servers," *Communications of the Association for Computing Machinery*, 16, 9, pp. 527-531 (1973).
2. Chin, K.E., *Networks of Queues with State Dependent Flows*, Ph.D. Dissertation. School of Industrial and Systems Engineering, Georgia Institute of Technology, Atlanta, GA (1986).
3. Egbelu, P.J., "The Use of Non-Simulation Approaches in Estimating Vehicle Requirements in an Automated Guided Vehicle Based Transport System," *Material Flow*, 4, pp. 17-32 (1987).
4. Graver, T.W. and L.F. McGinnis, "A Tool Provisioning Problem in an FMS," *International Journal of Fexible Manufacturing Systems*, To Appear (1989).
5. Kuhn, A., "Efficient Planning of AGVS by Analytical Methods," *Proceedings of the 2nd International Conference on Automated Guided Vehicle Systems and 16th IPA Conference* (1983).
6. Potts, R.B. and R.M. Oliver, *Flows in Transportation Networks*, Academic Press, New York and London (1976).
7. Quiroz, M.A., *Simulation of a Microload Automated Storage/Retrieval System*, Master Thesis, School of Industrial and Systems Engineering, Georgia Institute of Technology, Atlanta, GA (1986).
8. Solberg, J.J., "A Mathematical Model of a Computerized Manufacturing System", *Proceedings of the 4th International Conference on Production Research*, Tokyo (1977).
9. Suri, R., "Robutness of Queueing Networks Formulas," *Journal of the Association for Computing Machinery*, 30, 3, pp. 564-594 (1983).
10. Suri, R. and Y.C. Tay, "Error Bounds for Performance Prediction in Queueing Networks," *Association for Computing Machinery Transactions on Computer Systems*", 3, 3, pp. 227-254 (1985).
11. Toro Ramos, Zulma R., *Network of Queues and Decomposition Methods for Modeling Manufacturing Systems*, Ph.D. Dissertation, School of Industrial and Systems Engineering, Georgia Institute of Technology, Atlanta, GA (1988).
12. Toro Ramos, Zulma R. and L.F. McGinnis, "Performance Modeling for a Single Material Handling Device with Random Service Requests, Part I: Pure Blocking," Submitted to *IIE Transactions* (1989).
13. Yao, D.D. and J.A. Buzacott, "Modeling the Performance of Flexible Manufacturing Systems," *International Journal of Production Research*, 23, pp. 945-959 (1985).
14. Yao, D.D. and J.A. Buzacott, "Models of Flexible Manufacturing Systems with Limited Local Buffers," *International Journal of Production Research*, 24, 1, pp. 107-118 (1986).

SECTION FIVE
FLOW CONTROL AND AGVS

The flow of material through a manufacturing system represents one of the primary areas of interest in material handling. Providing the right amount of the right material at the right place at the right time, and in the proper condition, sequence, and orientation has been the classical domain of material handling. Recently the edict "...and nowhere else" has been added. The Just-In-Time philosophy developed from the need to control inventory. One of the simplest approaches to the control of material flow is the Kanban concept. Deceptively simple in concept, it requires a sophisticated analysis to ensure that the number of Kanbans in a system are sufficient, without creating unnecessary inventory. Askin and his colleagues provide guidance for multi-item systems.

Whereas most systems are laid out with some fixed pattern of production in mind, it is increasingly common to find system designers more concerned with flexibility of operation than with optimal performance for today's products. Manufacturing cells have become popular, wherein a number of different resources are grouped together, a mini production line that can be quickly and easily re-configured to produce a different product or component. This has been extended to the concept of a virtual cell, one that does not depend upon physical proximity, but can be created by software controlling the material handling and information systems. Thus, any collection of machines can be temporarily identified as a "cell" and function together to produce an item. The concept, and an approach to the control of such virtual cells, is discussed in the chapter by Drolet and Moodie.

Achieving flexibility requires material flow control systems that are capable of integrating flexible routing with changes in product mix and equipment availability. Taylor and Graves discuss the architecture of such systems using a printed circuit board assembly as an example.

The throughput of storage equipment used in conjunction with work-in-process or distribution systems is an important factor in computing material flow. Previously the throughput had to be estimated by simulation or the use of heuristics. Recent research by Foley and Frazelle have provided closed form analytical expressions for several of the more popular storage systems.

Automatically guided vehicles (AGVs) have proven a useful and flexible means of moving material in many environments. However, the failure of several large systems have shown that care must be exercised in the design and operation, or the system will not only not meet performance objectives but will disrupt the production process. Research is continuing to refine the design process and contribute to control development.

Evans, et. al. and Kim and Tanchoco discuss different aspects of planning the AGV System (AGVS). The first breaks the total design problem into a series of smaller, more tractable problems, and iterates on their solutions until convergence is reached. The resulting design is then ready for simulation and verification. The second addresses some of the issues involved when the path is not fixed to the floor, but is computer generated and can be changed at will. This added flexibility comes at the cost of a more complex control system.

Bohlander et. al. and Malmborg discuss the control of the vehicle in an AGVS. The first reviews the possibility of using the same workstation used to design the system to program the vehicle controller, thus providing the user with a quick and easy way to re-program to accommodate changes. The second describes a PC-based design aid which allows the use of the zone control model to help design systems before subjecting them to simulation and verification.

Ira W. Pence, Jr.

DETERMINING THE NUMBER OF KANBAN IN MULTI-ITEM JUST-IN-TIME SYSTEMS

Ronald G. Askin
M. George Mitwasi
Jeffrey B. Goldberg
The University of Arizona

ABSTRACT

Kanbans (cards) are one means of controlling just-in-time production systems. The Kanbans serve as production authorization orders and essentially become the information system. The presence of a production ordering Kanban for a specific part type constitutes an instruction to the workcenter to produce a "container" of that part. In this chapter we assume workcenters produce multiple part types and address the issue of the number of Kanbans needed for each part type. Our objective is to minimize the sum of inventory holding and backorder cost. A stochastic model is formulated. Steady-state results are derived for several operating environments and results and validated by simulation.

1 INTRODUCTION

Just-in-time (JIT) production systems have received much attention in recent years due to their role in improving the competitiveness of Japanese industry. Using cross-trained workers and simple information systems, JIT systems have proven effective at meeting production goals in environments with high process reliability, low setup times and low demand variability. Monden [4] describes the use of Kanbans (cards) as the within plant planning and scheduling systems for just-in-time manufacturing.

Several types of Kanban systems exist. To illustrate both the JIT concept and the use of Kanbans, we will describe a basic, single Kanban system. We envision a general assembly system. Raw materials are processed into parts, parts are processed into subassemblies and subassemblies into finished products. Any arbitrary number of levels, components and workcenters are permissible. The key is that raw materials and purchased parts "flow" through the system, visiting a sequence of workcenters on their way to becoming finished products. Completed parts at each workcenter are stored between that workcenter and the successor workcenter. When parts are needed at a workcenter, a worker goes to the previous inventory storage point and retrieves a container of parts. Each container houses a Kanban as well as the parts. When withdrawing the container of parts, the worker removes the Kanban and places it on a scheduling board to indicate the need to initiate a new order for this part type at its workcenter. When workers/machines become available at this previous workcenter, the Kanbans on the board are ordered to minimize setup time and production commences. For each item scheduled, production is the container size times the number of Kanbans. Upon completion, the containers with the parts and Kanban are moved to the inventory holding location. Kanbans thus serve as authorization for part orders. Fig. 1 depicts the system.

The issue addressed in this chapter is the number of Kanbans which should be kept in circulation for each part type. We envision a sequential production system where workcenters manufacture multiple parts. Each part type may be used in multiple final products. Since each full container of parts must contain a Kanban, the number of Kanbans multiplied by the container size (part capacity) place an upper bound on inventory. Average inventory will be the excess of the product of container size and the number of Kanbans over lead time demand at each workcenter.

The objective of a Kanban system is to respond to demand JIT. Eliminating unnecessary inventory and safety stock allows immediate detection of quality problems, reduction of inventory costs, and rapid response to minor changes in demand and product design. Inventory obsolescence is minimized. The system achieves these advantages with a very simple and inexpensive shop floor control system.

Kanban systems share some of the characteristics of base stock systems but are not equivalent. In a base stock system, a new item is ordered each time an item is demanded. Optimal results are available for certain simple demand and cost structures (Johnson and Montgomery [2]). In a Kanban system, items can be reordered only when parts are withdrawn. Outstanding orders cannot exceed the number of Kanbans.

Kimura and Terada [3] describe the operation of Kanban systems and examine the accompanying inventory fluctuations in a JIT environment. Rees et al. [6] studied empirical approaches for setting Kanban levels. Historical data was used to estimate lead times and enough Kanbans were added to cover demand in the

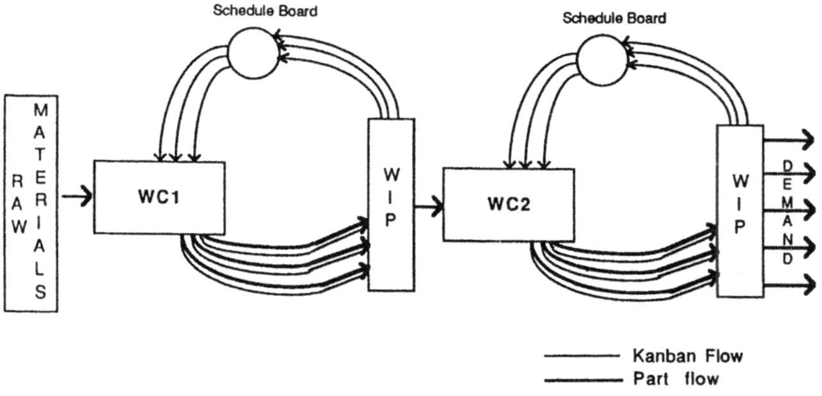

Fig. 1 Schematic of two-stage, single Kanban system

event of a long leadtime. Bitran and Chang [1] present a mathematical formulation of the Kanban determination problem. The formulation assumes planning periods of known length and finds the minimum feasible number of Kanbans. Each container can be used at most once a period and only one part type is allowed per workcenter. A heuristic is provided for selecting the number of Kanbans. Deleersnyder et al. [5] places the Kanban determination problem into the context of the overall push systems. A discrete time Markov process model is developed for an N-stage serial system subject to machine failures and demand uncertainty. The model assumes a single part type at each stage.

Much of the work to date has assumed assembly structures with a single part type per workcenter. In discussing such systems, Monden [4] noted that capacity is set by the slowest workcenter and noted that the ideal operating policy may be to simply produce a container whenever a container is taken, i.e., operate as in an assembly line without the necessity of Kanbans. In the current chapter we are interested in directly modeling the case where multiple part types are made at each workcenter and parts may be used in several workcenters.

When Kanbans are necessary, Monden describes the use of the model where k_i

$$k_i = \frac{(\tau_i D_i)(1+\alpha)}{n_i} \tag{1}$$

is the number of Kanbans for part type i, n_i is the container size, τ_1 is the lead time consisting of processing time, conveyance time, waiting time and Kanban collecting time, and D_i is the average demand rate for part type i. While the demand rate is known on average, some variability does exist due to the order sequence at final assembly and drifts in demand. The safety factor α is included in Equation (1) to handle variability. The problem in Equation (1) is the selection of α. Conventional wisdom has said that α is generally not more than 0.1. This chapter proposes an economic approach for selecting k_i, and accordingly α.

Kanban systems are often managed in an ad hoc manner. If parts are not available when needed, more Kanbans are added. If parts tend to sit in inventory, Kanbans are subtracted. Such a strategy suffers from the inefficiency of having to converge to the correct (optimal) solution by trial and error. Moreover, when random, short term fluctuations occur in demand rate, the manager may react by changing the number of Kanbans. If the fluctuation was a random event in a stationary, stochastic process, this action creates instability and serves to amplify production and inventory cycles. Such a phenomenon is similar to the machine operator who corrects parameter settings for every deviation from norm, whether statistically significant or not. Such overcorrections are known to increase, instead of dampen, variability. We aim for a decision model which allows selection of the proper number of Kanbans for the future, based on predictable demand patterns.

2 A MANY PARTS MODEL

In a Kanban system, the container size sets the minimum order size. The nature of the system is such that quick setups and reliable processes are assumed. The basic objective is to minimize the sum of inventory holding costs and shortages due to stockouts. Consider a workcenter that produces m part types. We let h_i be the cost per time to hold a full container of part type i and b_i the cost per time of shortages if the subsequent workstation is waiting on part type i.. The cost of lost capacity and idle labor can be included in b_i as well as any expediting cost. In practice, b_i is assumed large to prevent shortages. Our decision variables are the integer k_i values represented by $\underline{k} = (k_1, k_2, ..., k_m)$. We assume that the container size is fixed for each part type. Hence, inventory cost while processing (filling the container), and moving the filled container to inventory is unaffected by choice of $\underline{k}$. Let $p_i(x)$ be the probability of x full containers in inventory for part i, $x \leq k_i$. We allow the possibility of negative inventory to model the wait for containers by the next station. As demand is assumed to be externally generated, and must be eventually satisfied, backorders have no limit.

The assumption of external demand also permits the modeling of a multistage system as independent stages linked only by their proportional demand rates. Average demand rates for parts at each workcenter are determined by the demand for finished goods and their bills of materials. If mean demand changes for finished goods, this change will be felt by each individual workstation with a time delay equal to the sum of the lead times between that workstation and the final stage, i.e. the last production stage, N, feels the change in arrival of production requests immediately. The N-1th stage feels the delay τ_N later where τ_N is the lead time for production at stage N. This is true provided no workstation is asked to out produce its capacity. Capacity provides an upper limit for propagation of production orders to earlier stages. The minimum capacity at any stage sets the limit on flexibility to adjust to demand changes.

Production orders also require the availability of raw materials. For JIT systems, this implies availability of finished material at the preceding stage(s). Again, since total demand is externally determined, such availability does not affect the mean arrival rate of production orders at the subsequent workcenter. However, the higher order moments of the distribution of time between orders may be affected. We will assume the order of part type production is seemingly random among the part types waiting to be produced. Along with the implied environment of infrequent backorders, we can assume raw materials are always available. In practise, if parts are temporarily delayed we will produce one of the other scheduled part types which does not require the unavailable material.

2.1 Time Dependent Backorder Costs

We first consider the case where shortage costs are in effect for the duration of a backorder condition. The implication here is that expediting, overtime or subcontracting is occurring or perhaps the process is running at a faster than optimal speed. There may also be cost of idle labor at subsequent workcenters which are waiting for the product.

With this background, our problem is then

$$\underset{\underline{k}}{Minimize}\ E\,(cost/time) = \sum_{i=1}^{m} \left\{ h_i \sum_{x=0}^{k_i} x p_i(x) + b_i \sum_{x=-\infty}^{-1} p_i(x) \right\} \qquad (2)$$

A complete description of the system state involves the vector $\underline{x}$ of inventory for each part type. For our purposes, we need only determine the marginal $p_i(x)$ since inventory costs are additive in (2). It is understood that the $p_i(x)$ depend on $\underline{k}$. We will determine the $p_i(x)$ by use of the steady-state balance equations for the workstation. First however, we make several observations on the model.

Our assumption of external demand translates into a demand rate of D_i for all states. For instance, suppose backorder condition exists. Although continued requests for the part may cease to arrive, they implicitly continue. As containers are filled and delivered, these backorder requests immediately appear until production catches up with past demand. We further assume the service discipline in the workstation to be such that every Kanban scheduled has equal probability of being in process. A Kanban is "scheduled" if it is on the schedule board or associated with a container currently being filled. The expected number of Kanbans scheduled of part type i is denoted L_i. The total number of Kanbans scheduled at any time is , on average, $L = \sum_{i=1}^{m} L_i$. We assume, for now, that knowledge of the number of Kanbans scheduled of any one part type does not significantly effect L.

$$\text{Total arrival rate is } \sum_{i=1}^{m} D_i \text{ and average service time is } \mu^{-1} = \frac{\sum_{i=1}^{m} D_i t_i}{\sum_{i=1}^{m} D_i} \tag{3}$$

where t_i is the processing time per container of part type i. The variance of service time can be computed analogously. We use a simplistic model of the interarrival distribution. Each part produced at the successor workcenter has a certain probability of emptying a container and requiring a reorder. The random mixture of products prevents this from being a simple deterministic interval. The resulting geometric model for time to empty a container corresponds closely to an exponential interarrival time. We may then use an $M/G/1$ queueing model to estimate L. (If desired, a $GI/G/1$ model could be employed instead). The $M/G/1$ model yields

$$L = \rho + \frac{\lambda^2 V(t) + \rho^2}{2(1-\rho)} \tag{4}$$

where

$\lambda = \sum_{i=1}^{m} D_i$ is the total arrival rate,

$V(t)$ is the variance of the service time distribution, and

$\rho = \frac{\lambda}{\mu}$.

As any or all k_i increase, L_i is unaffected. Adding Kanbans at a workstation cannot increase demand at that workstation since orders must come from the subsequent production stage. Instead, adding Kanbans merely increase the average number of containers in inventory. (In a more complicated multistage model where backorders cease all work at the subsequent workstation and result in failure to meet demand, then the addition of Kanbans could modestly affect the demand seen by the workstation.) These remarks lead us to the recognition that expression (2) is separable in the part types i, i.e. $p_i(x)$ is unaffected by $k_j, j \neq i$. Thus, we can solve (2) separately for each k_i.

We now solve the steady-state balance equations for $p_i(x)$ where x indicates the number of containers in inventory. In so doing, we implicitly assume an exponential service time distribution for the workcenter. To simplify the presentation, we assume transit time is negligible. Thus, containers are either scheduled or in inventory. This assumption can be relaxed by indicating any fixed (average) transit time for containers. Using the standard *Rate Out* = *Rate In* balance equations we arrive at:

$$D_i p_i(x) = \frac{\mu}{L} p_i(x-1) \quad , x = k_i \tag{5.1}$$

$$\left(D_i + \frac{(k_i - x)\mu}{L}\right) p_i(x) = D_i p_i(x+1) + \left(\frac{(k_i - x + 1)\mu}{L}\right) p_i(x-1) \quad , 0<x<k_i \tag{5.2}$$

$$\left(D_i + \frac{k_i \mu}{L}\right) p_i(x) = D_i p_i(x+1) + \left(\frac{k_i \mu}{L}\right) p_i(x-1) \quad , x \leq 0 \tag{5.3}$$

The balance equations reflect the system transitions from state $p_i(x)$ to $p_i(x-1)$ at the rate D_i, (demand from successors). The system changes from state $p_i(x)$ to $p_i(x+1)$ when the workcenter completes service (fills a container). As we assume L jobs in process, service completions occur at the rate μ and the probability an arbitrary completion is of type i is equal to the proportion of in-process jobs which are type i. If x_i^+ is the actual on-hand inventory this proportion is $\frac{k_i - x_i^+}{L}$.

By imposing the constraint $\sum_{x=-\infty}^{k_i} p_i(x) = 1$ and recursive substitution we obtain the solution of equations (5.1-5.3) as

$$p_i(x) = p_i(k_i) \left(\frac{D_i L}{\mu}\right)^{k_i - x} \begin{cases} \frac{1}{(k_i - x)!} \; ; \; 0 \leq x \leq k_i \\ \frac{1}{k_i! k_i^{-x}} \; ; \; x < 0 \end{cases} \tag{6}$$

$$p_i(k_i) = \left[\sum_{x=1}^{k_i} \left\{ \left(\frac{D_i L}{\mu} \right)^{k_i - x} \frac{1}{(k_i - x)!} \right\} + \left(\frac{\mu k_i}{\mu k_i - D_i L} \right) \frac{1}{k!} \left(\frac{D_i L}{\mu} \right)^{k_i} \right]^{-1} \tag{7}$$

As we can solve for each k_i separately, we hereafter omit the subscript. To simplify notation let $\delta = \frac{DL}{\mu}$. For stability we assume $\delta < k$, i.e. the expected demand during a Kanban's expected waiting time does not exceed the number of Kanbans. Otherwise, we are besieged by backorders. Using δ along with expressions (6) and (7), the objective (2) becomes

$$\underset{k}{\text{Minimize}} \left[\sum_{x=0}^{k-1} \frac{\delta^x}{x!} + \left(\frac{k}{k-\delta} \right) \frac{\delta^k}{k!} \right]^{-1} \left[h \sum_{x=0}^{k} \frac{x \delta^{k-x}}{(k-x)!} + b \sum_{x=-\infty}^{-1} \frac{\delta^{k-x}}{k! k^{-x}} \right]$$

or equivalently, we wish to

$$\underset{k}{\text{Minimize}} \left[\sum_{x=0}^{k-1} \frac{\delta^x}{x!} + \left(\frac{k}{k-\delta} \right) \frac{\delta^k}{k!} \right]^{-1} \left[h \sum_{x=0}^{k} (k-x) \frac{\delta^x}{x!} + b \left(\frac{\delta}{k-\delta} \right) \frac{\delta^k}{k!} \right] \tag{8}$$

Problem (8) is a single variable optimization problem and can be easily solved by searching over $k \geq \delta$. Since incremental holding cost is strictly increasing in k, an upper bound on k is found when the incremental holding cost exceeds $b \sum_{x<0} p_i(x)$ for any k. Fig. 2 contains optimal values of k indexed for values of δ and the $\frac{b}{h}$ ratio. The $\frac{b}{h}$ ratio is shown for powers of 10 from 10^0 to 10^6. Values of the ratio less than 1 yield the same results as $\frac{b}{h} = 1$ in almost all cases. To use the figure, the ratio $\frac{D_i}{\mu}$ and total workcenter utilization must be known in addition to the $\frac{b}{h}$ ratio. Total workcenter utilization is used to find L, possibly using Equation (4). The product of these two values, $\frac{LD}{\mu}$, is indexed along the horizontal axis.

The diagonal line represents the minimal stable level whereby $k = \delta$. The vertical distance from the reference point to the optimality curves represents safety Kanbans. As expected, the number of extra Kanbans needed increases with the $\frac{b}{h}$ ratio.

An alternate objective would be to select the acceptable probability of shortage and determine the required k. If h is set to 0 in (8), and b to 1, the resulting

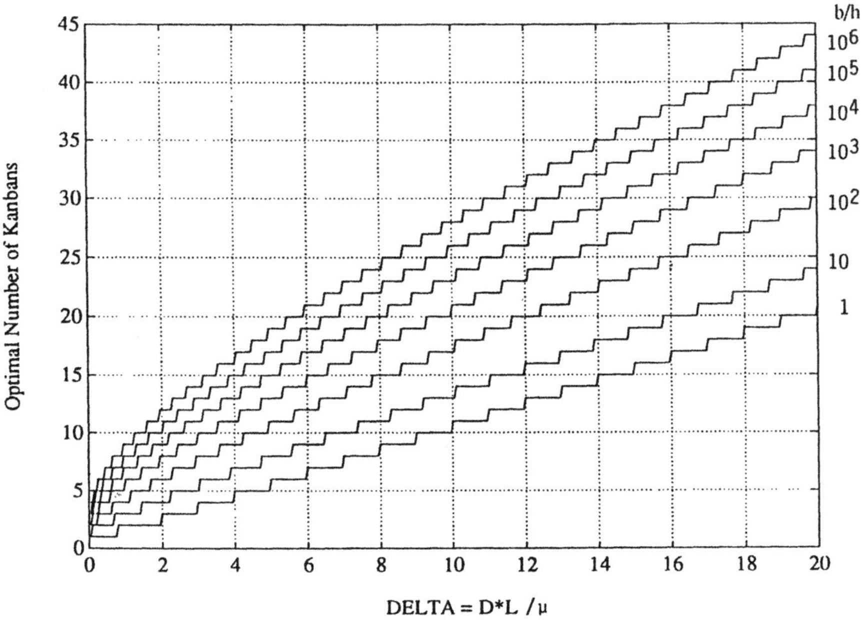

Fig. 2 Optimal Kanban for many parts model

expression is the probability of a stockout due to the replenishment cycle in the presence of a known long range demand rate but with variability in time between successive demand requests.

2.2 Occurrence Based Backorder Costs

Once a backorder condition starts, subsequent workstations are idle only for the time required to produce the first Kanban order. Thereafter, assuming matched production rates, the successor station is active with the just finished container of raw material. This next workstation is still waiting material, but it is no longer idle. In this case, b_i represents the cost to backorder customer demand for the final product for which part type i is destined. The same basic model can still be applied, however.

If entry into a backorder condition results in a change in schedule or manpower allocation, there may be an associated fixed cost. The rate of entry into a backorder condition is $D_i p_i(0)$. A fixed backorder condition cost term of the form $\pi D_i p_i((0)$ could be added to the model without affecting the solution procedure.

3 A FEW PARTS MODEL

When only a few part types are produced in a work cell, the above model may fail. The previous model assumed the number of containers of part type i in storage did not significantly affect the number of jobs in the production queue. If we drop these assumptions we can still develop the steady state balance expressions. The expression for state probabilities is more complex, however. Let L_{-i} indicate the expected number of containers in the production queue excluding those of part type i, i.e. $L_{-i} = \sum_{j \neq i} L_j$. For the current case, the steady-state balance Equations (5) become

Rate Out = Rate In

$$D_i p_i(x) = \frac{\mu}{L_{-i}+1} p_i(x-1), \; x=k_i \tag{9.1}$$

$$\left(D_i + \frac{(k_i-x)\mu}{L_{-i}+k_i-x}\right) p_i(x) = D_i p_i(x+1) + \frac{(k_i-x+1)\mu}{L_{-i}+k_i-x+1} p_i(x-1), \; 0<x<k_i \tag{9.2}$$

$$\left(D_i + \frac{k_i\mu}{L_{-i}+k_i}\right) p_i(x) = D_i p_i(x+1) + \left(\frac{(k_i\mu)}{L_{-i}+k_i}\right) p_i(x-1), \; x\leq 0. \tag{9.3}$$

Rearranging the terms in expressions (9.1-9.3), we obtain the recursive expressions

$$p_i(k_i-1) = \frac{D_i(L_{-i}+1)}{\mu} p_i(k_i). \tag{10.1}$$

$$p_i(k_i-j) = \left[\left(D_i + \frac{(j-1)\mu}{L_{-i}+j-1}\right) p_i(k_i-j+1) - D_i p_i(k_i-j+2)\right]\left(\frac{L_{-i}+j}{j\mu}\right), \; 2\leq j\leq k_i \tag{10.2}$$

$$p_i(-j) = \left[\left(D_i + \frac{k_i}{L_{-i}+k_i}\right) p_i(1-j) - D_i\, p_i(2-j)\right]\left(\frac{L_{-i}+k_i}{k_i\mu}\right), \; j>0. \tag{10.3}$$

In practice, we can place an upper bound j^-, on j in (10.3). The upper bound is the practical limit on the number of containers that can be backordered. Then, using the relation $\sum_{x=-j^-}^{k_i} p_i(x) = 1$, we can develop progressive relationships of the form $p_i(k_i-j)=a_j p_i(k_i)$ for the constants a_j obtained from expressions (10.1-10.3) and solve for the state probabilities.

The condition for stability in the few parts model is $\frac{D_i(L_{-i}+k_i)}{\mu}<k_i$. Figs. 3a, 3b and 3c show optimal numbers of Kanbans for the cases L_{-i} =1, 5 and 10. At first glance, these curves seem to imply significantly larger Kanban levels than Fig. 2. This is deceiving, however, as the workload due to part i is excluded.

We believe the format of Fig. 3 to be the more useful parameterization, however. For comparison, Fig. 4 illustrates the same results with $L_{-i} = 5$ indexed with k_i included in L. Using this upper bound on the number of Kanbans in process, we see that results are similar in form to those of Fig. 3 and indeed call for fewer Kanbans.

4 EXPEDITING

As a final model, suppose backordered items are given priority for production. Thus, when in state $p_i(x)$ for $x < 0$, the production rate of type i parts is μ. We are assuming here that either at most one part type can be backlogged at a time or other resources are obtained if several part types are in a backorder position. If inventory levels of part types are independent, the probability of being in a stockout (expediting) condition due to an item other than i is

$$Prob\ (Stockout) \equiv \beta_i = 1 - \prod_{j \neq i} \sum_{x=0}^{k_j} p_j(x) \tag{11}$$

The steady-state equations are then

$$D_i p_i(x) = \frac{(1-\beta_i)\mu}{L_{-i}+1} p_i(x-1)\ ,\ x=k_i \tag{12.1}$$

$$\left(D_i + \frac{(1-\beta_i)(k_i - x)\mu}{L_{-i}+k_i-x}\right) p_i(x) = \tag{12.2}$$

$$D_i p_i(x+1) + \left(\frac{(1-\beta_i)(k_i-x+1)\mu}{L_{-i}+k_i-x+1}\right) p_i(x-1)\ ,\ 0<x<k_i$$

$$\left(D_i + \frac{(1-\beta_i)k_i\mu}{L_{-i}+k_i}\right) p_i(x) = D_i p_i(x+1) + \mu_i p_i(x-1),\ \ x=0 \tag{12.3}$$

$$(D_i+\mu_i) p_i(x) = D_i p_i(x+1) + \mu_i p_i(x-1)\ ,\ x<0 \tag{12.4}$$

Expressions (12.1-12.4) assume part types without backorders are put on hold if a shortage occurs. Another possibility is to increase the production rate by using secondary resources (workers and/or processes). This case can be modeled by letting all β_i be 0 indicating that, even during a stockout of a part type, production of other part types proceeds normally. If parts without backorders are put on hold, the $\beta_i = 0$ solution can be used as a starting point to estimate β_i

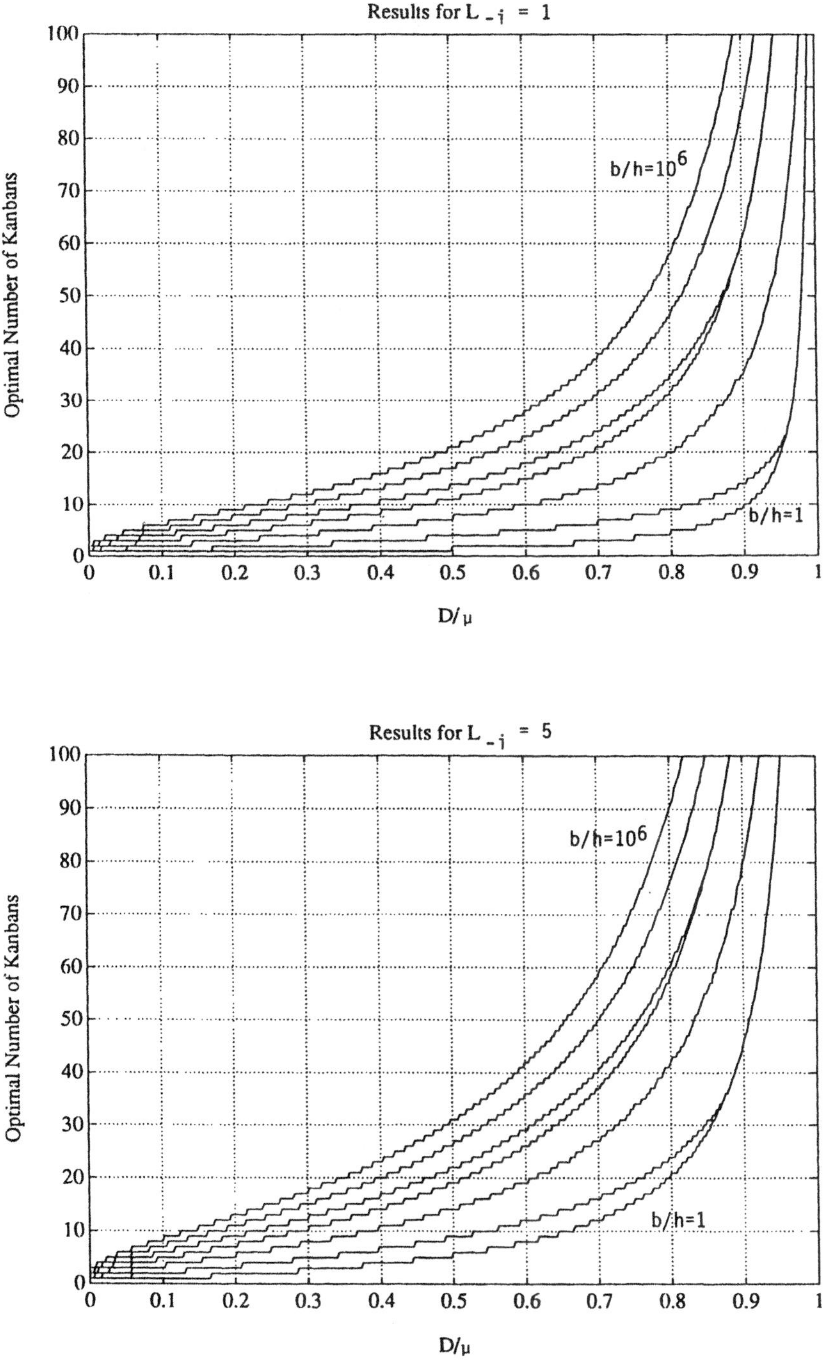

Fig. 3 Optimal Kanban for few parts model

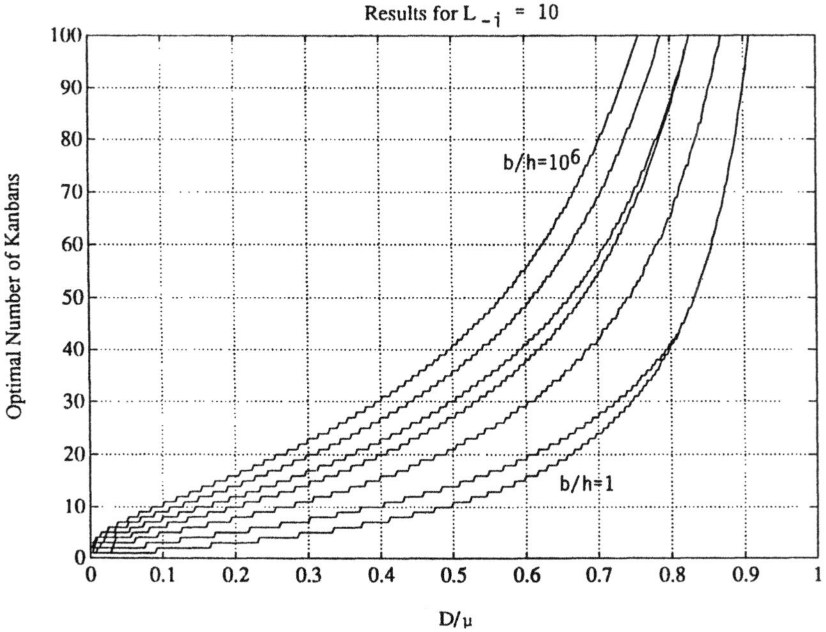

Fig. 3 (continued)

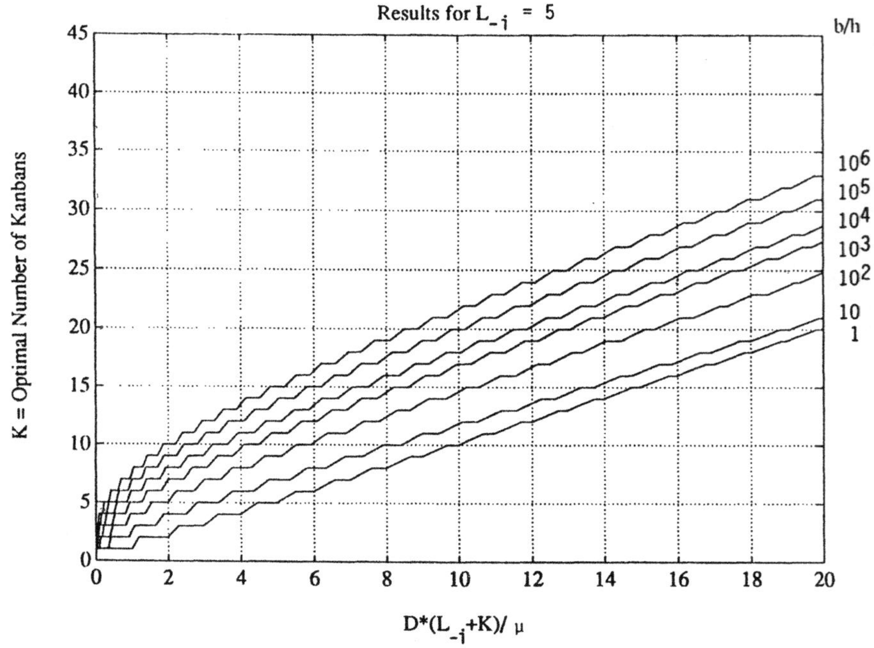

Fig. 4 Optimal Kanban for few parts model with K

in Equation (11). When part type i is in a backorder position ($x < 0$) we replace μ by $\mu_i \equiv t_i^{-1}$. The change is seen in equations (12.3) and (12.4). This modification implements our assumption that part type i is under constant production when its inventory level is negative.

Rearranging terms in the balance Equations (12.1-12.4) we obtain

$$p_i(k_i-1) = \frac{D_i(L_{-i}+1)}{(1-\beta_i)\mu} p_i(k_i) ; \qquad (13.1)$$

$$p_i(k_i-j) = \left\{\left[D_i + \frac{(1-\beta_i)(j-1)\mu}{L_{-i}+(j-1)}\right] p_i(k_i-j+1) - D_i p_i(k_i-j+2)\right\} \frac{L_{-i}+j}{(1-\beta_i)j\mu}, \qquad (13.2)$$

for $2 \leq j \leq k_i$;

$$p_i(-1) = \left\{\left[D_i + \frac{(1-\beta_i)k_i\mu}{L_{-i}+k_i}\right] p_i(0) - D_i p_i(1)\right\} \frac{1}{\mu_i} ; \qquad (13.3)$$

$$p_i(-j) = \left\{ (D_i+\mu_i)p_i(1-j) - D_i p_i(2-j) \right\} \frac{1}{\mu_i}, \; j \geq 2 . \qquad (13.4)$$

5 COMPARISON TO SIMULATION

The models included in this chapter assume exponential interarrival times and 100 percent availability of raw materials. In order to test the effect of these assumptions, a model was constructed in Simscript to simulate an hypothetical JIT system. The system consists of six workcenters in three levels as shown in Fig. 5. A small experiment was conducted by varying levels of utilization, demand interarrival distribution, b/h ratio and the number of parts per workcenter. The model was tested with both 2 and 10 parts per workcenter. These cases correspond to 2 and 10 finished products respectively, with each workcenter producing a distinct part type for each finished product and parts being assembled according to the structure of Fig. 5. Two parts were used to test the "Few Parts Model" and ten parts for the "Many Parts Model". Each part produced at a workcenter has an equal workload ($\mu_i = 1$ for all parts), demand was set to give the desired utilization. Utilization was tested at 80% and 95%. The number of Kanbans was selected from Figs. 2 and 3 to correspond to a b/h ratio of 10 and 1000.

The simulation model counts a backorder if all parts are not available at the time a Kanban *enters* the scheduling board queue. At the same time, raw materials are moved to the subsequent station for staging and the Kanbans are removed and sent to their scheduling board. The analytical model considers backorders only when parts are ready for processing, i.e. the successor Kanban reaches the head of the queue.

A comparison was made between the results predicted by the analytical models given in Equations (5) and (9) and the simulation. Comparisons were based on average number of full Kanbans in inventory and the probability of a part type being in a backorder state. Results are shown in Table 1 for exponential interarrival times for finished goods demand and in Table 2 for constant interarrival times. Values of L_M are computed using Equation (4) for exponential demand and an approximate theoretical expected time in queue expression for a GI/G/1 queue for Table 2. Expected waiting times were converted to average queue length using Little's Law (*Queue Length = Arrival Rate • Wait Time*). Average queue lengths found in the simulation are tabulated as L_S. Results shown for the simulation are the average at the middle level, i.e. workcenters 2 and 3 and the part types they produce.

Table 1 contains results for the case of exponential interarrival times for final demand. Average inventory and backorder predictions from the model were computed based on the theoretical L_M and $L_{-i,M}$ values. We note that in several cases the difference between simple queueing theory approximations and actual results cause the model to predict poorly, but this can be corrected by use of actual results from experience. For instance, for 2 parts, 95% utilization and 13 Kanbans, the simulation found an average inventory level of 5.85 Kanbans. The value of 3.77 reported in the table is based on a presumed queue length of 9.5. If the actual queue length of 7.15 is used instead, the analytical model computes an average inventory of 5.74. Likewise for the case of 24 Kanbans, using the simulation value of $L_{-i} = 8.58$ yields an average inventory of 15.35, which more nearly replicates the simulation result of 15.42. The simulation observes lower average queue lengths than predicted by the simple queueing model since, in reality, the queue length is constrained by the number of Kanbans. This bias diminishes as k increases. A finite population queueing model could be used to more accurately predict L for low values of k. In summary, the analytical model proposed in this chapter appears to predict well when an accurate estimate of the queue length at each workcenter is available.

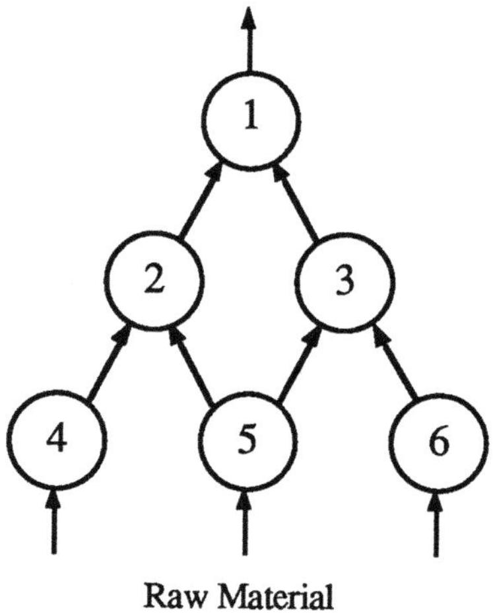

Fig. 5 Schematic model of simulated system

Table 1 Experimental results with exponential demand

No. Parts	ρ	L_M	L_S	k	Average Inventory Model	Average Inventory Simulation	Backorder Probability Model	Backorder Probability Simulation
10	0.8	4.00	3.81	2	0.84	1.62	0.026	0.002
10	0.8	4.00	3.80	4	2.83	3.62	0.000	0.000
10	0.95	19.0	15.1	4	1.85	2.49	0.070	0.034
10	0.95	19.0	16.6	7	4.84	5.34	0.001	0.015
No. Parts	ρ	$L_{-i,M}$	$L_{-i,S}$	k	Average Inventory Model	Average Inventory Simulation	Backorder Probability Model	Backorder Probability Simulation
2	0.8	2.00	1.76	4	2.14	2.28	0.120	0.065
2	0.8	2.00	1.95	10	8.01	8.05	0.001	0.005
2	0.95	9.50	7.15	13	3.77	5.85	0.230	0.168
2	0.95	9.50	8.58	24	14.53	15.42	0.001	0.056

Table 2 Experimental results with constant demand

No. Parts	ρ	L_M	L_S	k	Average Inventory Model	Average Inventory Simulation	Backorder Probability Model	Backorder Probability Simulation
10	0.8	2.05	5.06	2	0.92	1.49	0.006	0.000
10	0.8	2.05	5.08	3	1.92	2.49	0.000	0.000
10	0.95	9.51	13.60	3	1.49	1.64	0.036	0.040
10	0.95	9.51	12.80	5	3.48	3.72	0.001	0.001
No. Parts	ρ	$L_{-i,M}$	$L_{-i,S}$	k	Average Inventory Model	Average Inventory Simulation	Backorder Probability Model	Backorder Probability Simulation
2	0.8	1.02	1.29	3	1.78	1.71	0.122	0.023
2	0.8	1.02	1.19	8	6.66	6.80	0.001	0.000
2	0.95	4.76	5.81	8	3.03	2.19	0.190	0.322
2	0.95	4.76	5.25	17	11.81	11.75	0.001	0.018

Table 2 contains results for the case of a constant interarrival time for finished goods demand. In this case, using L_M in the analytical models yielded results as accurate as L_S. (Accurate here refers to closeness to the simulation results). The many parts model performed well with the high level of utilization. The few parts model performed better on the estimation of average inventory than on backorder probabilities. Outcomes differed by less than 4% in 3 of the 4 cases.

Unlike the exponential case, $L_M < L_S$. This occurs because interarrival times for demand were assumed to be constant. In practice however, the variability of processing time at the successor stage adds variability to the interarrival time for Kanbans. Thus, while queue lengths are reduced with respect to those in Table 1, they have not been reduced to the extent predicted with a constant interarrival time assumption.

6 SUMMARY AND CONCLUSIONS

A continuous time, steady-state Markov model has been developed for determining the optimal number of Kanbans to use for each part type at each workcenter in a JIT system. The model selects the proper safety factor in each case. Results can be used to shorten the trial-and-error cycle of Kanban insertion and extraction often used in JIT. Moreover, the approach allows the manager to determine the optimal number of Kanban to cope with foreseen shifts in demand thus reducing the time required by JIT systems to adapt to demand changes. These results provided a starting point for the integration of push information, supplied by a coordinating central controller, into a pull system.

As with traditional lot-sizing models, many different scenarios could be defined for the treatment of backorder costs as well as operating strategies for JIT. In each case a slightly different objective function and steady-state model respectively would apply. We have attempted to describe a general methodology for determining the optimal number of Kanbans for a given shortage cost function and operating policy. For several cases, graphs of optimal solutions to the problem are included.

7 ACKNOWLEDGEMENT

This chapter is based on work supported by the National Science Foundation under grant DDM 85-44993.

8 REFERENCES

1. Bitran, Gabriel R. and Li Chang, "A Mathematical Programming Approach to a Deterministic Kanban System", *Management Science*, Vol. 33, No. 4, pp. 427-441 (1987).
2. Johnson, Lynwood A. and Douglas C. Montgomery, *Operations Research in Production Planning, Scheduling and Inventory Control*, John Wiley & Sons, New York (1974).
3. Kimura, Osamu and Hirosuke Terada, "Design and Analysis of Pull System, A Method of Multi-Stage Production Control," *International Journal of Production Research*, Vol. 19, No. 3, pp. 241-253 (1981).
4. Monden, Yasuhiro, *Toyota Production System*, IE&M Press, Norcross, GA (1981).

5. Deleersnyder, Jean-Luc, Thom J. Hodgson, Henri Muller and Peter J. O'Grady, "Kanban Controlled Pull Systems: An Analytic Approach", *Management Science*, Vol. 35, No. 9, pp. 1079-1091 (1989).
6. Rees, Loren P., Patrick R. Philipoom, Bernard W. Taylor III, and Philip Y. Huang, "Dynamically Adjusting the Number of Kanbans in a Just-in-Time Production System Using Estimated Values of Leadtime", *IIE Transactions*, Vol. 19, No. 2, pp. 199-207 (1987).

AN ADVANCED AGVS CONTROL SYSTEM: AN EXAMPLE OF INTEGRATED DESIGN AND CONTROL

Ronald A. Bohlander
Wiley D. Holcombe
James W. Larsen
Georgia Institute of Technology

1 INTRODUCTION

The Automated Guided Vehicle System (AGVS) has attracted increased attention over the past decade from both the material handling user community and from academics, primarily due to its flexibility. From an academic perspective, the key challenges addressed have been 1) measuring or understanding AGVS performance and 2) extending its flexibility. Georgia Tech Material Handling Research Center programs, in this regard, have included:

- Development of an integrated path layout, fleet sizing and evaluation, and simulation tool, known as the AGVS Engineering Design Workstation, or AGVS EWS [1,2]. This tool is intended to be used by industrial engineers planning AGVS applications.

- Development of new guidance technologies incorporated in a software-guided vehicle which does not follow either a wire or stripe on the floor [3]. The vehicle still follows a prescribed path, but this is specified in software rather than by a continuous marker on the floor.

On the other hand, from a user's perspective, the principal challenge in the decade has been managing or mastering the complexity that comes with a highly flexible material transport system. The record of success has been mixed, with significant accomplishments and notable system failures [4]. Part of the difficulty has been the high cost of system software and the relative inaccessibility of the software to the user when changes need to be made. Relatively simple changes in the path network or location of pick-up/deposit

(P/D) points have typically resulted in significant contractor involvement and cost.

After we had established a reasonable level of guidance accuracy with our new guidance techniques [3], we reevaluated our R&D program. The most important motivation for the new guidance technology is to have a system which is easier to install and to reprogram. Merely having state-of-the-art guidance is not enough. It must be embedded in an environment for system planning and programming which allows the user to quickly make path and configuration changes. It occurred to us that the AGVS EWS, which we originally developed to plan and evaluate conventional AGV Systems, could also provide a system planning environment for a software-guided AGVS.

This chapter describes progress to-date on creating an integrated design and control environment, building on the developments of the AGVS EWS [1,2] and the software-guided vehicle [3].

2 A HIERARCHICAL SYSTEM

We must recognize that the AGVS EWS does not by itself provide a complete solution to designing an AGVS control system. It is not a software development tool; what it does do is help a person with the important task of *defining an AGVS configuration* by providing a highly interactive, graphical environment for this purpose. So we are, in effect, dividing the problem of designing controls into two parts:

1) First, we define the "members" of a control hierarchy and their responsibilities for the generic functions of an AGVS. We implement this as computer codes which will run the AGVS when the software is linked to data describing a particular installation.
2) To apply this in a specific site, we use the AGVS EWS to prepare data files describing the configuration (e.g., all allowed paths and control points) and all necessary control parameters.

A couple of requirements are relatively apparent. First, the workstation must give a sufficiently complete description of an installation. This ensures that there is a clean distinction between software procedures which are generic and the data which is special to each installation. We want to avoid the temptation to tune up the generic software itself for a given site. Second, we do not want to insert the data manually into the procedures each time we prepare for a new installation. A great deal of time is saved if the control codes automatically access the configuration information. The final requirement, which we will comment on here, relates to our research mission. We have defined the responsibilities of the blocks in the system hierarchy in a way that will support both conventional control rules and some advanced ones we are studying concerning dispatching and routing. It is interesting to note that, by keeping the AGV Control System modular, we can, in principle, apply this method of system preparation to both conventional and advanced-guidance systems.

We are focusing our initial efforts on developing a control system to use in experiments with two vehicles in our AGV Laboratory. We have been careful to make overall system architecture choices that permit us to expand control to a much larger number of vehicles, but two vehicles is a manageable number for early testing. While we recognize that an important part of AGV System software also deals with vehicle interactions with P/D stations, relatively little attention has been given to the details of this at the present stage.

About 1985, the Material Handling Research Center (MHRC) commissioned an 800 square-foot AGV Laboratory to support work on advanced guidance techniques. The Lab provides open floor space in which we can experiment with both conventional and advanced guidance vehicles. The layout of the room and typical AGV paths are shown in Fig. 1. Our present work focuses on two Litton Series-800 vehicles in this lab. The one pictured in Fig. 2 has been modified for software guidance, as described briefly in Section 5 and in more detail in [3].

Fig. 3 delineates the responsibilities of each of the principal computers used for design or control in the Laboratory; namely, the AGVS EWS, the AGV System Controller, and the computer(s) on-board each of the operating vehicles. It has been our basic philosophy to place responsibility at the lowest level at which a decision can be made. This is necessary to ensure that communication traffic is kept low so that a large number of vehicles can be controlled in the future. The following paragraphs explain the essential responsibilities shown in Fig. 3.

With the AGVS EWS [2], we prepare layouts of allowed paths, directions of travel, and positions of P/D stations. We can also analyze the throughput requirements of a given system in relation to the number of vehicles needed. The EWS provides a platform to generate simulations of detailed AGV performance. This is a useful precursor to full-scale experimentation, as well as a way to check the traffic and dispatching effects of having a larger fleet than is available in the AGV Lab. To achieve this, we are configuring the simulation generator to be able to mimic control rules that will actually be implemented in the Lab.

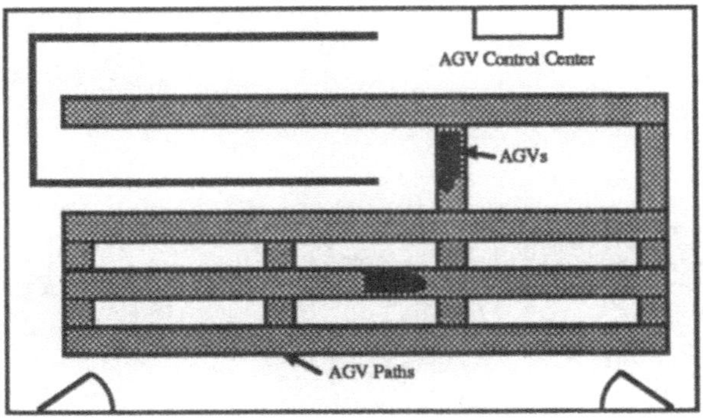

Fig. 1 Layout of the AGV laboratory at MHRC

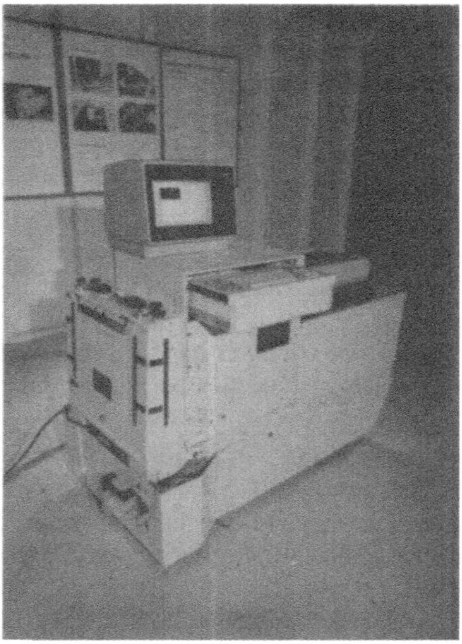

Fig. 2 Prototype, software-guided automated guided vehicle

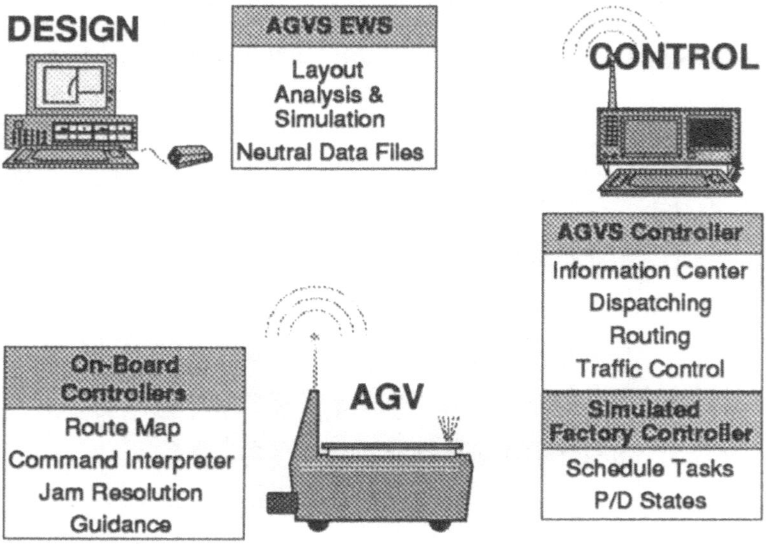

Fig. 3 Responsibilities for design and control of the AGVS

One of the main advantages in using the AGVS EWS as a foundation for configuration design is that it provides a standard format for describing an AGV System. All the modules of the AGVS EWS share a set of files called the Neutral Data Files [2] which describe the AGV System's path network and all other system parameters. We have adopted this same standard file set for use in our AGV Control System. We transfer the Neutral Data Files to the AGV System Controller (via a network connection or a floppy disk), and this computer, in turn, transmits them by radio (or floppy disk) to the computer in each vehicle, so that all components share a common image of the system. If we need to modify a design, the modifications are reflected in new Neutral Data Files which can be redistributed.

The AGV System Controller (AGVSC) augments the Neutral Data Files with additional information about the real-time system status. Fig. 3 refers to this combined set of data as an *Information Center*. These files serve as a blackboard through which the vehicles and the AGVSC communicate with each other about dispatching assignments, temporarily blocked path segments, and traffic flow.

The AGVSC has primary responsibility for *dispatching* and *idle vehicle management* but shares responsibility for *routing* with the controllers on board the vehicles. Each vehicle in our system carries enough information about the currently allowed paths so that they can do their own routing. The AGV System Controller need only tell the vehicle to pick up a load at station A and drop it at station B. The vehicle knows the details of the shortest route from A to B through its copy of the Neutral Data Files. However, one of our current research efforts concerns control of congested traffic, and in such cases, the AGVSC is able to intervene in the choice of a route. Further information on this will be given in Section 6.

The function titled *Simulated Factory Controller* is shown in the same block as the AGVSC in Fig. 3. This represents the activities normally associated with the plant host computer and ideally would be housed in a separate computer. However, for our current scale of experimentation, only a small extra set of procedures need be added to the AGVSC to simulate the responsibilities of this computer. We need only construct a rather arbitrary list of tasks for the AGVs to complete and maintain a file giving the status of each task and each P/D station.

Each AGV is responsible for all low-level functions like guidance. When it receives a high-level command from the AGVSC, it interprets this in terms of a route and constructs a sequence of primitive commands which it follows in carrying out the route. As it proceeds, it asks for control of each path segment it is entering. If it does not get this from the AGVSC, it goes into a blocked status. If this persists beyond a time-out limit, the AGV will attempt to take any available alternate path to the destination.

3 USE OF THE ENGINEERING DESIGN WORKSTATION

Details concerning the use of the AGVS EWS in the design of Automated Guided Vehicle Systems are covered in a companion article in the proceedings

[2], and will not be repeated here. What is worth elaborating is the central use of Neutral Data Files throughout the AGVS EWS and the control systems. Table 1 lists the kinds of information stored in the files. The table indicates files which were modified or added to support the AGVSC and vehicle controller needs. Not surprisingly, some of these have also been of subsequent use in the ongoing development of the simulator in the AGVS EWS.

The on-going development of the AGVS EWS and AGVSC has increased the level of detail covered in the Neutral Data Files. Since we are using the AGVS EWS in experimental work on control issues, we must try out changes in how we represent system data in these files. Up to this point, the Neutral Data Files have been flat text files, but it has become difficult to keep all software modules up-to-date when we make frequent changes in the referenced data files. Therefore, part of MHRC's current effort is a pilot conversion to a relational database in the AGVS EWS. A full database manager will probably not be used in the AGVSC, and certainly not on the vehicles, due to memory constraints. Rather, database utilities will be used to create "views" of the data needed by the AGVSC and the vehicles.

Table 1 List of neutral data files and the Impact of AGVSC design

File Type:	Vehicle Types
Purpose:	Contains the static information about the vehicle types in the system.
Impact:	Modified
File Type:	Vehicle Labels
Purpose:	Comprises the vehicle identification labels (or numbers) and the corresponding vehicle types
Impact:	Unaffected
File Type:	Control Point List
Purpose:	Contains control point information. The control point locations are specified by the user via the CAD program, but the actual control point IDs and coordinate location values are uniquely determined by the CAD program.
Impact:	Unaffected
File Type:	Segment List
Purpose:	Contains information about the guidepath segments of the AGVS network. A segment is uniquely identified by two control points. Its direction is implied as going from the first to the second control point (for unidirectional segments).
Impact:	Modified

Table 1 (continued)

File Type:	Landmark List
Purpose:	Contains the global locations of the navigation landmarks (or targets).
Impact:	Added
File Type:	P/D Station List
Purpose:	Associates an identification label (or number) with each P/D station, the associated work station label and the corresponding control point in the CAD program.
Impact:	Unaffected
File Type:	Work Station Information
Purpose:	Contains the static work station information. It associates a work station with its P/D stations.
Impact:	Unaffected
File Type:	Work Station Flows
Purpose:	Comprises the aggregated flow information between pairs of work stations.
Impact:	Unaffected
File Type:	Part Type Information
Purpose:	Contains information on part types, their generation distributions, and a pointer to their corresponding process routes.
Impact:	Unaffected
File Type:	Part Type Routes
Purpose:	Contains a list of work stations that describe the part routes (indicated by the value of the pointer in file type *Part Type Information*) and the corresponding set-up and processing times.
Impact:	Unaffected
File Type:	Shortest Paths
Purpose:	Comprises a shortest path table between each pair of control points. Each element in this square matrix specifies the next intermediate control point to travel from the top of the shortest path tree to the bottom. The indices for the control points are in increasing sequence.
Impact:	Unaffected
File Type:	Project File
Purpose:	This file associates all of the file types described above with their corresponding DOS file names.
Impact:	Unaffected

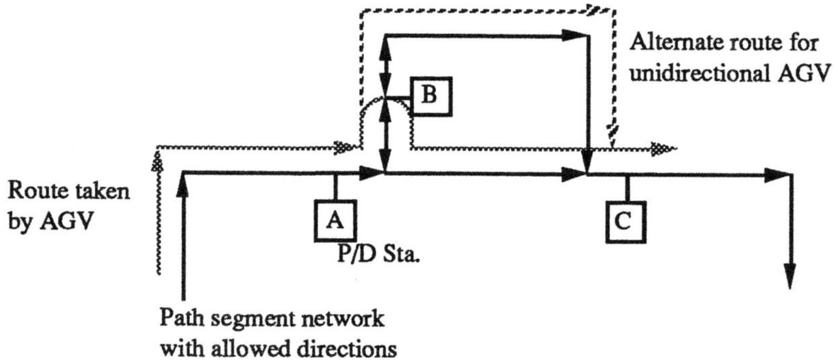

Fig. 4 Shortest routes depend on the type of vehicle

Our integrated projects on design and control have also helped us identify and avoid naive assumptions in the use of the AGVS EWS. An example is illustrated in Fig. 4 where we show part of a path network in which a bidirectional segment has been specified (near Station B). Suppose a material flow is assigned going from station A to station B, followed by another material flow from B to C. Given the bidirectional segment shown, the apparent shortest path is illustrated by the dotted AGV route in the figure. This route is feasible if the vehicles in the system are bidirectional since they can go from A to B and then reverse direction as shown. However, the existence of bidirectional path segments does not necessarily imply that bidirectional vehicles are being used. Today, when we use the AGVS EWS, we check to make sure the vehicles can actually execute maneuvers implied by shortest route logic. If the vehicles are not bidirectional, then an AGV flow like that shown by the dashed line would be the shortest feasible route.

4 AGV SYSTEM CONTROLLER

The AGV System Controller architecture is illustrated in Fig. 5. At the present time, we are implementing all of this within one 80386 PC and an MS-DOS operating system. Thus, the architecture is significantly affected by the need to organize the interrupt structure.

Interrupt routines are used to handle all communications. (In the case of the operator interface, the keyboard interrupt is actually managed within the programming language.) The interrupt routines manage the communications outside of the AGV System Controller Executive. Once an interrupt routine has completed a communications event, it sets a flag and writes the message to a global variable. If no flags are set, the AGV System Controller Executive rests in a busy loop. Once a flag is set, the AGV System Controller Executive passes the message to the appropriate module for action.

The AGV System Executive is responsible for the principal functions of the AGVSC as shown in Fig. 6. The dispatching and routing approaches which we are investigating are explained in Sections 6 and 7. Idle Vehicle Management covers procedures to assign a parking place to a vehicle if it is not needed for any task. The Interface block is used to manage the flow of information generated by interrupts, as shown in Fig. 5.

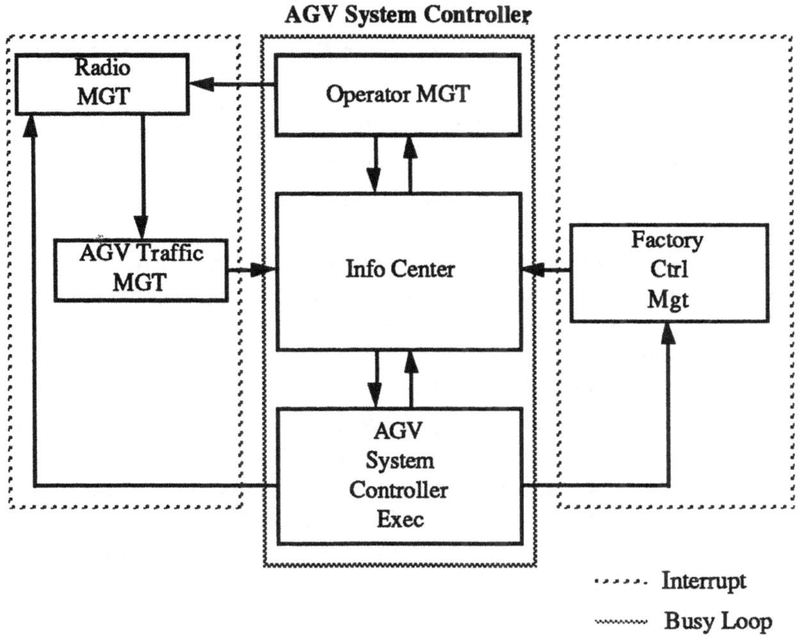

Fig. 5 AGV system controller architecture

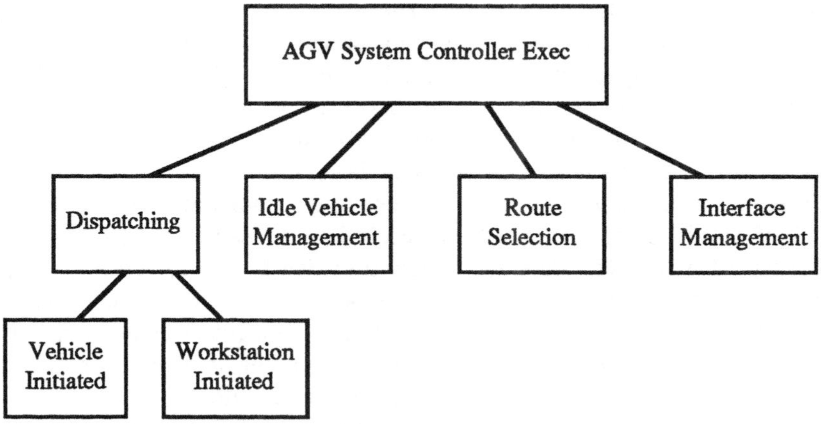

Fig. 6 AGV system controller executive

5 CONTROLS ON-BOARD THE VEHICLE

Details of the software-guided vehicle (shown in Fig. 2) have been published elsewhere [3] and will only be briefly summarized here. We started with a Litton Series 800 vehicle and modified it by adding a new control computer and additional sensors for navigation. The main on-board control is now provided by an 8086 PC which communicates

- with vehicle servo-motor controllers through a DAC and the AGV pendant port,
- with the original-equipment AGV computer through a serial port,
- with a vision-system computer through a parallel port, and
- with the AGVSC, through a serial port, modem, and mobile radio.

The original AGV computer is used to monitor safety systems like the contact bumpers. The vehicle has a tricycle wheel arrangement; the single front wheel is both steered and driven. The vehicle can back up, but not very accurately, so it is essentially a unidirectional vehicle.

Rapid servo-controlled guidance depends on inputs from wheel encoders on all three wheels including the steering angle of the front driven wheel. Wheel rotation measurements are translated by dead reckoning routines into estimated positions, which are compared with positions prescribed by the path segments the vehicle is following at the time. The servo-control loop operates to make the estimated position converge on the intended path or stopping point. Since dead reckoning becomes progressively less accurate, on account of wheels slipping and bouncing, the vehicle periodically measures its position with respect to markers in the room. Retroreflective targets on the ceiling are viewed by a low cost vision system mounted on the vehicle. Strobe lighting is used so that the vehicle can capture the images on the fly. This data is combined with wheel data through a Kalman filter to improve the running position estimate.

When we first got the software-guided vehicle running, we described the path it was to follow with a table of path segment parameters, like that shown in Table 2. This was typically typed in on the vehicle computer keyboard or loaded from a saved version on floppy disk. As we began to integrate the vehicle into an overall system controlled by the remote AGVSC, we clearly had to make a number of refinements in the on-board vehicle controls.

First, we needed a way for the vehicle computer to automatically generate a stack of primitive commands like this each time it received a move command from the AGVSC. This function is now provided by the vehicle's Command Interpreter which takes the command from the AGVSC, extracts the data stored in the on-board copy of the Neutral Data Files, and constructs the necessary stack of primitive path specifications. (If we were controlling another type of vehicle, this is the principal part of the code which would be rewritten.)

In addition, the main controller on board the vehicle now has a number of system responsibilities in addition to maintaining guidance, as mapped out in Fig. 7. In early work, the vehicle updated position estimates and servo-outputs

Table 2 Example path file; F(unction, S(traight), C(urve) entries

Segment Type	Segment Origin X-Value (meters)	Segment Origin Y-Value (meters)	Segment Origin Heading (radians)	Segment Velocity	Segment Length (meters)	Change in Heading (radians)
F	-0.002	2.798	0.0	0.0	1.82	1
S	2.919	5.736	0.0	0.2	3.66	0.0
C	2.919	9.4	0.0	0.2	-1	-1.57
S	3.919	10.4	4.71	0.2	0.5	0.0
C	4.419	10.4	4.71	0.2	-0.5	-1.57
F	0.006	1.362	-3.14	0.0	0.39	1
S	4.919	9.9	3.14	0.2	0.59	0.0
F	-0.01	2.599	-3.18	0.0	1.63	1
S	4.919	9.311	3.14	0.2	3.41	0.0
C	4.919	5.9	3.14	0.2	-0.5	-1.57
S	4.419	5.4	1.57	0.2	1	0.0
C	3.419	5.4	1.57	0.2	0.5	1.57
F	-0.003	1.856	-1.57	0.0	0.88	1
S	2.919	4.9	3.14	0.2	3	0.0
C	2.919	1.9	3.14	0.2	0.5	1.57
S	3.419	1.4	4.71	0.2	2.5	0.0
C	5.919	1.4	4.71	0.2	1	1.57
S	6.919	2.4	6.28	0.2	1.5	0.0
C	6.919	3.9	6.28	0.2	1	1.57
F	-0.023	1.003	-1.55	0.0	0.03	1
S	5.919	4.9	1.57	0.2	2.5	0.0
C	3.419	4.9	1.57	0.2	-0.5	-1.57
F	0.0	1.309	-4.71	0.0	0.34	1
S	2.919	5.4	6.28	0.2	0.34	0.0

every .11 seconds. The additional system responsibilities may occasionally mean that a guidance cycle would be missed and therefore we must provide for asynchronous control. As a result, a time-of-day field has been added to the stack tables. This allows the vehicle to compare its estimated position with calculated progress along the path. The on-board computer also uses a more refined desired path description based on splines. (Having the vehicle controller aware of time as well as position goals would make it possible, in principle, for us to experiment with AGV scheduling based on reservations of zones for time windows, as suggested by Tanchoco et al. [5], but we have no present plans to do so.)

Another way that time enters into the current system plan is when the vehicle determines whether it has been blocked for longer than some set limit. When the vehicle encounters a control point, it asks for the next zone and stops if this is occupied. If the next zone is blocked for longer than the set limit, there may be a problem on the network that will take maintenance personnel to solve. In

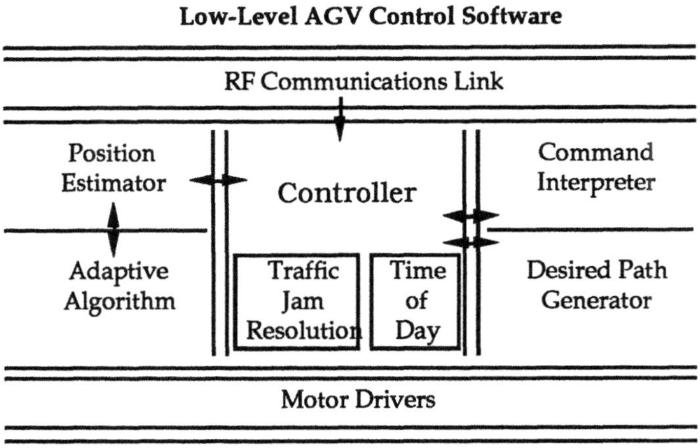

Fig. 7 AGV's on-board controller

this instance, the vehicle signals the AGVSC of the problem, selects an alternate route if a feasible one exists, and asks permission from the AGVSC for its new route and its next zone. This is an experimental concept which will be tested through simulation, along with others described in Section 6.

6 ADVANCED ROUTING AND TRAFFIC MANAGEMENT METHODS

As discussed previously, all controllers in the AGVS use the same shortest route table defined for the network of allowed paths. This is a conventional shortest-route-in-distance (SRID) table, which can be expressed quite compactly as a matrix of "next nodes" given a starting node on the left edge and a destination node on the top edge, as shown in Table 3. In this example, if want to go from node 2 to node 6, the table shows that from node 2 the next node is 9, and then dropping down to the row for this node, we find the next node is 10, and so on. The shortest-route-in-distance is then specified by visiting nodes 2, 9, 10, 12, 3, 13, 15, 14, and finally 6. As mentioned previously, during conditions of low congestion, a vehicle might simply be told to go from node 2 to 6, and the vehicle would get the information from the SRID table in the way we have outlined here. Our objective, however, is to make it possible for the AGVSC to intervene in selecting an alternate route when congestion is significant.

Most common traffic management methods consist only of zone blocking schemes (or queueing schemes) to assure that vehicles do not collide, e.g., by permitting no more than one vehicle per zone. No provisions are made to limit congestion on the network. We are also planning to employ conventional zone blocking and contact bumpers to prevent collisions, but we believe that congestion can be alleviated by being able to specify alternate feasible routes in the AGVS path network. (This of course applies only to networks which are relatively rich in possible paths between control points or network nodes. Systems with only a few simple loops may not be able to take advantage of

Table 3 A typical shortest route in distance table

	Destination																	
Origin	1	2	3	4	5	6	7	8	9	10	11	12	13	14	15	16	17	18
1	1	8	8	8	8	8	8	8	8	8	8	8	8	8	8	8	8	8
2	9	2	9	9	9	9	9	9	9	9	9	9	9	9	9	9	9	9
3	13	13	3	13	13	13	13	13	13	13	13	13	13	13	13	13	13	13
4	17	17	17	4	17	17	17	17	17	17	17	17	17	17	17	17	17	17
5	14	14	14	14	5	14	14	14	14	14	14	14	14	14	14	14	14	14
6	16	16	16	16	16	6	16	16	16	16	16	16	16	16	16	16	16	16
7	18	18	18	18	18	18	7	18	18	18	18	18	18	18	18	18	18	18
8	10	10	10	10	10	10	10	8	10	10	10	10	10	10	10	10	10	10
9	11	10	10	10	10	10	10	11	9	10	11	10	10	10	10	10	10	10
10	12	12	12	12	12	12	12	12	12	10	12	12	12	12	12	12	12	12
11	1	1	1	1	1	1	1	1	1	1	11	1	1	1	1	1	1	1
12	3	3	3	3	3	3	3	3	3	3	3	12	3	3	3	3	3	3
13	15	15	15	15	15	15	15	15	15	15	15	15	13	15	15	15	15	15
14	6	6	6	6	6	6	6	6	6	6	6	6	6	14	6	6	6	6
15	14	14	14	14	14	14	14	14	14	14	14	14	14	14	15	14	14	14
16	2	2	2	7	7	7	7	2	2	2	2	2	2	7	2	16	7	7
17	5	5	5	5	5	5	5	5	5	5	5	5	5	5	5	5	17	5
18	4	4	4	4	4	4	4	4	4	4	4	4	4	4	4	4	4	18

this possibility.) Consequently, we have designed our system to be able to explore some alternate route algorithms.

In the first method we will investigate, the AGVSC will construct a complementary *shortest-route-in-time* (SRIT) table. This will be derived from the *shortest-route-in-distance* (SRID) table and from a table of time delays for each of the control points in the network. As each route is assigned, an expected time delay will be added to each control point to be visited in the network. These time delays will impact the shortest route in time calculated for subsequent routes. Once a control point has been seized and released by a vehicle executing a route, the time delay for that node will be decremented.

Any control scheme that requires the transmission of route information from the system controller to the vehicles may jeopardize overall system responsiveness. The communications channel capacity can be limiting, especially in systems with a large number of vehicles. To avoid this, the AGVSC uses a compact specification of an alternate route; i.e., one which is not strictly according to the SRID table on-board the vehicle. The AGVSC simply decomposes the alternate route into a minimum number of sub-routes which *are* described by the SRID tables. The sub-routes are defined by the set of intermediate nodes between them. In this way, the AGVSC need only tell the AGVs these special intermediate nodes on the way to the ultimate destination; it does *not* have to broadcast *all* the nodes on the route.

These concepts and those described in the next section will be developed first on the AGVS EWS simulator, so that their value can be assessed before implementing them in the AGVSC.

7 ADVANCED DISPATCHING METHODS

Most common dispatching rules provide a method to select one of several tasks for a single vehicle (which has become idle) or to select one of several available vehicles for a single new job. These are based on the assumption that multiple idle vehicles and multiple waiting jobs hardly ever occur together. Generally quite simple rules are used, which either optimize service level or throughput. Assigning vehicles to oldest loads improves service level at the expense of greater empty vehicle travel time; assigning vehicles to nearest loads improves throughput but may strand a load in a low traffic corner of the network. We feel that overall performance improvements might be achieved by using a *predictive dispatching algorithm*. The AGVSC would predict when, in the near future, vehicles will become idle, and possibly also when new tasks will be created (e.g., when a process machine will complete its cycle). This would result in multiple idle vehicles and multiple waiting tasks in a planning window, thereby expanding assignment options. The dispatching algorithm could then make an optimum match between vehicles and tasks.

8 STATUS AND PERSPECTIVES

The developments described in previous sections are in various stages of completion. Many of the components are still in development and coding, some are in debug and testing, but we have enough experience with the integration of the system to make it worthwhile reporting on lessons learned to this point. Table 4 summarizes project status.

We are well-along in creating a capability to sit down at a CAD/CAE station and describe the configuration of an AGVS in preparation for an installation, or for a change in a system. A number of commercial AGVS firms now offer limited, similar capabilities. Some of these, for example, can facilitate planning for a simple change in the path to a P/D station, like that shown for station A in Fig. 8. Changes which shuffle the order of control points in a network, like the change to B in the figure, require more attention. The AGVS EWS can analyze the impact of such a change on system performance and will automatically update all system tables. When such a reconfiguration session is completed, then all that is necessary is to transmit new data tables to the vehicles and make any necessary changes in the "furniture". (If vehicles use conventional guidance, this would include laying new wires or stripes and, possibly, new communication lines.) In any case, it is a significant advance to be able readily to reconfigure the system software.

Our present implementation of the control system is focused on getting a rapid prototype working with realistic vehicle and system speeds. We are using some immediately available resources which are somewhat limiting in the short term; e.g., obsolete 8086 PC controllers on the AGVs, DOS operating systems, and

Table 4 Project status

Software-guided vehicle	• Operational, under limited system control • New on-board modules (in Fig. 7) fully coded & in tests
ADVS EWS	• Operational for preparing configuration and Neutral Data Files. • Relational database upgrades underway. • On-going evolution of the file contents,
Radio communications	• Partially coded & tested.
Automated Guided Vehicle System Controller	• Pseudo-code developed. • Detailed coding beginning.
Advanced Routing, Traffic Management & Dispatching	• Conceptual designs completed. • AGVS EWS Simulator undergoing modifications to test concepts.

procedure-based code, largely implemented in Turbo Pascal. Evolutionary upgrades are planned. We should, in time, incorporate multi-tasking and more use of faster and perhaps multiple microprocessors.

Our use of a modular, hierarchical organization may transition to an object-oriented control scheme if we can show that real-time speeds are feasible. Claims are being made by many that object-oriented disciplines are helpful in designing a system, preparing a system simulation, and then extending the simulation code to actual system control code. Since object-oriented code generates a lot of redundant overhead, its speed is an issue. Another is the use of data encapsulation in the object-oriented method to ensure modularity. In the design phase, competing relational data base approaches hold some advantages in flexible data sharing. It remains to been seen if it is possible to have the best

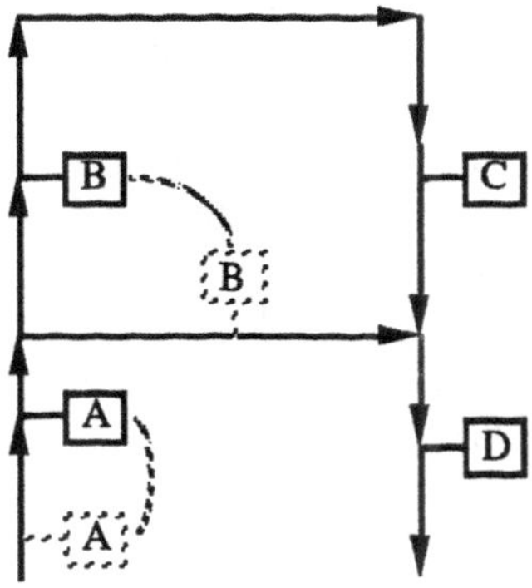

Fig. 8 Changing P/D locations

of both worlds since "object-oriented" and "relational" methods are often thought of as being at opposites ends of the data management spectrum.

MHRC is working on other, complementary CAE programs which will help in planning the interface of the AGVS to the rest of a plant. Establishing interface details has frequently been a major difficulty in new plant installations. Equally of interest is the overall architecture of the interfaces and control responsibilities. This is a key part of on-going research on integrated design and control systems.

9 ACKNOWLEDGEMENTS

We wish to thank a number of graduate students, past and present, for their contributions to the research progress summarized here. These include Omer Bakkalbasi, Larry Banta, Gene Durrence, Timothy Gfroerer, Dah-Chuan Gong, Mike Hegedus, Brett Lapin, Sang Ryong Lee, Petra Mayerhofer, Bill McKinney, Jesus Ortiz, John Parsons, Brett Peters, Pat Pudlo, and Thomas Tapp. We also thank our colleagues Marc Goetschalckx, Leon McGinnis and Stephen Dickerson for their leadership in the related developments of the AGVS EWS and the software-guided vehicle.

We gratefully acknowledge the support of the corporations and government agencies who fund the Material Handling Research Center and the CIMS Program at Georgia Tech, and IBM's gift of AGVs for our lab.

10 REFERENCES

1. Goetschalckx, M. and L.F. McGinnis, "Engineering Work Station Is a Design Tool for Computer-Aided Engineering of Material Flow Systems", *Industrial Engineering*, Vol. 21, No. 6, pp. 34-38 (June 1989). For a discussion of AGVS design methodology, see also: Omer Bakkalbasi and Leon F. McGinnis, "ABC's of Preliminary In-House Planning and Analysis of AGVS Applications", *Proceedings AGVS 89*, sponsored by MHIA, Philadelphia, 26-28 (September 1989).
2. McGinnis, L.F., "Computer-Aided Facilities Design Revisited: A Prototype Design Workstation for AGV Systems", *these proceedings*. See also: Marc Goetschalckx, "Computer-Aided Design of Industrial Logistics Systems", *these proceedings*.
3. Holcombe, W.D. , S.L. Dickerson, J.W. Larsen, and R.A. Bohlander, "Advances in Guidance Systems for Industrial Automated Guided Vehicles", *Proceedings of SPIE, Volume 1007: Mobile Robots III*, pp. 288-297, Cambridge, Massachusetts, (November 1988). See also, Larry E. Banta, *Advanced Dead Reckoning Navigation for Mobile Robots*, Ph.D. Dissertation, MHRC Report TD-86-16, Georgia Institute of Technology, Atlanta, GA.
4. Love, J.M., "AGVS in Automobile Assembly Applications: Trials, Tribulations, and Successes", *Proceedings Material Handling Focus '88*, sponsored by MHRC, Atlanta, (September 1988).

5. Kim, C.W. and J.M.A. Tanchoco, "Prototyping Integration Requirements of Free-Path AGV Systems", *these proceedings.*

A VIRTUAL CELL SCHEDULING ALGORITHM

Jocelyn R. Drolet
Universite du Quebec a Trois-Rivieres
Colin L. Moodie
Purdue University

ABSTRACT

In this chapter, we begin by defining a new manufacturing concept termed Virtual Cellular Manufacturing System (VCMS). The VCMS is an expansion of the virtual cell concept proposed by McLean and all at the NIST AMRF in the early 1980's [4].

Then, we will show how the virtual cellular concept can be a provider of flexibility for factories of the future. A time-phased linear programming algorithm which has been developed especially for controlling VCMS will be presented. Finally, a simulation comparison between a VCMS and a Group Technology Cell has been performed. Results obtained from the simulations demonstrate the superiority of VCMS over the GT Cell System.

1 THE EFFECT OF FACILITIES LAYOUT ON MATERIAL FLOW

Throughout the recent history of organized manufacturing, managers have sought to control material flow so as to minimize WIP, effectively utilize available machine tools, meet due dates and other manufacturing oriented criteria. One method for obtaining better material flow, with respect to these criteria, is to change the configuration (layout) of machine tools. Product demand seems to have a direct influence on the preferred configuration. A strong steady demand will favor one type of layout and a varying demand for low production volume products can favor another layout type. One bottom line consideration for choice of a layout is the return on capital invested in machine tools: for certain product demand-manufacturing situations one layout type will facilitate better material flow characteristics, low WIP, and superior productivity, than other layout configurations. In addition, current manufacturing philosophy favors reconfigurability: the ability for the manufacturing facilities to be rapidly reconfigured in response to changes in market, or other, demands to which they are exposed. A variation of the virtual cell concept, which will be described here, seems to provide good reconfigurability and flexibility features, and, with currently available material handling technology and computer control/integration capability, can now be achieved. Before describing the Virtual Cell Manufacturing System (VCMS), it will be instructive to review the material flow characteristics of other layout configurations.

Fig. 1 diagrams material flow in process (a), product (b), cellular (c), and virtual cellular (d) configurations. As stated above, certain manufacturing situations exist which will favor one layout over the other. For example, the process layout (job shop) is old but it still can be effective. Sule [7] notes that:

> "A job shop is a suitable mode for a company that produces many different products with a relatively small volume for each. The various items being produced to customer specifications require the collection of a variety of general-purpose equipment, highly skilled labor to operate this equipment and general-purpose tooling and fixtures. Similar machines can be grouped together within the plant to form a department."

The product layout, which has had a significant effect on reducing cost of production during this century, especially in the auto and home appliance industries, is effective for a production environment which is very different from

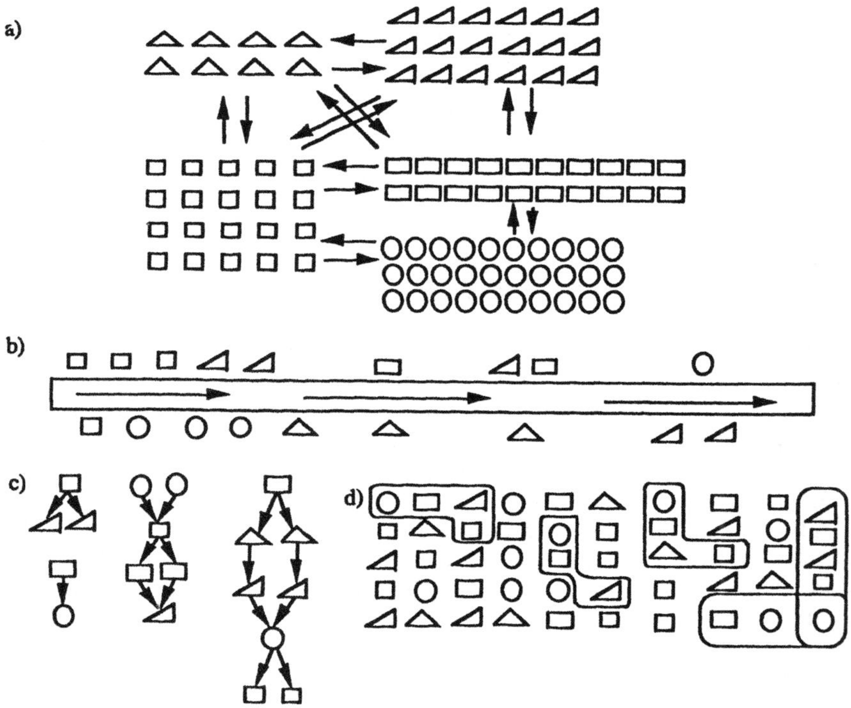

Fig. 1 Material flow in a) process, b) product, c) cellular, c) virtual cellular

that which favors the process layout. Tompkins and White [9] note this in their text:

> "Suppose a large, stable demand for a standardized product, like an engine block, is to be met by production. In such a situation, the workstations should be combined into a planning department so that all workstations required to produce the product are combined. The resulting product planning department may be referred to as production line department."

The cellular facility configuration, a post WWII development, is suitable for a manufacturing environment which seems to fall somewhere between that which favors either the process or product layout. Greene and Cleary [3] provide the following definition of cellular manufacturing:

> "Cellular manufacturing is the physical division of the manufacturing facility's machinery into production cells. Each cell is designed to produce a family of parts. A family of parts is defined as a set of parts that require similar machinery, tooling, machine operations, and/or jigs and fixtures. The parts within the family normally go from raw material to finished parts within a single cell. Usually, the manufacturing facility cannot be completely divided into specialized cells. Rather, a portion of the facility remains as a large, functional job shop which has been termed the Remainder Cell."

As noted above, the cellular layout tends to provide some of the material flow and low WIP characteristics identified with the product layout, but is suitable for product families where *individual* members of a family have low annual production volumes. A virtual cell is not identifiable as a fixed physical grouping of workstations but as data files and processes in a controller. When a job order needs a set of workstations to be "put together", a virtual cell controller takes over the control of these workstations and makes communication possible between them. A workstation is either a member of a pool of available workstations or a member of an existing virtual cell. This provides a manufacturing environment which is flexible and reconfigurable. This concept will be described in more detail in the next section of this chapter.

Modular facility layouts are also an alternative which makes reconfigurability practical. The spine concept is one type of modular layout which has found use in manufacturing environments where product life cycles are relatively short and reconfigurations are not uncommon. Tompkins and Spain [8] thoroughly describe this layout type. Fig. 2 shows a spine layout where the main material flow conduit is down the center with fabrication and assembly departments connected to the spine. Fig. 3 shows a modular layout (described by Bullinger [1]) which has three relatively independent sections. Subassemblies and components can be independently produced in the upper and lower modules while final assembly can take place in the central module. Each module can have its own flow pattern and can be expanded or contracted independently of the other modules.

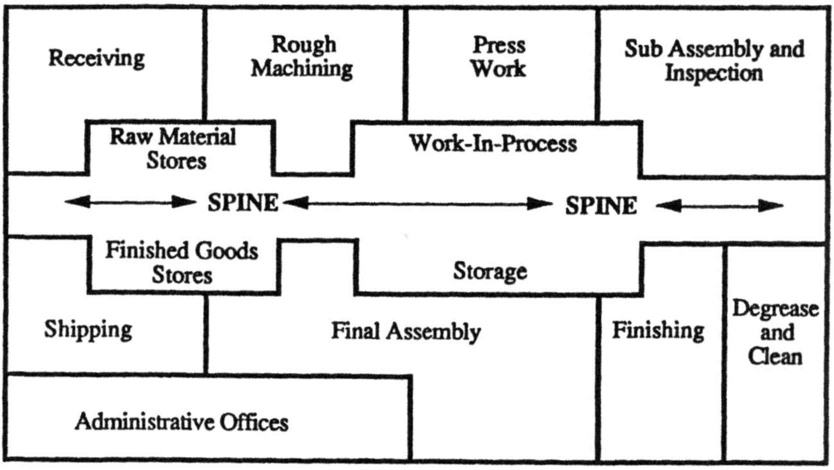

Fig. 2 Spine layout

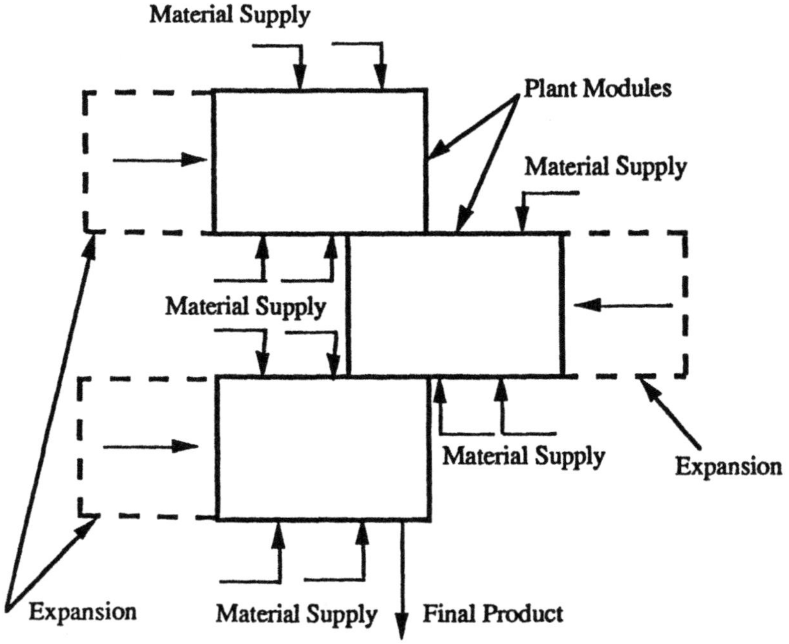

Fig. 3 Modular layout

2 THE VIRTUAL CELL AS A PROVIDER OF FLEXIBILITY

The facility layout configuration models described above are (except, perhaps, the virtual cell) available for, and are used in, contemporary factories. Looking ahead in time, however, Schonberger [5] states that metal processing factories of the future must be world class manufacturing systems, having no more than 20 highly productive machines with automated material handling systems. These factories will operate in JIT mode, under computer control. One might expect two or three similar machines and a handling device to be put together to form a workstation. In these workstations many activities can be going on in parallel: loading, unloading, processing, communicating, cleaning, are a few examples. Systems having up to 10 workstations working around the clock will not be uncommon [6]. These systems, although much smaller than current factories, will be much more productive.

Since cutting speed will drastically increase in the next decade, processing time will be much shorter compared with material handling time. Furthermore, the reduction of lot sizing down to the unit will dramatically increase demand on material handling even though WIP parts will generally visit fewer machines. Given this, it seems certain that cooperative workstations will have to be close to each other in such a situation. It is felt that the virtual cell layout strategy will become practical for scheduling parts to processors in this environment. The technology is available to achieve it.

A virtual cell is a logical grouping of workstations in a central controller. Fig. 4 illustrates a virtual cell which contains 15 workstations comprising five workstation types (differentiated symbolically by five different geometric shapes). This system currently has three active virtual cells and one shared workstation.

3 AN ALGORITHM FOR CREATING VIRTUAL CELLS

An efficient linear programming model has been developed for creating and scheduling virtual cells [2]. The solution satisfies the capacity constraints on all considered resources: tools, workstations, and fixtures. The methodology is best

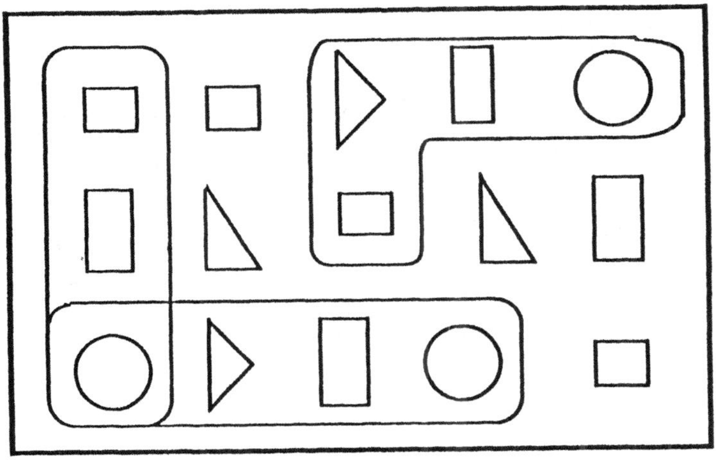

Fig. 4 Illustrative shop

explained, however, by a cell creation example which only seeks to minimize the flow distance between machine tools for a given job. The solution will consist of a list of workstations which will minimize the material flow for that job while being consistent with the requirement regarding the precedence and capacity constraints. Even though certain elements have been omitted, the general ideas are portrayed in Fig. 5.

The top left frame in the figure illustrates a facility containing 17 workstations. Each is expressed in terms of the following convention: the character represents the type of workstation, e.g. milling, turning, inspection, etc., the numeral identifies each replication of a given workstation type. For example, node "A4" denotes replication "4" of workstation type "A". The frame itself symbolizes the boundary of the facility. The axis system, located nearby, reminds us that the frame contains a scaled layout and thus, exact position of input/output workstations are known. From such a layout, one can calculate the inter-workstation distances (rectilinear or euclidean); these measures, along with the utilization profile of resources, are useful during virtual cell creation.

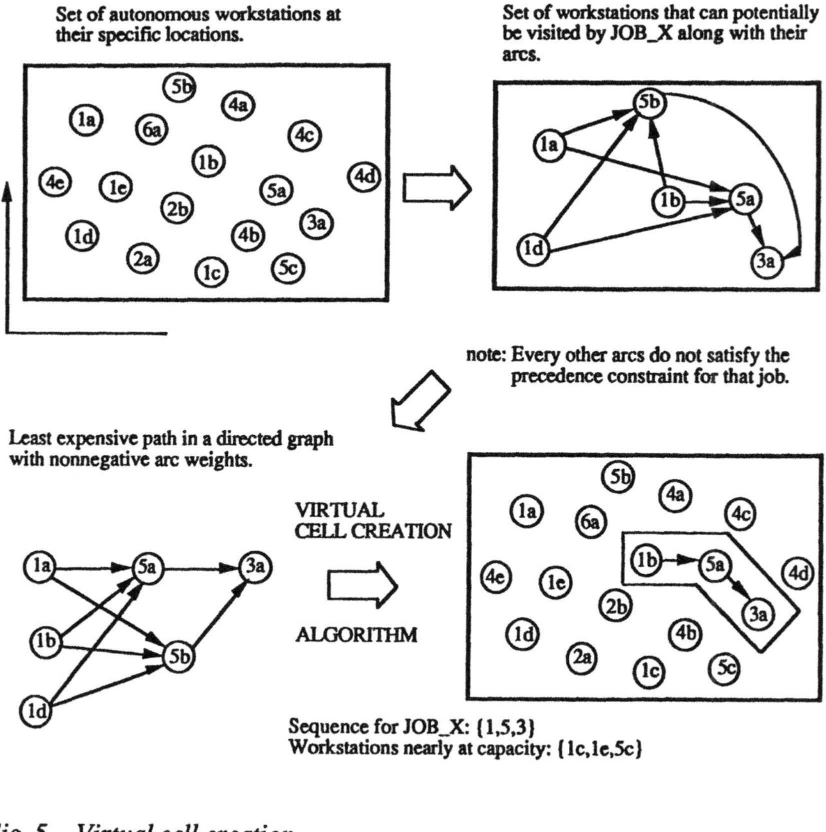

Fig. 5 Virtual cell creation

The second frame consists of the set of workstations, which can potentially be visited by the job. Workstations are connected by arcs and each arc represents an achievable flow of parts between two workstations. Only the arcs connecting successive workstations as restrained by the precedence constraint are shown.

The third frame depicts a restructuring of the second frame. Here one can easily distinguish the sequence of operations depicted by the precedence constraint of the job. The parts included in that job must go from a workstation of type A to a workstation of type E and then to a workstation of type C. The objective is to find the least expensive path from any workstation of type A to any workstation of type E via any workstation of type C while satisfying other constraints to be mentioned later. The output contains the list of workstations that minimizes the material flow for that job while being consistent with the requirement regarding the precedence and capacity constraints. The solution of this example is shown in the fourth frame. The set of workstations displayed in this frame represent the virtual cell just created for JOB_X. Along with it, a virtual cell controller will be created.

After virtual cell creation, the next phase consists of assigning resources such as tools and fixtures to the newly created cell. The resources are delivered just before the activation of the virtual cell and stay there as long as the job is not finished. The final phase aims at dispatching the parts within, and occasionally between, each virtual cell. In this phase the flow of parts and communication between cooperative workstations (members of the same virtual cell) is particularly important. The virtual cell controller has the algorithmic logic and the real time database to control the advance of processes within these workstations.

The objective function and constraints of the full linear programming model [2, 10] are listed below. Let us now define the decisional variable V_{jrp} which is critical to the understanding of the model. Variable V_{jrp} is the fraction of job j assigned to a virtual cell created at period p, and using resources on routing r. A Gantt chart which illustrates variable $V_{3,156,1}$ is represented in Fig. 6. As

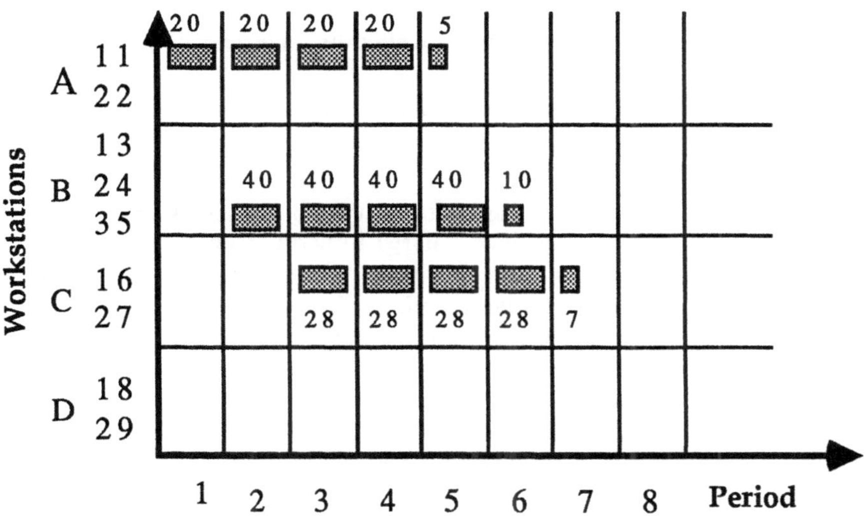

Fig. 6 A Gantt chart illustration of variable $V_{3,156,1}$

shown, job #3 begins processing at period 1 and its routing is A1 B3 C1. The bottleneck workstation determines the maximum introduction rate of parts. For sake of conciseness, the quantity of parts to be processed, their processing time and technological requirement are not shown. Such information is available in a data file for each job available for processing at the entrance of the system. The objective function creates and schedules virtual cells in a way that minimizes the total distance flow for all considered jobs. This is so as long as the lateness factor equals 0, otherwise, distance flow increases to permit the use of resources which are not as close. The first constraint set insures that each job will be produced in its totality. The first part of this constraint considers the detailed horizon in which one or more virtual cells will be created at some period in time for each job. The second part concentrates on the aggregate horizon in which a specific routing (by the same token, specific virtual cell) is not considered. The second set of constraints only applies to the detailed horizon, the third set only applies to the aggregate horizon.

p	=	A period.
t	=	A period.
h	=	Detailed scheduling horizon $(h>=0)$.
H	=	Aggregate scheduling horizon $(h<H)$.
j	=	A job.
r	=	A routing for some job j.
w	=	A resource type (workstation type, tool type, fixture type).
u	=	A unit of resource of some type w (a specific workstation, a tool, a fixture of some type w).
V_{jrp}	=	Fraction of job assigned to a virtual cell created at period p $(p<=h)$, and using resources on routing r.
W_{jp}	=	Fraction of job j assigned to an aggregate virtual cell created at period p $(h<p<=H)$.
N_j	=	Number of parts in job j.
D_r	=	Distance travelled by each part using routing r.
AD_j	=	Average distance travelled by each part of job j using any allowed routing r.
F_j	=	Flow time for job j (life cycle of any virtual cell for job j).
L_{jp}	=	Lateness of job j when its processing begins in period p $[\max(0,p-(H-F_j))]$.
M	=	Large cost incurred for each day of lateness on any job.
P_{twjrp}	=	Processing time required in period t on resource w when a virtual cell for job j using routing r and started in period p $(p<=h)$ has been created.
P_{twjp}	=	Processing time required in period t on resource w when an aggregate virtual cell for job j started in period p $(h<p<=H)$ has been created.
C_{wut}	=	Capacity of unit u of resource w in period t.
AC_{wt}	=	Aggregate capacity of all units of resource w in period t $(h<p<=H)$.

$$\text{Min}\sum_{p=0}^{h} \sum_{\forall j,r} V_{jrp} \ (N_j D_r + ML_{jp}) \ + \sum_{p=h+1}^{H} \sum_{\forall j} W_{jp} \ (N_j AD_j + ML_{jp})$$

S.T.

$$\sum_{p=0}^{h} \sum_{\forall r} V_{jrp} + \sum_{p=h+1}^{H} W_{jp} = 1 \qquad \forall j \qquad (1)$$

$$\sum_{\forall j,(w,u)\varepsilon r,(p:p<=t<=p+F_j,\ p<=h)} P_{twjrp} \quad V_{jrp} \ <= \ C_{wut} \qquad \forall(w,u),(t:t\leq h) \qquad (2)$$

$$\sum_{\forall j,(w,u)\varepsilon r,(p:t-F_j<=p<=h)} P_{twjrp}\, V_{jrp} \ + \sum_{\forall j,w\varepsilon r,(p:p<=t,<=p+F_j,h<=p<H)} P_{twjrp}\, W_{jp} \ <= \ AC_{wt} \qquad \forall w,(t:h<t\leq H) \qquad (3)$$

$$V_{jrp} >= 0 \qquad \forall jr,\ p<=h \qquad (4)$$

$$W_{jp} >= 0 \qquad \forall j,\ h<p<=H \qquad (5)$$

4 EXPERIMENTAL COMPARISON WITH A GT CELL ENVIROMENT

The virtual cell as defined here suggests a flexible machine configuration which can efficiently cope with a dynamic shop environment which produces small batch size orders of variable, engineered components. In order to test this hypothesis, a simulation comparison was made between a virtual cell shop and a shop which consists of the classical GT cells. It is not possible to make a GT cell environment completely equivalent to that of virtual cells; an effort was made to make the comparison fair, however. The two systems were evaluated for several criteria, such as **average makespan, number of parts waiting in queue, number of parts produced,** and **machine utilization.** For the modeled system, the virtual cell configuration proved to be superior. A portion of the comparison will be described in the following paragraphs and figures.

Fig. 7 illustrates the makespan for both systems. The plotted values represent an average across five simulation runs. The average makespan after 405 minutes of operations is 38.6 minutes for the VCMS and 48.08 minutes for the GT cell system; this is a significant difference. On the average, parts get through about 10 minutes faster in the VCMS. The second measure of performance to be evaluated with respect to the two systems was throughput. The graph presented in Fig. 8 shows the number of parts produced over time for each system. The plotted values represent an average across five simulation runs. The number of parts produced is significantly higher with the VCMS, in fact, the results indicate that a 20% increase in throughput is possible over the group technology

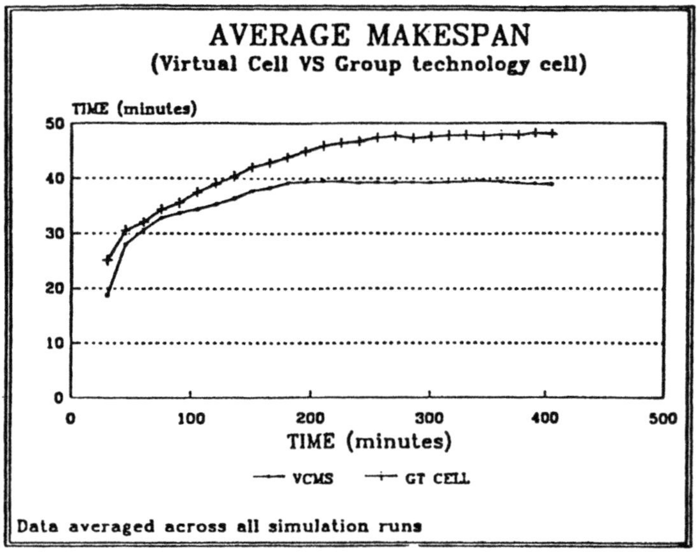

Fig. 7 Average makespan

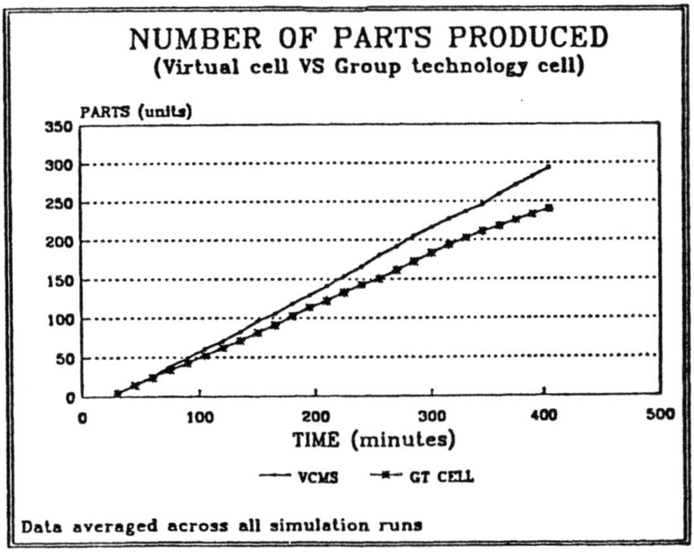

Fig. 8 Number of parts produced

cell system. This can mean increased profits if the market can absorb the extra production.

Machine utilization is another measure of performance considered in this study. The values plotted in Fig. 9 represent an average across 5 simulation runs. Similar to the throughput, machine utilization increases by almost 20%: from 63.3% with the GT cell system to 79.8% with the VCMS.

5 SUMMARY AND CONCLUDING REMARKS

A machine layout procedure which is based on the virtual cell concept has been presented. Cells are created in an optimal way based on machines needed by the job, availability of tools, distance between available machine tools, and other relevant criteria. Results obtained from an object oriented simulation show the superiority of virtual cellular manufacturing systems over group technology cell systems. Several aspects or measures of performance were considered: average makespan, throughput, machine utilization, queue length, number of parts in system, etc. The VCMS performed equally or better in each of these measures of performance. The potential benefits are high and it seems that further research should be encouraged. The good results presented here are based on a one simulation model: more are needed.

When considering what appears to be a significant advance in machine tool layout strategy, one must also consider the cost which can be required to achieve the potential advantages. First, there will be increased material handling. The material handling (transportation) might need advanced control systems in order to negotiate variable paths between machine tools within a virtual cell. Also,

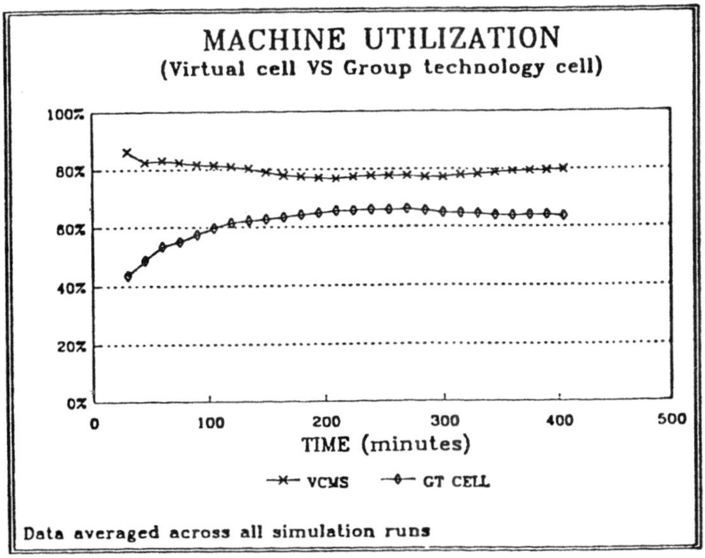

Fig. 8 Machine utilization

one must also consider the cell controller software for systems such as this; it will be complex.

The virtual cell does seem promising, however. The results presented here should encourage further study of this facility design strategy.

6 REFERENCES

1. Bullinger, H.J., H.P. Warnecke, and H. Lentes, "Toward the Factory of the Future," *Proceedings of 8th International Conference on Production Research* (August 1985).
2. Drolet, J.R., "Scheduling Virtual Cellular Manufacturing Systems," Unpublished Ph.D. Dissertation, Purdue University (June 1989).
3. Greene, T.J. and C.M. Cleary, "Is Cellular Manufacturing Right for You?", *Proceedings of Annual International Industrial Engineering Conference* (1985).
4. McLean, C.R., H.M. Bloom, and T.H. Hopp, "The Virtual Manufacturing Cell," *Proceedings of Fourth IFAC/IFIP Conference on Information Control Problems in Manufacturing Technology*, Gaithersberg, MD (October 1982).
5. Schonberger, R.J., *World Class Manufacturing: The Lessons of Simplicity Applied*, The Free Press, New York, NY (1986).
6. Solberg, J.J., D.C. Anderson, M.M. Barash, and R.P. Paul, "Factories of the Future: Defining the Target," A report of a research study conducted under National Science Foundation Grant MEA 82212044, Purdue University, West Lafayette, IN (January 1985).
7. Sule, D.R., *Manufacturing Facilities*, PWS-KENT Publishing Company (1988).
8. Tompkins, J.A. and J.D. Spain, "Utilization of Spine Concept Maximizes Modularity in Facilities Planning," *Industrial Engineering*, Vol. 13, No. 3, pp. 34-42 (March 1981).
9. Tompkins, J.A. and J.A. White, *Facilities Planning*, John Wiley and Sons, Inc. (1984).
10. Drolet, J.R and B. Montreuil, "An Efficient Model for Scheduling Virtual Cellular Manufacturing Cell Systems," Unpublished working paper, University of Quebec (January 1990).

A MODULAR ALGORITHM FOR PRE-SIMULATION DESIGN OF AUTOMATIC GUIDED VEHICLE SYSTEMS

Gerald W. Evans
Mickey R. Wilhelm
John S. Usher
University of Louisville

ABSTRACT

The AGVS design problem is extremely complex. The interrelationships among the many design variables make the development of a single, comprehensive analytical solution model practically impossible. Therefore, this chapter presents a modular modeling framework for breaking the AGVS design problem down into smaller, more tractable, sub-problems. The solutions resulting from each module act as input to the next sub-problem module. This process is iterated until convergence is reached.

1 INTRODUCTION

Automated guided vehicles (AGVs) are important components of many modern manufacturing systems. These AGVs are driverless vehicles used for transporting goods and materials throughout a facility, usually by following either a wire guidepath embedded on the floor, or a chemical guidepath painted on the floor [10].

An automated guided vehicle system (AGVS) includes not only the vehicle(s) and its guidance system, but also a set of pickup/delivery stations strategically located throughout the manufacturing facility, and a computer (or computers) for controlling the system. The computer determines vehicle destinations and routes, controls traffic in such a way so as to avoid collisions and minimize blocking, and schedules pickups and deliveries of loads to optimize some appropriate economic or productivity criteria.

AGVSs are inherently flexible in nature. Because of this flexibility, the problems associated with the design and operation of these systems are extremely complex. In particular, a large number of interdependent design variables, as well as various performance measures, must be considered in the design process. The complexity of the AGVS design problem has led many researchers to investigate the use of mathematical modeling techniques. Evidence of this is indicated by the fact that various simulation languages (e.g., SLAM-II [15] and SIMAN [14]) now include special facilities for the modeling of material handling systems such as AGVSs.

Although simulation remains the most popular modeling tool for solving design and operational problems associated with AGVSs, there are difficulties associated with this tool. In particular, simulation models are time consuming to build and run. Hence, several researchers (e.g., Suri [18] and Solberg [17]) have espoused the use of analytical models in manufacturing system analysis and design. These analytical models, based on such techniques as queuing theory, deterministic analyses, mathematical programming, and heuristics, retain many advantages over simulation models, especially in the early stages of design. For example, analytical models are usually not nearly as time consuming to build and run as simulation models. Hence, many more alternative designs can be investigated, which is especially important in the early stages of design.

Analytical models can be used in a complementary fashion with simulation models. That is, analytical models can be used as a pre-simulation tool to greatly reduce the number of designs to analyze with a simulation model (Fig. 1). This is the approach recently suggested by Haider et al. [9] in their work for IBM.

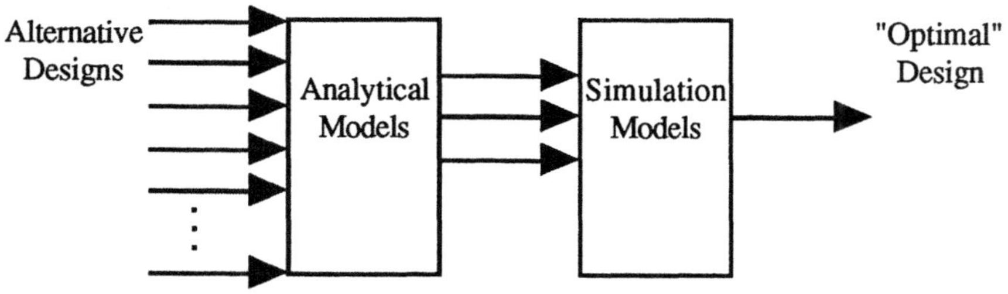

Fig. 1 Complementary use of analytical and simulation models for manufacturing system design

In this chapter, we discuss the development of such an analytical decision support system for AGVS design and operation. The system is composed of several modular components in order to facilitate the investigation of various alternative modeling approaches for each component.

In the next section of the chapter, we discuss the general application areas for AGVSs. In the third section, we present a brief literature review of analytical modeling approaches categorized according to the design and/or operational variable investigated. We also discuss the drawbacks of these approaches as a prelude to the fourth section of the chapter wherein we propose a decision support system. Finally, in the last section of the chapter, we present some observations and conclusions.

2 GENERAL AREAS OF APPLICATION FOR AGVS

The characteristic application areas for an AGVS can be categorized as warehousing, assembly, and job shop. An example guidepath layout for each of these applications is shown in Fig. 2.

In the warehousing application illustrated in Fig. 2a, the AGVs transport unit loads between receiving and storage, shipping and storage, or between receiving and shipping for backordered, or "hot," items. The operational problems associated with this application are not very difficult since the movement patterns, flows and pickup/delivery (P/D) station locations are well defined. Therefore, the routing and dispatching problems associated with this application are relatively easy to solve.

In the assembly application shown in Fig. 2b, the AGVs travel in one direction around the guidepath with work-in-process on board. The spurs labeled 1 through 5 represent assembly stations at which components are added to the product aboard the vehicle. Since travel is unidirectional, each vehicle may make several trips around the guidepath, stopping at a different assembly station each trip, depending upon the process routing for the product being assembled. Spurs 6 and 7 represent locations at which the vehicles may be resequenced during travel on the network due to production requirements or vehicle blocking considerations at assembly stations.

AGV assembly application problems are generally more difficult to model analytically than warehouse applications. However, they tend to be less difficult to formulate and solve than those for job shop applications.

In the job shop application illustrated in Fig. 2c, the AGV network is used to move materials among eight departments. The layout of the network is shown along with the direction of vehicle travel on each arc. Mathematical models can be used in this application to answer questions such as: (a) How many vehicles are required?; (b) What guidepath layout should be used?; (c) Where should the pickup and delivery (P/D) stations be located?; (d) How should AGVs be dispatched to pick up and deliver loads from or to the departments? Models used to address these problems for a job shop application tend to be very difficult to formulate and solve.

In the next section of the chapter, we give a brief overview of some of the analytical models developed for AGV's design and operation. The emphasis is on models which address job shop applications.

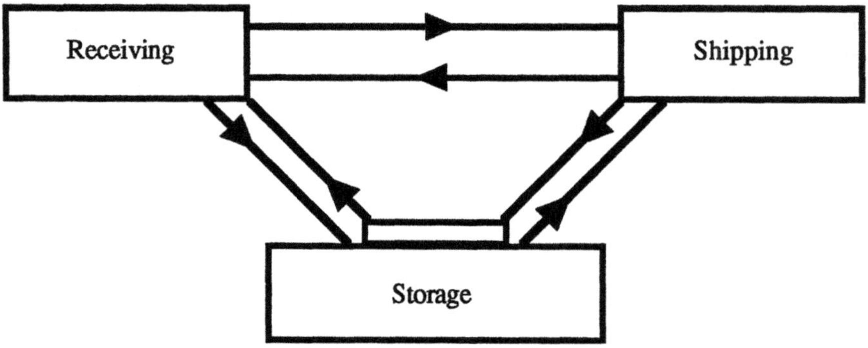

(a) Warehousing application example

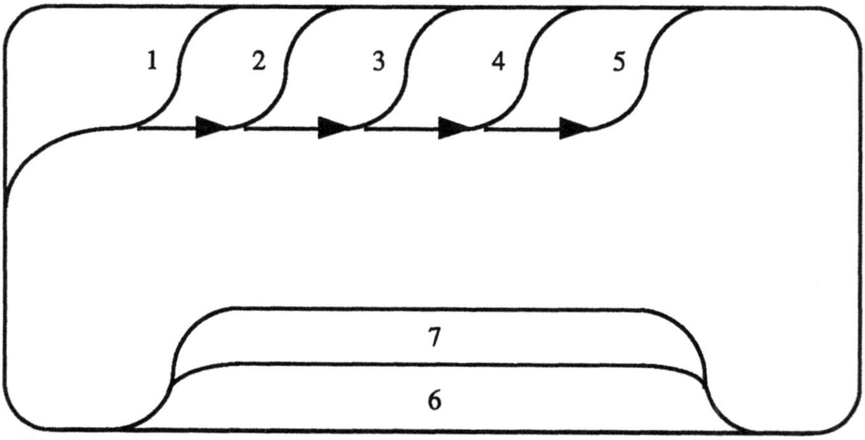

(b) Assembly application example

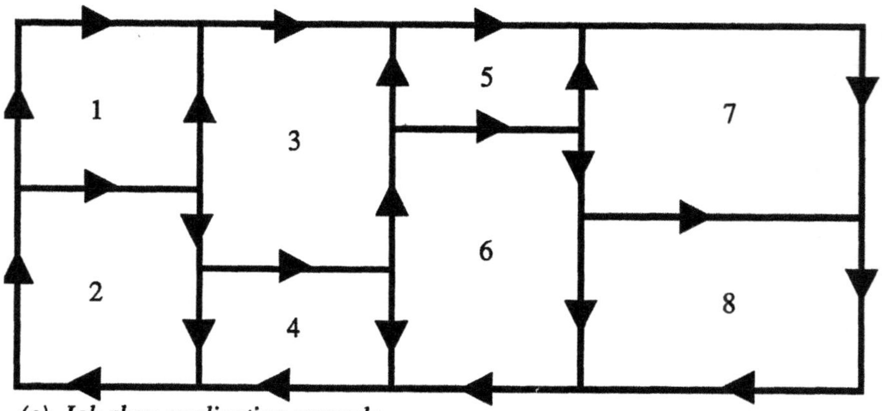

(c) Job shop application example

Fig. 2 Characteristic application areas for AGVSs

3 ANALYTICAL MODELS FOR AGVS DESIGN AND OPERATION

As noted earlier, one of the difficulties associated with the design and operation of an AGVS is that a large number of interdependent variables must be considered. The number of design and operational scenarios is typically so large that it may be impossible to simulate every single one of them. Hence, analytical models, which are typically easier to build and solve than simulation models, may be built to eliminate many of these scenarios from consideration prior to more detailed analysis with a simulation model.

In this section of the chapter, we discuss some of these analytical models. The discussion is organized according to the design variable which the model addresses. Reviews of a much more comprehensive nature can be found in either Wilhelm and Evans [21] or, more recently, in Malmborg [12]. As mentioned previously, the variables of interest in the design and operation of an AGVS include: The number of vehicles required for the system, the guidepath layout (direction and placement), the locations of the pickup and delivery (P/D) stations, and the dispatching and routing policies for the vehicles.

Maxwell and Muckstadt [13] appear to be among the first researchers to address the application of analytical modeling techniques to AGVS design and operation. They formulated a static analytical model for determining the minimum number of vehicles required for an AGVS and the expected blocking time for vehicles.

Following this work, Egbelu [4] presented four analytical procedures for estimating the number of vehicles required for an AGVS. These procedures were basically simple calculations based upon projected material flows between departments. Egbelu compared the results from his procedures with those provided by a simulation model employing combinations of the dispatching rules presented by Egbelu and Tanchoco [5]. His results indicated that the estimates obtained from the analytical procedures are consistently below the number of vehicles estimated by the much more detailed simulation procedure, which accounted for dispatching and routing policies.

Leung, Khator, and Kimbler [11] also addressed the problem of determining the required number of vehicles for a system, through the use of a mixed-integer linear programming formulation of the problem. Their formulation considered the criterion of minimization of total vehicle travel time.

With respect to descriptive modeling approaches which address the "number of vehicles problem," Tanchoco, Egbelu, and Taghaboni [19] compared the CAN-Q analytical model to a simulation model as aid for estimating the number of vehicles required in an AGVS. Their analysis showed that CAN-Q is a good approximating tool for determining a lower bound on vehicle requirements.

With respect to the problem of AGV routing, Blair [2] has developed heuristic approaches which attempt to minimize the distance of the longest vehicle route. In related work, Blair, Charnsethikul, and Vasques [3] presented a heuristic method for organizing a static set of material moves into tours with the objective of minimizing the maximum tour length. Their two-phased heuristic requires the solution of multiple traveling salesman problems. Since the traveling salesman problem is NP-complete, the performance of the heuristic degrades rapidly as problem size increases.

In the study of dispatching rules for AGVs, simulation modeling appears to be the most popular tool of choice. For example, Egbelu and Tanchoco [5] developed a dedicated simulation program, AGVSim, to test combinations of various dispatching rules for the two types of dispatching problems: workcenter initiated task assignment, and vehicle initiated task assignment problems.

In the area of guidepath design, Egbelu and Tanchoco [6] developed design strategies for single guidepath, bidirectional flow systems; and, Gaskins and Tanchoco [7] developed a 0-1 integer program for the design of unidirectional guidepath systems. The Gaskins and Tanchoco model used the objective of minimization of loaded vehicle travel.

Since the guidepath design depends upon the location of the P/D stations in the network (and vice versa), Usher, Evans, and Wilhelm [20] formulated a heuristic algorithm for simultaneously determining the unidirectional guidepath design and the locations of the P/D stations on the network. The algorithm basically involves two iterative phases, with the first phase involving the solution of the Gaskins and Tanchoco [7] integer program. The Usher, Evans, and Wilhelm heuristic also addresses the objective of minimization of loaded vehicle travel. Rabeneck, Usher, and Evans [16] extended this algorithm so that the criterion of total vehicle travel could be considered.

The models referred to above can be separated into two basic categories: descriptive or valuative models (e.g., simulation and queuing theory models) and prescriptive or optimization models (e.g., integer programming formulations). The descriptive approaches often contain the advantage of considering many of the design and operational variables simultaneously; however, these approaches contain no inherent optimization capability. The prescriptive procedures that have been developed do, of course, contain some optimization capability, but, they tend to analyze a small, selected subset (e.g, only one variable) of the entire set of design and operational variables associated with the overall problem. The other variable values are assumed to be ceteris paribus, and therefore represent input to the problem.

What is needed, therefore, is a prescriptive decision support system which allows the consideration of several design and operational variables simultaneously. Such a system should consist of various interacting modules which are designed to consider the relevant design and operational variables. Each module should address a restricted subset of the total set of variables using a specific modeling approach. Then, through this modular design, various modeling approaches could be tested for applicability and effectiveness in each module. This system could be used in greatly reducing the number of alternative designs prior to their evaluation by the use of a detailed simulation model, as shown in Fig. 1. Such a decision support system is discussed in the next section of the chapter.

4 THE MODULAR MODEL DESIGN FORMAT

As stated previously, the AGVS design problem is extremely complex in nature. The interrelationships among the many design variables make the development of a single, comprehensive analytical solution model nearly impossible. Even if such a model were developed, it would most certainly be intractable for reasonably sized AGVS design problems. As such, the use of heuristics in a modular format has been proposed as an effective approach for solving the AGVS design problem. In such a modular format, the overall problem is broken down into smaller, more tractable, sub-problems. Each of the sub-problems can be solved using optimal seeking or heuristic procedures depending on the problem. The resulting solutions then act as input to the solution of the next sub-problem. This process is iterated until convergence is reached.

Goetschalckx and McGinnis [8], present one type of modular design methodology for "... carrier based material flow systems in general, and for AGV systems in particular". Their resulting engineering work station (EWS), designed at Georgia Tech, allows the user to graphically develop an initial guidepath layout, estimate the required number of vehicles, refine the layout, and evaluate empty vehicle dispatching rules and general vehicle routing in an iterative fashion. The system calculates various system performance characteristics such as traffic intensity on given guidepath segments, total loaded and unloaded vehicle travel, etc. The system also allows the user to create a discrete simulation of the layout, and to perform some limited types of optimization. They have developed this PC-based system by integrating several existing and commercially available software packages with their own specialized software and a neutral database.

The Georgia Tech EWS is a state-of-the-art system. It offers the user an unmatched tool for the design and evaluation of an AGVS. It is however, more evaluation oriented than optimization oriented. That is, the system evaluates an initial design and, through the output of various performance measures, helps the user improve the design. As such, achieving a "good" final design relies heavily on the user's initial selection of key design parameters, such as flow path directions and P/D station locations, as well as extensive user expertise to find ways to improve the design. As a result, there still exists a need for a system capable of taking basic AGVS requirements and providing a near-optimal initial design. This design may then serve as the starting point for further evaluation and refinement via the Georgia Tech EWS, simulation, or other means.

Here we propose a modular modeling approach for analytically obtaining a good design of an AGVS prior to the use of evaluative techniques such as the

Georgia Tech EWS or simulation. The overall objective of the approach is to simultaneously find the flow directions along the guidepath network and the locations for the P/D stations such that the total vehicle travel distance (loaded and unloaded) is minimized.

The assumptions made by the model are as follows:

(a) AGV flow will be around the perimeter of a department served by the system,
(b) Unidirectional vehicle flow on each arc of the guidepath network,
(c) P/D stations may not be located at the intersections of aisles in the layout,
(d) The number of vehicles required and operational issues such as dispatching will be determined after directions of vehicle flow and P/D station locations have been found by this model.

In its current form, the heuristic approach, which is iterative in nature, is represented by four separate modules as seen in Fig. 3.

1. DATA INPUT - This module is to be designed to receive critical AGVS design information from the user. This data must consist of:
   - The block layout of the facility
   - The volume of loaded flow (trips/day, loads/year, etc.) between departments in the block layout
   - The initial locations of all P/D stations.

   This data is entered interactively by the user and stored in a common database consisting of two files. These files are:

   a) Total flow matrix - contains the total flow between departments, i.e., loaded and unloaded flow. Unloaded flow is calculated as the expected number of trips between departments resulting from the loaded flow information (see Rabeneck, Usher, and Evans [16]).
   b) Network description - contains the point-by-point information to describe the guidepath network layout in terms of intersections (or nodes), edges, edge lengths (or costs), and P/D station locations on the network. This portion of the common database is updated as iterations proceed and improvements in the overall design are found.

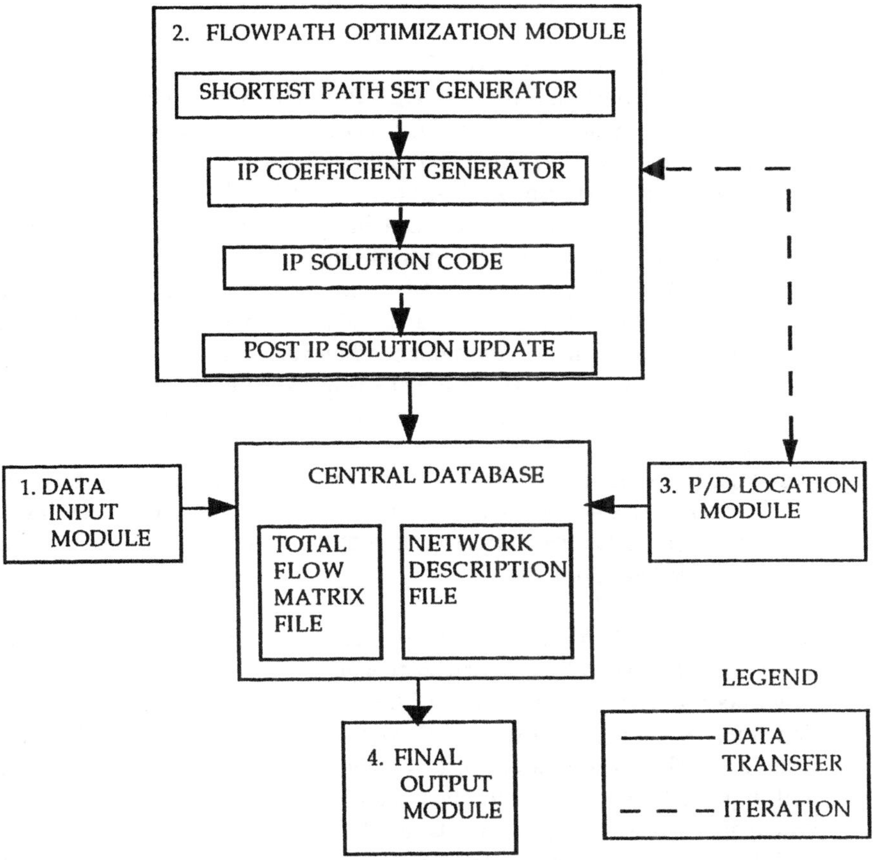

Fig. 3 Modular algorithm for AGVS pre-simulation design

2. FLOW PATH OPTIMIZATION - This module finds the directions of travel along each edge in the guidepath network, thereby converting edges to arcs, such that the AGV travel distance is minimized. Little work has been done to provide an optimal solution to this difficult problem. Gaskins and Tanchoco's [7] 0-1 integer program (IP) is currently used in the module. While this procedure has been found to yield possibly infeasible, and non-optimal solutions for certain cases, see Bakkalbasi and McGinnis [1], it does in most cases provide good solutions. It is, however, very cumbersome to apply since the integer program formulation becomes very large, even for

relatively small problems. Therefore, implementing this complex procedure is quite involved and requires the following sub-modules:

a) Shortest Path Set Generator - Gaskins and Tanchoco's method requires that the shortest path between each P/D station be calculated prior to the development of the objective function. This sub-module must read the network description file and find the shortest paths between all P/D stations. It must also check for non-feasibility that can result when a set of shortest paths are created in which flow is forced in both directions along one or more network edges.
b) IP Coefficient Generator - The function of this sub-module is to read the total flow matrix file and the shortest paths and generate the objective function and constraint coefficients for the IP.
c) IP Solution Code - This sub-module is designed to find the optimal 0-1 solution for the IP. Many choices of solution techniques are possible here.
d) Post-IP Solution Update - This sub-module is intended to check the optimal 0-1 solution to see whether an improvement has been found and, if so, update the current version of the flow path that resides in the network description file.

3. P/D STATION LOCATION - At this point in the approach, the directions of travel along the edges of the guidepath network (or arcs) have been determined based upon the assumed locations of the P/D stations. The next step is to find improved locations for the P/D stations. This module employs the random search heuristic developed by Rabeneck, Usher, and Evans [16], to find improved locations for the P/D stations. It requires reading the total flow matrix file as well as recalculating the shortest path distances between all P/D stations in the layout (only the distances are needed not the actual paths). This module checks whether an improvement has been found. If so, it will update the network description file and return to module 2 to search for improved flow paths.

4. FINAL OUTPUT - The objective of this module is to read the updated network description file and output the final results once the heuristic algorithms in modules 2 and 3 have converged.

5 RATIONALIZATION OF THE MODULAR APPROACH

There are two primary advantages provided by the modular approach to AGVS design proposed in this chapter: flexibility and ease of modeling enhancement.

The flexibility arises from the modularity incorporated into the design of the model. For example, the model contains a shortest path module for determining the network distances between P/D station locations, between P/D stations and guidepath intersection points (nodes on the network), etc. With the modularity of the model, it is quite easy to change shortest path algorithms as more efficient ones are discovered which may have special application to AGVS networks. Similarly, as more efficient integer programming codes are found, or as better heuristics are found for determining vehicle flow directions and P/D station locations, the model is flexible enough to accommodate new modules incorporating these codes and heuristics which will give better AGVS design solutions.

Some thought has also been given to enhancements of the modular approach to pre-simulation AGVS design. For example, the discussion in a preceding section detailed the rather extensive amount of input data required to execute this model. And, while the modules described previously aid the AGVS designer in determining a layout of the system network, directions of vehicular flow on each network arc, and P/D station locations on the network, the user must be very familiar with the operation of the model to insure that the input data is entered correctly. Thus, an obvious enhancement is an expert system to be used for gathering the input data from users, formatting the data for entry into the common data base, and initiating execution of the model.

Another enhancement which is envisioned at this time is the graphical portrayal of the design solutions output by the model. The output of the model could be linked to commercially available computer-aided drafting software such as AUTOCAD to display the design results to the user.

The model could also be enhanced with the addition of a module which computes the minimum number of vehicles required in the AGVS system. This module would operate on the output data after the directions of flow in the network and P/D station locations have been determined.

Finally, a post-processor module could be added to the model to ready the output for input to a simulation model. The simulation model would be used to either validate the design decisions made by the analytical model, or to determine where problems exist with the design during simulated operation of the system.

6 REFERENCES

1. Bakkalbasi, O. and L.F. McGinnis, "Technical Note: Flow Path Design for Automated Guided Vehicle Systems," submitted for publication to *Int. J. of Prod. Res.* (1989).
2. Blair, E.L., "Vehicle Routing to Support Automated Manufacturing," *Proceedings of the International Industrial Engineering Conference*, Institute of Industrial Engineers, pp. 254-256 (1985).
3. Blair, E.L., P. Charnsethikul, and A. Vasques, "Optimal Routing of Driverless Vehicles in a Flexible Material Handling System," *Material Flow* 4, 1&2, pp. 73-83 (1987).

4. Egbelu, P.J., "The Use of Non-Simulation Approaches in Estimating Vehicle Requirements in an Automated Guided Vehicle Based Transport System," *Material Flow* 4, 1&2, pp. 17-32 (1987).
5. Egbelu, P.J. and J.M.A. Tanchoco, "Characterization of Automatic guided Vehicle Dispatching Rules," *Int. J. of Prod. Res.* 22, 3, pp. 359-374 (1984).
6. Egbelu, P.J. and J.M.A. Tanchoco, "Potentials for Bi-Directional Guide-Path for Automated Guided Vehicle Based Systems," *Int. J. of Prod. Res.* 24, 5, pp. 1075-1097 (1986).
7. Gaskins, R.J. and J.M.A. Tanchoco, "Flow Path Design for Automated Vehicle Systems," *Int. J. of Prod. Res.* 25, 5, pp. 667-676 (1987).
8. Goetschalckx, M. and L.F. McGinnis, "Engineering Workstation is Design Tool for Computer-Aided Engineering of Material Flow System," *Industrial Engineering* 21, 6, pp. 34-38 (1989).
9. Haider, S.W., D.G. Noller, andT.B. Robey, "Experiences with Analytic and Simulation Modeling for a Factory of the Future Project at IBM," *Proceedings of the 1986 Winter Simulation Conference*, (Eds: J. Wilson, et al.), pp. 641-648 (1986).
10. Hodgson, T.J., R.E. King, S.K. Monteith, and S.R. Schultz, "Developing Control Rules for an AGVS Using Markov Decision Processes," *Material Flow* 4, 1 & 2, pp. 85-96 (1987).
11. Leung, L.C., S.K. Khator, and D.L. Kimbler, "Assignment of AGVS with Different Vehicle Types," *Material Flow* 4, 1&2, pp. 65-72 (1987).
12. Malmborg, C.J., "A Compendium of Archival Literature in AGVS Modeling," *Proceedings of the International Industrial Engineering Conference*, Institute of Industrial Engineers (1989).
13. Maxwell, W.L. and J.A. Muckstadt, "Design of Automatic Guided Vehicle Systems," *IIE Transactions* 14, 2, pp. 114-124 (1982).
14. Pegden, C.D., *Introduction to SIMAN*, Systems Modeling Corporation, State College, PA (1982).
15. Pritsker, A.A.B., *Introduction to Simulation and SLAM II, 3rd ed.*, John Wiley, New York (1986).
16. Rabeneck, C.W., J.S. Usher, and G.W. Evans, "An Analytical Model for AGVS Design," *Proceedings of the International Industrial Engineering Conference*, Institute of Industrial Engineers, pp. 191-195 (1989).
17. Solberg, J., "A Mathematical Model of Computerized Manufacturing Systems," *Proceedings of the Fourth International Conference on Production Research*, (1977).
18. Suri, R., "Modeling Flexible Manufacturing Systems Using Mean-Value Analysis," *Journal of Manufacturing Systems* 3, 1, pp. 27-38 (1984).
19. Tanchoco, J.M.A., P.J. Egbelu, and F. Taghaboni, "Determination of the Total Number of Vehicles in an AGV-Based Material Transport System," *Material Flow* 4, 1&2, pp. 33-51 (1987).

20. Usher, J.S., G.W. Evans, and M.R. Wilhelm, "AGV Flow Path Design and Load Transfer Point Location," *Proceedings of the International Industrial Engineering Conference* (1988).
21. Wilhelm, M.R. and G.W. Evans, "The State-of-the-Art in AGV Systems Analysis and Planning," *Proceedings of the AGVS '87 Forum*, Material Handling Institute, pp. 51-70 (1987).

RECENT ANALYTICAL RESULTS ON MINILOAD PERFORMANCE

R.D. Foley
E.H. Frazelle
Georgia Institute of Technology

ABSTRACT

We describe a basic model of miniload order picking systems. We also review selected analytical results on miniload performance.

1 INTRODUCTION

Miniload automated storage and retrieval systems are in the class of end-of-aisle or part-to-picker order picking systems. Miniload systems are popular for small parts order picking because they provide excellent space utilization, excellent item security, and accurate item picking [6]. In fact, over 500 miniload systems are currently in use in the U.S. [12]. However, at over $300,000 per aisle [15] and with limited ability to reconfigure, the initial system design must be accurate. In addition, system designers must make every effort to maximize system productivity in order to increase the return on a significant investment. System productivity can often be enhanced by taking advantage of the fact that a small minority of the items housed in the system generate a large majority of the bin retrievals. We show how this activity skew affects system throughput.

In this chapter, we summarize recent results for analytically determining miniload performance (i.e. throughput, picker utilization, and storage/retrieval machine utilization). We begin with a brief description of miniload system configuration and order picking operations.

2 PHYSICAL DESCRIPTION OF A MINILOAD AS/RS

A typical system (Fig. 1) is comprised of multiple aisles of storage racks, a S/R machine operating in each aisle, numerous modular storage bins for housing the items, and two load stands (or pick positions) at the end of each aisle to facilitate order picking.

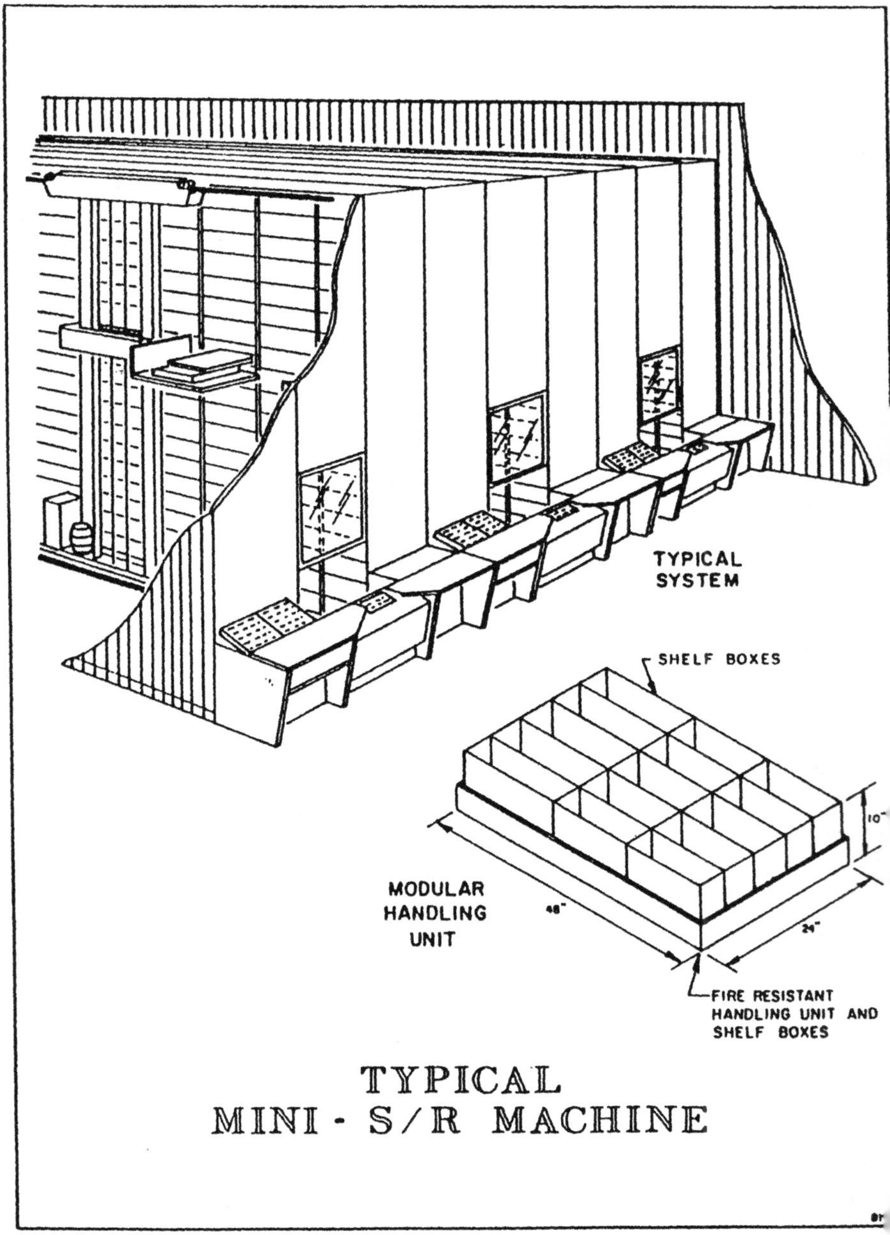

Fig. 1 Miniload AS/RS, reprinted from [17]

The load stands are arranged such that each aisle has a left and a right pick position. While the order picker is extracting items from the bin in one pick position, the S/R machine returns the container in the other pick position to its permanent, dedicated location in the rack, and returns with the next bin to be processed. If the picker finishes before the S/R machine returns with the next bin, the picker waits idle. If the S/R machine returns with the next bin before the picker is finished, the S/R machine waits idle.

The *dual command cycle time* is the length of time the S/R machine takes to pick up the bin from the pick station, return the bin to its home location, travel empty to the next bin location, retrieve the bin, travel to the I/O point, and deposit the bin on the empty table. The *pick time* is the length of time needed by the picker to process the bin. The processing activities may include extracting the items, documenting the tasks, counting and weighing items, interruptions, and restocking.

Not all miniload systems are configured like the one shown in Fig. 1. Systems may be configured with more than two pick positions per aisle, with a conveyor delivery system to deliver containers to "remote" order pickers, or with multiple input/output (I/O) points per aisle. However, it is the typical configuration, with one picker, two pick positions per aisle, and one I/O point located at the end of each aisle, that is modeled here. Since each aisle operates independently, we limit our analysis to a single aisle.

3 MODEL ASSUMPTIONS

In our analysis we assume retrievals are processed on a first-come-first-served (FCFS) basis and that there is an infinite queue of requests. Thus, we are determining the maximum possible throughput; i.e. the throughput capacity of the numbered aisle. We also assume the sequence of requested bin locations are independent and identically distributed. Therefore we are not allowing dependence induced by the fact that certain bins tend to be requested successively. For example, items in bin 2 may be associated with items in bin 8. Then, whenever, bin 2 is requested, bin 8 tends to be requested next [7].

We make three important assumptions concerning the characteristics of S/R machine travel within the aisle. First, we assume that the travel times are calculated with the Chebychev metric. Since one motor powers the mast horizontally back and forth along the guide rails, and another motor advances the shuttle table vertically up and down the mast, the assumption seems valid.

Our second assumption is that the S/R machine accelerates and decelerates instantaneously to/from its maximum travel speed. This is not the case in practice, however, new S/R machines are approaching instantaneous acceleration/deceleration.

We consider a variety of design scenarios. In each, the rack is assumed to be rectangular, and in some cases the rack is assumed to be square-in-time. A square-in-time rack is one in which the time required for the S/R machine to travel horizontally from the front of the rack to the back is equal to the time required to travel vertically from the bottom of the rack to the top. Since the horizontal travel speed is typically 2 to 4 times the vertical speed, the rack may be square-in-time but not square physically.

We have analyzed systems using both random and activity based storage. Under random storage activity, successive bin retrieval locations are assumed to be uniformly distributed over the rack face. Under activity based storage, we assume that the rack is divided into two regions for bin location assignments (Fig. 2). The region closest to the I/O point, region I, is assumed to house the high turnover items. Region I is defined on the normalized rack by the points (0,0), (a,0), (a,a), and (0,a). Region II occupies that portion of the rack not contained in region I. Within each region, we assume the random sequence of location visits are uniformly distributed. This is the case when items are classified as "fast" or "slow" movers and items are assigned to locations within a region without additional consideration of their turnover. For a more detailed discussion of these assumptions, see [4].

In some cases the distribution of pick times is assumed known, and in other cases only partial information on pick times is assumed available (e.g. mean). When only partial information is available, we have obtained tight upper and lower bounds on system performance.

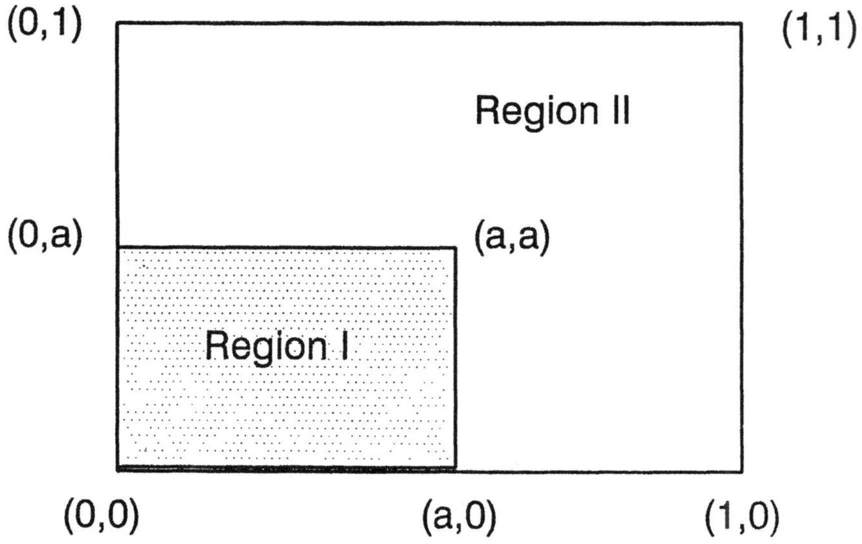

Fig. 2 Square-in-time rack with two activity zones

4 SUMMARY OF RESULTS

Several authors have conducted research on automatic storage and retrieval systems (AS/RS) travel time models. Gudehus [9] derived expressions for the expected single command travel time in an AS/RS as a function of the system's length, height, and horizontal and vertical travel speeds and acceleration. Hausman, Schwarz, and Graves [11] developed expressions for expected single command travel in square-in-time (SIT) racks with class-based dedicated storage. Graves, Hausman, and Schwarz [8] later developed an expression for dual command travel in SIT racks under class-based storage. Bozer and White [3] developed expressions for expected single and dual command travel time for racks of general shape in a variety of configurations. Finally, Han, McGinnis, Shieh, and White [10] derived an analytical expression for expected dual command travel time for a general rack shape, random storage, and nearest neighbor retrieval sequencing.

To our knowledge, the only published works which address the interaction of travel and processing time is that of Bozer and White [2] and Foley and Frazelle [4]. Bozer and White [2] developed approximate expressions for the expected rate at which bins are processed in a miniload AS/RS when bin processing times are assumed to be exponentially or uniformly distributed. They assume the dual command travel time is uniformly distributed with the same mean as the true distribution and with a variance which has been empirically determined to be close to the true variance.

Foley and Frazelle [4] develop two main results. First, they specify conditions for the long run average number of bins processed per unit time to converge to a number called throughput, and provide expressions for the throughput. In these expressions, the system throughput and other performance measures are unchanged when the pick time depends on the location of the bin currently being processed. The expressions use the distribution of dual command travel time. Second, they derived a closed form expression for the distribution of dual command travel time in a square-in-time rack with randomized storage.

In a later work, Foley and Frazelle [5] extended their earlier results in two directions. First, they derived a closed form expression for dual command travel time for square-in-time systems with two activity classes. This allowed the determination of the increase in performance available due to optimal two-class zoning as a function of the activity skew. Second, they derived bounds on system performance when only partial information about pick times is available (e.g. mean pick time or that all pick times lie in some closed interval). Six scenarios were considered:

(1) No pick time information,
(2) Mean pick time known,
(3) Mean and variance of pick time known,
(4) Mean and range of pick time known,

(5) Mean, variance, and range of pick time known, and
(6) Mean given and known NBUE (New Better than Used in Expectation).

Scharfstein and Foley [13] derived the cumulative distribution for dual command travel time in non-square-in-time racks with randomized storage. They also adapted a design algorithm for determining all system configurations which satisfy throughput and storage capacity requirements and size restrictions with a minimum number of aisles.

In [4,5, and 13] the expected number of picks per bin retrieval was ignored, and throughput was measured in bins processed per unit time. One way to improve system performance is to increase the number of picks per bin retrieval. Frazelle [7] developed procedures for assigning items to miniload bins to significantly increase picker utilization by increasing the number of picks per bin retrieval.

5 EXAMPLE

Example 1: Random and Storage

Consider a rack 30 feet high and 120 feet long serviced by an S/R machine which travels 90 feet per minute vertically and 360 feet per minute horizontally. Hence, the S/R machine takes 1/3 minute to travel from the front to the back and also from the bottom to the top of the rack. The expected processing time per container is 1.5 minutes, and the constant time for each pick-up or deposit is 0.15 minutes.

Since we have assumed that the S/R machine takes one unit of time to travel from the front to the rear of the back, it is easier to normalize the data to take this into account. Since the S/R machine takes 1/3 minute to travel from the front to the back, we assume that the units of time are 1/3 minutes. Converting the data to these time units results in the rack being one time unit long, one time unit high, the expected processing time being 4.5 time units, and each pickup or deposit requiring .45 time units. Since there are two pickups and two deposits per dual command cycle, c is $4 * .45 = 1.8$ time units.

The expected dual command travel time is $1.8 + c$ in [3,11] which in our example is 3.6 time units. We have not completely specified the pick time distribution, only its mean. Below we calculate the throughput for two particular cases: deterministic and exponentially distributed pick times.

For deterministic pick times with $2 + c < d < 3 + c$, the expected transaction time m is calculated from Equation (12) of [4]. With $c = 1.8$ time units and $d = 4.5$ time units, we have $m \approx 4.50064$ time units or 1.50021 minutes.

Table 1 Results for the random storage example

	Pick Times (mean of 1.5 minutes)	
	Deterministic	Exponentially Distributed
Picker Capacity	40/hour	
S/R Machine Throughput	50/hour	
Picker Utilization	99.9%	79.9%
S/R Machine Utilization	79.9%	63.9%
System Throughput	40/hour	32/hour

Hence, the throughput is $1/m \approx 0.6666$ containers per minute or roughly 40 containers per hour. The picker utilization is $E[P_n]/m \approx 4.5/4.50064 \approx .9998$ or 99.9%. Similarly, the S/R machine utilization is $E[C_n]/m \approx 3.6/4.50064 \approx 0.7999$ or 79.9%.

For exponentially distributed pick times the expected cycle time is calculated from Equation (14) of [4]. With $c = 1.8$ time units and $1/\lambda = 4.5$ time units or equivalently $\lambda = 2/9$, we have $m \approx 5.631$ time units or $m \approx 1.877$ minutes. Hence, the throughput is $1/m \approx 0.533$ containers per minute or roughly 32 containers per hour. The picker utilization is $E[P_n]/m \approx 4.5/5.631 \approx 0.799$ or 79.9%. The S/R machine utilization is $E[C_n]/m \approx 3.6/5.631 \approx 0.639$ or 63.9%. Note, that even though the mean pick time is the same for both cases, the variability of the exponential distribution has decreased the picker utilization, SR machine utilization, and system throughput. Also, note that if we analyzed the S/R machine in isolation, we would conclude that the S/R machine can process 50 containers/hour; similarly, the picker can process 40/hour. However, it would be hasty to state that the picker is the bottleneck and that the system throughput is 40 containers/hour. In the exponentially distributed case, the throughput is only 32/hour and both the picker and S/R machine are affecting the throughput. Decreasing either the pick time or the S/R machine travel time would increase throughput. From the results in [5], the throughput for any system with NBUE pick times with the same mean has to lie between the two cases described above.

Example 2: Activity Based Storage

We give some numerical examples of how the system throughput varies with zone size. We assume that the S/R machine takes 1/2 minute to travel horizontally from the front of the rack to the back. As in Example 1, we normalize time. In this example we use 1/2 minute as the normalized time unit. Henceforth, all system parameters will be given in normalized time units unless otherwise specified.

Since the fraction of items designated high-turnover increases as the size of the high turnover zone increases, we have assumed the following simple relationship. If the length of the high turnover zone is a, then a^2 of the items are stored in the high turnover zone. The amount of activity in the high turnover zone is

$$p(a^2)^\alpha$$

The parameter α determines how skewed the activity distribution is. If $\alpha = 1$, all items have the same activity and zoning will not increase the throughput. Normally, $\alpha < 1$. If one point on the curve is known, the parameter a can be determined. For example, if 80% of the activity is caused by only 20% of the items, then α is .139. In general, $\alpha = \log p / \log a^2$.

For this example we have selected a 90-10 activity skew. This activity skew is reasonable in many applications, and we have seen cases where 90% of the activity was concentrated on only 1% of the items. With such an extreme skew, it might be advisable to remove some of these items and store them in a more readily accessible location. We have set the fixed part of the dual command travel time to $c = 1/10$.

In Fig. 3, the expected dual command travel time is plotted as a function of the high turnover zone size a. With $a = 0$ or $a = 1$, we have essentially no zoning. As a increases, the expected dual command travel time decreases and the more skewed the distribution, the smaller the optimal zone size, and the more dramatic the decrease in expected dual command travel time. However, we need to include the pick time information to assess the effect of zoning on system performance.

Initially, we will consider both deterministic and exponential pick time distributions with a mean of 1. Fig. 4 shows the throughput as a function of the zone size for the 90-10 skew. Note, the greater the skew, the smaller the optimal zone size, and the greater the improvement in throughput. The increase in Fig. 4 for optimal zoning over no zoning is particularly dramatic: 62% for the exponential pick times and 71% for the deterministic pick times.

Similar results showing the effect of the manual pick time and the activity skew appear in [5]. In determining zone size, when the pick time distribution is unknown but the mean pick time is known and the pick time is NBUE, the safest procedure would be to use the optimal zone size from the exponential pick times. From Theorem 1 of [5], this guarantees the throughput will be at least as good as the throughput of the exponential case.

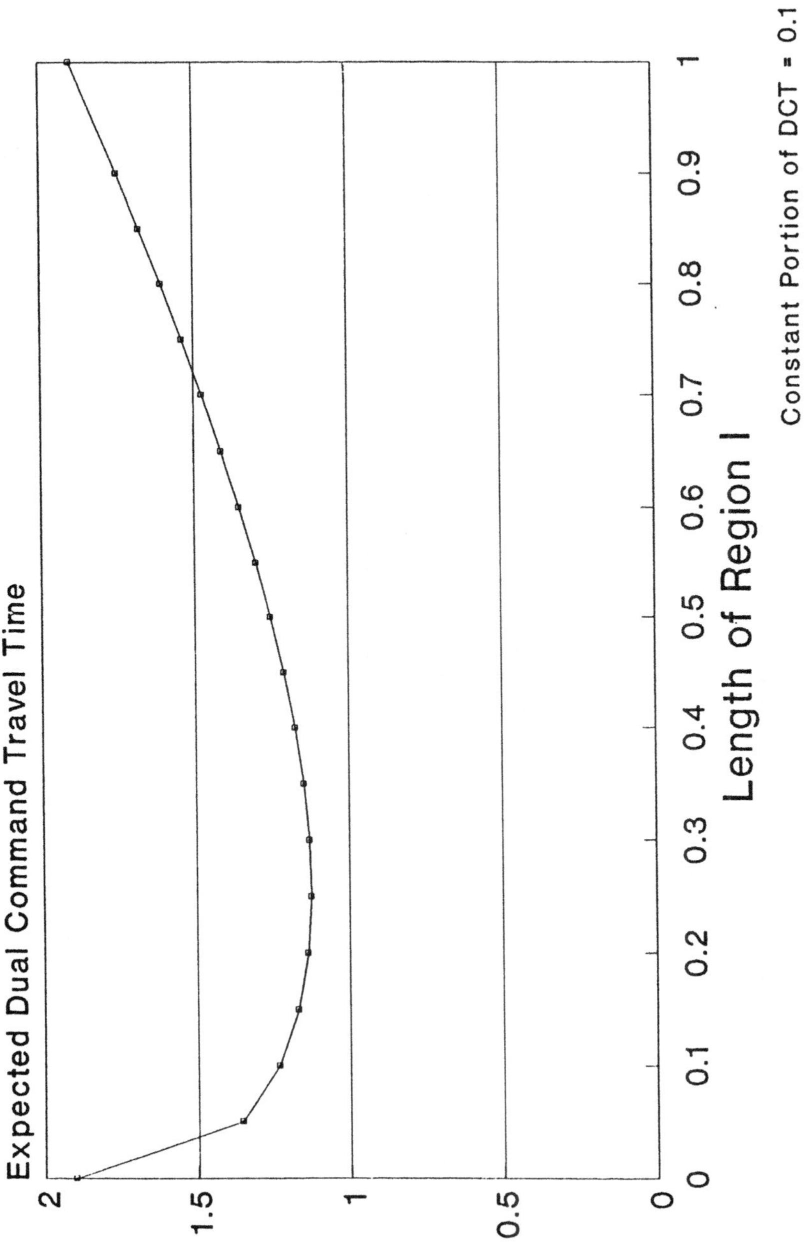

Fig. 3 Expected dual command travel time vs. zonesize

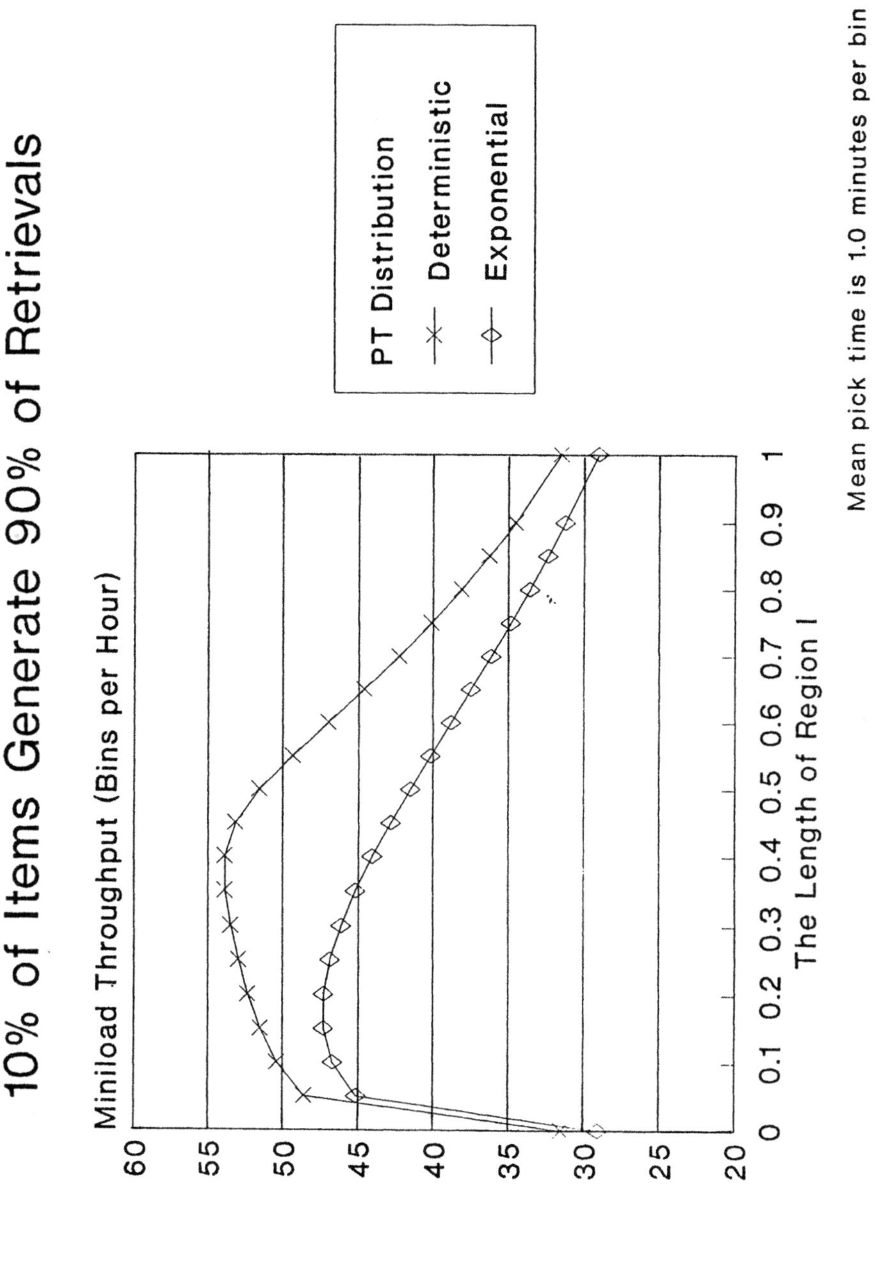

Fig. 4 Miniload throughput vs. zonesize (80-20)

Examples showing how these results can be used to design systems appear in [13]. Most of these results have been implemented in a software package for miniload system design and evaluation. The authors are empirically validating these results by comparison with historical performance of existing systems.

6 REFERENCES

1. Bozer, Y.A., "Optimizing Throughput Performance in Designing Order Picking Systems," PhD Dissertation, Georgia Institute of Technology, Atlanta, Georgia, (December 1986).
2. Bozer, Y.A. and J.A. White, "Design and Performance Models for End-of-Aisle Order Picking Systems," to appear *Management Science*.
3. Bozer, Y.A. and J.A. White, "Travel-Time Models for Automated Storage/Retrieval Systems," *IIE Transactions*, Volume 16, Number 4, pp. 329-338 (December 1984).
4. Foley, R.D. and E.H. Frazelle, "Analytical Results for Miniload Throughput and the Distribution of Dual Command Travel Time," to appear *IIE Transactions*.
5. Foley, R.D. and E.H. Frazelle, "Throughput Expressions and Bounds for End-of-Aisle Order Picking with Activity Based Storage," Technical Report J-90-10, School of Industrial & Systems Engineering, Georgia Institute of Technology, (1990).
6. Frazelle, E. H., "Small Parts Order Picking: Equipment and Strategy," Technical Report, Material Handling Research Center, Georgia Institute of Technology (October 1987).
7. Frazelle, E. H., "Stock Location Assignment and Order Picking Productivity," Ph.D. Dissertation, Georgia Institute of Technology, 1989.
8. Graves, S.C., W.H. Hausman, and L.B. Schwarz, "Storage-Retrieval Interleaving in Automatic Warehousing Systems," *Management Science*, Volume 23, Number 9, pp. 935-945 (May 1977).
9. Gudehus, T., "Grundlagen der Spielzeitberechnung" (1972).
10. Han, M., L.F. McGinnis, J.S. Shieh, and J.A. White, "On Sequencing Retrievals in an Automated Storage/Retrieval System," *IIE Transactions*, Volume 19, Number 1, pp. 56-66 (March 1987).
11. Hausman, W.H., L.B. Schwarz, and S.C. Graves, "Optimal Storage Assignment in Automatic Warehousing Systems," *Management Science*, Volume 22, Number 6, pp. 629-638 (February 1976).
12. Hill, John M., "Controlling Material Handling Systems in Manufacturing and Warehousing," Proceedings of the 1988 Material Handling Short

Course, Georgia Institute of Technology, Atlanta, Georgia, (April 14, 1988).

13. Scharfstein, D. and R.D. Foley, "Analytical Performance Measures for the Miniload Automated Storage/Retrieval System with Randomized Storage in Rectangular Racks", Technical Report, School of Industrial and Systems Engineering, Georgia Institute of Technology, Atlanta, Georgia (1990).
14. Schwarz, L.B., S.C. Graves, and W.H. Hausman, "Scheduling Policies for Automatic Warehousing Systems: Simulation Results," *IIE Transactions*, Volume 10, Number 3, pp. 260-270 (September 1978).
15. Zollinger, H.A., "Planning, Evaluating, and Estimating Storage Systems," presented at the Advanced Material Technology Seminar, Purdue University, Lafayette, Indiana (April 29-30, 1975).
16. "Warehouse Modernization and Layout Planning Guide," Department of the Navy, Naval Supply Systems Command, NAVSUP Publication 529 (March 1985).

IMPACTS OF FLEXIBLE ROUTING IN CONTROL ARCHITECTURES FOR MATERIAL FLOW SYSTEMS

G. Don Taylor
Robert J. Graves
University of Massachusetts

ABSTRACT

The degree to which routing flexibility can influence manufacturing system performance is the focus of this chapter. Of special interest is the performance of material flow control architectures which integrate flexible routing concepts (e.g., the conduct of fabrication/assembly operations in alternate sequences) with flexibility in the determination of product mix or machine level sequencing. The manufacturing context is that of diverse mixed technology circuit boards assembled using a cellular approach. The research methodology is that of simulation using the SIMAN language. Results indicate that the incorporation of flexible routing concepts into material flow architectures can significantly affect system performance, especially when integrated with other operational control mechanisms.

1 INTRODUCTION

In the United States, the concept of flexible manufacturing has largely been implemented in the form of flexible manufacturing systems (FMS) or flexible assembly systems (FAS). The FMS and FAS implementations generally represent computer integrated manufacturing systems consisting of a group of numerical control machining equipment pieces or flexible workstations in the form of cells connected by automated material handling equipment. Little effort, however, appears to have been devoted to the design of generalized processing systems, i.e. non-FMS or non-FAS, to incorporate flexibility concepts. This may be due to the absence of knowledge about the extent to which management and control system design could or should incorporate flexibility concepts. The current state of affairs therefore, is one of "islands of flexibility" where hardware flexibility in the context of design of unit processes has advanced further than system flexibility in the context of design of planning and control systems.

It is therefore important to study the incorporation of flexibility concepts into the planning and control systems for manufacturing. Those manufacturers using traditional processing hardware should be able to examine the potential impacts of flexible management and planning systems on the performance of their manufacturing systems. Also, those manufacturers using flexible manufacturing hardware should be able to assess the degree to which their management and planning systems are taking advantage of the available hardware flexibility. Those who design material handling systems for either of these manufacturing environments must understand the possible effects on the material handling system design when incorporating the dimension of routing flexibility, job mix flexibility, and sequencing flexibility into management and planning systems.

This chapter concentrates on the dimension of material routing flexibility and improved system performance relative to material flow goals. Routing flexibility in this context is defined to be the capability for the conduct of assembly operations in alternate sequences. The context for this concentration is a FAS, but the FAS is not the type of FAS as typically described in the research literature. This FAS environment is one of a number of unit processes which are under common computer control and are devoted to the simultaneous assembly of different types of printed circuit boards (PCBs). Flexible material routing heuristics are evaluated both with and without integrated flexible controls in work-in-process (WIP) sequencing at the machine level and in product mix determination at the system level.

2 RELATED LITERATURE

A great deal of research has concentrated on the product routing problem in a FAS. This section describes some of this work and introduces flexible routing as one of several building blocks in operational control architectures for WIP management. Section 2.1 introduces the use of heuristics to solve the dynamic routing problem and describes some of the recent approaches taken in an effort to make appropriate routing decisions at near real time. Section 2.2 describes the hierarchical breakdown of operational control associated with a general FAS, of which the routing problem is one level. Additionally, section 2.2 provides arguments for solution approaches which integrate decision making between control levels.

2.1 Heuristic Routing Methods

Some present solution approaches to the product routing problem perform dynamically at near real-time. Russo [16] states that, in a dynamic environment, improvements in material flow times are possibly by making routing decisions at the latest possible moment due to the fact that completion times are hard to predict. This opinion is supported by subsequent researchers as well, including Nof et al. [14]. Denzler and Boe [2] take a slight exception to this rule by stating that when minimal queues and reliable machines are involved, completion times can be estimated with more certainty, thus making Russo's observation less important. Perhaps automation can also add to the reliability of estimates, but the observation appears to be valid in general.

Nof et al. [14] studies and compares certain heuristic rules. They find that rules which perform poorly in a static environment may perform very well in a dynamic environment. In addition, the relationship of the routing problem to the product mix problem is developed. Nof et al. claim that having alternative orders or sequences for process steps actually increases productivity in some cases.

Buzacott [1] also demonstrates capacity improvements by making sequence decisions during production. These findings are consistent with Russo [16] as well as the results of the research presented in this chapter.

Other interesting approaches include the work done by Kimemia and Gershwin [7], [8]. These authors model the routing control problem as a network of queues and determine the optimal percentages of parts to follow each possible route with the objectives of minimizing congestion and delay. Additional detail regarding their approach to routing and dispatching is provided in Kimemia and Gershwin [9].

Donath [3] and Donath and Graves [4] describe a heuristic routing method (SCHEDULE) that can handle multiple products with operation times dependent upon the workplace selected. The heuristic assigns jobs to workplaces based upon cost estimates and works toward high workplace utilizations while treating job due dates as constraints. The authors further demonstrate that the heuristic performs quite satisfactorily at near real-time in Donath and Graves [5].

Maimon and Choong [10] have also developed an approach that attempts to achieve real-time load balancing in a stochastic environment. Their approach bases routings on the input buffer status of each work station. They attempt to maximize throughput and minimize WIP while maintaining balanced flow. In Maimon and Gershwin [11], the authors develop local operational rules to route products.

2.2 Hierarchical Production Control

The FAS is composed of a hierarchy of processing centers or levels of production; the total assembly process, specialized assembly cells within the process, and individual machines or processes within the cells. This hierarchy is a convenient way in which to view control levels as well as processing levels. The material routing problem is generally associated with the second (cell) level in the FAS control hierarchy. While each hierarchical level can be considered separately, a certain synergy may result from considering all levels of control in an integrated manner.

Significant literature concentrates on different hierarchical structures for operational planning in a FAS. Perhaps one of the most cited hierarchical design structures is offered by Nof et al. [14]. This chapter suggests the existence of three levels of operational control issues; the part mix problem which includes part type selection and ratio determination, the process selection or routing problem, and the part flow or sequencing task. Nof et al. suggest that these control issues are not independent and also suggest potential areas of improvements to the hierarchical system.

Other authors have suggested similar three level hierarchies. Buzacott [1] uses a three level hierarchy which includes a pre-release planner that determines which jobs to build, an input controller that determines sequence and timing, and a third level which matches jobs with machines. The work of Kimemia and Gershwin mentioned in section 2.1 involves a three level approach. On the top level, they determine product mix in the presence of machine failures. Secondly, the routing control problem is considered. Finally, the lowest level uses scheduling algorithms at specific operations. The work is later improved by Gershwin et al. [6] to reduce the off-line computations and help demonstrate the effectiveness of the hierarchical approach. Donath [3] decomposes the FMS scheduling problem into three components; the operation sequencing problem, the part-workplace assignment problem, and the part sequencing problem. Menga et al. [12] suggest that the hierarchy should be only two levels; a first level which attempts to route products to balance utilization, and a second level which schedules product lots to optimize an index of production performance while attempting to meet due dates.

In addition to suggesting hierarchical approaches to material flow problems in the FAS, researchers also suggest concentrating on total integrated solution procedures which consider the relationships between the control elements at the various hierarchical levels. In Muhlemann et al. [13], the authors suggest developing improved heuristics which work toward broader understanding of the total problem. In Buzacott [1], the author strongly suggests focusing upon the interactions between the levels of control. In Nof et al. [14], it is suggested that no one control issue can be considered while disregarding the others. The authors further note that complex interrelationships make intuitive approaches ineffective.

Even though integrated hierarchical approaches to operational planning may be beneficial, Buzacott [1] has pointed out that multi-media control is very complex, especially when considering dynamic control. The problems associated with dynamic control are noted by other authors as well. Muhlemann et al. [13] recognize the complexity of decision making in a dynamic environment and therefore suggest a concentration on non-optimizing procedures. Authors also recognize that sophisticated computer tools may be needed to help manage hierarchical control systems in a FAS. Vonderembse and Wobser [19] emphasize the importance of centralized computer control to manage real time routing decisions, load balancing, and production scheduling.

The solution approach taken in this chapter, which will be described in detail subsequently, has been influenced by the literature. Routing decisions are made at the latest possible moment in an effort to have maximum impact on material flow performance. The router is integrated with control decisions at other hierarchical levels. The solution approach taken involves the use of non-optimizing heuristics to keep computational complexity to a minimum.

3 EXPERIMENTAL SETTING

Since the primary focus of this effort is to examine the effect of routing flexibility as one component of flexibility in hierarchical control architectures for material flow systems, flexible material routing heuristics are examined both with and without integration to flexible control mechanisms at other levels. Routing is considered to be a cell level control mechanism and is therefore at the second level of control in the three level hierarchy. On the first (system) level, there is consideration of the product mix problem which determines what product to input, the release time for that input, and the initial lot size quantity to be released. On the third (machine) level, a WIP sequencer determines the processing order for jobs residing in each machine queue. This study examines four scenarios:

1) A control scenario that is characterized by a total lack of flexibility in operational decision making. This scenario is examined to establish a baseline for comparison with scenarios which exhibit some degree of flexibility in operational decisions.

2) A scenario identical to the control scenario with the exception of adding flexibility to the sequence of assembly operations required to produce a product. This scenario makes use of dynamic routing heuristics to control material flow.

3) A scenario identical to scenario 2 with the exception of adding flexibility in product mix decisions within the constraints of a tactical level planner. This scenario integrates dynamic routing heuristics with product mix rules to control both the input of material into the system and material routing.

4) A scenario identical to scenario 2 with the exception of adding the ability to control machine level sequencing dynamically. This scenario integrates flexible material sequencing heuristics with dynamic routing heuristics.

4 PRODUCTION ENVIRONMENT

The production basis for this study is a PCB assembly plant in Connecticut. The basis facility and the PCB industry in general are currently faced with challenges presented by very rapidly changing products and processes. The PCB industry hopes to manage this technology transition without the excessive capital expenditures associated with building multiple assembly systems. One clear strategy is to design and build highly flexible processes complete with the tools required to manage the flexibility. The basis facility is selected with these problems in mind. A discussion of the production environment follows in sections 4.1 and 4.2.

4.1 Product Issues

From an assembly viewpoint, the most critical change encountered by the PCB industry in recent years is the introduction of Surface Mount Technology (SMT). The resulting influx of new component packages leads to much greater product variability than previously encountered in assemblies populated by through-hole components only. As fine-pitch and Tape Automated Bonding (TAB) components are developed, the resulting mixed technology PCBs will have even greater component variability.

In an effort to better understand the evolution of mixed technology PCBs over the next few years, several 'experts' involved in new product development were consulted. These consultations resulted in the conceptual development of four products that are believed to be characteristic of the products that will develop as mixed technology PCB evolve. Table 1 presents the four products that these experts believe represent the changing nature of mixed technology PCB products over the next few years. These are the products selected for this study.

The first product is characteristic of one of the initial mixed technology PCBs that a manufacturing firm may attempt to build. It has SMT components on only one side and maintains a large percentage of through-hole devices. The second and third products demonstrate one possible continuation of mixed technology evolution leading to product four, a highly evolved mixed technology PCB. The fourth product design has eliminated all residual through-hole components, and includes active, passive, and fine-pitch components on both sides of the PCB.

Table 1 Generic products representing PCB evolution

Component type	Component count Product 1	Product 2	Product 3	Product 4
THORUGH HOLE	100	50	60	
SIDE 1 SMT:				
ACTIVES	20	70	50	40
PASSIVES/DISCRETES	150	75	95	150
FINE PITCH			5	10
SIDE 2 SMT:				
ACTIVES				15
PASSIVES/DISCRETES		100	200	220
FINE PITCH				5
TOTAL PART COUNT	270	295	410	440

4.2 Process Issues

If a separate process were specifically built for each of the four product types, four different processing systems would result. Most PCB manufacturers are not able or willing to build several processing lines to assemble PCBs with differing mixed technology needs. One design of an assembly system is to include the capability to build any of these four product types on one flexible line.

Fig. 1 depicts the assembly equipment used in this study to produce the four products. It is divided into six assembly cells:

1) SMT side one onsertion. (Also side two for product four.)
2) SMT side two onsertion.
3) Fine-pitch placement.
4) Automated through-hole insertion.
5) Non-standard through-hole insertion and cleaning.
6) Final assembly and shearing.

A great deal of flexibility is available for routings between these cells but within the assembly cells only flow shop routing is permissible. The process is operated as a "pull" process with finite queues capable of blocking stations upstream in the process.

Process reliability and capacity information is provided in Table 2. The placement rates and set-up requirements are shown for each processing step. Additionally, the mean time between failures (MTBF) and the mean time to repair (MTTR) are estimated. Production times are based on these machine rates and are proportional to the number of component placements at each process. Production times also include allowances for set-ups which may be performed concurrently with on-going assembly. An additional set-up time is required each time the part type of a lot differs from the part type previously run on a particular machine. This second type of set-up may not be performed concurrent to production and leads to additional processing delays beyond production time. The simplifying assumption is made that labor is always available to support assembly machinery.

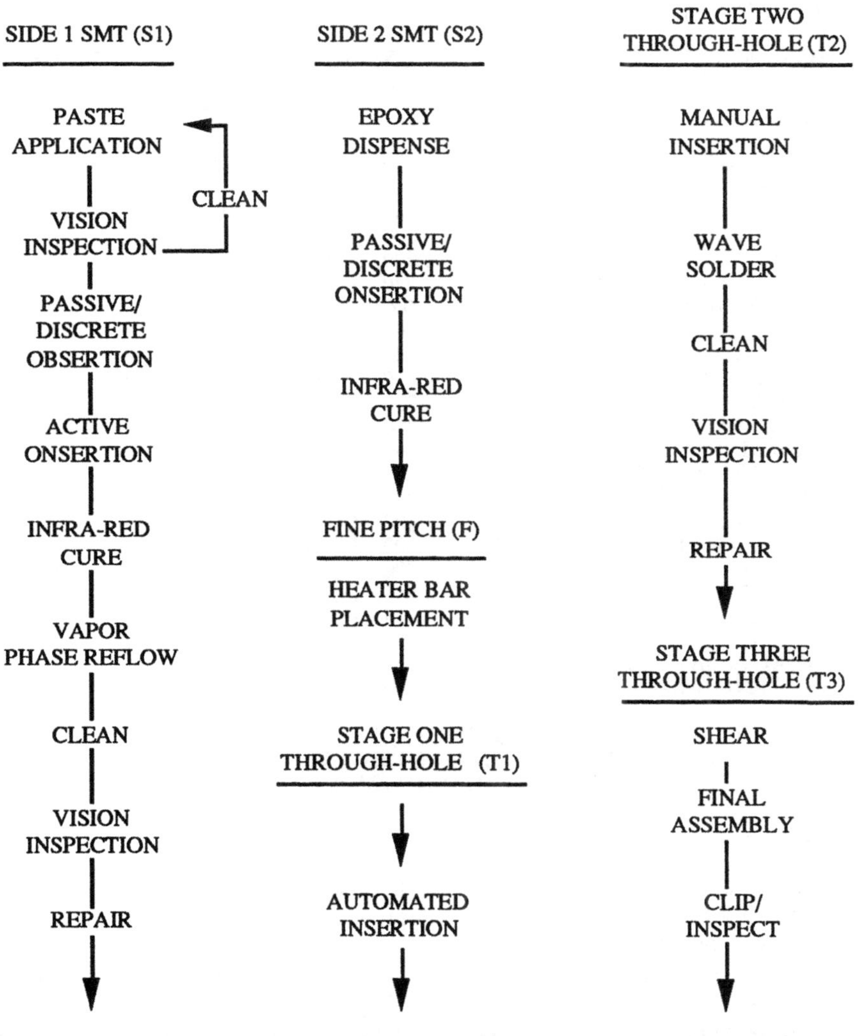

Fig. 1 Generic cells for assembly of mixed technology PCBS

Since each product stresses the assembly system in a different way, the mix of products released for processing will have a great affect upon the capabilities of the system. Even though many of the processes are similar, each product has different run times. The products also have different precedence constraints for inter-cell routings and this results in varying levels or routing flexibility. The precedence constraints for the four products previously presented are depicted graphically in Fig. 2.

An important ingredient in understanding dynamic system interaction is the time based availability of processes. In the case of equipment failures, it is impossible to predict the exact times that the process will be available. Therefore, the operational controller gathers dynamic process information to

Table 2 Process parameters

PROCESS STEP	# OF MACH. AVAILABLE	RUN TIME RATE/HR.		SET-UP PER-PCB*	SET-UP PER LOT*	RELIABILITY** MTBF	MTTR
PASTE APPLICATION	3	60	PCBs	.50	10.00	L	S
PASTE VISION	1	240	PCBs	.17	.50	M	M
PASTE REPAIR	1	60	PCBs	.00	..00	L	S
SIDE 1 PASSIVE/DIS.	1	8000	COMPs	.17	1.00	S	L
SIDE 1 ACTIVE	1	2500	COMPs	.25	1.00	M	M
SIDE 1 INFRA-RED	1	240	PCBs	.00	10.00	L	S
VAPOR PHASE	1	600	PCBs	.00	10.00	L	M
SIDE 1 CLEAN	1	300	PCBs	.00	10.00	L	S
SIDE 1 VISION	1	240	PCBs	.17	.50	M	M
SIDE 1 REPAIR	1	30	PCBs	.25	.00	L	S
EPOXY APPLICATION	1	8000	COMPs	.17	1.00	M	M
SIDE 2 PASSIVE/DIS.	1	8000	COMPs	.17	1.00	S	L
SIDE 2 INFRA-RED	1	240	PCBs	.00	10.00	L	S
FINE PITCH PLACEMENT	4	60	COMPs	.25	.50	M	S
AUTOMATED TH-HOLE	1	4000	COMPs	.25	2.00	S	M
MANUAL THROUGH-HOLE	7	180	COMPs	.25	.50	L	S
WAVE SOLDER	1	300	PCBs	.50	10.00	S	M
SOLDER CLEAN	1	300	PCBs	.00	10.00	L	S
SOLDER VISION	1	240	PCBs	.17	.50	M	M
SOLDER REPAIR	1	30	PCBs	.25	.00	L	S
SHEAR	1	60	PCBs	.17	.50	L	S
FINAL ASSEMBLY	1	120	PCBs	.17	.00	L	S
CLIP AND INSPECT	1	60	PCBs	.17	.00	L	S

* SET UPS ARE STATED IN MINUTES
** RELIABILITY INFORMATION (IN MINUTES)
MTBF: SMALL=1600, MEDIUM=2400, LARGE=4800
MTTR: SMALL=60, MEDIUM=120, LARGE=240

support decision making at each hierarchical control level so that processing flexibility may be used to achieve improved system performance, even under adverse conditions.

Both the characteristics of the current product mix and the process availability contribute dynamically to the creation of system level processing bottlenecks as well as local, cell level bottlenecks. Changes to mix or availability can cause significant changes in the location of these limiting processes. In some cases, such as machine failures, it is necessary to react to these changes. In other cases, it is possible to make operating changes to more directly control the system status and manage the bottlenecks as desired by the decision maker.

5 METHODOLOGY

Discrete event simulations are performed for a set of controlled experiments using the SIMAN simulation language. In addition, complicated control logic resides in FORTRAN subroutines. Static processing parameters are maintained in an external data base. The experimental frame for each experiment specifies that each control scheme be evaluated under identical system characteristics. Ten independent replications are run for each scenario; each representing 3 days (9 shifts) of production under steady-state conditions.

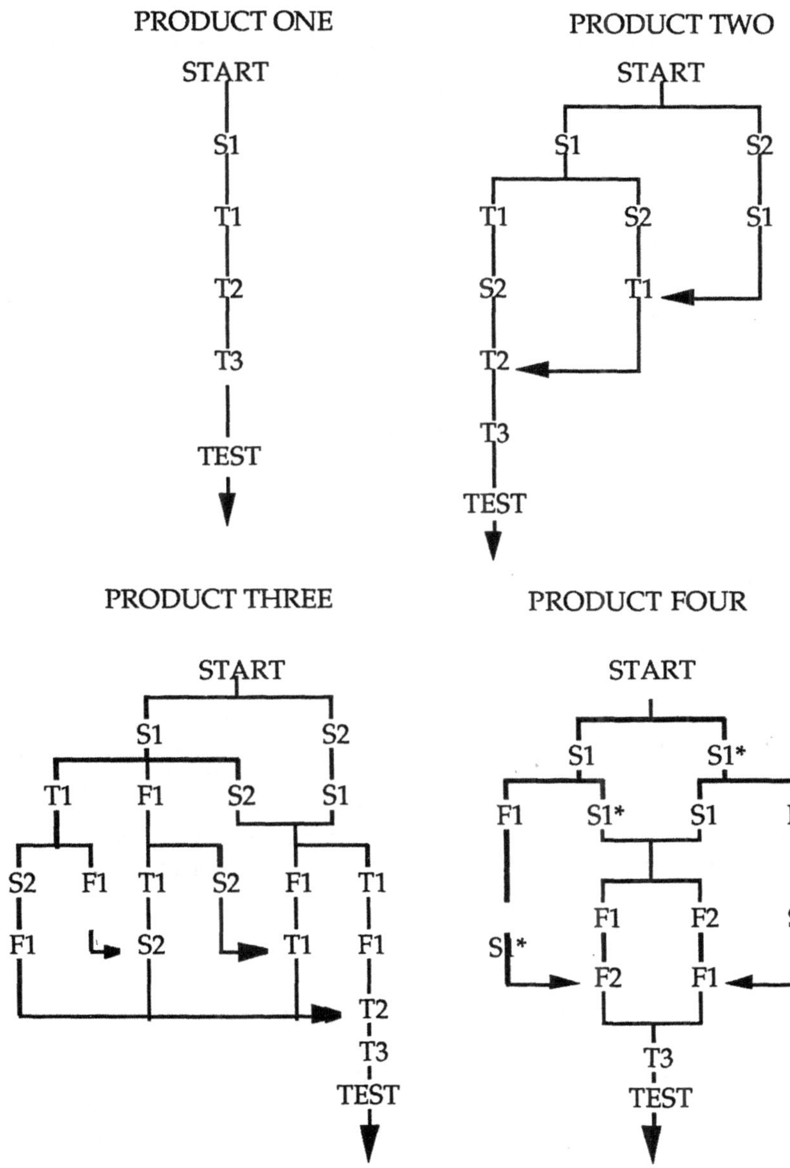

<u>LEGEND</u>

S1= SIDE 1 SMT
F1= SIDE 1 FINE PITCH
T1= STAGE 1 THROUGH-HOLE
T2= STAGE 2 TRHOUGH-HOLE
T3= STAGE 3 TRHOUGH-HOLE
S1*= SIDE 2 PROCESSING ON SIDE 1 CELL

S2= SIDE 2 SMT
F2= SIDE 2 FINE PITCH

Fig. 2 Precedence diagrams for generic products

The SIMAN output processor is used to perform statistical analysis. The INTERVALS command is used to generate confidence intervals at the 95% level. Additionally, the MEANTEST Command is used. Throughout this chapter, if a result is said to be "significantly" different from another, this indicates that the results have been found to be significantly different at the 95% confidence level using the MEANTEST Command. More information about this test may be obtained in Pegden [15].

5.1 Verification and Validation

The validity of results is partially supported using actual production data from similar products assembled at the basis facility in Connecticut. This is possible only for the baseline scenario and for limited test runs using routing flexibility. Integrated scenarios using two or more hierarchical levels of flexible control have not been implemented at the basis facility.

Internal validity of the SIMAN and FORTRAN code is analyzed by a series of test runs designed to evaluate each function. The SIMAN TRACE command is used to follow the flow of sample jobs throughout the simulated production facility for each scenario. These tests are extended to examine the proper functioning of all FORTRAN code and to ensure the effective transfer of data from external files.

5.2 Performance Criteria

The performance criteria used are the mean flow times for each product, the mean work-in-process levels, the average machine utilization at the bottle-neck (most heavily utilized) process, and the total throughput levels. The primary goals are to reduce mean flow times and work-in-process levels while meeting product demand. Therefore, bottleneck utilization is a secondary concern and the related throughput levels are only a concern if they are not adequate to support product demand.

6 RESULTS

This section describes the results obtained from the experiments described above. It is demonstrated that adding flexibility in product routings can be beneficial in reducing flow times and WIP levels, especially when integrated with control strategies from other levels of the hierarchy. Further, it is demonstrated that the degree of improvement possible is related to product and process issues specific to the particular FAS environment.

6.1 Effectiveness of Flexible Routing

In this study, routing flexibility refers to the ability to conduct assembly operations in alternate sequences, but does not extend to the ability to use alternative resources for specific assembly operations. Several routing heuristics are examined for use in this environment. The best routing heuristic observed in this experiment is one which routes products to the cell which offers the greatest opportunity for rapid processing. A shortest processing time (SPT) heuristic calculates the expected material flow time for the product in each potential path.

This calculation includes expected run time, expected set-up time, expected waiting time due to WIP congestion, expected time to repair for down processes, and an estimate of path reliability. The path with the shortest total expected material flow time is selected for processing.

In Fig. 3, the performance of the flexible routing scenario is compared to the performance of the baseline scenario which offers no flexibility in material flow decisions. Fig. 3 also shows the performance characteristics of the following two scenarios:

1) The flexible router in combination with the system level mixer.
2) The flexible router in combination with the machine level sequencer.

As shown in Fig. 3, the scenario featuring routing flexibility appears to provide better average flow times than the baseline scenario. Significantly better results than the baseline scenario, however, occur for product three only. This result is easily explained. As depicted in Fig. 2, product three offers the most opportunity for flexible routing. Product three has five decision points and eight potential paths. Comparatively, product one has no potential for flexibility in material flow, and product two has two decision points resulting in three potential paths. Product four has four decision points with six potential paths but loops back through the same machine cells twice, thus removing some of its potential for flexible material flow improvements. Therefore, for the product with the greatest potential for flexible routing, the heuristic performs better than for other products. The heuristic also performs significantly better than the baseline scenario for the same product. Fig. 4 shows that adding flexibility in routing also leads to improvement in work-in-process levels with a slight reduction in average bottleneck utilization.

In some circumstances, adding process routing flexibility can lead to more dramatic improvements. In Taylor and Graves [17], the authors describe such a scenario. The authors further demonstrate that the performance of a flexible routing methodology may be impacted positively by reductions in lot sizes; a strategy that moves the point of routing decision closer to the actual process. Three primary differences exist between the experiment described in this chapter and the experiment in Taylor and Graves [17]. First, the experiment in this chapter allows for flexibility between cells but not within cells; recall that flow processing occurs within the cells as shown in Fig. 1. In Taylor and Graves [17], opportunities exist for decision making within cells and this results in the ability to adapt or adjust decisions at even later times in processing. Secondly, the experiment in this chapter has a different process/product balance in the baseline environment. Each of the four products has product specific bottleneck in a different assembly cell. Because of this, the baseline process with no flexibility provides a stronger basis for comparison. Finally, the assembly system in Taylor and Graves [17] has buffers of infinite size. With infinite buffers, control strategies which effectively balance material flow perform very well by reducing large queues before bottlenecks. With finite queues, large WIP build-ups are not permissible by reason of system design.

6.2 Flexible Routing Integrated With Flexibility in Product Mix

The scenario which includes flexibility in both routing and product mix determination demonstrates the need to concentrate on resolving key areas of material handling interaction between control levels. In this case, the primary areas of integration involve product release times and the determination of

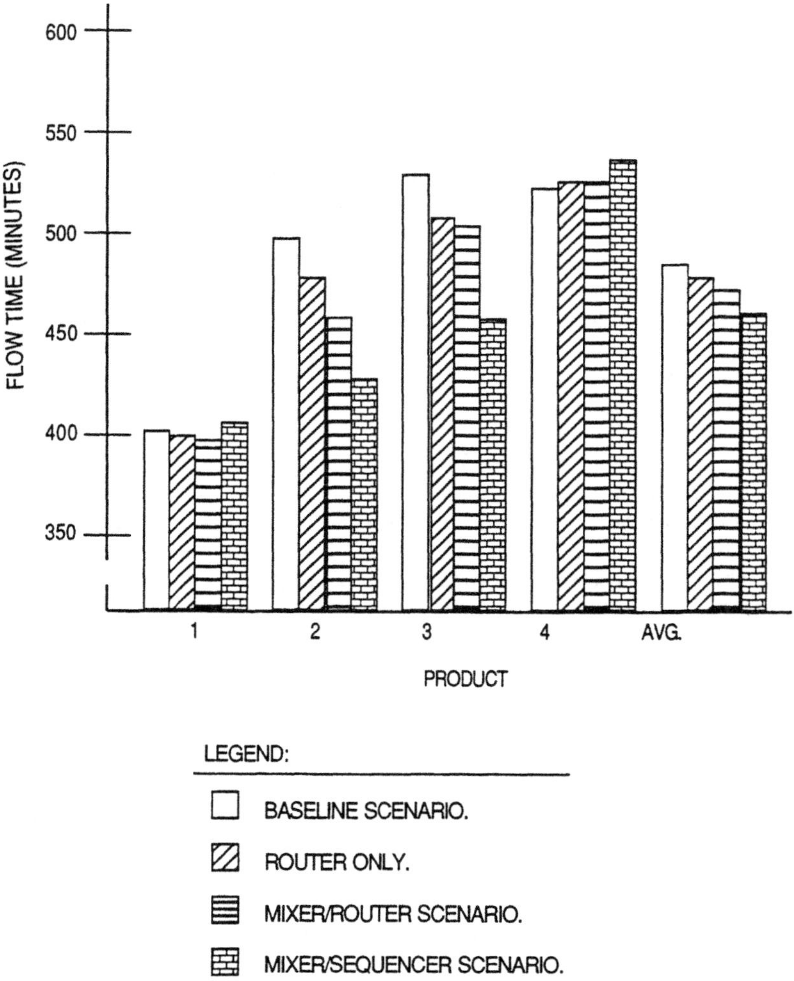

Fig. 3 Flow time comparison

desirable WIP levels. As stand-alone controls, the router and the mixer have different goals. The mixer desires to push products into the system in certain quantities, lot sizes, and release times based on current system status. It also tries to work within constraints established by a tactical level planner. The employed heuristic attempts to balance the WIP levels of each of the four products in the system. The router desires to pull products into the system when conditions are favorable for new jobs within a potential starting cell. A method of resolving this push-pull conflict is a necessary ingredient of the integrated model. Closely related to this problem is the maintenance of a favorable system WIP level, which is primary controlled by the mixer.

The method of problem resolution that is observed to perform best is one which makes use of an integrated periodic review system that receives input from the router, the mixer, and a dynamic data base consisting of both system level and cell level data. The extended SPT router described in section 6.1 is integrated with a mixer which seeks to balance the work load in each assembly cell by inputting appropriate jobs, while meeting throughput requirements. The resulting integrated model performs very well by effectively controlling WIP levels. Figs. 3 and 4 show that the integrated mixer/router performs better than both the baseline and the SPT router in terms of average flow time, bottleneck utilization, and work-in-process levels. Care is taken to ensure that the flexible mixer/router also performs adequately relative to throughput requirements.

To ensure that the integrated mixer/router model is performing well due to the concentration on integration, a control experiment is performed which includes both routing flexibility and flexible product mix decisions, but as stand-alone systems without the interfacing heuristics. In Table 3, the results of related significance tests are shown. The integrated mixer/router provides flow time results significantly better than those of the baseline scenario for products 2 and 3 as well as for average WIP levels. The non-integrated model, on the other hand, provides significant improvement only in average WIP levels. As noted earlier, products 2 and 3 provide greater opportunity for material flow improvement from flexible processing when compared to that of either product 1 or 4. The integrated model also performs better than either the stand-alone router or the stand-alone mixer.

A sample output using the MEANTEST command from the SIMAN output processor is presented in Fig. 5. The data and calculations presented are for a comparison of mean flow times for product 2. The sample output depicted is the comparison of the integrated mixer/router scenario with the baseline scenario.

6.3 Flexible Routing Integrated With Flexibility in Sequencing

This scenario demonstrates the need for using control strategies which not only concentrate on areas of problem resolution, but also those which are complementary with strategies at other hierarchical control levels. As stand-alone systems, the best observed router is the extended SPT model described in section 6.1. The best observed stand-alone sequencer is one which uses a truncated best set-up scenario. For sequencing at each station, the preceding finite-sized queue is searched for the first job which has desirable set-up attributes. This policy is abandoned when there are jobs in the queue which have been present for extended periods of time. The policy of changing each machine's sequencing rule from

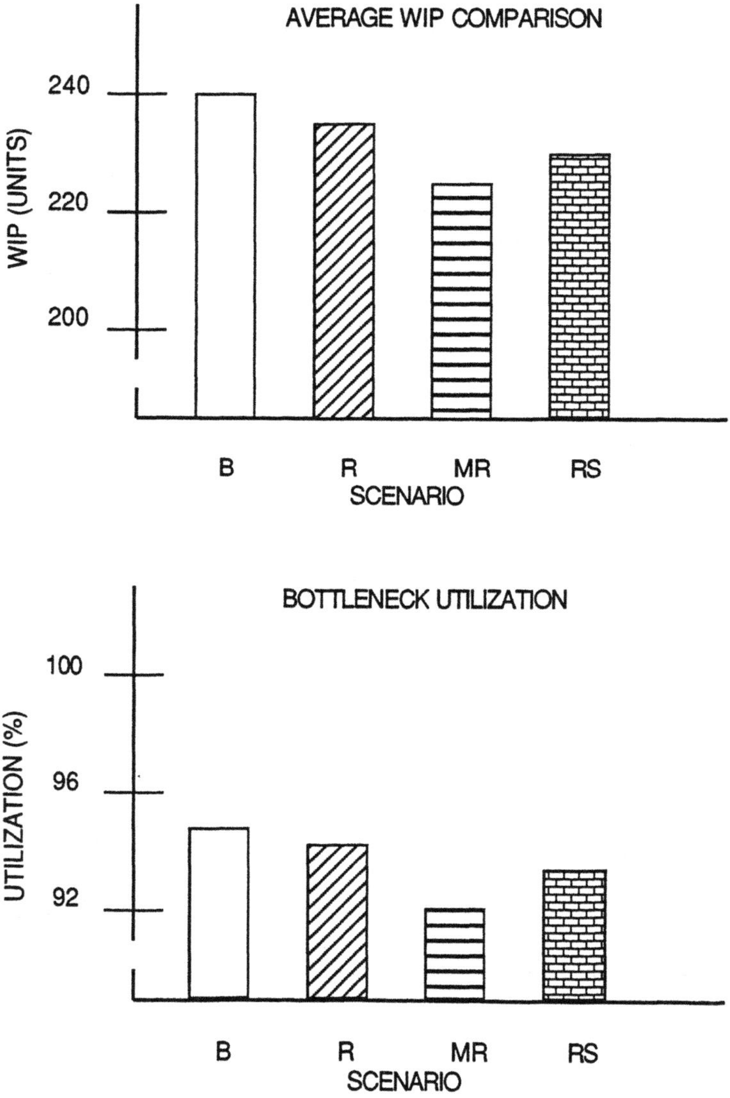

LEGEND:

B = BASELINE SCENARIO.
R = ROUTER ONLY.
MR= MIXER/ROUTER SCENARIO.
RS= ROUTER/SEQUENCER SCENARIO.

Fig. 4 Average WIP levels and bottleneck utilization

PRODUCT 2 AVERAGE FLOW TIMES FOR BASELINE SCENARIO.

RUN 1	472.72
RUN 2	486.67
RUN 3	480.62
RUN 4	498.07
RUN 5	452.96
RUN 6	490.87
RUN 7	509.20
RUN 8	520.40
RUN 9	501.45
RUN 10	486.91
MEAN	490

PRODUCT 2 AVERAGE FLOW TIMES FOR MIXER/ROUTER SCENARIO.

RUN 1	449.60
RUN 2	450.19
RUN 3	452.72
RUN 4	473.77
RUN 5	440.04
RUN 6	482.02
RUN 7	453.23
RUN 8	547.85
RUN 9	462.61
RUN 10	447.04
MEAN	457

MEANSTEST :

IDENTIFIER	ESTD. MEAN DIFFERENCE	STANDARD DEVIATION	0.95 C. I. HALF-WIDTH	MINIMUM VALUE	MAXIMUM VALUE	NUMBER OF OBS
0.30-0.31	33.1	7.25	15.5	453.	520.	10
				440.	482.	10

REJECT H0 - MEANS ARE NOT EQUAL AT 0.05 LEVEL

MEAN-DIFFERENCE INTERVALS :

```
                     0                             33.1
0.30-0.31            |-----------------(---------------X---------------)
                                       17.6                             48.5

 -----------------------------------------------------------------------------
 | | = TEST ( 0 OR 1 )   ( = LOWER 95% CL   X = AVERAGE   ) = UPPER 95% CL |
 -----------------------------------------------------------------------------
```

Fig. 5 Sample mean test comparison

first-in-first-out to a policy which helps to avoid unnecessary set-ups, effectively alters the lot size dynamically at each station and results in improved system performance. Although the router and sequencer heuristics perform very well independently, they perform very poorly together. Table 3 shows that the non-integrated router/sequencer performs very poorly relative to the baseline scenario. Average flow times are larger than the baseline and two products have mean flow times which are significantly worse.

The course of integration pursued is one which coordinates the heuristic decision making at each level so that similar decisions are guiding policy at each level in the operational control hierarchy. The best of such approaches observed is one in which both the router and the sequencer use an SPT heuristic for decision making. The router uses the extended SPT model to choose the cell with the shortest expected processing time. The sequencer at each machine within a cell also selects that job from its finite input queue which can be processed in the shortest amount of time.

This approach works very well. Fig. 3 reveals that average flow times for the integrated router/sequencer scenario are better than those for any other scenario. Furthermore, Fig. 4 shows that the average work-in-process level and the average bottleneck utilization are lower than those of the baseline and the stand-alone

Table 3 Results of significance tests

Parameter of Comparison	Scenario: Router Only	Mixer/Router Non-Int.	Mixer/Router Integ.	Router/Sequencer Non-Int.	Router/Sequencer Integ
Flow Time:					
Product 1				⊖	
Product 2			⊕		⊕
Product 3	⊕		⊕	⊖	⊕
Product 4					
WIP Level		⊕	⊕		⊕

⊕ Denotes results with mean values significantly better than the baseline scenario at the 95% confidence level.

⊖ Denotes results with mean values significantly worse than the baselin scenario at the 95% confidence level.

router. As shown in Table 3, flow times are significantly better than those of the baseline scenario for Products 2 and 3. Work-in-process levels are significantly lower than the baseline as well.

7 CONCLUSIONS

These results show that incorporating routing flexibility as a component in material flow control architectures can lead to improvement in performance relative to a number of system performance criteria. The results further show that integration of routing flexibility with other control levels in the FAS hierarchy can result in even greater improvement. Integrated control strategies can produce synergistic results which are better than those of individual control mechanisms acting separately.

Using a heuristic approach, it is possible to make near real-time material flow decisions. This approach should support decision making for either automated or manual material handling systems. Decision outputs which determine product mix, inter-cell routing, or buffer sequencing can be presented for implementation by manual means or by automatic means.

Regardless of the method of implementation employed, a key component of a material flow system must be a sophisticated data collection system. This system must be capable of collecting dynamic process data and must have the capability to present the data to the decision maker in a recognizable format.

The material handling system to support process flexibility must be able to make use of this dynamic data. In a short time interval, the flexible material handling system must be able to identify a workpiece at a specific location. It must respond to instructions from the decision maker to route or re-route a workpiece to a processing location based on current status. The material handling system must also make efficient use of limited buffering availability to avoid both traffic flow problems and unnecessary delays.

Additional research concentrates on the integration of material flow decisions at all control levels of the FAS and is reported in Taylor [18]. The work is extended to determine which general characteristics of the FAS may improve the performance of this approach. Heuristic approaches are tested for a larger variety of product mixes, process reliability parameters, set-up requirements, and buffer configurations.

8 REFERENCES

1. Buzacott, J.A., "'Optimal' Operating Rules For Automated Manufacturing Systems," *IEEE Trans. Automatic Control*, AC-27, pp. 80-86 (Feb. 1982).
2. Denzler, D.R., and W.J. Boe, "Experimental Investigation of FMS Scheduling Decision Rules," *International Journal of Production Research*, Vol. 25, No. 7, pp. 979-994 (July 1987).
3. Donath, M., "A Scheduling Methodology for Flexible Manufacturing Systems," Ph.D. Dissertation, Department of Industrial Engineering and Operation Research, University of Massachusetts, Amherst, MA (1985).
4. Donath, M. and R.J. Graves, "Flexible Assembly Systems: An Approach for Near Real-Time Scheduling and Routing of Multiple Products," *International Journal of Production Research*, Vol. 26, No. 12, pp. 1903-1919 (1988).
5. Donath, M. and R.J. Graves, "Flexible Assembly Systems: Test Results for an Approach for Near Real-Time Scheduling and Routing of Multiple Products," *International Journal of Production Research and Development*, Vol. 29, No. 4, pp. 392-400 (July 1985).
6. Gershwin, S.B., R. Akella, and Y.F. Choong, "Short-term Production Scheduling of an Automated Manufacturing Facility," *IBM Journal of Research and Development*, Vol. 29, No. 4, pp. 392-400 (July 1985).
7. Kimemia, J.G. and S.B. Gershwin, "An Algorithm for the Computer Control of Production in a Flexible Manufacturing System," *IEEE Annual Transactions*, pp. 628-633 (1981).
8. Kimemia, J.G. and S.B. Gershwin, "An Algorithm for the Computer Control of a Flexible Manufacturing System," *IIE Transactions*, Vol. 15, No. 4, pp. 353-362 (December 1983).
9. Kimemia, J.G and S.B. Gershwin, "Flow Optimization in Flexible Manufacturing Systems," *International Journal of Production Research*, Vol. 23, pp. 81-96 (1985).
10. Maimon, O.Z. and Y.F. Choong, "Dynamic Routing in Reentrant Flexible Manufacturing Systems," *Robotics and Computer Integrated Manufacturing*, Vol. 3, No. 3, pp. 295-300 (1987).
11. Maimon, O.Z. and S.B. Gershwin, "Dynamic Scheduling and Routing for Flexible Manufacturing Systems That Have Unreliable Machines," *IEEE Annual Transactions*, pp. 281-287 (1987).
12. Menga, G., G. Bruno, R. Conterno, and M.A. Dto, "Modeling FMS by Closed Queueing Network Analysis Methods," *IEEE Transactions on*

Components, Hybrids, and Manufacturing Technology, Vol. CHMT-7, No. 3, pp. 241-248 (Sept. 1984).

13. Muhlemanm, A.P., A.G. Lockett and C.K. Farn, "Job Shop Scheduling Heuristics and Frequency of Scheduling," *International Journal of Production Research*, Vol. 20, No. 2, pp. 227-241 (1982).
14. Nof, S.Y., M.M. Barash, and J.J. Solberg, "Optimal Control of Item Flow in Versatile Manufacturing Systems," *International Journal of Production Research*, Vol. 17, No. 5, pp. 479-489 (1979).
15. Pegden, C.D., "Introduction to SIMAN with Version 3.0 Enhancements," *Systems Modelling Corporation*, State College, PA (1986).
16. Russo, F.T., "A Heuristic Approach to Alternate Routing in a Job Shop," Master's Thesis, M.I.T., USA (June 1965).
17. Taylor, G.D. and R.J. Graves, "An Examination Of Routing Flexibility for Small Batch Assembly of Printed Circuit Boards," In Review for publication in *International Journal of Production Research.*
18. Taylor, G.D., "A Dynamic Production Control Methodology for Flexible Assembly of Mixed Technology Circuit Boards," Ph.D. Dissertation, Dept. Of Industrial Engineering and Operations Research, University of Massachusetts, Amherst, MA (1990).
19. Vonderembse, M.A. and G.S. Wobser, "Steps for Implementing a FMS," *Industrial Engineering*, pp. 38-48 (April 1987).

PROTOTYPING THE INTEGRATION REQUIREMENTS OF A FREE-PATH AGV SYSTEM

Chang Wan Kim
J. M. A. Tanchoco
Purdue University

ABSTRACT

Automated guided vehicle systems (AGVS) are gaining more attention from both academic and industrial research organizations. However, there exist some basic problems that still remain unsolved which prevent widespread installation of AGV systems. System integration is such an issue.

This chapter describes an on-going research on Intelligent Transport Systems (ITS) at Purdue University. It is an AGV-based factory transport system composed of experimental free-path vehicles, a vehicle simulator, a transport system controller and a shop simulator. Discussions include issues related to the integration requirements of a free-path AGV system. The main focus is on the design of the control architecture, division of functions between system components, data communications considerations, etc. The concept of a virtual transport system is also presented.

1 INTRODUCTION

An automated guided vehicle (AGV) is a driverless, autonomous vehicle used to transport materials within a factory or warehouse. Despite the large investment costs, AGV systems offer several attractive features which are not easily obtained by conventional transport methods such as conveyors or lift trucks. They provide greater routing flexibility, and they can be integrated with other computerized systems. They also provide other advantages such as less product damage, ability to operate in hazardous or clean environments, etc.

While most existing AGVs move along physical guide path networks defined by embedded wires or painted stripes or tapes on the floor, more research efforts are now being focused on free-path AGVs [16] [6] [2] [7] [15]. Free-path vehicles follow virtual flow path networks defined only in the computer memory. Since the need for installing guide paths is eliminated, free-path AGVs provide even greater flexibility. However, a much more complicated control system is required to achieve the promised advantages of such a system.

In the literature, most of the research efforts in the AGV area are devoted to the advancement of vehicular technology. Few works are reported on the system's operation and integration issues. Although a free-path AGV system is a more attractive transport method than conventional AGVs, conveyors, and fork trucks, the lack of good control systems is a major hurdle that must be overcome if such a system is to gain widespread use. This chapter discusses the major issues related to the design, development and integration of a control system for a free-path AGV system. The discussions include the control architectures, division of functions between system components, data communication considerations, etc. The discussions are based on a prototype system called Intelligent Transport System, under development at Purdue University.

2 THE INTELLIGENT TRANSPORT SYSTEM

The proposed Intelligent Transport System is based on a fleet of small, autonomous vehicles with the ability to provide a high degree of flexibility and to reduce work-in-process inventories and flow times, as opposed to conventional large unit-load carriers. The movement of the vehicle is controlled by a vehicle controller which resides on an on-board microprocessor. The routing and scheduling of the vehicles are done by a supervisory controller which resides on a workstation-level computer and communicates with the vehicles through a common radio channel. A shop simulator running on a central computer generates simulated events for the transport system and collects statistics. The vehicle simulator can replace the actual vehicle when testing the control software or doing experiments for analysis purposes. Fig. 1 shows the current configuration of the Intelligent Transport System. More detailed descriptions are given below.

2.1 The Experimental Vehicle

The experimental vehicle is a tricycle-type free-path AGV equipped with a navigation system based on a dead-reckoning method [7]. The mechanical design of the vehicle is shown in Fig. 2. The front wheel drives and steers the vehicle. Encoders attached to the rear wheels send sensory information needed to estimate the position and orientation of the vehicle to the vehicle controller. The vehicle is equipped with a micro-level computer as its on-board processor. Information exchange with the supervisory controller is achieved through wireless modems.

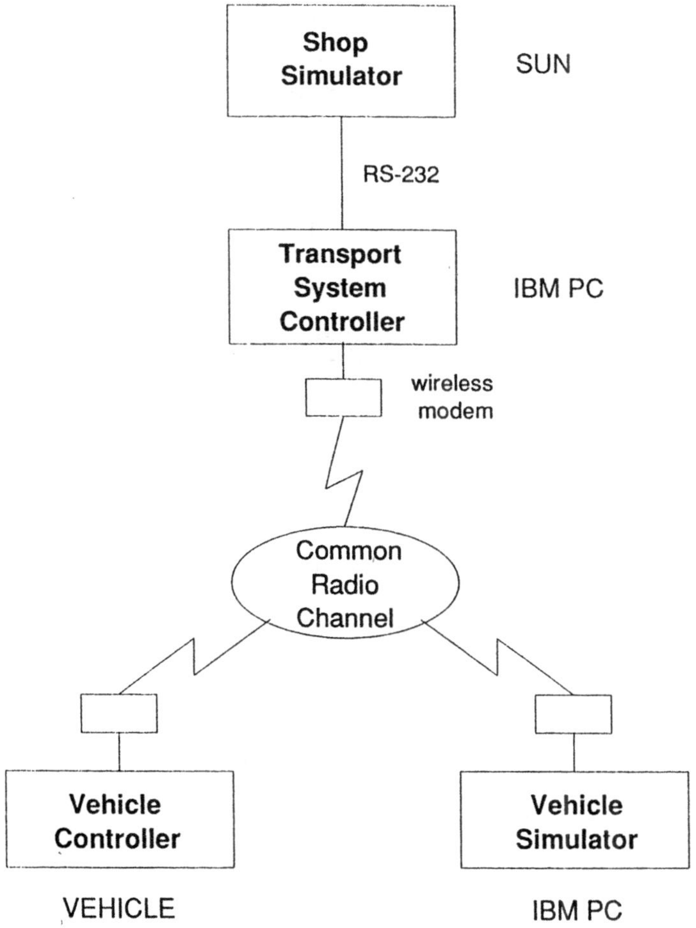

Fig. 1 ITS configuration

2.2 The Vehicle Simulator

The vehicle simulator is a virtual vehicle which behaves as closely to the actual vehicle as desired. When the vehicle simulator is running in place of an actual vehicle, the supervisory controller does not recognize the change since it follows the same procedures for data exchange between the vehicles and the supervisory controller. Though it does not receive real sensory inputs, it uses the same routines for navigation and maneuver control, based on the estimated counter values. It runs on any microcomputer of the same type as the on-board processor. Fig. 3 shows the graphic display generated by the vehicle simulator.

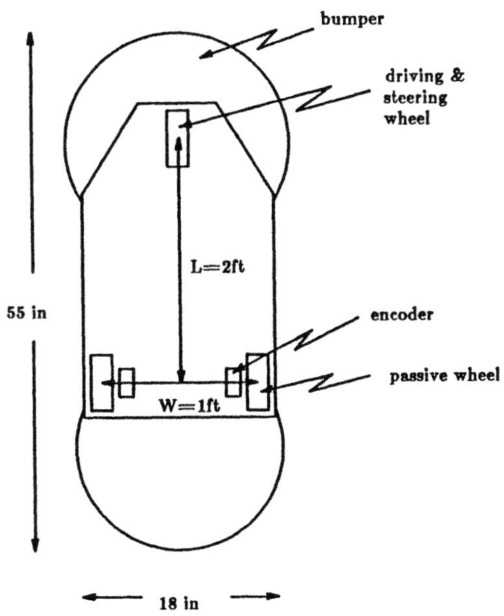

Fig. 2 Mechanical design of the experimental vehicle

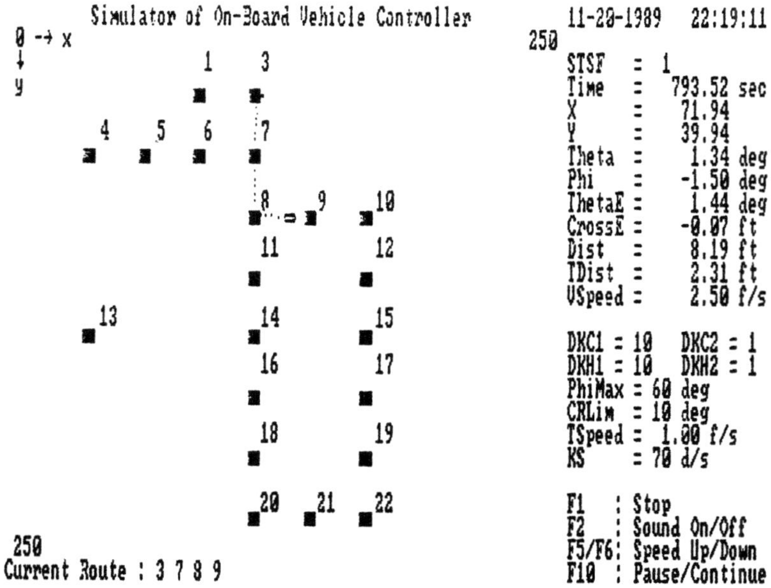

Fig. 3 Graphic display of vehicle simulator

The vehicle simulator plays a very important role in the design, development, and analysis of the control system for ITS. It can be used (1) to test the navigation control technique employed, (2) to test the operation procedures of the system, (3) to test the communication protocols, and (4) to perform other experiments for systems analysis purposes.

2.3 Transport System Controller

The transport system controller running on a microcomputer is a C version [13] of the LISP-based controller described in [12] for a free-path automated guided vehicle system. The controller is designed to operate in real-time, and make dispatching decisions based on the current status of the vehicles and the flow path network. It is capable of handling both unidirectional and bidirectional traffic flows.

The tasks of the controller are performed by several cooperating modules. When a transport request is received from the shop simulator, the dispatcher module of the controller, with the help of the route planner and the scheduler, selects a vehicle which will arrive at the pickup station in the shortest time. In doing this, the route planner and scheduler generate a tentative travel schedule for each vehicle accounting for its current travel schedule and the status of the whole flow path network. After a vehicle is selected, the route planner and the scheduler find a congestion-free route and schedule which will satisfy the request in the shortest time. To generate congestion-free routes and schedules, the controller maintains a list of reservation times of vehicles at each node. Both the loaded and unloaded movements are congestion-free.

2.4 Shop Simulator

AGVSim2 [4] is a special-purpose discrete-event simulator designed to evaluate the control strategies of supervisory controllers for a free-path AGV system used in a job shop environment. It provides the capability to model a variety of system configurations, and allows for complex free-path vehicle maneuvers such as vehicle take-over, bidirectional flow, multiple vehicle types, and multiple loads on a vehicle. A system trace and performance statistics are reported by AGVSim2.

Instead of modeling the supervisory control logic within the simulator, AGVSim2 is directly interfaced with the supervisory controller described in the previous section. The simulator emulates the system and creates the events such as transport requests that are transmitted to the AGV system supervisory controller. The responses from the supervisory controller are used by AGVSim2 to update the status of the system, collect statistics, and generate new events.

2.5 Operation of ITS

For the sake of simplicity, the discussion below assumes there is only one vehicle in the system. The execution of each component software goes through three phases: initialization, operation and termination. In the initialization phase, each component performs the following functions:

(1) Establish Connection with Other Components

Each component opens a serial communication port of the computer and checks if the other party is running.

(2) Synchronize Clocks

The higher-level component sends the current time since midnight. The lower-level component compares it to its own time and records the difference.

(3) Download Layout

The layout file is downloaded from the shop simulator to the transport system controller and to the vehicle in that order.

After initialization, the system waits for the shop simulator to initiate the operation phase. A simplified procedure of operation is illustrated in Fig. 4. The shop simulator generates simulated events according to an input file. A transport request is generated when a unit load is ready to go to the next workstation. Then the transport request is transmitted to the transport system controller in a predefined format. The transport system controller, upon receipt of the transport request, calculates the route and downloads it to the vehicle. The vehicle then moves according to the route received. The vehicle reports its location and status back to the transport system controller whenever it stops. When the vehicle completes a route, the transport system controller reports this to the shop simulator. Then the shop simulator continues the simulation and generates another transport request. Meanwhile the shop simulator collects statistics.

3 INTEGRATION ISSUES

3.1 Control Architecture

The prototype system described above assumes a hierarchical control structure. At the top of the hierarchy is the shop simulator AGVSim2 which generates simulated events such as transport requests for the AGV system. Then the transport system controller takes these events as input and

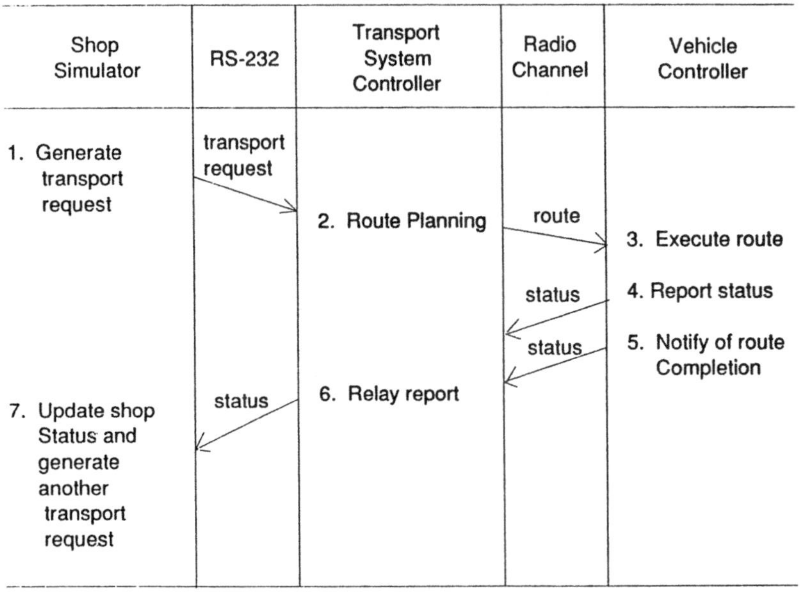

Fig. 4 ITS operation

generates routes for the vehicle. The vehicle controller residing on the on-board processor of the vehicle receives these routes and controls the movements of the vehicle according to the routes received. During the movements, the vehicle controller reports the status of the vehicle to the system controller. Then the transport system controller relays these status reports to the shop simulator. The shop simulator collects statistics based on the status reports (refer to Fig. 1).

The Automated Manufacturing Research Facility of the National Institute of Standards and Technology (formerly the National Bureau of Standards) is a typical example of a hierarchical control based system. In the concept proposed by NIST AMRF, a higher-level controller sends commands to and receives status reports from its subordinate controllers. Long-range tasks enter the system at a higher level and are broken down into subtasks to be executed as procedures at that level or to be sent as commands to the next lower level [8]. Fig. 5(a) shows a typical configuration of an AGV system under hierarchical control. In this architecture, the central controller sends movement commands in the form of travel itineraries and the vehicle reports its location and status to the central controller. The vehicle controller performs only minimal functions such as navigation and maneuver control, load transfer control, and battery monitor and recharge control, etc. The central controller makes global decisions such as vehicle routing, vehicle

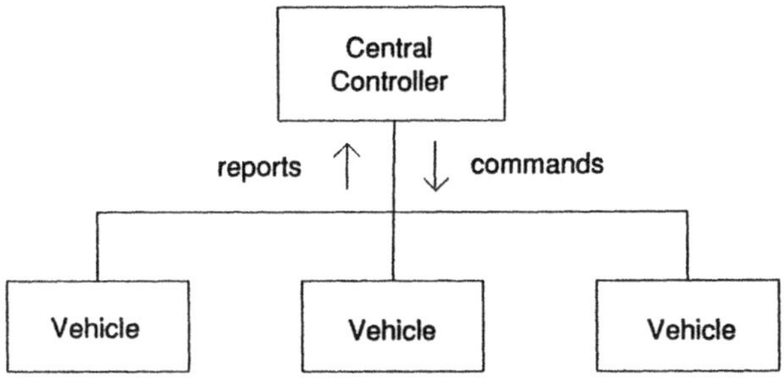

(a) Hierarchical Control

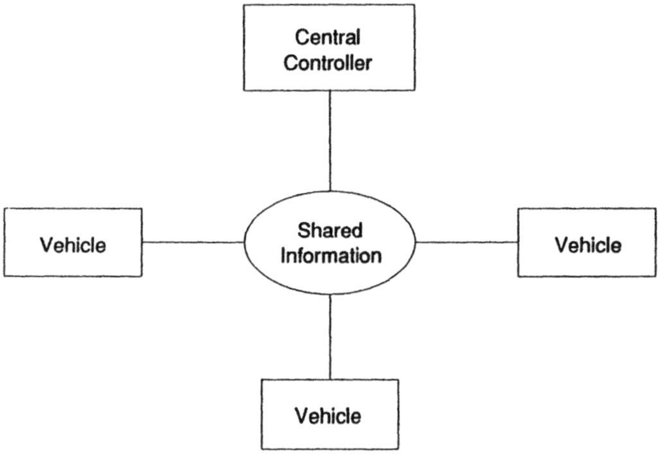

(b) Heterarchical Control

Fig. 5 Control architecture

dispatching, and traffic control, etc. The central controller maintains global information and the vehicle controller keeps local information. Thus, the control is centralized.

Most existing AGV systems are configured more or less in a hierarchical architecture. The hierarchical architecture of a centralized control system resembles the structure of many organizations in human society. The underlying philosophy is well understood to the people involved in development and operation of such a system. Thus, it is easy to implement and brings a more efficient human synergy. However, the system is vulnerable to the failure of the central controller; if the central controller fails, then the whole transport system becomes useless. A backup computer is typically necessary. Since most computations are done by the central controller, heuristic rules are preferred to time-consuming optimizing algorithms even though a high-performance computer is used. The amount of information transmitted to the vehicle from the central controller is also substantial, which requires faster and more reliable data communication devices and protocols. Finally, it is less adaptable to vehicle or communication failure.

On the contrary, the concept of cooperative systems [14] or heterarchical architecture was opened by recent advances in distributed computing and open systems architectures of local area networks. In a cooperative, heterarchical system, all the intelligent, autonomous participant subsystems have: (1) equal right of access to resources, (2) equal mutual access and accessibility to each other, (3) independent modes of operation, and (4) strict conformity to the protocol rules of the overall system [5]. The central controller, if one exists, coordinates the operation of participants to achieve the global goals. The participants make most of the important decisions based on shared information and cooperate with each other. They should be intelligent and autonomous enough for the system to be operable even when the central controller fails. Examples of cooperative heterarchical systems can be found in ant or bee societies, computer networks, etc.

Fig. 5(b) illustrates an AGV system configured based on heterarchical architecture. The vehicles and the central controller exchange information through a common communication channel and maintain identical copies of global information on the flow path network and the current locations of the vehicles, etc. Each vehicle can also have individual information dedicated for its own use such as vehicle length, width, maximum speed, etc. Most of the functions performed by the central controller under a hierarchical control such as vehicle routing and traffic control can be delegated to the vehicle controller. The central controller performs minimal functions such as information exchange with the larger CIM environment, assignment of transport tasks to vehicles, system monitoring, records keeping, etc. If only one communication channel is assigned to the entire CIM system, then the vehicles can operate in complete autonomy without the help of the central controller which serves as a mere monitoring station.

In a heterarchical mode of operation, relatively high-performance processors are required on vehicles since most of the computational burdens are loaded on the vehicle controllers. Since this concept is currently not well understood, more time and efforts are required in the planning and development stages before implementation. However, the system can readily adapt to partial failures of vehicles or communication devices since the operating system does not make assumptions on the number of vehicles in the system or the degree of the reliability of communication (the system performance still depends on these). Thus, the system becomes more robust. The amount of information to be transmitted is significantly reduced, which makes the communication more reliable. Optimizing algorithms can be implemented by distributing the computational burden to the on-board vehicle processors.

When implemented, the control structure of a specific system will lie at some point in the spectrum between the strict hierarchy and the strict heterarchy. The hybrid structure proposed in [11] is a combination of the two strategies, aimed to maintain the capabilities of abstraction and tight coupling provided by hierarchical structure and the flexibility provided by heterarchical structure. The transport functions are isolated from the rest of the CIM system by introducing an interface that remains constant and independent of any probable changes in the transport system. Transport services are available to the cell controllers and the shop controller via the transport bulletin board provided in the interface.

Defining and allocating the functions between the vehicle and AGV system controllers are closely related to the control structure employed. Generally, functions related to overall planning and control are performed by the system controller and functions related to mechanical control of the vehicles are performed by the vehicle controller. As the system becomes more heterarchical, the vehicle controller becomes more intelligent and autonomous, and assumes more responsibilities. Of course, there is no single general solution which gives the best results to every system. But the best control architecture should be decided based on the analysis of the specific system under study.

3.2 Data Communications

Integration of the vehicle controller and the transport system controller described above requires a fast and reliable communication method between the vehicles and the central computer. Two possible communications media are radio and infrared. Wideband radio local area networks with spectrum allocation are believed to be the most suitable [7]. Though there are some commercial free-path AGVs with radio communication capabilities [6] [15], and commercial wireless modems available in the market, designs and techniques for robust indoor radio local area networks are not currently available.

Since infrared beams propagate in only line-of-sight (LOS) directions in space, communication using infrared requires installing several communication points where the vehicles have to stop to communicate with the central controller. Possible locations of the communication points are the load transfer stations, battery charge points, and the parking lots. Synchronizing the communications between vehicles is not desirable though possible. A central communications server is required to store and forward messages between vehicles.

Selection of an adequate media access control (MAC) method is another important issue. Three standard MAC methods are recommended by IEEE Project 802 for local area networks [10]; CSMA/CD (IEEE 802.3) for random access, token bus (IEEE 802.4) and token ring (IEEE 802.5) for controlled access. Several other schemes are also reported in the literature for both random and controlled access. Since a controlled access method requires a central controller to manage the right of access of other nodes, a hierarchical architecture is the natural strategy to organize the transport system. A random access method is suitable for complete heterarchy or autonomy.

3.3 Design and Development Aid Tools

The problem of deciding the best system design and architecture cannot be formulated mathematically. Simulation techniques are generally used to solve this problem. AGVSim [3] is a useful tool to decide better heuristic rules for an AGV system operation. AGVSim2 [4] can be used to evaluate the performance of a transport system controller. Simulation is also used in the design and development of control software [9]. The vehicle simulator described in this chapter has proven to be an essential tool in the development of the vehicle controller and the integration of ITS. Though the vehicle simulator represent a single vehicle, the same idea can be extended to develop a virtual transport system containing multiple vehicle simulators. As mentioned earlier, it can be used to test the transport system controller and the communication protocols in the development stage and to perform simulation experiments in the later analysis stages. An object-oriented language, featuring inheritance and data abstraction [2], is a possible implementation tool of a virtual transport system.

4 FINAL REMARKS

AGV systems are more attractive and advantageous alternatives to other material transport methods such as conveyors or fork trucks. Free-path AGVs add more flexibility to conventional AGV systems. However, the promised advantages can be achieved only when good control systems are available. While most of existing AGV systems are operated under a hierarchical control, there exist other modes of operation ranging from the complete heterarchy to the strict hierarchy, and some hybrid methods between these two. The functions performed by individual controllers are defined and allocated based on the control architecture employed. The algorithms and procedures to solve specific problems should be selected considering the architecture. Selection of the communications media and media access control method is also interrelated with the issue of control architecture. Though these issues are the very fundamental questions to be answered in the design stage, few researches are reported in the literature. What makes it worse is that there is no general strategy which is the best suited to all systems, nor is there any formal methodology to decide the best strategy for a specific system. The virtual transport system described in this chapter can be used to analyze the performance of various strategies without performing expensive time-consuming experiments on actual systems.

5 ACKNOWLEDGEMENT

The work reported in this chapter was partially supported by the NSF/Purdue University Engineering Research Center for Intelligent Manufacturing Systems under Grant No. CDR 8803017.

6 REFERENCES

1. Cox, Brad J., *Object-Oriented Programming An Evolutionary Approach*, Addison-Wesley, Reading, MA (1987).
2. Cox, I., "Blanche: Model-Based Autonomous Navigation," *Material Handling Focus '88*, MHRC, Georgia Institute of Technology, Atlanta, GA (1988).
3. Egbelu, Pius J. and J.M.A.Tanchoco, "Designing The Operations of Automatic Guided Vehicle System Using AGVSim," *Proc. 2nd Int. Conf. AGVS*, pp 21-30 (1983).
4. Gaskins, R.J. and J.M.A. Tanchoco, "AGVSim2 - A Development Tool for AGVS Controller Design," *Int. J. Prod. Res.*, Vol. 27, No. 6, pp. 915-926 (1989).
5. Hatvany, J., "Intelligence and Cooperation in Heterarchic Manufacturing Systems," *Robotics & Computer-Integrated Manufacturing*, Vol. 2, No. 2, pp. 101-104 (1985).
6. *Material Handling Engineering*, "SGV's Find Their Way by Reading Bar Codes," October (1987).
7. McGillem, C.D., J.D. Callison, and T.S. Rappaport, "Autonomous Roving Vehicle Navigation, Control and Communication," *Proc. 6th Int. Conf. Automated Guided Vehicle Systems*, pp. 179-188, IFS(Publ.) Ltd, UK, (1988).
8. McLean, C., M. Mitchell, and E. Barkmeyer, "A Computer Architecture for Small-Batch Manufacturing," *IEEE Spectrum* (May 1983).
9. Meinberg, Uwe, "Effective Design of Control Software for Highly Flexible AGV Systems," *Proc. 10th Int. Conf. on Automation in Warehousing*, pp. 301-304 (1989).
10. Schwartz, M., *Telecommunication Networks: Protocols, Modeling and Analysis*, Addison-Wesley, Reading, MA (1987).
11. Solberg, J.J. and J.A. Heim, "Managing Information Complexity in Material Flow Systems," *Proceedings of NATO Advanced Research Workshop on Advanced Information Technologies in Industrial Material Flow Systems*, Grenoble, France (1988).
12. Taghaboni, F. and J.M.A. Tanchoco, "A LISP-based Controller for Free-Ranging Automated Guided Vehicle Systems," *International Journal of Production Research*, Vol. 26, No. 2, pp. 173-188 (1988).

13. Ting, L. and J.M.A. Tanchoco, "C-Version of The Taghaboni-Tanchoco AGV Supervisory Controller," Technical Report, School of Industrial Engineering, Purdue University, West Lafayette, IN 47907 (1989).
14. Vamos, Tibor, "Cooperative Systems - An Evolutionary Perspective," *IEEE Control Systems Magazine*, pp. 9-14 (August 1983).
15. Van Brussel, H., C.C. Van Helsdingen, and K. Machiels, "FROG Free Ranging on Grid: New Perspectives in Automated Transport," *Proc. 6th Int. Conf. on AGVS*, pp. 223-232, IFS(Publ.), UK (1988).
16. Walker, S.P., S.K. Premi, C.B. Besant and A.J. Broadbent, "The Imperial College Free-Ranging AGV(ICAGV) and Scheduling System," *Proc. 3rd Int. Conf. on AGV*, pp. 189-198, IFS (Publications), UK (1985).

A PC BASED IMPLEMENTATION OF THE CONTROL ZONE MODEL FOR AGVS DESIGN

Charles J. Malmborg
Rensselaer Polytechnic Institute

ABSTRACT

A computerized decision support system is introduced for implementation of the zone control model for AGVS design. A hypothetical design problem is introduced and defined in terms compatible with the system database. The application of the computer model is illustrated through a series of six adjustments in the original design profile corresponding to variation in AGVS design variables including guidepath layout, vehicle routing, vehicle dispatching, and load transfer point location. For each variation, the package is used to study the performance of the proposed system including gridlocking risk factors and the minimum vehicle fleet size needed to meet a fixed throughput requirement.

1 INTRODUCTION

In a recent paper (see Malmborg [20]), the control zone analytical model for the design of AGV systems was developed to extend the range of design variables simultaneously used in the prediction of system throughput capacity. In contrast to most earlier analytical models dealing with AGVS design, (such as those reported in Maxwell et al. [1], Egbelu [5], Tanchoco et al. [6], Gaskins et al. [11], Blair et al. [9], and others), the control zone model explicitly measures five variables considered basic to the specification of a zone control AGVS (Wilhelm et al. [12]). These five variables include the guidepath layout (including the designation of vehicle control zones and buffering areas), load transfer point location, vehicle fleet size, vehicle dispatching rules, and vehicle routings. Based on these values, the control zone model predicts AGVS throughput capacity and shop locking risk factors associated with guidepath gridlock. The general approach used by the model is to generate explicit, quantitative estimates of vehicle interference levels. The availability of the control zone model raises the possibility of expanding the traditionally limited role played by analytical models in the AGVS design process by providing a tool for more in-depth analysis of system alternatives prior to the simulation phase of the design process. By providing the system designer with guidance in the very early phases of the design process (i.e., prior to fixing the levels of the six basic variables) the potential exists for extensive prescreening of design profiles. Thus, the designer could potentially identify a reduced set of promising design alternatives prior to committing to the development of a simulation model for validation.

To realize the potential benefit described above, an implementation tool for the control zone model is needed. The objective of this chapter is to describe a PC based package developed to serve as this implementation tool. In essence, the system provides a protocol through which the user can make adjustments in design alternatives, receive almost instantaneous feedback on the effects of the proposed system changes, and subsequently make further adjustments in a design alternative. The features of the package are illustrated through application to a test facility for which variations in system design profiles are evaluated. The next section of this chapter thus overviews the zone control model for evaluation of AGVS performance. (Readers interested in a quantitative description of the details of the model should see Malmborg [20].) This is followed by a description of the package as applied to a sample design problem. The final section offers a summary and conclusions.

2 OVERVIEW OF THE ZONE CONTROL MODEL

The zone control model for AGVS design follows a series of five steps in the prediction of system performance based on the five design variables including: guidepath layout, load transfer point location, workstation storage capacity, vehicle fleet size, vehicle dispatching, and vehicle routing. Each of these steps captures the full or partial effects of one or more of these decision variables on the performance of the AGVS. The steps of the procedure are summarized below.

Step 1. Computation of unobstructed, interstation travel times. This step uses data specified by the user in the form of linear guidepath segment endpoints, vehicle travel parameters, load transfer point locations, and vehicle routings to compute the "unobstructed" travel time between pairs of workstations served by an AGVS. The term unobstructed refers to travel times in the absence of guidepath contention. The travel times are generated using the vehicle travel speed, linear segment lengths, the time necessary for negotiating a transfer from one linear segment to the next while traveling through a control zone or between zones, the sequence of control zones traversed in traveling between a pair of workstations (i.e., routings), and the guidepath locations of the appropriate load transfer point at each workstation. In the model, a control zone is defined as a collection of adjacent linear guidepath segments in which vehicle collisions are avoided by allowing only one vehicle at a time in the zone.

Step 2. Estimation of empty vehicle travel. The estimation of empty vehicle travel in the control zone model is based on the vehicle fleet size and dispatching rules used in the AGVS. A dispatching rule as used within an AGVS governs the process of vehicle recirculation. AGV dispatching rules generally involve two components referred to as a workstation initiated and a vehicle initiated component (Egbelu et al. (2)). The workstation initiated component of an AGV dispatching rule determines which vehicle is selected to service a materials handling transaction originating at a workstation when idle vehicles are available. Examples of workstation initiated dispatching rules include nearest vehicle (NV), least utilized vehicle (LUV), random vehicle (RV), farthest vehicle (FV), longest idle vehicle (LIV), etc. The vehicle initiated component of an AGV dispatching rule is invoked when there is a pending queue of materials handling transactions and no vehicles immediately available. When a vehicle does become available, the vehicle initiated component of the rule combination determines which transaction is served by that vehicle. Examples of workstation initiated rule components include shortest travel time (STT), minimum outgoing remaining queue space (MORQS), random workstation (RW), modified first come first served (MFCFS, where FCFS is used and each workstation is allowed to have only one transaction in the queue), unit load shop arrival time (ULSAT), etc. Thus, a major decision variable in the design and operation of an AGVS is the dispatching rule combination selected for vehicle recirculation (e.g., MFCFS/RV, RW/NV, ULSAT/LUV, etc.)

The control zone model generates estimates of empty vehicle travel within an AGVS by first predicting the proportion of time that the workstation initiated component of the dispatching rule is invoked relative to the vehicle initiated component. A rough estimate of this proportion is obtained by representing the AGVS as an M/M/N queuing system where N is the number of the vehicles selected by the decision maker. In this queuing system, the service time is represented by the average, unobstructed travel time for a loaded transaction. This average is computed by the cumulative sum of the products of the unobstructed travel time and relative frequency of travel between all workstation pairs. Although this estimate is biased downward for the average travel time (since it does not account for vehicle blocking) it provides an initial rough estimate. The rate of arrivals used within the M/M/N model is given by the sum of the material flow matrix elements describing the number of transactions between each pair of workstations per unit time. Given this service rate, arrival rate, and fleet size (i.e., number of parallel servers), the steady state probability that there are fewer than N materials handling transactions in the system can be computed. This probability estimates the proportion of time that the workstation initiated component of the dispatching rule combination is invoked. Similarly, the complement represents the proportion of time that the vehicle initiated component of the dispatching rule is invoked.

Once the proportion of time that each rule component is invoked has been estimated, the corresponding loaded travel base for computing the volume of empty vehicle travel can be determined. For the workstation initiated component, the loaded travel base from which empty vehicle travel is given by the product of the probability of invoking this component and the material flow matrix representing the total volume of loaded travel per unit time. Analogously, the loaded travel base for the vehicle initiated component is determined by the product of the complement of this probability and the material flow matrix. In each case, the loaded travel base provides a volume of loaded travel between each pair of workstations that will result in an estimate of empty travel for vehicle recirculation that is determined by the dispatching rule used.

As an example, suppose that NV is the workstation initiated dispatching rule. In this extreme case, the volume of recirculation travel is obtained through the solution of the transportation problem where vehicle arrivals to each workstation (in the loaded travel base for the workstation initiated component) represent supply nodes and vehicle departures represent demand nodes. The cost coefficients are given by the travel times between each pairwise combination of workstations. When an alternative dispatching rule is used, the user must approximate the point in the range between the extreme cases of the NV (i.e., solution to the minimization transportation problem) and the FV (i.e., solution to the maximization transportation problem) rule components associated with the effects of the dispatching rule selected. For example, if the user selects the RV rule, an "alpha" value of 0.5 may be used. In this case, the volume of empty vehicle travel is approximated by the product of 0.5 (i.e., alpha) and the matrix associated with the solution to the minimization transportation problem and the product of 0.5 (i.e., one minus alpha) and the matrix associated with the solution to the maximization transportation problem. Obviously, the alpha value must be treated as a sensitivity parameter with a range of values considered for the complete analysis of a given design profile.

An analogous procedure using a second alpha sensitivity parameter is used to estimate the volume of empty vehicle travel resulting from the vehicle initiated component of the dispatching rule. This volume is obtained from the loaded travel base for this rule component. The vehicle initiated empty vehicle travel estimates are added to the empty travel estimates obtained from the workstation initiated component of the dispatching rule to estimate the total volume of empty vehicle travel. Thus, the second modeling step produces a modified material flow matrix representing both loaded and unloaded vehicle travel.

Step 3. Estimation of Arrival Rates to Each Control Zone. In this step of the modeling process, the maximum average number of vehicle arrivals per unit time is estimated for each control zone. These estimates are obtained by first computing the relative frequency of arrivals to each zone in the AGVS. The relative frequency of arrivals to a zone is obtained by the ratio of the total number of routings per unit time which include vehicle travel through the zone, and the sum of these values for all control zones. (These frequencies are used by the model to determine the probability that a randomly selected transaction will involve a given control zone.) In parallel with the frequency values, the probability that a transaction will involve a load transfer operation is also computed for each control zone. These values are estimated by the number of transactions through a zone which represent the start or completion of a routing, to the total number of transactions through the zone from all the routings among workstation pairs. Using these probabilities, the transit time through each zone including load transfer is given by the sum of the unobstructed travel time and the product of the load transfer probability and load transfer time.

The control zone frequencies and transit times are applied by the model to estimate the maximum rate of arrivals to a zone. The model estimates this rate using the product of the relative frequencies for each zone and the ratio of the number of vehicle hours available per unit time in the AGVS, and the sum of transit times (including load transfer) through each zone. The total number of vehicle hours available per unit time is given by the product of the number of vehicles and the system efficiency rate (which accounts for maintenance, battery charging, etc.). The ratio of the number of vehicle hours available per unit time in the AGVS and the sum of transit times through each zone is used in this calculation to approximate the "average" arrival rate to a zone. This estimated maximum arrival rate is an approximation of course, since the zone transit times do not yet account for vehicle blocking.

Step 4. Estimation of Vehicle Blocking Effects. The control zone model estimates the effects of vehicle blocking resulting from guidepath contention using the arrival rates described above to generate the probability distribution of the number of vehicles using, and waiting to use, a control zone. Specifically, maximum average vehicle arrival rates and the unobstructed transit times are used in modeling each control zone as a separate queuing system. However, since the transit time through each zone is essentially a constant, a simple M/M/1 model cannot be used to generate the probability distribution of the number of vehicles using and waiting to use a given control zone. Instead, this probability distribution is obtained through modeling each zone as a Markov process where system transitions correspond to vehicle departures from the zone (i.e., service completions) and system states correspond to the number of vehicles traveling through and waiting to travel through the zone. Transition probabilities in this Markov chain are obtained by assuming a probability distribution for the arrival rates to control zones (e.g., Poisson) and computing the probability of alternative numbers of arrivals to a zone during a service interval. These probability calculations take account of the vehicle fleet size and current system state (e.g., if N=5 and there are two vehicles in the zone and buffer, then the arrival rate is only 3/5 of what it would be if there were no vehicles in the zone). The resultant probabilities are then used to formulate the transition probabilities between the system states associated with the control zone. The probability distribution of the number of vehicles using and waiting to use a zone is obtained by solving for the vector of steady-state probabilities associated with the Markov chain.

Estimates of vehicle blocking times are obtainable directly from the probability distribution of the number of vehicles using and waiting to use a control zone through simple expectation. The control zone model estimates the travel time through a zone including blocking using the sum of the unobstructed travel time and the expected waiting time. The expected waiting time is obtained by the product of probability that a given number of vehicles is waiting to use the zone, and the average waiting time that would correspond to this number of vehicles waiting. These products are summed for all feasible values of the number of vehicles waiting. The average waiting time corresponding to X vehicles waiting to use a zone is estimated by the model as the product of (X-1/2) and the unobstructed travel time. The reasoning in using this calculation is that on the average, the vehicle using the zone will be half completed when a new vehicle arrives at the queue for that zone.

The probability distribution of the number of vehicles using and waiting to use a zone is also applied is estimating the likelihood that the number of vehicles waiting for access to a zone will exceed the vehicle buffering capacity allocated to that zone. This calculation provides the user with some insight into the probability that a gridlock situation will occur at various sites along the guidepath before the final guidepath layout is fixed.

Step 5. Estimation of Maximum AGVS Throughput Capacity. Using the estimates of travel times through each control zone including vehicle blocking, revised estimates of the travel time between each pairwise combination of workstations are obtained by the model. The travel time estimates which include vehicle blocking are then used by the model to compute the number of vehicle hours per unit time necessary to complete the volume of vehicle travel captured in the modified material flow matrix, (i.e., the matrix which includes loaded and empty vehicle travel). The number of vehicle hours needed to service each pairwise combination of workstations is obtained by the product of the corresponding travel frequency in the modified material flow matrix, and the travel time including blocking between the workstation pair. The total vehicle hours available is obtained by summing these products for all pairwise combinations of workstations. If the total vehicle hours necessary exceeds the product of the number of vehicles and the efficiency factor, the number of vehicles is inadequate. Otherwise, the number of vehicles is automatically reduced until the minimum fleet size for which vehicle hours available exceeds vehicle hours necessary is determined. The resultant fleet size is the minimum number of vehicles for which the model predicts the AGVS will have sufficient throughput capacity.

3 APPLICATION OF THE CONTROL ZONE MODEL

In this section, the application of the PC based package for the control zone model is described for a test problem. The package is written in (compiled) BASIC. It reads a data file created by the user to initialize the design process and describe the facility served by the AGVS. To create these files, the user must impose a cartesian coordinate system on the facility to specify the location of all intersections of linear guidepath segments, the specific intersections defining each segment and the segments included in each control zone. To alter the guidepath, the user interactively makes adjustments to this input file. The package also reads in the material flow matrix describing the required volume of loaded vehicle travel in the facility and the routings between each workstation pair. The specification of the routings defines the location of load transfer points (i.e., at the terminal of the last guidepath segment in the corresponding routing).

The user enters initial values including the scale of the coordinate system, the vehicle travel speed, efficiency factor, load transfer time, time to negotiate guidepath segment changes, load transfer time, and vehicle fleet size. This can be done interactively or through an input file serving as the initial database. The selection of a dispatching rule by the user is communicated interactively through the separate specification of alpha values for the workstation and vehicle initiated portion of the vehicle dispatching rule combination. (The system provides for an automated sensitivity analysis of these values.) The database used by the system was designed to make modification of inputs as a result of changes in a design profile as straight forward as possible. Thus, to study the effects of design changes, the user merely edits the input files describing a design profile (which would be the quickest alternative for the computer literate user) or makes adjustments to parameters interactively, by selecting from a menu. In each case the user reruns the main analysis package to study the effects of changes in a design profile.

Using the procedures described in the second section of the chapter, the package generates outputs related to vehicle dispatching. These include the estimated frequency with which the workstation initiated and vehicle initiated components of the rule combination is invoked, and the material flow matrix modified for empty vehicle travel. The package also outputs the steady state probability vector describing the number of vehicles in each control zone, travel times between workstation pairs (with and without blocking) and the total number of vehicle hours available and required per unit time. The user can also compute the gridlock probability at each control zone based on the vehicle buffering capacities. Based on the outputs provided, the user edits the input database (either on-line or off-line) in attempting to attain a feasible design profile that most effectively satisfies the relevant performance criteria.

To illustrate the application of the control zone model, a database correspon-ding to the sample problem depicted in Fig. 1 was created. As illustrated in the figure, the facility in question consists of five workstations where the initial guidepath layout has 11 linear guidepath segments defining nine control zones. The data describing the original guidepath layout, the load transfer point locations, material flow volume requirements, control zone buffer capacities, and fixed parametric data are summarized in Tables 1 and 2. (For simplicity in the example problem, it is assumed that pickup and delivery load transfers occur at the same location at each workstation and that routings are symmetric. These are not limitations imposed by the package.) Based on this starting point, the package was used to interactively study the effects of six variations in the original design profile. This entire process, including user modification of the database was completed on a standard IBM PC XT in approximately one hour. The system response time for doing all computations to determine the throughput capacity for the alternative design profiles studied averaged

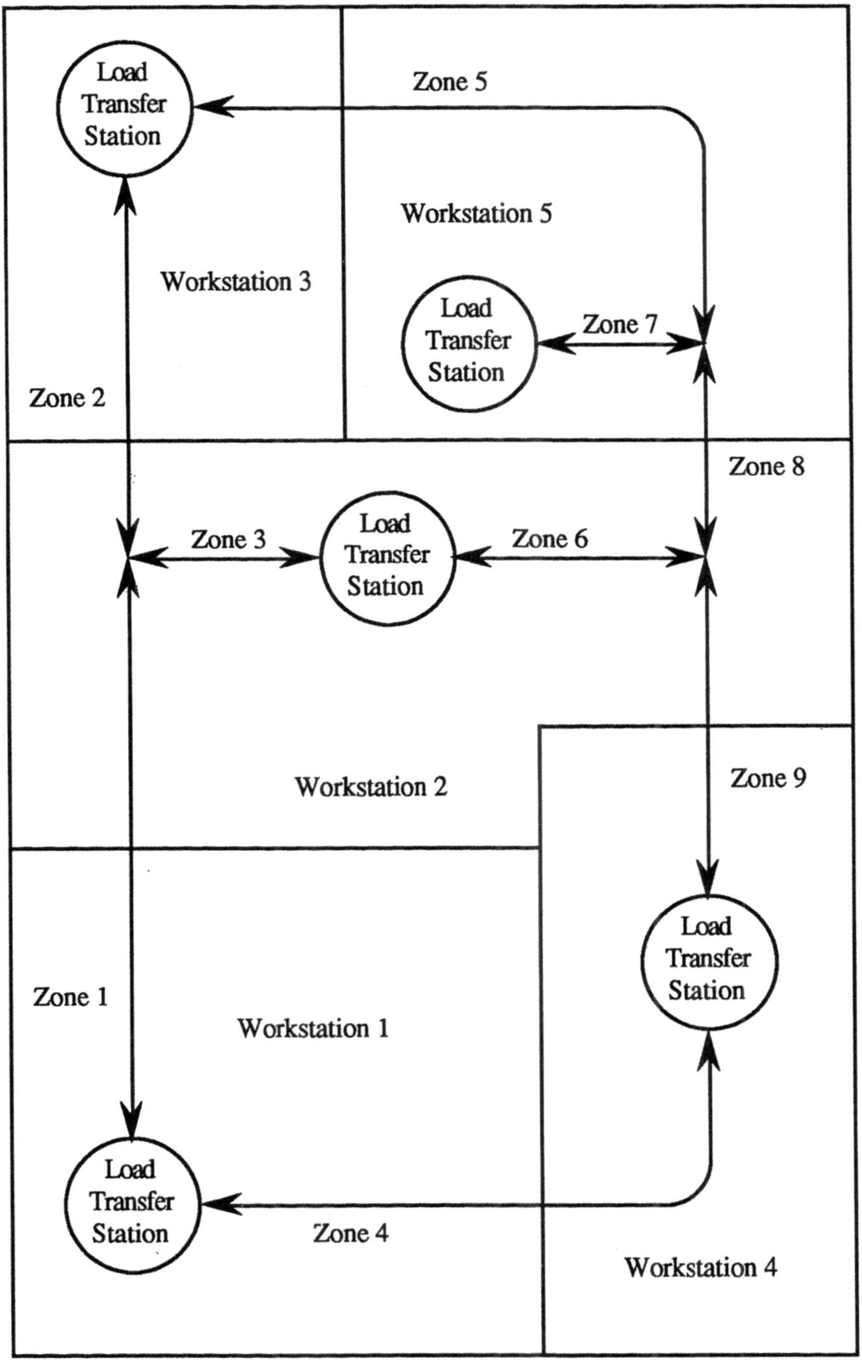

Fig. 1 Initial guidepath layout

Table 1 Summary of the input database describing materials handling requirements and the initial guidepath layout

Material Flow Matrix in Required Hourly Unit Loads Between Workstation Pairs (Pickup and Deposit Load Transfers Share the Same Location at Each Workstation):

		Workstation				
		1	2	3	4	5
	1	-	1	0	2	2
	2	3	-	2	3	1
Workstation	3	4	0	-	2	0
	4	0	4	3	-	1
	5	2	0	1	4	-

Guidepath Segment Intersections:

Intersection	Coordinates	Intersection	Coordinates
1	(1, 1)	6	(6, 1)
2	(1, 6)	7	(6, 3)
3	(1, 9)	8	(6, 6)
4	(3, 6)	9	(6, 8)
5	(4, 8)	10	(6, 9)

Intersection Combinations Defining Endpoints of Linear Guidepath Segments:

Segments										
1	2	3	4	5	6	7	8	9	10	11
(1,2)	(2,3)	(2,4)	(1,6)	(3,10)	(4,8)	(5,9)	(6,7)	(7,8)	(8,9)	(9,10)

Linear Guidepath Segments Included in Each Control Zone:

Control Zones								
1	2	3	4	5	6	7	8	9
(1)	(2)	(3)	(4,8)	(5,11)	(6)	(7)	(8)	(9)

Vehicle Buffer Capacities Assumed for Each Control Zones (in Vehicles):

Control Zones:	1	2	3	4	5	6	7	8	9
Buffer Capacity:	(3)	(3)	(2)	(3)	(3)	(2)	(1)	(1)	(2)

Table 2 Summary of the input database describing fixed parameters and four alternative vehicle routings (assumed symmetric)

Fixed Parametric Data:

Vehicle Efficiency Factor:	0.85
Coordinate Scale:	1" = 40 ft.
Vehicle Speed:	60 ft./min.
Load Transfer Time:	10 seconds

Initial Load Transfer Point Locations:

Workstation	Coordinates
1	(1,1)
2	(3,6)
3	(1,9)
4	(6,3)
5	(4,8)

Routings Represented by Sequences of Control Zones Between Workstation Pairs:

Routing 1		Routing 2		Routing 3		Routing 4	
Stations	Sequence	Stations	Sequence	Stations	Sequence	Stations	Sequence
1,2	4-9-6	1,2	1-3	1,2	11-1-3	1,2	11-1
1,3	4-6-9-3-2	1,3	1-2	1,3	11-1-2	1,3	11-1-2
1,4	1-3-6-9	1,4	4	1,4	11-10	1,4	11-10
1,5	1-3-6-8-7	1,5	4-9-8-7	1,5	4-9-8-7	1,5	4-9-8
2,3	6-8-5	2,3	3-2	2,3	3-2	2,3	2
2,4	3-1-4	2,4	6-9	2,4	6-9	2,4	1-10
2,5	3-2-5-7	2,5	6-8-7	2,5	6-8-7	2,5	2-5
3,4	2-1-4	3,4	5-8-9	3,4	2-1-10	3,4	2-1-10
3,5	5-7	3,5	5-7	3,5	5-7	3,5	5
4,5	4-1-2-5-7	4,5	9-8-7	4,5	9-8-7	4,5	9-8

approximately 170 seconds per design profile. The minimum number of vehicles required was determined using the package for each of the six design variations summarized below:

Case 1 The original guidepath layout (Fig. 1) and load transfer point locations are used in conjunction with the second set of routings, (summarized in Table 2), and the FV/LTT dispatching rule (i.e., both the workstation and vehicle initiated alpha values equal to zero).

Case 2 The original guidepath layout and load transfer point locations are used in conjunction with the second set of routings and the NV/STT dispatching rule (i.e., both the workstation and vehicle initiated alpha values equal to one).

Case 3 The original guidepath layout and load transfer point locations are used in conjunction with the first set of routings and the RV/RW dispatching rule (i.e., both the workstation and vehicle initiated alpha values equal to 0.5).

Case 4 The original guidepath layout and load transfer point locations are used in conjunction with the second set of routings and the RV/RW dispatching rule.

Case 5 A cutover is added to the guidepath as shown in Fig. 2 resulting in a revised guidepath with 13 linear segments and 11 control zones. The third set of routings summarized in Table 2 is used in conjunction with the RV/RW dispatching rule.

Case 6 The modified guidepath layout described in case 5 is used in conjunction with relocation of the load transfer points for workstation 2 from its original location to the intersection of control zones 3, 2 and 1. The load transfer point for workstation 5 is relocated to the intersection of control zones 6, 8 and 9. (These changes are illustrated in Fig. 2) The fourth set of routings summarized in Table 2 is used in conjunction with the RV/RW dispatching rule.

The outputs provided by the package for each of the six cases are summarized in Tables 3 through 8. As these tables show, minimum vehicle requirements are highly sensitive to the design variables. Based on the outputs in Tables 3 and 4, it is evident that changing from the extremes of the dispatching rule spectrum can nearly double minimum vehicle requirements. Similarly, routings that result in high congestion levels at specific control zones can have a similar effect. This is illustrated by the contrast in outputs between cases 3 and 4 where the average number of vehicle minutes per hour due to vehicle blocking is respectively 22.1 and 45.3 minutes. (See Tables 5 and 6.) Correspondingly, minimum vehicle requirements go from 4 to 8 as the second set of routings is replaced by the first set of routings in Table 2. Comparing the outputs associated with cases 3 and 5, it is evident that the effects of changes in the guidepath layout can be difficult to predict. In this case, the combined effect of the corresponding guidepath and routings modification is not enough to actually change the minimum vehicle requirements, but these adjustments do result in additional guidepath contention that very nearly results in the need for an additional vehicle. Thus, the effect of expanding the guidepath to create a cutback actually increases the vehicle minutes required. This result emphasizes the difficulty of predicting the effects of changes in the design variables.

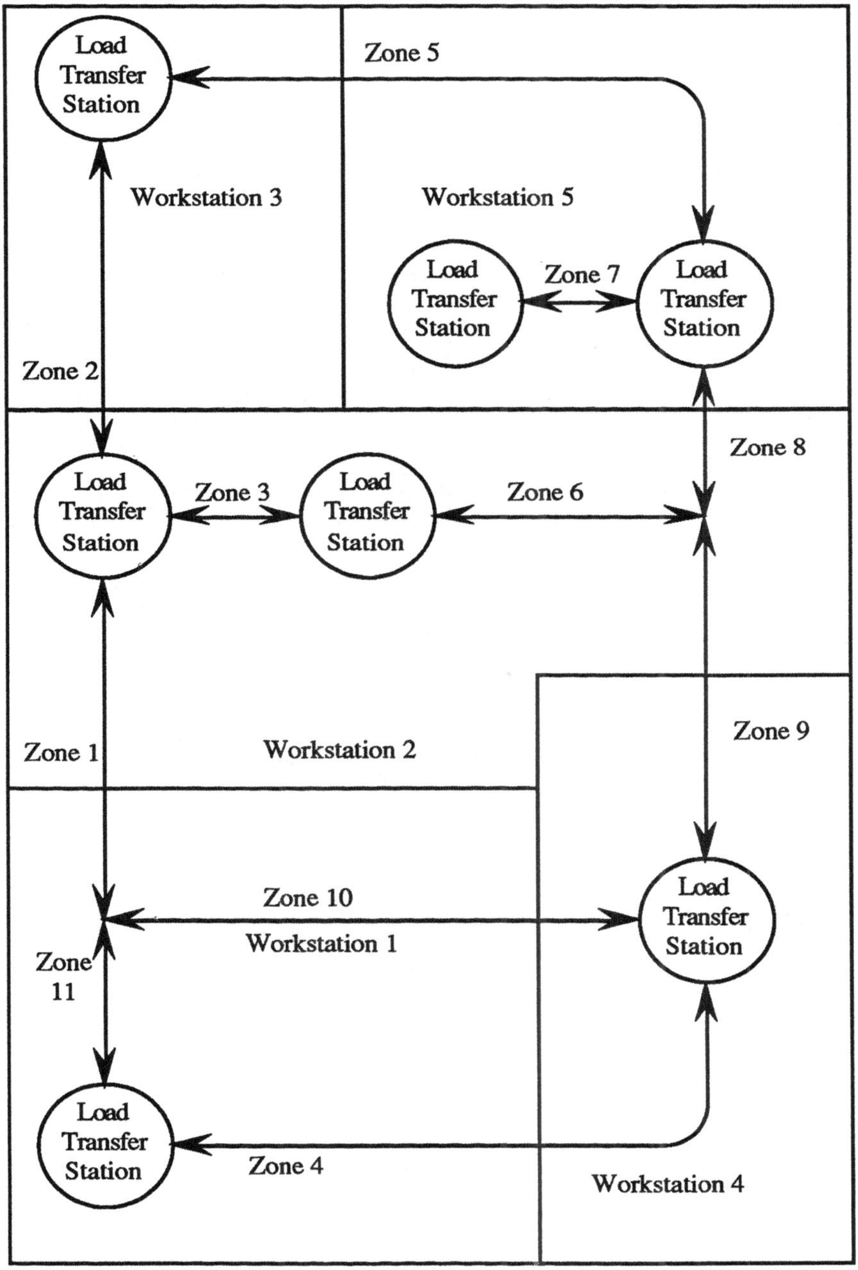

Fig. 2 Modified guidepath layout (Case 5 and 6) and relocated load transfer points for workstations 2 and 5 (Case 6)

Table 3 Summary of outputs for case 1

Minimum Number of Vehicles Computed by the Model	10
Total Vehicle Minutes Available Per Hour	510
Total Vehicle Minutes Required Per Hour	469.7

Steady State Probability Vectors for Each Control Zone:

	System States - Number of Vehicles Using Zone and Waiting								
	1	2	3	4	5	6	7	8	9
Zone 1	0.81	0.17	0.02	0.00	0.00	0.00	0.00	0.00	0.00
Zone 2	0.91	0.09	0.00	0.00	0.00	0.00	0.00	0.00	0.00
Zone 3	0.93	0.07	0.00	0.00	0.00	0.00	0.00	0.00	0.00
Zone 4	0.67	0.27	0.05	0.01	0.00	0.00	0.00	0.00	0.00
Zone 5	0.85	0.14	0.01	0.00	0.00	0.00	0.00	0.00	0.00
Zone 6	0.91	0.08	0.00	0.00	0.00	0.00	0.00	0.00	0.00
Zone 7	0.96	0.04	0.00	0.00	0.00	0.00	0.00	0.00	0.00
Zone 8	0.98	0.02	0.00	0.00	0.00	0.00	0.00	0.00	0.00
Zone 9	0.95	0.05	0.00	0.00	0.00	0.00	0.00	0.00	0.00

Material Flow Matrix Including Empty Vehicle Travel in Unit Loads/Hr.

	Workstation				
Workstation	1	2	3	4	5
1	0.00	1.00	5.00	2.00	2.00
2	6.00	0.00	2.00	9.00	1.00
3	4.00	6.00	0.00	2.00	0.00
4	0.00	7.00	4.00	0.00	5.00
5	2.00	2.00	1.00	9.00	0.00

Interstation Travel Times in Minutes Without Blocking:

	Workstation				
Workstation	1	2	3	4	5
1	0.00	5.00	6.83	5.00	5.42
2	5.00	0.00	4.33	5.33	5.08
3	6.83	4.33	0.00	5.67	3.25
4	5.00	5.33	5.67	0.00	8.58
5	5.42	5.08	3.25	8.58	0.00

Interstation Travel Times in Minutes With Blocking:

	Workstation				
Workstation	1	2	3	4	5
1	0.00	5.67	7.58	5.31	5.72
2	5.67	0.00	4.58	6.15	5.37
3	7.58	4.58	0.00	6.51	3.45
4	5.31	6.15	6.51	0.00	9.63
5	5.72	5.37	3.45	9.63	0.00

Table 4 Summary of outputs for case 2

Minimum Number of Vehicles Computed by the Model	6
Total Vehicle Minutes Available Per Hour	306
Total Vehicle Minutes Required Per Hour	262.1

Steady State Probability Vectors for Each Control Zone:

	System States - Number of Vehicles Using Zone and Waiting								
	1	2	3	4	5	6	7	8	9
Zone 1	0.86	0.13	0.01	0.00	0.00	0.00	0.00	0.00	0.00
Zone 2	0.96	0.04	0.00	0.00	0.00	0.00	0.00	0.00	0.00
Zone 3	0.95	0.05	0.00	0.00	0.00	0.00	0.00	0.00	0.00
Zone 4	0.81	0.17	0.02	0.00	0.00	0.00	0.00	0.00	0.00
Zone 5	0.95	0.05	0.00	0.00	0.00	0.00	0.00	0.00	0.00
Zone 6	0.94	0.05	0.01	0.00	0.00	0.00	0.00	0.00	0.00
Zone 7	0.97	0.03	0.00	0.00	0.00	0.00	0.00	0.00	0.00
Zone 8	0.98	0.02	0.00	0.00	0.00	0.00	0.00	0.00	0.00
Zone 9	0.97	0.03	0.00	0.00	0.00	0.00	0.00	0.00	0.00

Material Flow Matrix Including Empty Vehicle Travel in Unit Loads/Hr.

	Workstation				
Workstation	1	2	3	4	5
1	5.00	1.00	0.00	2.00	2.00
2	3.00	5.00	2.00	7.00	1.00
3	4.00	0.00	6.00	2.00	0.00
4	1.00	4.00	3.00	7.00	1.00
5	5.00	0.00	1.00	4.00	4.00

Interstation Travel Times in Minutes Without Blocking:

	Workstation				
Workstation	1	2	3	4	5
1	0.00	5.00	6.83	5.00	5.42
2	5.00	0.00	4.33	5.33	5.08
3	6.83	4.33	0.00	5.67	3.25
4	5.00	5.33	5.67	0.00	8.58
5	5.42	5.08	3.25	8.58	0.00

Interstation Travel Times in Minutes With Blocking:

	Workstation				
Workstation	1	2	3	4	5
1	0.00	5.33	7.21	5.21	5.63
2	5.33	0.00	4.43	5.77	5.20
3	7.21	4.43	0.00	6.11	3.32
4	5.21	5.77	6.11	0.00	9.10
5	5.63	5.20	3.32	9.10	0.00

Table 5 Summary of outputs for case 3

Minimum Number of Vehicles Computed by the Model	4
Total Vehicle Minutes Available Per Hour	204
Total Vehicle Minutes Required Per Hour	181.9

Steady State Probability Vectors for Each Control Zone:

	System States - Number of Vehicles Using Zone and Waiting								
	1	2	3	4	5	6	7	8	9
Zone 1	0.96	0.04	0.00	0.00	0.00	0.00	0.00	0.00	0.00
Zone 2	0.97	0.03	0.00	0.00	0.00	0.00	0.00	0.00	0.00
Zone 3	0.98	0.02	0.00	0.00	0.00	0.00	0.00	0.00	0.00
Zone 4	0.96	0.04	0.00	0.00	0.00	0.00	0.00	0.00	0.00
Zone 5	0.97	0.03	0.00	0.00	0.00	0.00	0.00	0.00	0.00
Zone 6	0.97	0.03	0.01	0.00	0.00	0.00	0.00	0.00	0.00
Zone 7	0.97	0.03	0.00	0.00	0.00	0.00	0.00	0.00	0.00
Zone 8	0.97	0.03	0.00	0.00	0.00	0.00	0.00	0.00	0.00
Zone 9	0.92	0.07	0.01	0.00	0.00	0.00	0.00	0.00	0.00

Material Flow Matrix Including Empty Vehicle Travel in Unit Loads/Hr.

	Workstation				
Workstation	1	2	3	4	5
1	2.5	1.00	2.50	2.00	2.00
2	4.50	2.50	2.00	8.00	1.00
3	4.00	3.00	3.00	2.00	0.00
4	0.50	5.50	3.50	3.50	3.00
5	3.50	1.00	1.00	6.50	2.00

Interstation Travel Times in Minutes Without Blocking:

	Workstation				
Workstation	1	2	3	4	5
1	0.00	2.83	3.17	2.83	5.42
2	2.83	0.00	2.17	2.50	2.92
3	3.17	2.17	0.00	4.33	3.25
4	2.83	2.50	4.33	0.00	2.92
5	5.42	2.92	3.25	2.92	0.00

Interstation Travel Times in Minutes With Blocking:

	Workstation				
Workstation	1	2	3	4	5
1	0.00	2.88	3.22	2.89	5.54
2	2.88	0.00	2.19	2.56	2.96
3	3.22	2.19	0.00	4.42	3.29
4	2.89	2.56	4.42	0.00	2.99
5	5.54	2.96	3.29	2.99	0.00

Table 6 Summary of outputs for case 4

Minimum Number of Vehicles Computed by the Model	8
Total Vehicle Minutes Available Per Hour	408
Total Vehicle Minutes Required Per Hour	362.7

Steady State Probability Vectors for Each Control Zone:

	System States - Number of Vehicles Using Zone and Waiting								
	1	2	3	4	5	6	7	8	9
Zone 1	0.83	0.15	0.01	0.00	0.00	0.00	0.00	0.00	0.00
Zone 2	0.93	0.06	0.01	0.00	0.00	0.00	0.00	0.00	0.00
Zone 3	0.94	0.06	0.00	0.00	0.00	0.00	0.00	0.00	0.00
Zone 4	0.74	0.23	0.03	0.00	0.00	0.00	0.00	0.00	0.00
Zone 5	0.90	0.09	0.01	0.00	0.00	0.00	0.00	0.00	0.00
Zone 6	0.93	0.07	0.00	0.00	0.00	0.00	0.00	0.00	0.00
Zone 7	0.96	0.04	0.00	0.00	0.00	0.00	0.00	0.00	0.00
Zone 8	0.98	0.02	0.00	0.00	0.00	0.00	0.00	0.00	0.00
Zone 9	0.96	0.04	0.00	0.00	0.00	0.00	0.00	0.00	0.00

Material Flow Matrix Including Empty Vehicle Travel in Unit Loads/Hr.

	Workstation				
Workstation	1	2	3	4	5
1	0.00	5.00	6.83	5.00	5.42
2	5.00	0.00	4.33	5.33	5.08
3	6.83	4.33	0.00	5.67	3.25
4	5.00	5.33	5.67	0.00	8.58
5	5.42	5.08	3.25	8.58	0.00

Interstation Travel Times in Minutes Without Blocking:

	Workstation				
Workstation	1	2	3	4	5
1	0.00	5.00	6.83	5.00	5.42
2	5.00	0.00	4.33	5.33	5.08
3	6.83	4.33	0.00	5.67	3.25
4	5.00	5.33	5.67	0.00	8.58
5	5.42	5.08	3.25	8.58	0.00

Interstation Travel Times in Minutes With Blocking:

	Workstation				
Workstation	1	2	3	4	5
1	0.00	5.49	7.38	5.26	5.67
2	5.49	0.00	4.50	5.95	5.28
3	7.38	4.50	0.00	6.30	3.38
4	5.26	5.95	6.30	0.00	9.35
5	5.67	5.28	3.38	9.35	0.00

Table 7 Summary of outputs for case 5

Minimum Number of Vehicles Computed by the Model	4
Total Vehicle Minutes Available Per Hour	204
Total Vehicle Minutes Required Per Hour	202.9

Steady State Probability Vectors for Each Control Zone:

	System States - Number of Vehicles Using Zone and Waiting										
	1	2	3	4	5	6	7	8	9	10	11
Zone 1	0.94	0.06	0.00	0.00	0.00	0.00	0.00	0.00	0.00	0.00	0.00
Zone 2	0.97	0.03	0.00	0.00	0.00	0.00	0.00	0.00	0.00	0.00	0.00
Zone 3	0.99	0.01	0.00	0.00	0.00	0.00	0.00	0.00	0.00	0.00	0.00
Zone 4	0.96	0.04	0.00	0.00	0.00	0.00	0.00	0.00	0.00	0.00	0.00
Zone 5	1.00	0.00	0.00	0.00	0.00	0.00	0.00	0.00	0.00	0.00	0.00
Zone 6	0.98	0.02	0.00	0.00	0.00	0.00	0.00	0.00	0.00	0.00	0.00
Zone 7	0.98	0.02	0.00	0.00	0.00	0.00	0.00	0.00	0.00	0.00	0.00
Zone 8	0.98	0.02	0.00	0.00	0.00	0.00	0.00	0.00	0.00	0.00	0.00
Zone 9	0.95	0.04	0.01	0.00	0.00	0.00	0.00	0.00	0.00	0.00	0.00
Zone 10	0.96	0.04	0.00	0.00	0.00	0.00	0.00	0.00	0.00	0.00	0.00
Zone 11	0.94	0.06	0.00	0.00	0.00	0.00	0.00	0.00	0.00	0.00	0.00

Material Flow Matrix Including Empty Vehicle Travel in Unit Loads/Hr.

	Workstation				
Workstation	1	2	3	4	5
1	2.50	2.51	0.00	2.00	2.99
2	6.48	2.50	2.00	5.00	2.01
3	4.00	0.00	3.00	5.01	0.00
4	1.99	4.99	6.00	2.00	1.00
5	3.02	0.00	1.00	7.99	2.00

Interstation Travel Times in Minutes Without Blocking:

	Workstation				
Workstation	1	2	3	4	5
1	0.00	2.83	3.17	2.08	5.42
2	2.83	0.00	2.17	2.50	2.92
3	3.17	2.17	0.00	4.92	3.25
4	2.08	2.50	4.92	0.00	2.92
5	5.42	2.92	3.25	2.92	0.00

Interstation Travel Times in Minutes With Blocking:

	Workstation				
Workstation	1	2	3	4	5
1	0.00	2.89	3.24	2.12	5.51
2	2.89	0.00	2.19	2.54	2.94
3	3.24	2.19	0.00	5.02	3.26
4	2.12	2.54	5.02	0.00	2.96
5	5.51	2.94	3.26	2.96	0.00

Table 8 Summary of outputs for case 6

Minimum Number of Vehicles Computed by the Model	3
Total Vehicle Minutes Available Per Hour	153
Total Vehicle Minutes Required Per Hour	150.6

Steady State Probability Vectors for Each Control Zone:

	System States - Number of Vehicles Using Zone and Waiting										
	1	2	3	4	5	6	7	8	9	10	11
Zone 1	0.91	0.09	0.00	0.00	0.00	0.00	0.00	0.00	0.00	0.00	0.00
Zone 2	0.97	0.03	0.00	0.00	0.00	0.00	0.00	0.00	0.00	0.00	0.00
Zone 3	1.00	0.00	0.00	0.00	0.00	0.00	0.00	0.00	0.00	0.00	0.00
Zone 4	1.00	0.00	0.00	0.00	0.00	0.00	0.00	0.00	0.00	0.00	0.00
Zone 5	0.97	0.03	0.00	0.00	0.00	0.00	0.00	0.00	0.00	0.00	0.00
Zone 6	1.00	0.00	0.00	0.00	0.00	0.00	0.00	0.00	0.00	0.00	0.00
Zone 7	0.00	0.00	0.00	0.00	0.00	0.00	0.00	0.00	0.00	0.00	0.00
Zone 8	1.00	0.00	0.00	0.00	0.00	0.00	0.00	0.00	0.00	0.00	0.00
Zone 9	1.00	0.00	0.00	0.00	0.00	0.00	0.00	0.00	0.00	0.00	0.00
Zone 10	0.96	0.04	0.00	0.00	0.00	0.00	0.00	0.00	0.00	0.00	0.00
Zone 11	0.94	0.06	0.00	0.00	0.00	0.00	0.00	0.00	0.00	0.00	0.00

Material Flow Matrix Including Empty Vehicle Travel in Unit Loads/Hr.

	Workstation				
Workstation	1	2	3	4	5
1	2.50	1.51	0.00	2.00	4.00
2	5.99	3.51	2.00	5.49	1.00
3	4.00	0.00	3.00	5.01	0.00
4	0.00	4.99	6.00	3.99	1.00
5	5.51	0.00	1.00	5.50	2.00

Interstation Travel Times in Minutes Without Blocking:

	Workstation				
Workstation	1	2	3	4	5
1	0.00	2.08	3.17	0.33	4.25
2	2.08	0.00	1.42	2.08	2.50
3	3.17	1.42	0.00	3.17	1.42
4	0.33	2.08	3.17	0.00	2.08
5	4.25	2.50	1.42	2.08	0.00

Interstation Travel Times in Minutes With Blocking:

	Workstation				
Workstation	1	2	3	4	5
1	0.00	2.17	3.27	0.33	4.33
2	2.17	0.00	1.43	2.17	2.52
3	3.27	1.43	0.00	3.27	1.42
4	0.33	2.17	3.27	0.00	2.12
5	4.33	2.52	1.42	2.12	0.00

Finally, case 6 demonstrates the effects of relocating load transfer points at workstations 2 and 5. The net effect of these changes combined with corresponding changes in the routings is to effectively eliminate control zones 3, 6 and 7. Although this results in a slight increase in vehicle blocking in some of the remaining control zones, there is an overall reduction in vehicle minutes required due to reduced travel times between the workstation load transfer points.

5 SUMMARY AND CONCLUSIONS

A computer package for implementation of the control zone model for AGVS design is described. Results from implementation of the control zone model are obtained for a series of design profiles representing variations in the guidepath layout, vehicle routings, vehicle dispatching rules and load transfer point locations associated with a hypothetical design problem. These variations are typical of the "what-if" type of analysis involved in designing a zone control AGVS. Preliminary results suggest the difficulty of intuitively predicting the effects of changes in design profiles defining control zone AGV systems. However, the availability of an implementation tool for the zone control model enables the user to expand the range of alternatives that can be investigated prior to the simulation phase of the design process for zone control AGV systems. In this way the user can more effectively identify a set of promising design alternatives for which construction of detailed simulation models is warranted and subsequently increase the effectiveness of the design process.

6 ACKNOWLEDGEMENTS

This work was supported in part from a grant from the New York State Center for Advanced Technology in Automation and Robotics.

7 REFERENCES

1. Maxwell, W.L. and J.A. Muckstadt, "Design of Automatic Guided Vehicle Systems," *IIE Transactions*, Vol. 14, No. 2, pp. 114-124 (1982).
2. Egbelu, P.J. and J.M.A. Tanchoco, "Characterization of Automatic Guided Vehicle Dispatching Rules," *International Journal of Production Research*, Vol. 22, No. 3, pp. 359-374 (1984).
3. Newton, D., "Simulation Model Helps Determine How Many Automatic Guided Vehicles are Needed," *Industrial Engineering*, Vol. 17, No. 2, pp. 68-79 (1985).
4. Egbelu, P.J. and J.M.A. Tanchoco, "Potentials for Bi-Directional Guidepath for Automated Guided Vehicle Based Systems," *International Journal of Production Research*, Vol. 24, No. 5, pp. 1075-1097 (1986).

5. Egbelu, P.J., "The Use of Nonsimulation Approaches in Estimating Vehicle Requirements in an Automatic Guided Vehicle Based Transport System," *Material Flow*, Vol. 4, No. 1, pp. 17-32 (1987).
6. Tanchoco, J.M.A., P.J. Egbelu, and F. Taghaboni, "Determination of the Total Number of Vehicles in an AGV Based Material Transport System," *Material Flow*, Vol. 4, No. 1, pp. 33-52 (1987).
7. Wysk, R.A., P.J. Egbelu, C. Zhou, and B.K. Ghosh, "Use of Spreadsheet Analysis for Evaluating AGV Systems," *Material Flow*, Vol. 4, No. 1, pp. 53-64 (1987).
8. Leung, L.C., S.K. Khator, and D.L. Kimble, "Assignment of AGVS With Different Vehicle Types," *Material Flow*, Vol. 4, No. 1 (1987).
9. Blair, E.L., P. Charnsethikul, and A. Vasques, "Optimal Routing of Driverless Vehicles in a Flexible Material Handling System," *Material Flow*, Vol. 4, No. 1 (1987).
10. Hodgson, T.J., R.E. King, S.K. Monteith, and S.R. Schultz, "Developing Control Rules for an AGVS Using Markov Decision Processes," *Material Flow*, Vol. 4, No. 1 (1987).
11. Gaskins, R.J. and J.M.A. Tanchoco, "Flow Path Design for Automated Guided Vehicle Systems," *International Journal of Production Research*, Vol. 25, No. 5, pp. 667-676 (1987).
12. Wilhelm, M.R. and G.W. Evans, "The State-Of-The-Art in AGV Systems Analysis and Planning," *Proceedings of the AGVS `87*, Pittsburgh, PA (October 1987).
13. Taghaboni, F. and J.M.A. Tanchoco, "A LISP Based Controller for Free Ranging Automatic Guided Vehicle Systems," *International Journal of Production Research*, Vol. 26, No. 2, pp. 173-188 (1988).
14. Usher, J.S., G.W. Evans, and M.R. Wilhelm, "AGV Flow Path Design and Load Transfer Point Location," *Proceedings* of the International Industrial Engineering Conference, Orlando, FL, pp. 174-179 (1988).
15. Ozden, M., "A Simulation Study of Multiple Load Carrying Automatic Guided Vehicles in a Flexible Manufacturing System," *International Journal of Production Research*, Vol. 26, No. 8, pp. 1353-1366 (1988).
16. Malmborg, C.J., "A Compendium of Archival Literature in AGVS Modeling," *Proceedings*, of the International Industrial Engineering Conference, Toronto, Ontario, pp. 185-190 (1989).
17. Rabeneck, C.W., J.S. Usher, and G.W. Evans, "An Analytical Model for AGVS Design," *Proceedings*, of the International Industrial Engineering Conference, Toronto, Ontario, pp. 191-195 (1989).
18. Malmborg, C.J., "An Analytical Strategy for Minimum Fleet Size Determination in Automated Guided Vehicle Systems," Decision Sciences and Engineering Systems Report Series, Rensselaer Polytechnic Institute, Troy, NY (June 1989).

19. Malmborg, C.J., "A Model for the Design of Zone Control Automated Guided Vehicle Systems," Decision Sciences and Engineering Systems Report Series, Rensselaer Polytechnic Institute, Troy, NY (July 1989).

GPSR Compliance

The European Union's (EU) General Product Safety Regulation (GPSR) is a set of rules that requires consumer products to be safe and our obligations to ensure this.

If you have any concerns about our products, you can contact us on ProductSafety@springernature.com

In case Publisher is established outside the EU, the EU authorized representative is:

Springer Nature Customer Service Center GmbH
Europaplatz 3
69115 Heidelberg, Germany

Batch number: 10844669

Printed by Printforce, the Netherlands